PHYSICS
Structure and Meaning

NEW EDITION

LEON N COOPER

Thomas J. Watson, Sr., Professor of Science
Brown Univerisity

PHYSICS
STRUCTURE AND MEANING

NEW EDITION

BROWN UNIVERSITY

Published by UNIVERSITY PRESS OF NEW ENGLAND

Hanover and London

Brown University Press
Published by University Press of New England, Hanover,
NH 03755
© 1968, 1970 by Leon N Cooper
Prologue to the New Edition © 1992 by Leon N Cooper

The original edition of this book was published by Harper
& Row in 1968 under the title *An Introduction to the
Meaning and Structure of Physics*. A short edition,
upon which this volume is based, was published by Harper & Row in 1970.

Printed in the United States of America 5 4 3 2 1

CIP data appear at the end of the book

To Kay, my wife

CONTENTS

APPENDICES

PREFACE

This book is a shorter version of *An Introduction to the Meaning and Structure of Physics* which was published in 1968. Some of the more difficult technical variations of the original book have been deleted and a considerable number of problems have been added. It is hoped that this new edition, while retaining the approach of the original, will require less effort of the reader and will be more generally accessible.

Concerning the approach of the original, a number of questions have been raised so that (although I have little patience with auto-exegesis of any text) perhaps a few words would be useful.

Physics teaching, like orchestral music, has been dominated until recently by the masterpieces of the nineteenth century, as though to be more than one century behind the times is embarrassing and antique. I feel, with many of my colleagues, that to meaningfully introduce physics one must enter the twentieth century and seriously discuss relativity and quantum physics. Since I am attempting to speak to those who have little background or intuitive feeling for mathematical symbols, and who are thought to be more at home with words than with equations, concepts, as much as possible, must be put into words or pictures and one must avoid using mathematical relations as explanations for physical ideas. This requires considerable attention to what is really necessary technically as opposed to what is only traditional and can be deleted, as well as an occasional new approach.

In so short a space there can be no attempt to develop in detail all physical theories. The overall selection of subject material is guided first by the desire to look at physical theories from various points of view: as they are developing, in a classical, formal or axiomatized stage, and the extent to which they can be developed to provide a picture of the world. No attempt has been made to include everything; the only subjects included of necessity are those required to arrive at today's problems. The idea is to give the reader a solid example of a physical theory as a structure, then to rely on this understanding when developing more briefly the ideas of other theories. The pace of the book changes intentionally. Some parts move very quickly, attempting only to sketch the background of ideas; others more slowly, where that part of the structure must be understood in order to go on to what follows.

A great deal of attention is given to the formation of concepts. The reason for displaying their development is partially to emphasize that they are not obvious or even easy (something clear enough to the student) and further that often they are not necessary—that is, they are not the only way the particular experience might be organized. A question often very puzzling to students is why such a thing was done at such a time. Frequently the answer can only be given in the milieu of the time—the problems that seemed important, the opinions of the people involved. Since the reader in a sense is duplicating the process of the original author (he is struggling to arrive at the concept), he can be helped and reassured by seeing part of the struggle, examining the alternatives, and having underlined what is evident to him—that the concept he is trying to understand is not at all obvious. The original material used in the attempt to place the various themes in context (although I have made a great effort to be accurate) is intended more as pedagogy than history. (My approach to history in any case is closer to Herodotus than Thucidites.)

There is relatively careful attention to the use of words, especially those with philosophical overtones, a continuing attempt to distinguish what is observed from the varying interpretations of those observations, and some emphasis on the game-like aspects of physics in an attempt to loosen the mind from the grip of everyday conventions and prepare it for the changes that occur periodically in the subject. I have found that a little attention to these matters can avoid a great deal of trouble with questions such as, "Does the electron really exist?" It can prepare students to accept with relative equanimity the changes in thinking required to understand relativity and quantum physics as well as to give them a fairly solid understanding of what the subject is about. Although a certain number of the literary or philosophical allusions may escape the reader, they will hardly injure his comprehension and might possibly arouse his interest and direct his attention to the many connections between physics and other endeavors.

I have tried, in all, to present physics as one attempt made by human beings to organize their experience—different in technique, but not totally different in outlook from that of the painter, for example, whose canvas is often his organization of his experience of light and color. It seems to me that important physics, as important painting, imposes the vision of the scientist/artist on the raw data, in principle available to everyone. A generation or two later the world appears to us as that vision.

I am pleased to acknowledge once again the help of all those who aided in the production of the original text. For this shorter version, I would like to express my gratitude as well to George Telecki, of Harper & Row, for his patience, quiet competence, and his ability to introduce order into the everyday chaos called normal.

Leon N Cooper

PROLOGUE TO THE NEW EDITION

We are told that more scientists are alive today than in all previous history. Just look at the literature that comes into my office. No possibility of reading everything. One might guess that more has been published in the last twenty years than ever before. Almost every day, it seems, some new and remarkable discovery or conjecture is made. It is no surprise, then, that this book, first published in 1968, is no longer up-to-date. Perhaps, however, it is not out-of-date.

Viewed from the perspective of a century or two from now, what has occurred since 1968 will very likely appear just a small addition to what happened previously. The great theories are still in place—quantum mechanics, statistical mechanics, gravitation, and electromagnetic theory, just as one might have described them twenty or even fifty years ago. From the eternal viewpoint, everything moves slowly. But for those participating, the events seem extraordinary. Thus, though there is little in the previous chapters I am inclined to change, perhaps a few more words are justified. Hence this prologue to the new edition.

What has happened in twenty years? Let me list a few highlights.

We have seen the unification of electromagnetic and weak interactions into what we now call electro-weak interactions. Thus, the four great forces have been reduced to three. A strong candidate for unification of electro-weak and quantum chromodynamics (quantum theory of nuclear interactions) has appeared, and it is possible that a final unification of all interactions, including gravity, awaits.

There was high-temperature superconductivity found, most improbably, in ceramics, metallic oxides, and most recently, in large molecules with geodesic-like structures called Bucky balls. A certain embarrassment to those of us who encouraged research into novel forms of superconductivity; we pointed out that there were no fundamental reasons why the temperature of transition had to be so low. However, no one, to my knowledge, suggested looking at ceramics. We didn't even know Bucky balls could exist.

And then there were non-events. For example, the great cold fusion uproar—the proposal that fusion could take place contrary to the arguments presented on pages 406 and 413 with no clear indication of how the Coulomb barrier was being breached. Because of its enormous technological implications, this raised very high hopes. The normal human desire to appear in the papers, to announce results first, to hold the first press conference, led to sloppy work. I recall the amusement and skepticism my colleagues showed for some of the early work in cold fusion, work in which individuals all over the world employed delicate and hard-to-use instruments with which they had little experience and within days reported results to the newspapers in hastily called press conferences, only to retract these results a week or two later. It was an insight into the day-by-day working of science that scandalized some but, to my mind, provided those who don't do science an interesting window into what actually happens in the subject.

Among the most striking developments have been new techniques for observation and computation. As we have seen in the past, we often know what we would like to do but are unable to do it. The famous experiment measuring the charge-to-mass ratio of the electron was done first by Thompson, possibly because he had the one technician who could create a sufficient vacuum.

Among the more striking technological advances in the past twenty-five years have been the use of satellites for observation and the use of computers in scientific computation. Satellites have enabled us to make observations that are difficult or impossible from the surface of the earth. Computers have opened a new dimension in our ability to analyze, to visualize, and to extract consequences from a set of hypotheses.

It's not possible in this brief prologue to review, even superficially, all of these events. However, there has been a change of focus that I would like to discuss. After many years in the wings, astronomy, astrophysics, and cosmology have once more come to center stage. We appear to be witnessing that combination of cosmology and physics in which cosmological events can be explained by the "laws of nature," as we discover them here on earth, reminiscent of the incredible period of Kepler, Galileo, and Newton. Among physicists, there is now general agreement on a scenario for the origin of the universe. This is known as the Standard Model ("an ill phrase, a vile phrase," as Polonius once said—worse, unimaginative). It goes as follows.

In the beginning, there is a Big Bang, an enormously energetic fireball. The universe itself—space, time, and energy—is concentrated in some very small, perhaps infinitesimal volume, held together by gravitation, but expanding rapidly due to the very large internal pressure.

After $t = 0$ (the origin of time—so far no opinion on the prior countdown), the primordial fireball expands very rapidly. In a striking demonstration of the modesty of theoretical science, we do not claim to be able to say very much about times before 10^{-43} seconds. Before this time, when temperatures exceeded 10^{32} kelvins, we cannot trust extrapolations of current theories since we are in a region where we do not know their consequences or whether they are consistent. (One problem is that we do not yet have a quantum theory of gravity.) Under these extreme conditions, it is thought, the four forces of nature—gravitation, nuclear, electromagnetic, and weak—are unified. So there we have it: 10^{-43} seconds of almost total ignorance.

At 10^{-43} seconds, the temperature 10^{32}K, a kind of phase transition is thought to have occurred, in some respects similar to the transition between water and ice, in which the gravitation force is believed to have condensed as a separate force. Now its properties are very different from the other three, which remain unified. During this period (the Grand Unified Era), there is no distinction between nucleons, quarks, electrons, and leptons. These particles transform easily into one another.

At 10^{-33} seconds, the universe having expanded considerably, the temperature now having dropped to 10^{27}K, another phase transition occurs during which the strong or the nuclear forces condense. Soon after, we enter what is called the hadron era, in which nucleons and electrons separately fill the universal soup. In this, we see leptons, antileptons, photons, nucleons, and antinucleons colliding with one another and exchanging energy.

When the universe has reached the advanced age of a millionth of a second, the temperature has cooled to about 10^{13}K. Now the vast majority of hadrons—nucleons—disappear. The even balance between particles and antiparticles results in the annihilation of most of these nucleons, while the temperature has dropped so that nucleon-antinucleon pairs can no longer be created by photons. However, as we can see by looking about us, a few nucleons must have remained. Various conjectures have been made as to why there is an excess of nucleons. Present theoretical preference is that some time after the Big Bang there was a slight excess of quarks over antiquarks. This resulted in an excess of nucleons over antinucleons, and finally in the matter we see about us today. Early in the hadron era, there were about as many nucleons as photons. When the era ends, there is only one nucleon for every 10^9 photons, and this ratio has persisted until today.

After about one full second of life, the temperature has cooled to roughly ten billion kelvins. Still, it's enough energy for photons to create electron-positron pairs, so there is equilibrium between them, and there are about as many electrons as photons. But a few seconds more and the temperature has dropped sufficiently that electrons and positrons can no longer be created by photons. So, as with the nucleons before them, the electrons and positrons continue to annihilate, and no more can be formed. Almost all disappear, except for a small excess of electrons. (Eventually, when temperatures have cooled sufficiently, these will combine with protons and other nuclei to form atoms that are not immediately broken up.) This period is now called the radiation era—there are a few electrons, a few photons, and other nucleons, but mostly photons and weakly interacting neutrinos.

Within minutes, fusion begins to occur. The temperature is now about a billion kelvins. Nuclei strike each other and begin to form heavier nuclei by fusion. During this period, we have production of deuterium, helium, and some lithium. A few minutes later, the temperature has dropped and these processes can no longer occur. Further synthesis of nuclei will occur much, much later in stars.

Some hundred thousand years later, the universe now having cooled to about three thousand kelvins, atoms can be formed without being broken apart immediately. Once atoms can be formed, most free electrons disappear, being captured by protons to form atoms. And now the photons are freed, since the electrons from which they scattered previously are captured in atoms; they can now travel across the entire universe with minimum absorption. As the universe continues to expand, the now almost unimpeded radiation cools further and remains until this day. Do we see such a thing?

After this, between about a million years and the present age of the universe (something like ten to fifteen billion years), galaxies begin to form, probably by gravitational condensation from clouds of hydrogen. These clouds in turn attract one another by gravitational forces; with local clumping in some regions, the hydrogen becomes rather dense, eventually forming stars; and then begin the various complicated cycles through which stars go. Typically, as the hydrogen cloud contracts under gravitational forces, the average speed of the atoms increases (gravitational potential energy becoming kinetic energy), and the average temperature increases. When the center of the hydrogen cloud (now a star) becomes hot enough, the Coulomb barrier can be penetrated and fusion (hot) begins to occur. The tremendous release of energy in this fusion process is what produces light and other radiation coming from stars, and it can, for a while, halt the gravitational collapse of matter that forms stars.

As the hydrogen burns (fuses) to form helium, the helium accumulates in the core of the star. When the hydrogen left in the core no longer provides sufficient energy (in a typical star, this takes about ten billion years), the core again begins to collapse under gravitational forces. After somewhat complicated cycles involving a stage in which the star grows in volume and appears reddish (red giant), the further evolution of the star depends critically on its mass. A star of less than 1.4 solar masses eventually becomes a white dwarf, continues to lose energy, decreases in temperature, becomes dimmer and dimmer, and finally it flickers and goes out.

For a more massive star, the gravitational forces are so great they finally force the collapse of the nuclei themselves, until they form what is essentially a giant nucleus—a neutron star. The collapse of the core is thought to be accompanied by a great explosion—a type of supernova. During such explosions, most of the heavy nuclei of the periodic table are formed. The presence of heavy nuclei on earth is thought to indicate that we are part of the debris of such a supernova.

For even more massive stars, the gravitational forces are so strong that eventually even light cannot escape. (The energy of the photon becomes smaller than the gravitational potential energy.) The star then collapses to become possibly the most exotic of all galactic objects—a black hole.

On top of all this activity the universe continues to expand. We may ask: Will this go on forever? Will there be a collapse (the Big Crunch)?

Arrogance, fantasy, or science fiction? Perhaps all of the above. How is it that a group of otherwise seemingly reasonable people, gathering in conferences at various spas all over the world, can talk seriously to each other in such a language, can argue about details, can discuss for days complex calculations that elucidate one or another variation? How, in short, can anyone believe such a thing?

The base on which everything stands is that most fundamental assumption: the same "laws" of physics apply throughout the entire universe for all time. What we can glean from a twentieth century tabletop in our reasonably temperate climate applies equally to the hellish environment at the first millionth of a second after the creation of the universe. On this base rise two pillars that support the entire structure. The first is the deduction that the universe is expanding—that other galaxies are moving away from us, and that the further away they are, the faster they are moving. The second is the observation of uniform background microwave radiation.

The idea that the universe is expanding was first proposed by Hubble in 1929. It is based on observations of Doppler shifts of light emitted from stars in various galaxies. One can compare, for example, the spectrum of hydrogen as we see it here on earth with the spectrum of hydrogen as seen from distant stars. The basic observation is that the further a star is from us (how we know how far away a star is, is a good question; see below), the more the spectrum is shifted toward the red. Since the amount of shift toward the red is related to the speed with which the galaxy is moving away from us, Hubble concluded that distant galaxies are moving away from us, and the further away they are, the faster they are moving. Hubble's Law, as it is known, is given by the following, extraordinarily simple equation:

$$y = Hd,$$

where H is Hubble's constant, about 15 km/second/million light years of distance.

What does it mean to say that the distant galaxies are moving away from us? No, we have not surreptitiously returned to a geocentric theory—the earth at the center of the universe, with everything moving away from us. Possibly the easiest way to envision this is to imagine the universe as the surface of an expanding sphere. From any point on the sphere, other points would seem, on average, to be moving away; the further from us they are, the faster they move. The situation is made more complex because objects move randomly with respect to one another on the surface of the sphere, in addition to moving due to expansion of the sphere. Close by, the random motion overwhelms motion due to the expansion of the sphere, but at greater distances the expansion can be seen above the random motion. It follows that everyone in the universe would observe the same expansion. Thus, the earth is not in a preferred position; what we see is the common expansion of the sphere seen in the same way by everyone.

The background radiation was first observed in 1964, when Arno Penzias and Robert Wilson began to run into problems with static in their radio telescope. The radio telescope is designed to search the heavens for radio emissions (the point being that various stellar objects emit radio waves in addition to or perhaps rather than light). In attempting to adjust their radio telescope, they had difficulty eliminating some static. It must have been as aggravating as attempting to eliminate static from a favorite music station. Eventually, they realized that the static itself is a profound emission, not of individual galaxies, but of the universe itself. Making precise measurements at a wavelength of 7.35 cm, which is in the microwave region of the electromagnetic spectrum, they found that the intensity of this radiation does not vary by day, night, time of year, nor direction; that is to say, it comes at all times from all parts of the universe with equal intensity. It is as though the universe were uniformly filled with 7.35 cm radiation moving in all directions at all times with equal intensity.

Radiation at other wavelengths has also been found to have intensities unaffected by direction or time. The relative magnitude of the radiation at different wavelengths falls on the curve shown in fig. 31.2, precisely the curve of black-body radiation measured at the end of the nineteenth century and explained by Max Planck. The peak of the radiation, near 10^{-1} cm, corresponds to a temperature of 2.7K. It is as though the universe were a container filled with radiation corresponding to that of a black body at 2.7K. What would be a possible source of such radiation?

We might be inclined to guess that stars or other objects emit radiation of this kind. But it is very difficult to account for its extraordinary uniformity. In addition, it's clearly remarkable that the intensity of the radiation as a function of wavelength follows the black-body curve so closely. We realize after the fact that this was predicted. George Gamow and his collaborators had calculated in the late forties that such radiation would be one result of the Big Bang. The radiation trapped in the container that is the universe cools as the universe expands. For the size of the universe today, one would expect that the temperature would be roughly 2.7K.

It is always remarkable how we can speak with such confidence about events so distant in time under conditions so different from those we experience ourselves. In fact, our reasoning is based on several very precise ideas:

1. The fundamental assumption that the same laws of physics apply throughout the entire universe throughout all time (although it must be said that, according to recent ideas, while the fundamental laws are the same, their manifestations may appear quite different).

2. Astute conjectures about individual situations.

3. Observations.

Consider the following example: How do we estimate how far away a distant galaxy is? It is crucial to be able to do this. Otherwise, how could we say that the red shift increased as galaxies were further and further away from us? The distances we are speaking of are much too large to be measured directly, for example by parallax (using the slightly different angle the line from the earth to the galaxy makes at the extremes of the earth's orbit about the sun). The following is one procedure that is employed.

Observe the galaxy through a large telescope. If possible, pick out individual stars. Focus attention on the brightest star.

Astute conjecture one: assume that the brightest star in this galaxy is no more or less bright than the brightest star in others, our own included. Why? Other evidence indicates that different galaxies are composed of populations of stars, some young, some old, and most middle-aged; but the stars themselves are no different from one galaxy to another. With this conjecture, we can, by observing nearby stars, estimate the absolute amount of light coming from this brightest star. (We determine how much light comes from the brightest stars nearby and assume that the same amount of light is coming from the brightest star in the distant galaxy.)

Astute conjecture two: assume that this light is not appreciably absorbed in intervening space. This seems to be consistent with our observations of nearby stars and galaxies and is consistent with our general ideas about what constituents fill space. Then, a measurement of the amount of light reaching us per square centimeter immediately gives us the distance to this faraway star.

Now, if the universe is in fact expanding as proposed by Hubble, this galaxy on average should be rushing away from us with a speed, and thus a red shift, that increases with its distance.

Another example: although times such as 10^{-43} seconds seem short, and temperatures such as 10^{32}K seem high, beyond this first brief period of almost total ignorance, the picture we obtain of the origin of the universe is quite consistent with existing theory. The rate of expansion is a consequence of gravitational theory applied to a violently expanding gas. The calculation in principle is no different from the calculation of the expansion of any fireball following an explosion. Although the temperatures seem high, below 10^{32}K we are in a region where we can hopefully deduce consequences from existing theories (assuming there is no surprising new physics somewhere between the energies we have been able to observe and the energies in the early fireball). The temperatures are translated into the average kinetic energy of the constituents. For example, from the dynamics of the expansion we conclude that after about one second the fireball has cooled to about ten billion kelvins. At this temperature, the average kinetic energy of the constituents is about one million electron volts. Shortly thereafter the average kinetic energy has dropped below a million electron volts, so that a photon can no longer create an electron-positron pair.

There are various exotic objects that might be expected according to the scenario above. Probably the first big news event to elude the initial publication date of this book was the discovery of a new type of stellar object, the pulsar. It was late 1967 when J. Bell and A. Hewish, with their collaborators at Cambridge University, identified an object in the heavens that emitted short, rapid bursts of radiation at regularly timed intervals. This and other pulsars, as they came to be known, produced extremely sharp radio pulses typically separated by intervals of a second or less. This discovery inspired the usual flurry of theoretical activity and the usual wild conjectures—for example, that some form of life in outer space was attempting to signal us. (I don't mean to intimate that these conjectures are entirely out of the question. Someday one of them may prove to be correct. But it's amusing to observe how quickly the same people dust them off and put them forward when any new observation is made.) It is now thought that the explanation of the source of this radiation is what is known as a neutron star. In such stars, matter has become so tightly packed that an ordinary nucleus such as helium is no longer possible. Under immense gravitational forces the nuclei collapse into each other. Since neutrons have no electron charge, their most stable form is that of a star consisting solely of neutrons held together by gravitational forces. Such stars are expected to be very small, a few kilometers in diameter, and extraordinarily dense, packing more than the mass of the entire sun into this small volume. It is now believed that these very tiny dense objects rapidly rotate while emitting a narrow beam of radiation, somewhat like the beacon of a lighthouse. More than a thousand pulsars have been catalogued. Since there must be many others that either radiate in the wrong direction or have ceased radiating, more neutron stars undoubtedly exist than have been seen. It is estimated that perhaps one-tenth of one percent of all stars are neutron stars.

A particular pulsar with a very short period, known as pulsar NP0531, has been discovered in the Crab Nebula. Its period is 0.033 seconds. This Crab pulsar is probably the most carefully studied of all known pulsars. It emits radiation at all observable wavelengths, from radio to visible light to the highest X rays. All of these wavelengths produce the same pulse period. This pulsar was formed a little over nine hundred years ago in a supernova explosion observed in A.D. 1054. Thus it is very young. As it grows older, the Crab pulsar will slow down and its radiation will become limited to the radio region of the spectrum.

Another type of pulsar has been discovered that emits radiation at X-ray wavelengths. These are known as X-ray pulsars and they differ from the radio pulsars in that they always occur in two-star (binary) systems. An example is Hercules X-1. This first X-ray source was discovered in the constellation of Hercules. It is a companion star to the visible star called HZ Hercules. Every 1.7 days, the visible star eclipses the X-ray source. In addition to this 1.7-day period of the binary star system, Hercules X-1 displays a 1.24-second period due to the rotating beam of X rays coming from the pulsar. It is thought that the X rays are generated on neutron stars by the transfer of mass from their companions. This is primarily due to the extremely strong gravitational attraction of the neutron star. In order for this to occur, the gas must attain a temperature of about 10^6K.

Thus it is believed that neutron stars in the form of pulsars have been observed. Their exotic offspring, the black hole, has so far only been hypothesized and perhaps observed. If the mass of the neutron star is as large as several solar masses, the gravitational force becomes so strong that eventually light cannot escape. In effect, we cannot see this object directly. Therefore the name—black hole.

Although (by definition) a black hole does not emit light, it is by no means unobservable, since nearby bodies would be affected by its gravitational forces. Potential black holes have been detected, and it is possible that large numbers are scattered throughout the universe. In addition, there is evidence that suggests there may be a large black hole with a mass estimated at millions of times that of the sun at the center of our galaxy. There may even be large black holes at the center of many, if not all, galaxies. It could be that this huge gravitational force is what is required to assemble the stars into a galaxy.

One possible indicator of a black hole would be its gravitational interaction with some nearby object. Suppose, for example, there were a binary system, one of whose members was a black hole. Gaseous material from the luminous star falling into the black hole would be greatly accelerated and would be expected to emit characteristic X rays. There is at present some evidence for such a black hole in the binary star Cygnus X-1.

It is by such combinations of deduction, conjecture, and observation that our picture of the universe and its evolution is formed. But not everything fits together. Even within this grandiose and seemingly consistent scenario there are a few small problems. Let me mention two.

Among the observations that do not fit so neatly into the picture outlined above are objects called quasars (quasisteller objects). These objects are as bright as nearby stars but display very large red shifts, suggesting, according to Hubble's Law, that they are very distant. If they are this far distant, they must be incredibly bright—thousands of times brighter than normal galaxies. On the other hand, in spite of their large red shifts they might be much closer than expected. In either case, we have a problem. If so far away, why so bright, or, if not so far away, why such a large red shift? Among the various suggestions made is one related to the idea that large black holes occur at the center of galaxies. If those black holes occasionally capture large numbers of nearby stars, enormous amounts of energy would be given off, thus appearing incredibly bright.

And there is the problem of what is called dark matter or missing mass. There are reasons to believe that the actual amount of mass distributed throughout the universe is much larger than that we can see. One reason for this point of view is theoretical hope that the universe has sufficient mass so that gravitational forces eventually slow the expansion, but not so much as to produce a Big Crunch. This would require that the amount of dark or non-luminous matter be approximately ten times what we see. There have been many attempts to measure the actual density of matter in the universe. An indication that dark matter exists is given, for example, in observations of rotating galaxies. A detailed analysis of these rotations suggests that they actually contain much more matter than is luminous. Using ordinary gravitational theory, we can calculate the mass of the central sun by measuring the acceleration of a planet about this central sun and knowing the distance of the planet from it. When such calculations are done for rotating galaxies (including our own Milky Way), we find that the mass required is about ten times larger than what we actually see.

As usual, there are many suggestions as to what this dark matter might be, including weakly interacting massive particles (wimps), black holes, small faint stars, massive neutrinos, etc. At present, there are many reasons to believe that much of the mass of the universe is not luminous and has not yet been seen, though there are no firm conclusions.

Even if and when we attain a fuller understanding, it will appear astonishing to many to speak of this vision of the creation with times like 10^{-43} seconds and temperatures of billions of degrees. We must recall that 10^{-12} seconds can be long for a nuclear event. Closer to our own experience, the mayfly lives for a day, the tortoise perhaps for 1000 years—a factor of 10^5. Similar questions must have been raised when Galileo proposed that the planets were made of earthy material, when Newton proposed that all objects were subject to the same gravitational force, when it was proposed that the stars were so far distant that one would not see any effect of parallax in the course of the earth's orbit, or when Copernicus proposed that somehow, in spite of a motion of 1000 miles an hour at the surface of the earth, or 18 mi/sec about the sun, we felt no effects of such motion. What is important is not the extraordinary shortness of the times or the extremes of conditions, but rather our belief that we can make precise statements and connect these statements with consequences that can be observed. As Galileo said, to make what is said depend on what was said before.

After the first 10^{-43} seconds, we can in fact make fairly precise connections between the conditions, as remote as they seem, and what we observe today. Perhaps the most striking example is that of background radiation. This radiation is one of the predictable consequences of the Big Bang scenario and, as far as we can tell, seems to be there precisely as would be expected. It is, in fact, very difficult to understand where this radiation would come from, if one did not have an origin in some sort of a Big Bang.

Thus, our situation from a logical point of view is not that different from that of Copernicus, Kepler, and Newton. In order to understand the peculiar motions of five spots of light and the sun and the moon, all of Greek cosmology and physics had to be revised. The incredible explanatory power of physics has always been its ability to go from sometimes remote phenomena through long chains of argument to make contact with the world as we see it. The arguments sometimes seem very direct, and other times remarkably distant. But what seems direct today often seemed distant to our ancestors. The sequence of argument that leads from Coulomb's Law and the quantum theory to the property of materials would have seem ludicrously extended conjecture to scientists of the late nineteenth century. Today, they are almost accepted as fact. In the end, the existence of a chair, an electron, or the occurrence of the Big Bang are assumptions that we make to explain what we experience.

Are we up-to-date? Who can say? I haven't yet read tomorrow morning's papers. But I can't go through this every season. So let's try to write a few things that will keep this volume up-to-date for the next twenty-five years. This gets us into prediction, I know. And I realize I'm no better than anyone else at predicting the future. (I can barely recall the past.) In addition, past predictions of what the future of science would be are littered with colossal failures. One example: Rutherford is reputed to have said that the idea that the nucleus could ever be a source of energy was moonshine. Another: at the turn of the century, Michelson is reputed to have made the famous comment that all that was left for physics was the filling in of the sixth decimal place. He is also reputed to have regretted ever saying such a thing. And in recent years we have heard talk of the end of history, the end of science. As with all good things, perhaps someday science and history will end. As I write, the Communist Party is close to becoming illegal in Russia. Let's see what tomorrow brings.

To my mind, one of the most exciting developments of the last twenty years, a development likely to dominate our thinking for many years to come, is what has come to be known as string theory. This is a subject on which there are passionate differences of opinion, appropriate for a new subject whose development is far from complete.

For many years, physicists have recognized that the extension of our theories to geometrical points involved difficulties. This can easily be seen in the simplest case of a gravitational or the Coulomb force, forces that vary as $1/R^2$ and become infinite as R goes to zero. This has led to a variety of difficulties. The obvious way out of such difficulties, one explored by many physicists for many years, has been in effect to say the obvious. There is no reason to believe that particles are points; they're very likely extended in space, so perhaps they're little spheres. In one way or another, such ideas solve some of the difficulties but have always led to cumbersome theories that seem to add nothing. In addition, no evidence of non-pointlike behavior has yet been found in the case of electrons and light.

The ingenious innovation of string theory is to propose what should almost have been obvious. If zero dimensions doesn't work, try one dimension. In effect, string theory might begin as line theory. The point is generalized to a line. By this single stroke, many of the problems involving infinities are removed and, if the line is generalized to an arbitrary one-dimensional curve, we have what is called a string. Since the string can vibrate, this can give rise to the internal states that have been sought for so long.

Theories of this type have been developed extensively in the last twenty years; we're now presented with baffling and frustrating combinations of circumstances. Qualitatively, the idea seems right; it is just the generalization of field theory that seems to have been required. But the full mathematical properties of these theories are so complex that they have not yet been worked out, and some of the simplest things that can be said in the previous theories cannot yet be said for string theories. Thus, certain physicists have joined mathematicians in working out the complex new mathematical structures that are required. We hear the usual complaints that physics has become too mathematical and too far removed from experimental results, that this is a fruitless direction which will lead nowhere. Some or all of the above may be true.

My own suspicion is that this is the generalization of field theory we have been waiting for; it is the natural way to remove infinities from field theories and to produce internal states that correspond to the different elementary particles. In addition, there is the intriguing possibility that gravitation may for the first time be united in a natural way with the rest of field theory. It may be, as Witten says, that string theory provides us with a peek at physics of the twenty-first century in the twentieth century. We will have to wait for the mathematics to be worked out, but I suspect there is a great new physical principle that eludes us, possibly something of the order of the quantum, and this may emerge from the wings in the next twenty-five years.

Since I am now clearly in the mood for great conjectures, let me continue. I suspect that such a new physical principle will tell us something about the deep problems that lie at the heart of the quantum theory. All physicists who have thought about the quantum theory realize that, although the theory provides a structure by which we can understand what is seen in the world, the interpretation of the theory is, at best, peculiar. Some of the problems have been discussed on pages 353–62.

Though I am an advocate of the idea that a physicist works as an artisan (he or she doesn't have to know what a theory means in order to use it with great success), I am still allowed to conjecture that the resolution of problems related to the mathematics of string theory may also address problems that arise out of the strict linearity of the quantum theory. Please regard the above statement with a great deal of suspicion, since one is always inclined to attempt to explain one mystery by invoking another. However, it could be true.

We are on the verge of a revolution in medicine. Genetic engineering and other manipulation that will lead to production of proteins and highly directed probes may be the answer to viral diseases and other maladies. Molecules that will interact with other biological molecules in such a way as to reinforce or eliminate their activity are now being designed, in part using calculations and computer simulations based on everything we know about the quantum theory of molecules under electrical interactions. We can reasonably expect to arrive at a new stage in the control of diseases, which will be as remarkably different from what we have today as the pre- and post-antibiotic eras.

One of the most remarkable innovations in the years to come will be the increasing use of high-speed computers that are more and more intelligent. We are on the verge of building machines that, to a greater and greater degree, will begin to reason the way we reason and be capable of participating with us in complex decision-making processes. This inevitably produces a great deal of anxiety as people fantasize themselves being in the hands of machines that, like Frankenstein monsters, will destroy them. It all depends on us. We always have the problem of controlling the machines we build. My own opinion is that some of the political and bureaucratic monsters we've built are more cumbersome, more difficult to manage, and in general worse for our health than any machine we will design.

Thus, I believe that one of the major innovations of the next century will be the increasing presence of complex reasoning machines that interact with us in making scientific, economic, and natural resource decisions. In the future, rather than fearing these machines, we will look back and ask ourselves how we could have survived without them. They will seem to have arrived just in time to help us solve the very complex environmental and economic problems that elude us at present. Inevitably, such machines will help us in the evolution of new scientific theory into areas too complex to deal with today. We may, by the end of the twenty-first century, be able to deal with problems such as turbulence, complex economic systems, and others that are presently beyond our ability.

When we consider the scientific problems that have been attacked, we realize that they are a small subset of all possible problems, a subset that has been chosen because it is capable of analysis by our size brain. With a smaller brain we could not have arrived at Newtonian theory. With a larger brain we might have solved turbulence. In many ways, the problems we have solved are like the games we construct—complicated enough to be interesting, but not so complicated that we don't have a chance to work our way through them. Tic-tac-toe is trivial; two-dimensional chess taxes the human mind to the limit; seven- or eight-dimensional chess, as far as I know, has never seriously been attempted by human beings.

Among the great changes that have and will continue to take place are changes in what might be called the sociology of science. We now have scientific commentators, watchdog committees, and fraud-detection commandos. We will soon have ethnic as well as women's science. Hopefully, the real work will continue. And though it is not easy for those outside the enterprise to distinguish true metal from dross, serious from frivolous, those of us on the firing line have no such problem.

Perhaps the most serious change is the increasing need for money. As systems become more and more complex, more and more people, equipment, and therefore money are required for each new result. Naturally, people hark back with sentimentality to the good old days when results could be obtained on a tabletop. The fact is, some results are still obtained on tabletops, but the tables are getting larger and the tops more expensive. More and more results come from huge collaborations demanding enormous resources. And this brings us inevitably to the questions: Who pays? How does one pay? And why should one pay? Should money be parcelled out to various states? Is this money a gift, political pork, or is it a precious investment in our future to be used, as other investments, with a maximum of intelligence?

One problem with fundamental science is that its results are largely unpredictable. It is a problem faced by government agencies when funds are tight, and it is a problem faced by corporations when they have to report to their shareholders. Everyone will agree that the invention of radio, radar, the transistor, penicillin (let me not list them all) have a value that is hard to overestimate for the quality of our lives as well as the gross domestic product. Further, it is generally agreed that these could not have come about without advances in fundamental research (electromagnetic theory, quantum theory, and the microbe theory of disease) that came shortly before. But who pays for the fundamental research? If one has to decide as a stockholder whether or not to support research, one has to decide whether or not to spend one's own money for a program whose consequences and whose benefits are hard to predict.

I recall, at a seminar organized by the Army Research Office, making the point that research on superconductivity in which I participated had been financed by the Army and, as a major consequence, led to the development of what is called the superconducting quantum interference device (the SQUID, which is used to measure magnetic fields in a very sensitive fashion). Now, the prime military user of the SQUID, as far as I know, is the Navy, which is very interested in the measurement of magnetic fields. I'm sure that the Army doesn't begrudge its sister service this great benefit of research that the Army financed, but it does make clear what the problem is. In a market economy, we expect to be paid for what we sell and we expect to enjoy the fruit of our investments. When one invests in research (other than intellectual pleasure, for which resources are somewhat limited), one invests statistically, based on history that tells us the benefits will be enormous. In times of fat budgets, perhaps this is not too much of a problem; but when budgets are constricted, one always tends to cut the future for the present, to look for short-term goals as opposed to long-term. To my mind, this is the underlying problem involved in supporting the type of fundamental research that is required if science is to progress.

A solution, to my mind, would be to invest in fundamental research as a separate line item (separated from all development projects) as some fixed percentage of the GNP and thought of as part payment (a type of royalty) on the economic worth of fundamental ideas of the past as well as an investment for the future. This payment should be regarded in the same way as any other obligation (interest on the national debt, for example). Obviously, there are many arguments pro and con, but one interesting benefit could be increased thought by people who do fundamental research on how to convert their results into actual increases in the GNP.

The French have an expression, "Plus ça change, plus ça reste la même chose." Science and history have ended before; they will end again. We will hear similar sentiments at the end of the twenty-first and twenty-second centuries. One day, no doubt, just as repeated predictions of flood, disaster, and the end of the world, it will finally happen; but, dear reader, that day has not yet come.

PHYSICS
Structure and Meaning

NEW EDITION

ON THE PROBLEM OF MOTION

Take two balls of lead (as the eminent man Jean Grotius, a diligent investigator of Nature, and I formerly did in experiment), one ball ten times the other in weight; and let them go together from a height of 30 feet down to a plank below—or some other solid body from which the sound will come back distinctly; you will clearly perceive that the lighter will fall on the plank, not ten times more slowly, but so equally with the other that the sound of the two in striking will seem to come back as one single report. And the same thing happens with bodies of equal magnitude, but differing in weight as ten to one. Wherefore the alleged proportion of Aristotle is foreign to the truth.

SIMON STEVIN (1548–1620)

The downward movement of a mass of gold or lead, or

of any other body endowed with weight, is quicker in

proportion to its size. . . .

ARISTOTLE (384–322 B.C.)

Here is something absolutely false, and something we can better test by observed fact than by any demonstration through logic. If you take two masses greatly differing in weight, and release them from the same elevation, you will see that the ratio of times in their movements does not follow the ratio of the weights, but the difference in time is extremely small; so that if the weights do not greatly differ, but one, say, is double the other, the difference in the times will be either none at all or imperceptible.

JOHN PHILOPONUS (533 A.D.)

1
THE TOWER OF PISA

If we can believe Diogenes Laërtius, Aristotle "spoke with a lisp . . . , his calves were slender, his eyes small and he was conspicuous by his attire, his rings and the cut of his hair." He was born in Stagirus in 384 B.C., came to Athens when he was 17, and remained there for 20 years studying—among others, with Plato. After Plato's death Aristotle left Athens and at the age of 42, ready for a serious position, agreed to head a school in Pella for the particular benefit of the 13-year-old son of Philip of Macedonia. When, six years later, Philip suddenly died, the school was closed; Alexander, the new king, needed no further instruction.

In 335 B.C. Aristotle returned to Athens and founded the Lyceum, outside the city. Alexander is said to have given him 800 talents, an enormous grant for the time, and to have ordered fishermen and hunters of the kingdom to inform Aristotle if they ran across any matters that might interest him. Setting a pattern for the modern college, Aristotle laid down rules of conduct, established common meals for students, and probably created a museum. In addition, anticipating the graduate school, he held a symposium once a month and organized research on a large scale. With Alexander's death (323 B.C.), the political situation in volatile Athens became somewhat uncertain for Aristotle, a known favorite of the conqueror. When the traditional charge of impiety was lodged against him, he left the city, thus saving Athens (with the memory of Socrates still fresh) the disgrace of a second philosophocide. A year later, he died.

To his son Nicomachus he advised the golden mean. The function of poets, he said, is not to repeat experience, but to provide its organization. He introduced a system of logic still in use today and attempted to disentangle the scientific subjects from one another and to exclude the possibility that theorems belonging to one branch might be demonstrated by first principles belonging to another. For each science—geometry, arithmetic, etc.—one has the individual postulates. The factors found in all, Aristotle recognized as axioms, meaning common sayings, agreements as to what the language means. He wrote voluminously, and often with great accuracy, on subjects of natural history; he was a moralist, political scientist, literary critic, physicist, biologist, practicing naturalist, logician, and teacher, as well as a philosopher. He devised methods, coined words, made observations, collected specimens, gathered together, criticized, and summarized almost all of what had been done before; he seems to have been willing to offer an opinion (or the opinion of an-

3

other) on almost any subject that arose. It was possibly his misfortune to have compiled in his lectures an encyclopedia of ancient thought so rich in its diversity, so complete in its world view, that he came to overwhelm the thinking of Europe before the great Renaissance.

After Aristotle's death, his notebooks, preserved for a while in caves near his home, were sold to the library at Alexandria, which became, after the decline of the Greek city states, the center of what learning and scholarship continued. After the second century A.D., few original works were created in the West. What was produced were encyclopedias and commentaries on what had gone before, and by the time of the Moslem invasion, in the seventh century, scholarly activity had largely ceased. During the centuries that followed—the time we now call the Dark Ages —even what had been known was largely forgotten, except at the few monasteries where copies of the ancient documents were recopied from generation to generation. In this process documents were lost and altered. Of Euclid's works there remained only an incomplete Latin translation prepared by Boethius in the sixth century, of Ptolemy's almost nothing, and of Aristotle's only a few of his writings on logic.

Between the tenth and twelfth centuries, amidst the general increase in activity, various ancient texts were rediscovered in Arabic and translated into Latin. The universities of Europe began as informal gathering places where the new documents could be discussed; there Aristotle's texts, translated into Latin, were taught again. In the course of these translations and retranslations, errors were made, resulting in a great deal of confusion and in a new profession: that of the scholar who spent his time trying to decide what was actually contained in these ancient texts. In their amazement over this newly reclaimed heritage, such scholars devoted tremendous effort and energy to understanding what it was about, and in the process they committed themselves very heavily to its content.

During this time, the attitude of the Church, the dominant political and cultural force in Western Europe, changed quite radically. At first the newly introduced ancient writings were considered suspect. In Paris the teaching of Aristotelian physics was prohibited; the ancient texts and Christian doctrine did not fit together very well. But through the effort of Christian scholars, the most important of whom was St. Thomas Aquinas (1225–1274), Aristotle was reconciled with Christian doctrine, and when this was done, the Aristotelian conception of the universe—his compendium of his own research, criticism, his repetitions of hearsay, and the ideas of others—became part of the Christian drama of salvation; an attack thereafter on Aristotle was an attack on the Church itself.

ARISTOTLE ON THE MOTION OF BODIES

The earth was the center of Aristotle's universe and physics. Heavy things fell toward it and light things rose. He wrote:

> I call absolutely light that whose nature is to move always upward, and heavy whose nature is to move always downward, if there is no interference.

Further:

> . . . the natural motion of the earth as a whole, like that of its parts, is towards the centre of the Universe: that is the reason why it is now lying at the centre . . . light bodies like fire, whose motion is contrary to that of the heavy, move to the extremity of the region which surrounds the centre.[1]

His universe was closed, bounded at its outer edge by the celestial sphere, populated with air, earth, fire, water, or celestial substance. Between the celestial sphere at the periphery and the earth at the center were the spheres of all the planets, the sun, and the moon. Each of these followed its natural circular motion about the earth, and in between was the plenea (that substance presumably thin as air), which filled all space.

In addition to material bodies—whose natural motions were down or up—he introduced the idea of celestial material (the stuff out of which the stars and planets were made) whose natural motion was in a circle about the center of the universe. All motions were in this way divided into two classes: natural motion, motion consistent with the nature of the body and requiring no external influence; and forced or violent motion, motion not consistent with the nature of the body, requiring an external force.

The elements out of which Aristotle built his cosmos in fact differed from one another not primarily in material substance but in their natural movements and their tendency to occupy different places. Whenever a body does not move with its natural motion, Aristotle has to find an external influence. For example, a cart, when drawn along a road, is not in its natural motion toward the center—it is the horse that provides the force. However, the stars and the moon moving around the earth require no forces, since they are celestial material and are following their natural motion. The place an object occupies thus has an absolute significance; the center of the universe is different from its periphery, providing a fundamental connection between the geometry of space and the motion of bodies.

The physical theories of Aristotle were a systemization of some of the beliefs of his times, as well as in agreement with the facts as he understood them. The experiences to which he referred were common things: a horse pulling continually to move a cart on a level road, or a stone falling to the bottom of a pond. For this variety of experience it seems to be an immediate qualitative fact that a force is required to pull a cart or that a heavy body falls faster than a light one. The only motions that seem to proceed without help are those of objects falling (if they are heavy) or rising (if they are light) or moving about the earth in circles (if they are celestial stuff). An object moving along a straight line with uniform speed (such as a cart) has to be pushed. He thus never classified what we now call resistance or friction as a force separated from the movement. The separation, when it was finally made, resulted in the natural motion of inertia and the modern organization of the motion of bodies.*

---

* The Greeks probably never had the experience—now common—of moving so uniformly (in a plane or ocean liner) that one realizes that, without looking out a window, it is not possible to say whether or not one is moving. Such an experience gives an immediate perception of what is later to be called the law of inertia—one

The details of Aristotle's physics, once they were available, were questioned by many of his Scholastic critics. In the fourteenth century Nicole Oresme, a member of a Parisian school, disputed Aristotle's argument for the earth's uniqueness. He proposed that the motion of the stars cannot be deduced from their apparent motion viewed from the earth because these two motions are relative; the stars might be at rest while the earth moves just as easily as the earth could be at rest while the stars move—either would produce the same effect. "Just," he said, "as it seems to a man in a moving boat that the trees outside the boat are in motion." Aristotle's theory of projectile motion came under continual criticism. It was difficult to understand why a projectile continued moving upward (not its natural motion) once the impelling force was removed. The Scholastic critics of Aristotle proposed a theory in which such objects acquire an impetus (we would now call it momentum) that keeps them moving.

This process of criticism and revision of the Aristotelian views gradually brought the revisors into conflict with the majority of the scholarly world, who professed to regard what they thought Aristotle had written as the last word—in fact, the only word. Although for several hundred years the inadequacies of Aristotle's physics were known—some extremists (such as Pierre La Ramée) had declared that everything in Aristotle's physics was unscientific—still for the scholastics what "*ille philosophus*" had written had become the ultimate body of knowledge. Learning for them consisted in studying and knowing Aristotle and the commentaries interpreting Aristotle. (On the more obscure passages there could be much commentary.) Questions, even questions of fact, could be resolved by consulting Aristotle. It was a comfortable world, safe from sudden discoveries and dilemmas. When one studied Aristotle, reconciled with Christianity, one did not fear for one's soul. And if there were matters that Aristotle had not considered—possibly, probably, they were not worth considering.

Whether Aristotle would have approved their attitude we do not know, but it is known that he treated none of his predecessors with such deference. "*Amicus Plato, sed magis amica veritas*" ("Plato is dear to me, but dearer still is the truth"), he is supposed to have said. The doctrines of his teachers, and all who had gone before, he attacked with generous impartiality. Our knowledge of many of the early Greek writers comes from Aristotle, who states their opinions so that he may refute them. In matters of the intellect he has been compared to an Oriental prince who must eliminate his brothers before he feels secure. And Plato complained of his greatest, his most famous, pupil that he behaved like a young colt kicking his mother.

But for those who were attempting to do what Aristotle himself had done—to see the way the world was and to try to understand it—it was clear that Aristotle, like other men, had many faults; he made mistakes,

---

does not have to feel a push or pull to continue in uniform motion. The closest they came was possibly in the interior of a moving ship on a very calm sea. But the Aegean was not famous for its tranquillity. (For another view of Aristotle's physics, see Kuhn.[2])

repeated erroneous opinions, talked too much and too loosely, and, what was more important, simply had not considered many of the relevant questions. Perhaps they were wiser; perhaps they were less patient; perhaps they could not read or understand Aristotle as well as their learned contemporaries but, in any case, to make themselves heard they had to penetrate the inertia of tradition and belief.

In attempting to demonstrate the impossibility of a void ("void is a place where nobody happens to be") Aristotle's infamous argument on falling bodies appears.

> We see that bodies which have a greater impulse either of weight or of lightness, if they are alike in other respects, move faster over an equal space, and in the ratio which their magnitudes bear to each other. Therefore, they will also move through the void with this ratio of speed. But that is impossible; for why should one move faster? (In moving through *plena* it must be so; for the greater divides them faster by its force. For a moving thing cleaves the medium either by its shape, or by the impulse which the body that is carried along or is projected possesses.) Therefore, all will possess equal velocity. But this is impossible.[3]

The speed of a naturally falling body, he argues, is retarded by the resistance of the medium. Through a void, infinitely thin, without any substance, and thus offering no resistance, all bodies will fall at the same rate (possibly he believed infinitely rapidly). This he rejects, since he wishes to associate different natural motions with each different element —fire is light, earth is heavy. Therefore, he concludes, there can be no void; a variation on this view was stated succinctly by Lucretius:

> All things that fall [*cadunt*] through the water and thin air, these things must needs quicken their fall [*casus celerare*] in proportion to their weights [*pro ponderibus*], just because the body of water and the thin nature of air cannot check each thing equally, but give place more quickly when overcome by heavier bodies. But, on the other hand, the empty void cannot . . . support anything; . . . wherefore all things must needs be borne on through the calm void, moving at equal rate with unequal weights.[4]

Whether it is more appropriate to conclude that Aristotle felt that bodies fall in proportion to their weight, that a vacuum was impossible, or that given a vacuum all bodies would fall at the same rate is not easy to say. The argument is not mentioned again in his treatment of motion in the later books of his physics.

It was, however, when interpreted in its worst light, one of the weakest links in that enormous chain that both he and his commentators had forged, and it was toward this that his critics directed a good part of their fire. They became somewhat harsh, as when Galileo quotes Aristotle as saying "that an iron ball of 100 pounds falling from a height of 100 cubits reaches the ground before a one pound ball has fallen a single cubit," for what was needed was a confrontation dramatic enough to shake the lethargy of centuries of exegesis. Galileo and Simon Stevin by that time were attacking not Aristotle but rather those of their scholarly contemporaries and learned predecessors (something quite obvious to their colleagues) who for 500 years had confined their thinking, their observations, and their knowledge to what was written in Aristotle.

REDEFINITION

But finally, as though the long years of restraint on intellectual activity, the long years of submission to printed authority had created an unbearable pressure, the boundaries of that anthropomorphic medieval world of magic and animism burst. Years of criticism gradually had achieved a redefinition of the problem of the motion of bodies which would enable a new and immensely fruitful attack. And there appeared those figures who as the old world crumbled were creating already that rational world which was to dominate Western thought thereafter.

The universe itself changed. Copernicus put the sun at the center, leaving for earth the reduced status of one among other planets. Giordano Bruno burst the starry sphere; apostle of infinite space (burned for his heresy) he made the universe endless, earth and sun lost among countless other planets and suns. Galileo, looking at the heavens through his telescope, discovered in celestial matter those imperfections so characteristic of the earth. Motion became relative to other bodies—not referred to space, since space is uniform and no point is preferred to any other. Oresme: "I suppose that local motion can be perceived only when one body alters its position relative to another." The universe, no longer centered about a unique, fixed point, was no longer filled, and no longer necessarily finite. Objects moved through this space uniformly from point to point because the space on all sides is the same.

Returning to an atomism that came from Democritus and Lucretius, Gassendi* proposed that the world was made of atoms and their arrangement in a void. No longer would stones, fire, or celestial material strive to find their natural place; no longer was the world of objects, animate or inanimate, charged with meaning and purpose. God might be the first cause of all things, but after that initial push everything proceeded according to forces and laws, just as a giant machine.

It was cold in Bavaria on November 10, 1619; to get warm René Descartes shut himself up in a heated room. There, he says, he had three visions, saw flashes of light, and heard thunder. When he emerged, he had formulated analytic geometry and had conceived the idea of applying mathematical methods to philosophy. Rejecting all doctrines and dogmas, putting aside all authorities, especially Aristotle, he would start with a clean slate and doubt everything.

> . . . I ought to reject as absolutely false all opinions in regard to which I could suppose the least ground for doubt. . . .[5]

How does one begin if everything is to be doubted? Can one find something

---

* Among his students were Molière and Cyrano de Bergerac; his private opinion about the Scholastic philosophers is possibly reflected in Molière's *Le Mariage Forcé*, his mechanism in Cyrano's fantastic account of his *Voyages to the Sun and the Moon.* His heresies were as extreme as those of Bruno, but he was much quieter. Durant suggests that because he was so amiable a youth, so modest in his conduct, and so regular in his religious duties, no one seems to have thought of burning him.

> . . . so certain and of such evidence that no ground of doubt,
> however extravagant, could be . . . capable of shaking it . . . ?[6]

This something, the first principle of his philosophy, is: *Je pense, donc je suis. Cogito, ergo sum.* "I think; therefore I am." Not intended as a syllogism, but an immediate and irrefutable experience, this is the clearest and most distinct idea that we can ever have. Other ideas should be considered true in proportion as they approach this primal intuition, this direct perception in clarity and distinctness. Descartes' new method of philosophy was to analyze complex conceptions into their constituents until the irreducible elements are simple, clear, distinct ideas. Rather than believing everything that is printed or that one has been taught, he attempted to begin by believing a few things which are immediate and intuitively obvious.

> . . . I concluded that I might take, as a general rule, the principle,
> that all the things which we very clearly and distinctly conceive
> are true. . . .[7]

From this foundation, very much as the geometer he was, he would attempt to construct his view of the world.

He saw the whole universe, except perhaps God and the human soul, as mechanical. God created matter and endowed it with motion; after that the world evolves by the laws of mechanics without interference. And from this world viewed as a machine, composed of material corpuscles, following mechanical laws, he attempted to deduce the entire structure of the Copernican universe, the universe as it is observed:

> Thereafter, I showed how the greatest part of the matter of this
> chaos must, in accordance with these laws (laws of nature), dispose
> and arrange themselves in such a way as to present the appearance
> of the heavens; . . . an earth . . . planets . . . comets . . . how
> the mountains, seas, fountains, and rivers might naturally be formed
> . . . metals produced . . . plants grow. . . .[8]

The vision was vast, as was the magnitude of the enterprise: to deduce all of experience by applying mechanical laws to moving corpuscles. But the effort, not as complete as the vision, brought objections: Privately, Hobbs

> . . . was wont to say that had Descartes kept himself wholly to
> geometry . . . he had been the best geometer in the world, but
> that his head did not lie for philosophy.

Fontenelle:

> It is Descartes . . . who gave us the new method of reasoning,
> much more admirable than his philosophy itself, in which a large
> part is false, or very doubtful according to the very rules that he
> has taught us.

But if the connection between the behavior of his corpuscles and the world of experience was more a wish than a fulfillment, having begun this way he was led to the question: How do the corpuscles themselves behave? He proposed, not liking any influences that did not seem immediate, that the corpuscles of matter exert no influence on each other,

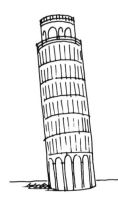

except when they collide. In this universe the question arises: How does a single corpuscle move if there is no interference? What happens to a moving corpuscle in a space where there is no center and where no point is different from any other? Although he never accepted the idea of empty void, suggesting, for example, that "a vacuum existed nowhere except in Pascal's head," he proposed that such a corpuscle would continue to move at constant speed in a straight line—the natural motion of the new physics—the law of inertia, Newton's first law of motion.

Thus, by the end of the sixteenth century there were available alternatives to the universe as conceived by Aristotle. Rather than closed, the universe was proposed to be open; rather than filled, it was empty. Space, rather than having a unique point, was the same in all directions, peopled with particles that do not fall or rise but remain in uniform motion unless they collide. Out of this world came Galileo. That he may never have climbed the stairs leading to the top of the Tower of Pisa was not to prevent him from changing forever our conception of the motion of bodies.

2

A VERY NEW SCIENCE

THIRD DAY

CHANGE OF POSITION. *[De Motu Locali]*

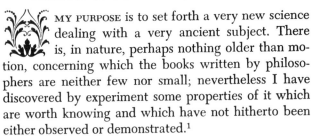 MY PURPOSE is to set forth a very new science dealing with a very ancient subject. There is, in nature, perhaps nothing older than motion, concerning which the books written by philosophers are neither few nor small; nevertheless I have discovered by experiment some properties of it which are worth knowing and which have not hitherto been either observed or demonstrated.[1]

The words are those of Galileo; the year is 1638; and in the dialogue that follows, Salviati, Galileo's spokesman, Sagredo, a neutral and open-minded observer, and Simplicio, who presents the traditional view, lay the foundations for the science of mechanics.

Some superficial observations have been made, as, for instance, that the free motion [*naturalem motum*] of a heavy falling body is continuously accelerated;* but to just what extent this acceleration

---

* "Natural motion" of the author has here been translated into "free motion"—since this is the term used today to distinguish the "natural" from the "violent" motions of the Renaissance. [*Trans.*]

occurs has not yet been announced; for so far as I know, no one has yet pointed out that the distances traversed, during equal intervals of time, by a body falling from rest, stand to one another in the same ratio as the odd numbers beginning with unity.

It has been observed that missiles and projectiles describe a curved path of some sort; however no one has pointed out the fact that this path is a parabola. But this and other facts, not few in number or less worth knowing, I have succeeded in proving; and what I consider more important, there have been opened up to this vast and most excellent science, of which my work is merely the beginning, ways and means by which other minds more acute than mine will explore its remote corners.

This discussion is divided into three parts; the first part deals with motion which is steady or uniform; the second treats of motion as we find it accelerated in nature; the third deals with the so-called violent motions and with projectiles.

Uniform Motion

In dealing with steady or uniform motion, we need a single definition which I give as follows:

Definition

By steady or uniform motion, I mean one in which the distances traversed by the moving particle during any equal intervals of time, are themselves equal.

Caution

We must add to the old definition (which defined steady motion simply as one in which equal distances are traversed in equal times) the word "any," meaning by this, all equal intervals of time; for it may happen that the moving body will traverse equal distances during some equal intervals of time and yet the distances traversed during some small portion of these time-intervals may not be equal, even though the time-intervals be equal.[2]

We can define a ratio of the distance traveled to the time taken and call this the speed:

$$\text{speed} = \frac{\text{change in distance}}{\text{time interval}} \qquad (2.1)$$

For uniform motion the speed will be constant, since the distances traveled by the moving particle during *any* equal intervals of time are themselves equal. (Note Galileo's caution.) In addition, it is implied that the direction of motion is constant. This uniform motion—constant speed, unvarying direction—is the natural motion of the new physics. Any non-uniformity will be attributed to a force.

The speed does not change in time, so it is said that the speed (as a function of time) equals a constant (v_0), or

$$v(t) = \text{constant} = v_0 \qquad (2.2)$$

and this dependence of the speed as a function of time can be displayed as a graph. We observe, as a special case of a general rule to be developed

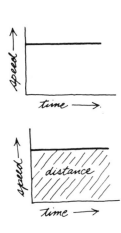

soon, that the distance traveled by a body moving at a **uniform** speed is equal to the area under the curve (line in this case).

$$\text{distance} = \text{speed} \times \text{time} = \text{height} \times \text{base} = \text{area} \qquad (2.3)$$

Examples are completely trivial. A car moving uniformly at 30 miles/hr goes 15 miles in ½ hr. A bullet with a speed of 1000 ft/sec goes a mile in 5.28 sec; a rocket moving 12,000 m/sec moves 60,000 m in 5 sec.

Motion in nature is rarely uniform; the speed of a body, usually not constant, changes with time. During any time interval the average speed may be defined as

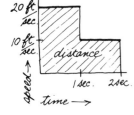

$$\text{average speed} = \frac{\text{distance traveled}}{\text{time interval}} \qquad (2.4)$$

Thus, if a body moves 30 ft in 2 sec, its average speed over that 2-sec interval is

$$\text{average speed} = \frac{30\ \text{ft}}{2\ \text{sec}} = 15\ \frac{\text{ft}}{\text{sec}} \qquad (2.5)$$

However, it may have moved 20 ft during the first second and 10 ft during the second second. Thus the average speed during the first second would be

$$\text{average speed (sec)}_1 = \frac{20\ \text{ft}}{1\ \text{sec}} = 20\ \frac{\text{ft}}{\text{sec}} \qquad (2.6)$$

and the average speed during the second second is

$$\text{average speed (sec)}_2 = \frac{10\ \text{ft}}{1\ \text{sec}} = 10\ \frac{\text{ft}}{\text{sec}} \qquad (2.7)$$

If the motion was uniform during the first interval and uniform again (although with different speed) for the second interval, the graph of speed as a function of time would look as shown, and, again, the distance traveled would be the area (shaded) under the steplike curve.

In general, we could imagine the speed of an object changing in an almost completely arbitrary manner, yielding a graph of the speed as it depends upon time that might look as shown in Fig. 2.1. Such a curve

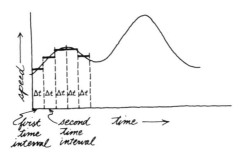

FIG. 2.1. Speed as a function of time.

might be obtained by taking the average speed over smaller and smaller time intervals, and plotting its dependence upon time (the horizontal steplike lines above). To do this, it is convenient to introduce the notation

Δt a time interval that is supposed to be small (as small as we wish) *but never zero* and

Δd the distance change in that time interval
—again presumably small—possibly zero,
since the body might be standing still

Then the average speed[*] in the time interval Δt is defined as

$$\text{average speed (time interval } \Delta t) = \frac{\text{distance traveled}}{\text{time interval}} = \frac{\Delta d}{\Delta t} \quad (2.8)$$

Both Δd and Δt can be thought of as ordinary numbers—that is to say, 2 in. or $\frac{1}{20}$ sec, so that

$$\Delta d = (\text{average speed}) \times \Delta t \quad (2.9)$$

To the extent that the steplike curve approximates the smooth curve, the area under the steps approximates the area under the curve itself. We might guess that as the intervals Δt are made smaller and smaller, the approximation can be made as accurate as we like. This, with some restrictions on the curves allowed, is exactly true. Thus, as the distance traveled by the body is the sum of the distances it travels in the first, second, third, etc., time intervals,

$$d = (\Delta d)_1 + (\Delta d)_2 + (\Delta d)_3 + \cdots$$
$$= (\text{area of rectangle})_1 + (\text{area of rectangle})_2 + \cdots \quad (2.10)$$

the total distance it travels is the total area of the rectangles, which, as the interval grows small, approaches as closely as we wish the area under the curve.

To begin his analysis of nonuniform motion, Galileo fixes his attention on one useful and important case. He considers

Naturally Accelerated Motion

and writes

. . . accelerated motion remains to be considered. And first of all it seems desirable to find and explain a definition best fitting natural phenomena. For anyone may invent an arbitrary type of motion and discuss its properties . . . we have decided to consider the phenomena of bodies falling with an acceleration such as actually occurs in nature and to make this definition of accelerated motion exhibit the essential features of observed accelerated motions. And this, at last, after repeated efforts we trust we have succeeded in doing. In this belief we are confirmed mainly by the consideration that experimental results are seen to agree with and exactly correspond with those properties which have been, one after another, demonstrated by us . . . hence the definition of motion which we are about to discuss may be stated as follows: A motion is said to be uniformly accelerated, when starting from rest, it acquires, during equal time-intervals, equal increments of speed.[3]

Just as the speed can be defined as

$$\text{average speed} = \frac{\text{change in distance}}{\text{time interval}} = \frac{\Delta d}{\Delta t} \quad (2.11)$$

[*] What is often called the "instantaneous" speed is just the average speed for a very (infinitesimally) small time interval (see the Appendices, pp. 483–487).

so the acceleration can be defined as

$$\text{average acceleration} = \frac{\text{change in speed}}{\text{time interval}} = \frac{\Delta v}{\Delta t} \qquad (2.12)$$

From the definition of uniformly accelerated motion—"It acquires, during equal time intervals, equal increments of speed"—it follows that for that motion the acceleration is constant and that during equal time intervals the speed increases by equal amounts:

$$\text{change in speed} = \Delta v = (\text{uniform acceleration}) \times \Delta t \qquad (2.13)$$

Galileo's uniformly accelerated motion is motion in a straight line, for which the speed, as a function of the time, increases in proportion to the time elapsed:

$$v(\text{speed}) = v_0(\text{speed at time } t = 0) + a(\text{uniform acceleration}) \times \text{time}$$
$$= v_0 + at \qquad (2.14)$$

where a is the uniform acceleration and v_0 the speed at time $t = 0$. The graph of the speed as a function of time is shown in Fig. 2.2. For more

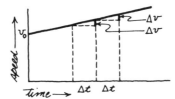

F I G. 2.2. Uniformly accelerated motion.

complicated motions the acceleration need not be constant and the speed as a function of time might have the arbitrary shape shown in Fig. 2.1.[*]
Now Sagredo responds,

> Although I can offer no rational objection to this or indeed to any other definition, devised by any author whomsoever, since all definitions are arbitrary, I may nevertheless without offense be allowed to doubt whether such a definition as the above, established in an abstract manner, corresponds to and describes that kind of accelerated motion which we meet in nature in the case of freely falling bodies. And since the Author apparently maintains that the motion described in his definition is that of freely falling bodies, I would like to clear my mind of certain difficulties in order that I may later apply myself more earnestly to the propositions and their demonstrations.[4]

Salviati then proceeds to try to convince Sagredo and the others that the motion of a body released from rest near the surface of the earth is uniformly accelerated. Earlier Salviati has been in disagreement with Aristotle on this point and has made the most of it, for he knows he is right. Then there follows a discussion of some of the properties of uniformly accelerated motion. Let us concentrate on one of particular interest. Given that a body is accelerating uniformly (that is, its speed is given by $v = v_0 + at$), through what distance does it move in a given time?

---

[*] In common usage, there is some confusion between fast motion and accelerated motion. Acceleration means only that the speed is changing—it need not be fast. Anyone who has gone by rail between Florence and Rome on the *Accelerato* would not forget the difference.

For uniform motion, the distance $= v_0 t$ is just the area under the curve: speed versus time. This idea has been generalized so that we can say the distance traveled for uniformly accelerated motion (or any motion for that matter) is equal to the area under the curve: speed as a function of time. To calculate the distance traveled by a body that is uniformly accelerated we have only to calculate the area under the speed-versus-time curve. And this is just the area of the triangle and the rectangle shown in Fig. 2.3. *The area of the rectangle* is $v_0 t$—the distance traveled

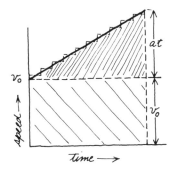

F I G. 2.3. Distance is the area of the rectangle plus the area of the triangle.

by a body moving uniformly with speed v_0. *The area of the triangle* is $\frac{1}{2}$(base) × (altitude). But the altitude is the *increase* in speed, in the time t, due to the acceleration a. The speed as a function of time is

$$v = v_0 + at \qquad (2.15)$$

which gives an increase (the final speed minus the original speed)

$$v - v_0 = at \qquad (2.16)$$

Thus the area of the triangle is

$$\text{area} = (\tfrac{1}{2}\text{ base} \times \text{altitude}) = \tfrac{1}{2}t \times at = \tfrac{1}{2}at^2 \qquad (2.17)$$

Therefore the distance traveled by a uniformly accelerated body in the time t is

$$\text{distance} = \text{area of rectangle plus area of triangle}$$
$$= v_0 t + \tfrac{1}{2}at^2 \qquad (2.18)$$

which is the result obtained by Galileo.

We can now use this to obtain the relation Galileo mentions in his introduction. Suppose a body is dropped from rest. In this case its initial speed v_0 is zero; so we have

$$\text{distance} = 0 + \tfrac{1}{2}at^2 = \tfrac{1}{2}at^2 \qquad (2.19)$$

Table 2.1 indicates how distance depends on time. The distance fallen in the first second is $\frac{1}{2}a$; in the second second it is $\frac{1}{2}a \cdot 4$ minus the distance

TABLE 2.1

Time, sec	Distance	Distance fallen in each second
1	$\frac{1}{2}a$	$\frac{1}{2}a$
2	$\frac{1}{2}a \cdot 4$	$\frac{1}{2}a \cdot 4 - \frac{1}{2}a = \frac{1}{2}a \cdot 3$
3	$\frac{1}{2}a \cdot 9$	$\frac{1}{2}a \cdot 9 - \frac{1}{2}a \cdot 4 = \frac{1}{2}a \cdot 5$
4	$\frac{1}{2}a \cdot 16$	$\frac{1}{2}a \cdot 16 - \frac{1}{2}a \cdot 9 = \frac{1}{2}a \cdot 7$

already fallen in the first second, $\frac{1}{2}a$, which equals $\frac{1}{2}a \cdot 3$; in the third second it is $\frac{1}{2}a \cdot 9$ minus the distance fallen in the first and second seconds, $\frac{1}{2}a \cdot 4$, which equals $\frac{1}{2}a \cdot 5$. Thus the ratios of the distances fallen in the equal time intervals are

$$\frac{\text{second second}}{\text{first second}} = \frac{3 \cdot \frac{1}{2}a}{1 \cdot \frac{1}{2}a} = \frac{3}{1} \qquad (2.20)$$

$$\frac{\text{third second}}{\text{second second}} = \frac{5 \cdot \frac{1}{2}a}{3 \cdot \frac{1}{2}a} = \frac{5}{3} \qquad (2.21)$$

or, as Galileo says, ". . . in the same ratio as the odd numbers beginning with unity. . . ."

Simplicio now objects:

> I am convinced that matters are as described, once having accepted the definition of uniformly accelerated motion. But as to whether this acceleration is that which one meets in nature in the case of falling bodies, I am still doubtful; and it seems to me, not only for my own sake but also for all those who think as I do, that this would be the proper moment to introduce one of those experiments—and there are many of them, I understand—which illustrate in several ways the conclusions reached.[5]

And Salviati responds:

> The request which you, as a man of science, make, is a very reasonable one; for this is the custom—and properly so—in those sciences where mathematical demonstrations are applied to natural phenomena, as is seen in the case of perspective astronomy, mechanics, music, and others where the principles once established by well-chosen experiments, become the foundations of the entire superstructure. I hope therefore it will not appear to be a waste of time if we discuss at considerable length this first and most fundamental question upon which hinge numerous consequences of which we have in this book only a small number, placed there by the Author, who has done so much to open a pathway hitherto closed to minds of speculative turn. So far as experiments go they have not been neglected by the Author; and often, in his company, I have attempted in the following manner to assure myself that the acceleration actually experienced by falling bodies is that above described.[6]

He goes on to describe the experiments that Galileo has done to demonstrate that freely falling bodies are in fact uniformly accelerated.

In an experiment described by Salviati, the distance a body falls is measured as a function of time. Thus Galileo attempts to demonstrate that the motion is uniformly accelerated by verifying that one of the consequences of uniformly accelerated motion,

$$\text{distance} = v_0 t + \frac{1}{2}at^2 \qquad (2.22)$$

is in agreement with observation. We can essentially repeat what Galileo did but take advantage of modern equipment to photograph a falling

PLATE 2.1. Flash photograph of a falling golf ball. The time intervals between successive positions of the ball are equal. (F. W. Sears and M. W. Zemansky, *University Physics*, Addison-Wesley, Reading, Mass., 1952)

TABLE 2.2

Interval number	Interval length Δx, cm	Average speed $\Delta x/\Delta t$, cm/sec	Change in speed Δv, cm/sec	Acceleration $\Delta v/\Delta t$, cm/sec^2
1	7.70	231	32	960
2	8.75	263	31	930
3	9.80	294	32	960
4	10.85	326	34	1020
5	11.99	360	33	990
6	13.09	393	32	960
7	14.18	425	32	960
8	15.22	457	32	960
9	16.31	489	35	1050
10	17.45	524	32	960
11	18.52	556		
				Av. 980

SOURCE: Physical Science Study Committee, *Physics.* D. C. Heath, Boston, 1967.

golf ball (Plate 2.1), each picture taken at an equal time interval. In Table 2.2 is an analysis of a similar motion (a falling billiard ball this time). The time interval between recordings is $\frac{1}{30}$ sec. The calculated values of the acceleration are constant within the limits of accuracy of the measurements. (The last figure in the Δx column is quite uncertain; it is just a reasonable estimate of a fraction of a millimeter.) We can verify for ourselves, if we wish, within the accuracy of the measurements, that the acceleration is constant or that the distance fallen varies with the time as distance $= v_0 t + \frac{1}{2}at^2$. Or perhaps say with Simplicio:

I would like to have been present at these experiments; but feeling confidence in the care with which you performed them, and in the fidelity with which you relate them, I am satisfied and accept them as true and valid.[7]

The constant acceleration near the earth's surface, denoted by the special symbol g, is gravity

$$g = 980 \text{ cm/sec}^2 = 32 \text{ ft/sec}^2 \qquad (2.23)$$

Thus an object released from rest falls 16 ft or 490 cm in 1 sec, 64 ft in 2 sec, and so on.

THE MOTION OF PROJECTILES

On the motion of projectiles, Aristotle—who proposed that if a heavy body is released, it falls toward the center of the earth unless a force impels it otherwise—was particularly weak. For if a stone is released from a sling, where is the force that impels it to move upward before it falls down again? The air itself, Aristotle suggested, thrust out before the stone, rushes behind to push. But he did not seem any more satisfied by this explanation than those who followed him.

To attack this problem, Galileo suggested that a stone thrown horizontally is given a certain horizontal speed—this speed remains uniform (neglecting the resistance of the air) since there are no forces acting in the horizontal direction. There is a force in the vertical direction—the one that causes (?) bodies to fall toward the earth with a uniform acceleration. The motion of a projectile, Galileo suggested, is compounded of a uniform speed in the horizontal direction superimposed upon a uniform acceleration in the vertical direction. Perhaps Plate 2.2 will make this clear.

The important and not intuitively obvious point follows that an object thrown horizontally falls in just the same way as one dropped at the same time with no horizontal velocity. The only difference is that the body thrown moves uniformly in the horizontal direction as it falls with a uniform acceleration in the vertical direction.

$$\text{magnitude of horizontal component of displacement} = v_0 t \quad (2.24)$$

$$\text{magnitude of vertical component of displacement} = \tfrac{1}{2}gt^2 \ (2.25)$$

The displacement is composed of both these motions simultaneously and, as Galileo showed, the curve described by the projectile is, as a consequence of these assumptions, a parabola.

Sagredo is taken aback:

One cannot deny that the argument is new, subtle and conclusive, resting as it does upon this hypothesis, namely, that the horizontal motion remains uniform, that the vertical motion continues to be accelerated downwards in proportion to the square of the time, and that such motions and velocities as these combine without altering, disturbing, or hindering each other. . . .[8]

In the classic illustration, a cannonball is dropped from the mast of a moving ship (Fig. 2.4). From the point of view of an observer on shore,

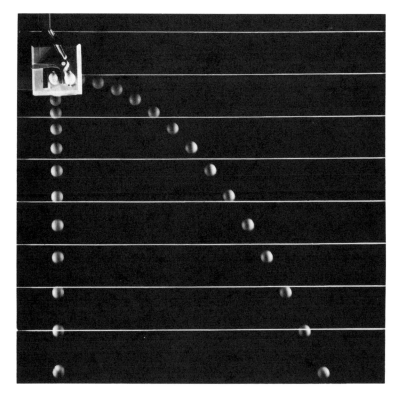

PLATE 2.2. Flash photograph of two golf balls, one projected horizontally at the same time that the other was dropped. The strings are 6 in. apart, and the interval between flashes was $\frac{1}{30}$ sec. (Physical Science Study Committee, *Physics*, D. C. Heath, Boston, 1967)

the ball falls with a uniform acceleration, while moving with constant speed in the horizontal direction. A sailor on the ship, however, is moving with the same horizontal speed as the ball, so that from his point of view the cannonball appears to fall straight down; it lands at the base of the mast.

It is the language that is particularly striking in the *Dialogue Concerning Two New Sciences*: for the first time it is modern. Excluding

FIG. 2.4. A falling cannonball from several points of view.

changes in fashion that inevitably occur in a period of 300 years, Galileo speaks to us in words and about problems that are immediately recognizable—forces, accelerations, uniform motions, and inertia. And when he gropes for a concept, such as the motive or the impelling force, he gropes in a direction that will eventually lead to a fruitful definition. When in the course of the arguments Salviati points out that the acceleration of the same body sliding down a smooth inclined plane varies with the incline of the plane and is less than that which it would experience in free fall, it is proposed that the force causing (?) the body to slide down the plane, also related to the incline, is proportional to the acceleration of the body. (This will become Newton's second law.)

In these dialogues he has applied the method of Euclid to the problem of bodies moving in space. He has introduced definitions and postulates and developed their consequences. In doing so he has developed a structure of bodies in motion, very much as Euclid developed a structure of relations in space. With the structure created by Galileo, the next generation of physicists would pose the question of what causes a body to accelerate uniformly near the surface of the earth, because, given Galileo's result, its movement as a function of time follows. Thus his observations of the actual motion of bodies under various circumstances near the surface of the earth, and his analysis of these motions, opened the way and provided the means by which, as he predicted with a certain modesty, "other minds more acute than mine will explore its remote corners."

3

WHAT IS A FORCE?

The use of the word "law" in scientific literature is not fortunate. We know of no legislation or decree that established the rules we describe; it is probably more accurate to think of them as having been invented rather than discovered. Historical records reveal the origin of some; others are concealed, because the man who first proposed them has not let us know how they came to his mind. But the invention of rules occurs continually in current scientific enterprises.

Let us now, ourselves, invent the law of forces.

In the absence of forces a body remains at rest or in uniform motion. But how do we identify a force? We would prefer not to be limited to saying that a force acts when the motion of a body changes, for then the law of inertia would have no empirical content: merely stating that a body continues in uniform motion unless it does not. But even this is not entirely meaningless. If we accept the law of inertia as a convention, we shall then look for forces whenever a body changes its motion.

We have convinced ourselves that sugar in a bowl does not spill itself. Therefore, if we find sugar spilled all over a table, we look for the pair of hands that did it. But our initial supposition is based on the observation that in the past we have been able to find who spilled the sugar. We believe it so firmly that, if we found no one who spilled the sugar we would still rather suppose that the someone had disappeared than suppose that the sugar spilled itself.

Fortunately, there are not too many forces in nature, and they can be identified; they generally feel like a push or a pull. Although we shall eventually refine the concept, we can begin with this intuitive identification of a force. Often a push or a pull on a body can be associated with some object in contact with it—a hand, a rope, water, or air. In addition, we all experience the heaviness of our bodies and of objects about us, which may be interpreted as due to a force between all bodies and the earth.

Let us agree that we can identify forces in the physical world. How does one invent mathematical objects that can be associated with them? The objects we invent will depend on what properties of force we are interested in characterizing. A positive integer is sufficient to characterize a number of pebbles. A positive number (not an integer) is sufficient to characterize a length. Will a number do for force? The pushes and pulls we experience have the quality of largeness or smallness, that is, magnitude. Such a quality could possibly be characterized by a number. However, forces also have the quality of direction. If we want to characterize both (and these are the two properties of greatest initial interest), we shall need a mathematical object that has at least these two properties.

Let us introduce an object which we call a force arrow or, for short, an arrow, which we denote by a capital letter with an arrow above it:

$$\vec{A}$$

A picture of the arrow is more informative. It can be drawn as a line which begins at a point (call it P) and terminates in an arrowhead. Such an arrow has at least the two properties that we require. It has magnitude (its length) and it has direction. We therefore say that a force acting on a point P is to be associated in some way with an arrow whose length corresponds to the magnitude of the force (say 1 in. for 1 lb) and whose direction is the direction of the force:

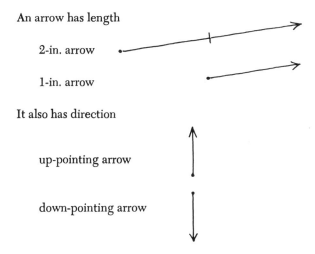

An arrow has length

    2-in. arrow

    1-in. arrow

It also has direction

    up-pointing arrow

    down-pointing arrow

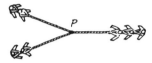

Now that we have the arrows, we need the rules. They must not only be consistent (contradictions are not allowed), but they must also allow us to make a correspondence between the arrows and the forces of the natural world.

What are the observed properties of the forces in the natural world? The most primitive is that of combination. Two, or more, forces can act on a point. Is this the equivalent of some other single force? We know that two forces can be balanced by a single force. Consider the forces acting here on the point P. Or consider a three-team tug of war in a momentary stalemate. The point P does not move, so according to the principle of inertia there must be no force acting on it. Therefore, in some sense, the forces acting on P (there are three forces acting on P in each case above) through the ropes or strings must combine to give no force.

The principle of inertia—the decision to classify uniform motion as the natural motion of the theory—thus places requirements on the structure of the forces that alter natural motions. Another choice of natural motions (Aristotle's, for example) would necessitate different rules for the forces. The ultimate requirement is that a theory which is in correspondence with the facts of experience can be constructed. In principle many theories are possible, but it is considered difficult to obtain even one.

We must therefore define an operation in the world of arrows which can correspond to the combination of forces in the physical world. We call this operation arrow addition and designate it by the symbol + (plus), which looks like ordinary addition but is not the same. The first rule of addition is

$$\vec{A} + \vec{B} = \vec{C} \qquad \text{(Postulate I)} \qquad (3.1)$$

The addition of two arrows gives us another arrow or, translated into the physical world, the combination of two forces gives a third force. Such a rule is not always true among either mathematical or physical objects. The combination of two liquids may give a solid, the addition of two fractions an integer, or the combination of two poisons a harmless substance —consider sodium and chlorine. Thus, although the statement that the combination of two arrows gives an arrow seems natural enough, it is not what one would call self-evident; it is a rule devised to correspond to the empirical nature of forces.

The second rule of arrow addition we propose is

$$\vec{A} + \vec{B} = \vec{B} + \vec{A} \qquad \text{(Postulate II)} \qquad (3.2)$$

That is, the order of addition of arrows is irrelevant or, in the world, the order of combination of forces is irrelevant. Adding force $\vec{A}$ to force $\vec{B}$ is the same thing as adding force $\vec{B}$ to force $\vec{A}$:

$$\vec{A} + \vec{B} = \vec{B} + \vec{A}$$

Again, this is not true among all mathematical objects or among events in the world. Aiming, then shooting, gives a different result from shooting, then aiming; the hairpiece must be put in place after the sweater has been pulled on. We are less familiar with mathematical objects that do not obey Postulate II, but they exist and are sometimes useful.

We propose further that there exists in the world of arrows a zero arrow (denoted by $\vec{0}$, an arrow with zero length), which is to be associated with a zero force and which has the property that added to any arrow it leaves that arrow unchanged.

There exists $\vec{0}$ such that for any arrow $\vec{A}$,

$$\vec{0} + \vec{A} = \vec{A} \qquad \text{(Postulate III)} \qquad (3.3)$$

$$\vec{0} + \vec{A} = \vec{A}$$

$$+ \ \nearrow \ = \ \nearrow$$

Further, we propose that for every arrow $\vec{A}$ there is also a negative arrow $-\vec{A}$, so that the sum

$$\vec{A} + (-\vec{A}) = \vec{0} \qquad \text{(Postulate IV)} \qquad (3.4)$$

This corresponds to the observation that for every force in nature there is always another force that will balance it.

The final rule we propose states that the order of the sum of three arrows is irrelevant:

$$(\vec{A} + \vec{B}) + \vec{C} = \vec{A} + (\vec{B} + \vec{C}) \qquad \text{(Postulate V)} \qquad (3.5)$$

which is to say, if we add first $\vec{A}$ and $\vec{B}$ and then add $\vec{C}$, the result is the same as adding first $\vec{B}$ and $\vec{C}$ and then adding $\vec{A}$.

We now ask a more specific question. Given the two arrows $\vec{A}$ and $\vec{B}$, given that they are combined (added), what is the arrow that results? That is to say, given the magnitude and direction of both $\vec{A}$ and $\vec{B}$, what should be the magnitude and direction of their sum? The postulates proposed above are the mathematical expressions of various qualitative properties of systems of forces, which we could recall if our attention was properly directed to them. How can we invent a rule for arrow addition which satisfies all these postulates and can be associated with physical forces?

There are at least two procedures. One is to look at the mathematical objects and invent a rule that is consistent, elegant, and aesthetically pleasing. There is absolutely no guarantee that the forces of the physical world will combine to follow the rule. However, it is a faith (not always justified) of theoretical physics that if man proposes what is sufficiently elegant, nature, pleased and flattered, will say yes. Or we might go into the laboratory (the world) and combine forces, one after another, measuring $\vec{A}$, $\vec{B}$, and their sum to see if there was any consistent pattern which we could discover in the sum $\vec{A} + \vec{B}$. *Door Meten tot Weten* ("knowledge through measurement") is the motto of the Kamerlingh-Onnes low-temperature laboratory (Leiden, The Netherlands). What are called physical laws have been invented almost always by doing both.

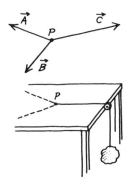

[To make such a measurement, we might subject point P to the forces $\vec{A}$ and $\vec{B}$ and then determine what $\vec{C}$ is required to keep the point from moving. But now we must make more precise how a force in the world is to be associated with an arrow. Imagine a table, a frictionless* pulley, and a weight on a string. The string is under tension due to the weight, and we assert that the string exerts a force on point P equal in magnitude to the magnitude of the weight, and in the direction of the string. Thus a 2-lb weight produces a 2-lb force, a 3-lb weight a 3-lb force, etc. Further, if we agree to ½ in. for 1 lb, the force systems on the real table would be associated with arrows, as shown in Fig. 3.1. (This method is neither

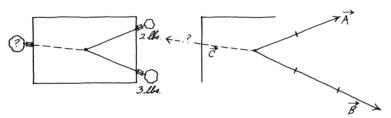

F I G. *3.1.* A method for associating forces with arrows.

unique, absolute, or even the best; it is simple and quick. Any success it achieves will depend—as with other conventions—on properties of the world that will be made more precise later. We ask only: Identifying forces and associating them with arrows as shown, can we construct a system of relations among the arrows that can be made to correspond to the relations among the forces?)]

From the principle of inertia (if a body is at rest or in uniform motion there is no force acting on that body), we conclude that the sum

$$\vec{A} + \vec{B} + \vec{C} = \vec{0} \qquad (3.6)$$

That is, the three forces add up to the zero force (no force). From this, adding $-\vec{C}$ to both sides and using the fact that

$$\vec{C} + (-\vec{C}) = \vec{0} \qquad \text{(III and IV)} \qquad (3.7)$$

we have

$$\vec{A} + \vec{B} = -\vec{C} \qquad (3.8)$$

or, the result of adding $\vec{A}$ to $\vec{B}$ is a force which when combined with $\vec{C}$ gives zero.

The object of making such measurements is to accumulate enough results so that we might see if some rule is evident that relates the magnitude and direction of the observed $\vec{C}$ to those of $\vec{A}$ and $\vec{B}$ (see Fig. 3.2). The process is not unlike the examination question that reads: "A sequence of numbers follows, of which the first are 1, 4, 9, 16, 25, . . . What is the next number?" In both cases we presume a rule can be made. Once proposed we use it to "predict" an event not already observed (that is, we test the rule).

* By frictionless we mean, for the moment, a well-oiled pulley that turns very easily.

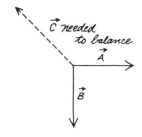

FIG. 3.2. The force, $\vec{C}$, that will balance the sum, $\vec{A} + \vec{B}$.

With or without these observations, in the end we must speculate, because no observations, no matter how complete, ever directly give us a rule. We hope that, using the observations and/or recalling to ourselves the properties of forces we know from our daily experience, we can limit the numbers of possible rules so that among the few that remain we might, possibly on esthetic considerations, pick one that works.

It is perhaps plausible to assume that if the forces are equal in magnitude and if they face in opposite directions, the total force should add to $\vec{0}$ (that is, no force at all). Thus, if the arrow for $\vec{A}$ is

$$\longrightarrow$$

the arrow for $-\vec{A}$ is

$$\longleftarrow$$

and

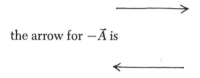

Now imagine two forces running in the same direction. Let us propose that they add just as numbers do; this seems to be what one observes. Two boys of the same strength, pulling in the same direction, seem to exert twice the force that one alone does. Thus

$$\vec{A} + \vec{A} = 2\vec{A} \tag{3.9}$$

or perhaps

$$2\vec{A} + 3\vec{A} = 5\vec{A} \tag{3.10}$$

$$\bullet\!\!-\!\!+\!\!\rightarrow \; + \; \bullet\!\!-\!\!+\!\!-\!\!+\!\!\rightarrow \; = \; \bullet\!\!-\!\!+\!\!-\!\!+\!\!-\!\!+\!\!-\!\!+\!\!\rightarrow$$

Thus we might be inclined to agree that arrows lying along the same line add as do ordinary numbers—just as two rulers, for example, placed end to end: the resultant magnitude the sum of the original magnitudes, the resultant direction just the original direction.

With this agreement we can define the meaning of the multiplication of an arrow by a number:

$$2\vec{A} = ? \tag{3.11}$$

We agree that

$$2\vec{A} = \vec{A} + \vec{A} \qquad (3.12)$$

is the arrow

magnitude: twice that of $\vec{A}$
direction: that of $\vec{A}$

Suppose now the two arrows are as shown; they are not opposite to one another, nor in the same direction.

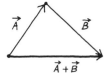

What is $\vec{A} + \vec{B}$ in this case? What is a general rule that gives

in the special cases? Here the warriors of theoretical physics fall as those heroes before Ilium, mist darkening their eyes. We cannot pretend that any logical steps will lead us to this rule, because many rules will do. We are forced to gamble and take the risk. In science, reputed to be the child of logic and cold reason, we find perhaps with surprise that the midwife must perform a creative act.

The gamble follows: We propose that two force arrows $\vec{A}$ and $\vec{B}$ combine so that $\vec{A} + \vec{B}$ is as shown. This now-famous rule (also called the triangle or parallelogram rule for the composition of forces) satisfies all the postulates proposed previously for arrow addition. It was introduced first by Simon Stevin (who dropped two balls from a height), an engineer with an immediate practical need for this knowledge.

Obeying all these rules, the arrows now become the prototype for what are called *vectors*,* that is, objects that obey the rules listed above. Just as different material things (apples, stones, and people) can be counted with the same numbers, different physical objects (forces and others to be introduced later) can be associated with vectors.

But now we ask if this rule is "true" or "correct." Logically, it is perfectly all right. A + B give another arrow C and A + B equal B + A. In the same way all the other postulates (I–V) are satisfied. We can assert that the structure that results is consistent. However, it would be possible to write down a large number of other rules of addition that also would not lead to contradictions. When we ask if the rule is correct, we do not ask only whether it is mathematically consistent (it is required to be so), but also whether vectors combine according to the rule in such a way that the physical force corresponding to A combined with the physical force corresponding to B results in. a force that is represented by the vector C. The only way to answer such a question is to combine actual

* From this point on, the conventional boldface type will be used to indicate vectors.

forces in the real world in an attempt to verify that they do in fact combine according to the rule stated above It is to obtain such verification that physicists go into the laboratory and do experiments.

(Such an experiment is not difficult to imagine or to do. If one has forces of magnitudes **A** and **B** acting on the point P in the directions shown in Fig. 3.3, one can measure the magnitude and direction of the

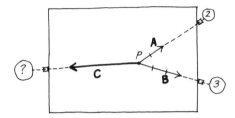

FIG. 3.3. Testing the rule.

force **C** required to keep the point P at rest—as was done previously. These measured magnitudes and directions would then be compared with those calculated using the rules. Presumably they should agree. It is a fact that they do.)

Thus by combining symbols on paper we can "predict" just what force will be necessary to balance two other forces in the real world. The equations we write seem to contain in themselves the working of the world; by their manipulation it seems one can manipulate nature itself. The most remarkable aspect is psychological. When a relation such as the rule of vector addition is first suggested as the rule for the combination of forces, the suggestion is tentative; but when buildings and bridges have been constructed for years following the rule, it becomes a law of nature. And we hear: "Forces are vectors." Of course, forces are not vectors—they are forces; but the association of forces with vectors has been so successful that the distinction becomes blurred. It is the almost magical power one has, sitting and writing in an office, to predict precisely what strength of steel cable will be necessary to support a strut in a bridge not yet built that produces the conviction that leads one to believe finally that a force is a vector.

SOME OF THE FORCES IN THE WORLD

In a sense one could say that the underlying problem of post-Galilean physics would be to identify the various forces of the world. We list a few.

1. The force attracting all heavy bodies toward the earth. Its magnitude is by definition the weight of the body on earth, and it is directed toward the center of the earth. Thus a 2-lb object has exerted on it a force of 2-lb magnitude directed down.

2. Various contact forces: pushes and pulls, let us say, transmitted by a string, a hand, or by any two palpable bodies in contact. These are probably the easiest to accept intuitively. Descartes, for example, would accept only contact forces between corpuscles when they collided. Consider, for example, a body pulled by a string. The magnitude of such a force depends on how hard the pull is, and it is directed in the direction of the string. The origin might be, for example, a weight hung over a pulley or a coiled spring.

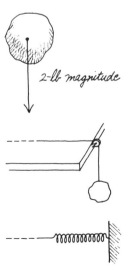

2-lb magnitude

3. Forces that arise due to the resistance of media, such as the resistance of air to a projectile or to an airplane, the resistance of water to the passage of a ship, or the resistance of one surface moved over another.

The identification of such as forces in their own right is one of the distinctions between Aristotelian and post-Galilean physics. For example, Aristotle would say it required a force to keep a body moving in uniform motion in a straight line (say a bale of hay along a level road). There is no question that it does require a force—one of the reasons farmers have horses—but if one identifies an opposing force between the bale and the road (how hard one has to pull depends upon the condition of the road), one can say that the total force is zero.

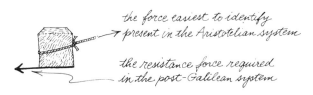

the force easiest to identify
present in the Aristotelian system

the resistance force required
in the post-Galilean system

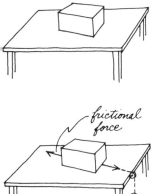

frictional
force

All these latter types are called frictional forces. Their origin is complex, but they have the general property that their magnitude depends upon the speed with which the body moves through the medium, and they are always directed to oppose the motion. Thus a body at rest on a table experiences no frictional force. However, if we attempt to pull it, we find a (resisting) force opposed to the direction of motion, whose magnitude depends in a very complicated way on the nature of the surfaces, and so on. It is a force we can sense directly. We know that it requires less push to move a sled across an icy lake than across a lawn.

SOME OTHER PHYSICAL QUANTITIES ASSOCIATED WITH VECTORS

The structure we have invented has an existence independent of forces. We could consider the consequences of the rules of vector addition even in a world where no forces existed. Just as for numbers and, as we shall see, for geometry, the same structure can be associated with a variety of physical situations.

Displacement in Space

One of the simplest physical things that can be associated with a vector is the operation of a displacement. Consider two points a and b in a plane (or two cities on a plain). If we displace an object or a person from a to b (independent of the actual path that was followed) we propose to associate this displacement with the vector **A** which begins at a and terminates at b. Its magnitude and direction are thus

> magnitude: distance between a and b
> direction: direction of line from a to b

(In doing this we have decided to retain an interest only in the initial and final positions. If we were interested in the actual path that was followed, the displacement vector would not be sufficient.)

If we now make the further displacement from b to c, this would be associated with the vector **B**. From this it is clear that the displacement from a to c, associated with the vector **C**, is related to the displacements from a to b and from b to c by

$$\mathbf{A} + \mathbf{B} = \mathbf{C} \tag{3.13}$$

We thus have another physical realization of the mathematical object—the vector. We have interpreted the vector and vector addition as a displacement and combination of displacements in space. It is not hard to verify that all the postulates for vector addition listed above are satisfied.

Galileo's analysis of projectile motion can be restated now as a decision to view the combination of displacements in different directions as the addition of vectors, and to regard the displacement in the horizontal direction as unaffected by the displacement in the vertical direction, as shown in Plate 3.1.

Velocity

Speed was defined as change in distance/time and uniform motion as motion with constant speed in a straight line. This definition can be restated in a compact way using the concept of the vector. In analogy with speed, which was defined as

$$v = \frac{\text{change in distance}}{\text{time interval}} = \frac{\Delta d}{\Delta t} \tag{3.14}$$

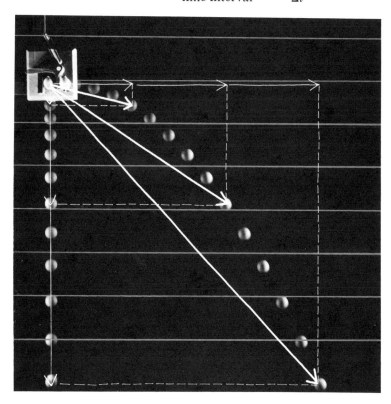

PLATE *3.1.* The displacements in the vertical direction are the same for both balls. (Photograph from Physical Science Study Committee, *Physics*, D. C. Heath, Boston, 1967)

we define **velocity** as

$$\mathbf{v} = \frac{\text{change in } \mathbf{displacement}}{\text{time interval}} = \frac{\Delta \mathbf{d}}{\Delta t} \qquad (3.15)$$

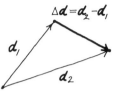

$\Delta \mathbf{d}$ = change in **displacement** = $\mathbf{d}_2 - \mathbf{d}_1$

(Note that $\mathbf{d}_1 + \Delta \mathbf{d} = \mathbf{d}_2$ from the definition of vector addition.)

From this definition it follows that the **velocity** is a vector whose magnitude is the speed and whose direction is the direction of motion

$$\textbf{velocity:} \begin{cases} \text{magnitude:} & \text{speed} \\ \text{direction:} & \text{direction of motion} \end{cases}$$

which means, for example, that the **velocity** of a body changes if *either* its speed or its direction of motion changes. Thus for motion in a straight line with changing speed or motion with unchanged speed on a curved path the **velocity** would not remain constant.

The **acceleration** vector is defined as

$$\textbf{acceleration : a} = \frac{\text{change in } \mathbf{velocity}}{\text{time interval}} = \frac{\Delta \mathbf{v}}{\Delta t} \qquad (3.16)$$

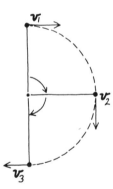

Thus a body may accelerate by, for example, increasing its speed along a straight line, or by changing the direction of its motion without changing its speed. A body at the end of a string moving at constant speed in a circle accelerates because its velocity changes direction. (Since it accelerates, a force is required—provided in this case by the tension in the string.)

4

"THE LION IS KNOWN BY HIS CLAW"

Isaac Newton, born the year that Galileo died, grasped the tools, the insights, the knowledge that had set the seventeenth century scientific world in a ferment, and adding his own inventions created the first great modern physical theory, a structure so remarkable that it dominated the landscape of human thought for two centuries. In the *Mathematical Principles of Natural Philosophy* (published in 1687) Newton restated and generalized

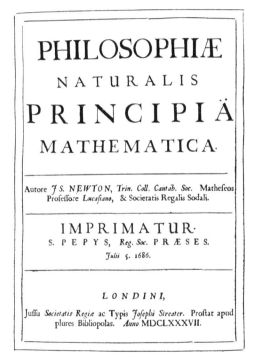

PHILOSOPHIÆ

NATURALIS

PRINCIPIÄ

MATHEMATICA·

Autore *J* S. *NEWTON*, *Trin. Coll. Cantab. Soc.* Mathefeos Profeffore *Lucafiano*, & Societatis Regalis Sodali.

IMPRIMATUR·
S. PEPYS, *Reg. Soc.* PRÆSES.
Julii 5. 1686.

LONDINI,

Juffu *Societatis Regiæ* ac Typis *Jofephi Streater.* Proftat apud plures Bibliopolas. *Anno* MDCLXXXVII.

FIG. *4.1.* Title page of Newton's *Principia.*

Galileo's findings in the form of two laws, added a third, and proposed further, that bodies attract one another according to a rule, now called the *law of gravitation.* He then developed the consequences of these postulates into a system of the world so comprehensive, from the motion of the planets to the ebb and flow of the tides, so detailed, including the precession of the earth's equinox (a small rotation of the earth's axis completed once every 26,000 years), that men like Alexander Pope would be dazzled:

> Nature and Nature's laws lay hid in night.
> God said: "Let Newton be"; and all was light.[1]

His powers as creator and user of mathematics and physics were so great that he has partially eclipsed many of his contemporaries, such as Hooke or Huygens, who contributed to and understood much of what was about to happen. To his fellows, Newton, "the Lion," was "known by his claw."

MATHEMATICAL PRINCIPLES
OF
NATURAL PHILOSOPHY

DEFINITIONS

DEFINITION I

The quantity of matter is the measure of the same, arising from its density and bulk conjointly.[2]

The quantity of matter is today called the mass, denoted usually by the letter m. The mass, Newton says, to relate it to things we know, may be set equal to a density multiplied by the volume of material.

$$\text{mass} = \text{density} \times \text{volume}$$

Thus twice the volume of the same material has twice the mass, while the same volume of different materials (for example, lead and aluminum) has different masses, owing to the different densities.

DEFINITION II

The quantity of motion is the measure of the same, arising from the velocity and quantity of matter conjointly.[3]

We now call the "quantity of motion" the momentum, and denote it by the vector **p**:

$$\textbf{momentum} = \text{mass} \times \textbf{velocity} \qquad (4.1)$$

$$\mathbf{p} = m\mathbf{v} \qquad (4.2)$$

Thus the momentum vector is

> magnitude: mass × speed
> direction: direction of motion

For a body of constant mass the momentum will be changed if either the speed or the direction of motion changes. Therefore, a body moving due north with a speed of 50 miles/h will experience a change in momentum if its speed is increased to 60 miles/h or if a turn is made so that it moves east at 50 miles/h.

DEFINITION IV

*An impressed force is an action exerted upon a body, in order to change its state, either of rest, or of uniform motion in a right line.**

> This force consists in the action only, and remains no longer in the body when the action is over. For a body maintains every new state it acquires, by its inertia only. But impressed forces are of

---

* In order not to change Newton's own words unnecessarily, we retain his use of "right line" (meaning straight line). This archaic English usage is preserved in modern French—*tout droit* means "straight ahead," and *une ligne droite* ("a right line") means a straight line.

different origins, as from percussion, from pressure, from centripetal force.[4]

Force for Newton (as will become clear later) was to be treated as a vector; he is trying to distinguish force from inertia—two concepts not clearly separated previously.

Then follow what he calls the axioms or laws of motion. They are the postulates of his system.

LAW I

Every body continues in its state of rest, or of uniform motion in a right line, unless it is compelled to change that state by forces impressed upon it.

> Projectiles continue in their motions, so far as they are not retarded by the resistance of the air, or impelled downwards by the force of gravity. A top, whose parts by their cohesion are continually drawn aside from rectilinear motions, does not cease its rotation, otherwise than as it is retarded by the air. The greater bodies of the planets and comets, meeting with less resistance in freer spaces, preserve their motions both progressive and circular for a much longer time.[5]

This is the law of inertia and defines the natural motion of the new physics.

LAW II

The change of motion is proportional to the motive force impressed; and is made in the direction of the right line in which that force is impressed.

> If any force generates a motion, a double force will generate double the motion, a triple force triple the motion, whether that force be impressed altogether and at once, or gradually and successively. And this motion (being always directed the same way with the generating force), if the body moved before, is added to or subtracted from the former motion, according as they directly conspire with or are directly contrary to each other; or obliquely joined, when they are oblique, so as to produce a new motion compounded from the determination of both.[6]

This is the famous second law. It proposes an answer to the question: How does the motion change under the action of an impressed force? The change of motion (momentum), as Galileo had suggested, is to be proportional to the motive force (from the way Newton used this, the force multiplied by the time it acts) and in the direction of that force. Thus

$$\mathbf{F}\,\Delta t = \Delta \mathbf{p}$$

In this one equation is condensed the underlying maxim of the Newtonian world.

The final law of motion is really a statement about the nature of the forces one finds in the world.

LAW III

To every action there is always opposed an equal reaction: or, the mutual actions of two bodies upon each other are always equal, and directed to contrary parts.

Whatever draws or presses another is as much drawn or pressed by that other. If you press a stone with your finger, the finger is also pressed by the stone. If a horse draws a stone tied to a rope, the horse (if I may so say) will be equally drawn back towards the stone; for the distended rope, by the same endeavor to relax or unbend itself, will draw the horse as much towards the stone as it does the stone towards the horse, and will obstruct the progress of the one as much as it advances that of the other. If a body impinge upon another, and by its force change the motion of the other, that body also (because of the equality of the mutual pressure) will undergo an equal change, in its own motion, towards the contrary part. The changes made by these actions are equal, not in the velocities but in the motions of bodies; that is to say, if the bodies are not hindered by any other impediments. For, because the motions are equally changed, the changes of the velocities made towards contrary parts are inversely proportional to the bodies. This law takes place also in attractions, as will be proved in the next Scholium.[7]

There then follow several corollaries; among these, the first makes explicit his decision to view forces as vectors.

COROLLARY I

A body, acted on by two forces simultaneously, will describe the diagonal of a parallelogram in the same time as it would describe the sides by those forces separately.

If a body in a given time, by the force M impressed apart in the place A should with an uniform motion be carried from A to B, and by the force N impressed apart in the same place, should be carried from A to C, let the parallelogram ABCD be completed, and, by both forces acting together, it will in the same time be carried in the diagonal from A to D. For since the force N acts in the direction of the line AC, parallel to BD, this force (by the second Law) will not at all alter the velocity generated by the other force M, by which the body is carried towards the line BD. The body therefore will arrive at the line BD in the same time, whether the force N be impressed or not; and therefore at the end of that time it will be found somewhere in the line BD. By the same argument, at the end of the same time it will be found somewhere in the line CD. Therefore it will be found in the point D, where both lines meet. But it will move in a right line from A to D, by Law I.[8]

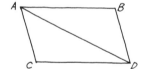

(A. Motte, trans., *Sir Isaac Newton's Mathematical Principles of Natural Philosophy and His System of the World*, University of California Press, Berkeley, 1962)

What he asserts, rephrased a bit, is that the result of the action of two forces on a body is related to their separate actions according to the parallelogram he has constructed, or that two forces **M** and **N**, acting on

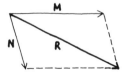

a body separately for a time interval Δt, give the same result as another force **R** acting for the same time interval. Thus the two forces combine so that

$$\mathbf{M} + \mathbf{N} = \mathbf{R}$$

But this is just the rule for vector addition proposed in Chapter 3.

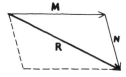

Next he explains a bit where all this has come from.

SCHOLIUM

Hitherto I have laid down such principles as have been received by mathematicians, and are confirmed by abundance of experiments. By the first two Laws and the first two Corollaries, *Galileo* discovered that the descent of bodies varied as the square of the time (*in duplicata ratione temporis*) and that the motion of projectiles was in the curve of a parabola; experience agreeing with both, unless so far as these motions are a little retarded by the resistance of the air. When a body is falling, the uniform force of its gravity acting equally, impresses, in equal intervals of time, equal forces upon that body, and therefore generates equal velocities; and in the whole time impresses a whole force, and generates a whole velocity proportional to the time. And the spaces described in proportional times are as the product of the velocities and the times; that is, as the squares of the times. And when a body is thrown upwards, its uniform gravity impresses forces and reduces velocities proportional to the times; and the times of ascending to the greatest heights are as the velocities to be taken away, and those heights are as the product of the velocities and the times, or as the squares of the velocities. And if a body be projected in any direction, the motion arising from its projection is compounded with the motion arising from its gravity. Thus, if the body A by its motion of projection alone could describe in a given time the right line AB, and with its motion of falling alone could describe in the same time the altitude AC; complete the parallelogram ABCD, and the body by that compounded motion will at the end of the time be found in the place D; and the curved line AED, which that body describes, will be a parabola, to which the right line AB will be a tangent at A; and whose ordinate BD will be as the square of the line AB. On the same Laws and Corollaries depend those things which have been demonstrated concerning the times of the vibration of pendulums, and are confirmed by the daily experiments of pendulum clocks. By the same, together with Law III, Sir *Christopher Wren*, Dr. *Wallis*, and Mr. *Huygens*, the greatest geometers of our times, did severally determine the rules of the impact and reflection of hard bodies, and about the same time communicated their discoveries to

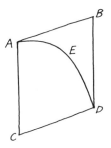

(A. Motte, trans., *Sir Isaac Newton's Mathematical Principles of Natural Philosophy and His System of the World*, University of California Press, Berkeley, 1962)

the *Royal Society*, exactly agreeing among themselves as to those rules. Dr. *Wallis*, indeed, was somewhat earlier in the publication; then followed Sir *Christopher Wren*, and, lastly, Mr. *Huygens*. But Sir *Christopher Wren* confirmed the truth of the thing before the *Royal Society* by the experiments on pendulums, which M. *Mariotte* soon after thought fit to explain in a treatise entirely upon that subject.[9]

The fundamental object in Newton's system is the quantity of motion (momentum):

$$\textbf{momentum} = \text{mass} \times \textbf{velocity}$$
$$\mathbf{p} = m\mathbf{v} \tag{4.3}$$

and the fundamental question is: How does this quantity of motion change under the action of impressed forces? When there are no forces on a body, its quantity of motion (momentum) remains unchanged (first law). Where there is a force on a body, the momentum changes so that the force, multiplied by the time it is applied, is equal to the change in momentum:

$$\mathbf{F}\,\Delta t = \Delta(m\mathbf{v}) \tag{4.4}$$

Dividing by the small time interval this becomes

$$\mathbf{F} = \frac{\Delta\mathbf{p}}{\Delta t} = \frac{\Delta(m\mathbf{v})}{\Delta t} \tag{4.5}$$

And if the mass of the body being pushed does not change (an important special case) we may write

$$\mathbf{F} = m\,\frac{\Delta\mathbf{v}}{\Delta t} = m\mathbf{a} \tag{4.6}$$

since $\Delta\mathbf{v}/\Delta t$ is by definition the acceleration. When $\mathbf{F} = 0$ it follows from Eq. (4.6) that

$$m\,\frac{\Delta\mathbf{v}}{\Delta t} = 0 \tag{4.7}$$

or

$$\Delta\mathbf{v} = 0 \tag{4.8}$$

and so the body moves uniformly. The first law of motion is therefore a consequence of the second: that case in which the force is zero. Newton listed them separately, however, and the significance of the first law is emphasized by this repetition. It states the decision to view motion from the point of view of forces and changes of momentum.

An inspection of Eq. (4.6) reveals immediately some of its consequences. When the force is zero the velocity is constant, in agreement with the first law. The change of velocity is in the direction of the force. Both sides of the equations are vectors, which is necessary, because it is not possible to write an equation consistently in which a vector is said to be equal to a number. If the force is constant, the acceleration is constant; thus the motion of bodies falling near the surface of the earth might be attributed to the action of a constant force.

If a force acts on a body that body changes its motion in the direction of the force.

1. If the body was at rest initially, it begins to move in the direction of the force:

2. If the body was moving initially with a velocity v_0:

a. If the force is in the direction of v_0, the speed is increased:

b. If the force is opposed to v_0, the speed is decreased:

c. If the force is at some angle to v_0, the direction and magnitude of v_0 may be changed:

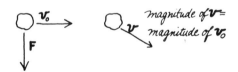

It is possible to change the speed of a body without changing its direction of motion by applying a force in the direction of v_0 or opposed to it. Is it possible, we might ask, to change the direction of motion without changing the speed? The answer is yes; we must apply a force that has no component parallel to the direction of motion—a force perpendicular to the velocity:

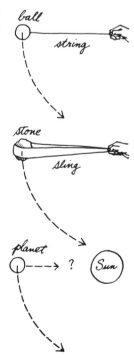

The velocity vectors have the same magnitude but changing direction. Drawn from the same center they go around a circle. There is a small trick here, however; the force must continually turn so as to remain perpendicular to the velocity. But this is not difficult to achieve: a ball on a string, a stone at the end of a sling, or a planet moving about its sun in a circular orbit are all examples.

UNITS OF FORCE AND MASS[*]

To make some quantitative statements, and to compare consequences of Newton's postulates with experience, it is convenient to introduce various units in which force and mass can be measured.

[*] See also the Appendices, pp. 493-495.

Length and time are commonly measured in meters, centimeters, or feet, and seconds. If mass is measured in $\left(\begin{array}{c}\text{grams}\\\text{kilograms}\end{array}\right)$,[*] force is measured in $\left(\begin{array}{c}\text{dynes}\\\text{newtons}\end{array}\right)$.

A $\left(\begin{array}{c}\text{dyne}\\\text{newton}\end{array}\right)$ of force is by definition that force which gives a body of 1 $\left(\begin{array}{c}\text{gram}\\\text{kilogram}\end{array}\right)$ mass an acceleration of 1 $\left(\begin{array}{c}\text{centimeter}\\\text{meter}\end{array}\right)$ /sec^2:

$$1 \left(\begin{array}{c}\text{dyne}\\\text{newton}\end{array}\right) = 1 \left(\begin{array}{c}\text{gram}\\\text{kilogram}\end{array}\right) \times \frac{1 \left(\begin{array}{c}\text{centimeter}\\\text{meter}\end{array}\right)}{\text{sec}^2} \qquad (4.9)$$

Although the justice of using Newton's name to describe a unit of force can hardly be disputed, and although this unit is a part of the convenient metric system, it brings no immediate sense of its magnitude to those of us used to English measure. In the English system, ironically, Newton is given no mention; the unit of force is the pound. Those of us born and bred in the United States have an immediate sense of the force exerted on us by an object that weighs 1 lb; we know how hard to expect to lift to pick up a package that weighs 1 lb. An object that is said to weigh 1 lb is an object on which the earth pulls with a force of 1 lb; to lift it, we must pull it away from the earth with a force at least as great as 1 lb—a force approximately equal to 4.5 newtons (450,000 dynes).

What is commonly known as 1 lb of butter, however, is clearly not a force. It is that quantity of butter on which the earth pulls with a 1-lb force. On the moon the same quantity (mass) of butter would feel less heavy (it would no longer weigh 1 lb); in outer space it would weigh nothing at all; however, this quantity of butter, which affects scales so differently depending upon what planets are nearby, can be used to cook as many omelets in one place as the other.

Suppose now we choose a mass so that 1 lb of force pulling it will give it an acceleration of 1 ft/sec^2. (This would be considered a convenient unit of mass in the English system.)

$$1 \text{ lb} = (1 \text{ English mass unit}) \times \frac{1 \text{ ft}}{\text{sec}^2} \qquad (4.10)$$

The system that has no object denoted by the unit newton calls its mass unit a slug:

$$1 \text{ lb} = 1 \text{ slug} \times \frac{1 \text{ ft}}{\text{sec}^2} \qquad (4.11)$$

That quantity of matter which is a slug weighs about 32 lb near the surface of the earth. The mass of the average heavyweight boxer is therefore about 6 slugs. A force of 1 lb on this boxer will impart to him an acceleration of $\frac{1}{6}$ ft/sec^2. To accelerate the average automobile (say one that weighs 3200 lb and thus has a mass of 100 slugs) to 60 miles/hr in

---

[*] These standard quantities of matter can be found in Paris, Washington, and under glass jars in most physics laboratories; near the surface of the earth they weigh about $\left(\begin{array}{cc}0.0022 & \text{lb}\\2.2 & \text{lb}\end{array}\right)$.

TABLE *4.1*

Fundamental units	CGS	MKS		English	
Length	centimeter	meter:	there are 100 cm in 1 m	foot:	there are 30.5 cm in 1 ft
Mass	gram	kilogram:	there are 1000 g in 1 k	English mass unit (slug):	there are 14,590 g in 1 slug
Time	second	second:	there is 1 sec in 1 sec	second:	there is 1 sec in 1 sec
Force	dyne	newton:	there are 10^5 dynes in 1 newton	pound:	there are 444,823 dynes in 1 lb

10 sec (88 ft/sec in 10 sec, an acceleration of 8.8 ft/sec^2) requires a force of 100 slugs × 8.8 ft/sec^2 = 880 lb.

The mass of a bullet might be 1 g; a force of 2×10^7 dynes will give it an acceleration of 2×10^7 cm/sec^2. Starting from rest, in 10^{-3} sec it will acquire a speed of 2×10^4 cm/sec. In that time it will move a distance

$$d = \tfrac{1}{2}at^2 = \tfrac{1}{2} \times 2 \times 10^7 \text{ cm/sec}^2 \times (10^{-3} \text{ sec})^2 = 10 \text{ cm} \quad (4.12)$$

Thus if a gun barrel is about 10 cm (3.94 in.) long and the explosion of the gunpowder provides an approximately constant force of 2×10^7 dynes (about 45 lb) over the length of the barrel, the bullet will leave the barrel with a speed of 2×10^4 cm/sec (about 655 ft/sec).

UNIFORM CIRCULAR MOTION

Let us now pose the question, so important historically: What force is required to keep a body moving in a circle at constant speed? Aristotle would have said: No force is required for celestial bodies, because their natural motion is a circle about the center of the universe. But if the celestial bodies are made of earthly material, their natural motion in the post-Galilean physics is in a straight line with uniform speed. Therefore, some force is required if they are to move in a circle.

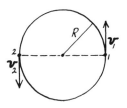

We first ask: What is the acceleration of a body moving in a circle of radius R, at the constant speed v? The velocity is not constant (it is directed up at 1 and down at 2) so, by definition, the body accelerates.

Consider, for example, the acceleration over half a circle (positions 1 and 2 in the diagram). The average acceleration of a body in the time t is, by definition,

$$\text{average acceleration} = \frac{\text{change in velocity}}{\text{time interval}} \quad (4.13)$$

which in this case is

$$\text{Change in } \mathbf{velocity} = \mathbf{v}_2 - \mathbf{v}_1 = \Big\downarrow \text{-} \Big\uparrow$$

The magnitude of $\mathbf{v}_2 - \mathbf{v}_1$ is $2v$, the direction of $\mathbf{v}_2 - \mathbf{v}_1$ is down ($\downarrow$), and the time to traverse half the circle = distance/speed = $\pi R/v$. Thus the average acceleration has the magnitude

$$\frac{2v}{\pi R/v} = \frac{2}{\pi} \frac{v^2}{R} \tag{4.14}$$

and the direction is down.

Here we are asked to calculate the change in velocity of a particle moving through space.

We perform the vector sums by moving the vectors about in space (putting the tail of $-\mathbf{v}_1$ on the head of $\mathbf{v}_2$). This is consistent with the rules in the same sense that adding 30 apples in the barrel near the door to 50 apples in the barrel near the stove gives

$$30 \text{ apples} + 50 \text{ apples} = 80 \text{ apples}$$

(Where? In the store, presumably.)

When we add $\mathbf{v}_2$ to $-\mathbf{v}_1$ we do not ask where in space the particles had these velocities. Rather, we ask how the velocity of the particle changed (as it was moving through space). Thus we might have said

$$\mathbf{v}_2 \text{ (at the point in space 2)} - \mathbf{v}_1 \text{ (at the point in space 1)} = \\ \mathbf{v}_2 - \mathbf{v}_1 = \Delta\mathbf{v} \text{ (as above)} \tag{4.15}$$

Forces with which we introduced the concept of a vector sometimes act at a point (the apples are all in the same barrel), and there the question does not arise.

The result (4.14) is crude but gives an idea of what the precise result will be. To do better, we can calculate exactly as above, except that we allow the time interval to grow very short. We hope that as the time interval grows very small, the acceleration that we compute will become independent of the precise size of Δt as long as Δt is very small. This hope turns out to be justified. The result is

$$\text{instantaneous acceleration} = \frac{v^2}{R} \text{ (magnitude)} \tag{4.16}$$

The direction is from the body to the center of the circle.

Except for the factor $2/\pi$ (4.16) is the same in form as the average acceleration over one half of the circle. A body moving in a circle with

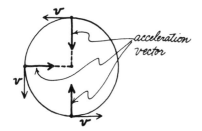

acceleration vector

constant speed is thus accelerated toward the center, the magnitude of the acceleration being v^2/R. We note that the acceleration is perpendicular to the velocity. This result, so important in the theory of planetary motion, was obtained by Huygens[10] and Newton.[11] It is an exact consequence of the definition of acceleration in this situation of uniform circular motion.

We find the force required to keep a body moving in a circle immediately by using Newton's second law:

$$\mathbf{F} = m\mathbf{a} \tag{4.17}$$

$$F \text{ (magnitude)} = \frac{mv^2}{R} \tag{4.18}$$

The direction is from the body to the center of the circle. Thus from this point of view, the force required to keep a planet moving in a circular orbit around the sun at constant speed must be directed from the planet toward the sun.

These are some of the tools with which Newton made his attempt to understand the motion of the planets. This problem, which had occupied astronomers' attention for 2000 years, had become the consuming scientific question of Newton's time. It was the ideal mate to the laws of motion proposed by Newton, and the solution of the problem of the planets via Newton's theory was the supreme achievement of seventeenth-century science. For on earth the question of forces is tedious and difficult; there are many forces and they tend to be complicated: friction, air resistance, gravity, pushes, pulls, tugs, and so on. In the absence of a really clear-cut problem, it might be difficult to say that anything had been achieved. Newton's laws of motion, his view of the world, in the absence of any resulting clarity would not be readily distinguishable from, let us say, Aristotle's view, which also had a certain success. We turn, therefore, to the problem of the planets and to its ultimate formulation in the hands of Kepler.

Courtesy National Aeronautics and Space Administration

5

THE MUSIC OF THE SPHERES

One time the earth rested on an elephant that stood on the back of a tortoise who required no further support. In Thales' time the earth was hemispherical and fixed, supported by pillars. The moon and sun moved above, crossing the sky at the appointed time, disappearing and somehow returning to the proper starting point to appear again at the proper hour. Anaximander, probably the first man ever to do so, conceived of the earth unsupported, floating in space, but the center of the universe, so it did not fall. The Pythagoreans, possibly Pythagoras or Parmenides, contributed the idea that the earth was spherical, believing as they did that the sphere was the most perfect physical form. In a very short time the earth had become spherical, unsupported, and poised in the middle of the universe. Thus began that great debate concerning the earth's place in the cosmos, which would continue for over 2000 years.

The Pythagoreans surrounded their spherical earth by eight giant, transparent, concentric spheres which bore all the objects in the sky and which revolved around the earth on different axes and at different speeds. These spheres were meant first to explain the most striking fact about the heavens—that in spite of the nightly turning of the heavenly bowl, the stars (as though they were tacked on the inside of a great revolving sphere and we were in the center of this sphere watching) always maintain the same relations to one another (see Plate 5.1). The Big Dipper, turning and moving, sometimes pouring, sometimes upright, always looks like the Big Dipper.

Among the fixed stars move the sun, the moon, and those points of light we know as the planets (from the Greek word for wanderer). They

PLATE 5.1. One-hour time exposure taken with the camera pointed at the North Star. The circular arcs show the apparent (?) motion of the stars. (Yerkes Observatory photograph)

are distinguished from the stars partially by their brightness but primarily because they do not remain fixed with respect to their celestial neighbors, appearing to wander erratically about the heavens. This erratic motion attracted the attention of early astronomers to them. (There were five known to ancient man: Mercury, Venus, Mars, Jupiter, and Saturn, as well as the sun and the moon, also considered to be planets.) It was noted that they seemed brighter than the stars, and that their distance from the earth seemed to change. Soon they became associated with various human endeavors (Venus, love; Mars, war—astrologers would see the future in the planetary positions of the present), as though they formed an intermediary between the immutable perfection of the stars and the restless imperfection of the earth.

The first major problem of astronomy was to explain the apparently peculiar motions of the planets (see Fig. 5.1) in some reasonable way. The Pythagoreans, by putting the sun, moon, and planets on different spheres, each with its own uniform rate of rotation, attempted to account for their motion with respect to the stars. However, one Pythagorean, Philolaus,

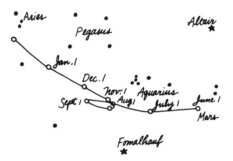

F I G. *5.1.* Progress of the planet Mars with respect to the fixed stars. At various intervals, Mars appears to reverse its direction of motion. (After R. H. Baker, *Astronomy*, D. Van Nostrand, Princeton, N.J.)

proposed that the earth was just a planet like all the others, revolving around a central fire and joined in its revolutions by all the other heavenly objects, including the sun. This idea was immensely controversial and provocative (so much so that Plato, who eventually rejected it, is reported to have paid about $2500 for a copy of Philolaus' book).

Plato decided finally that the earth was immobile. However, the spheres conceived by the Pythagoreans were not sufficient to explain the irregular paths of the planets. According to long tradition,* Plato set a problem to astronomers ("those studying such things") to find "which uniform and ordered movements must be assumed to account for the apparent movement of the planets."

Plato and his Greek contemporaries had their preconceptions about "ordered and uniform" movements. Because uniform motion in a circle was the most perfect, the most symmetric of motions, it was the most suitable for a celestial body. (And the disposition to attribute highly symmetric properties to the fundamental objects of the natural world is one, as we shall see, that is still very much with us.) So the problem became: What are the combinations of perfect circular paths along which the planets really move?—possibly the first recorded thesis problem. It occu-

---

* Simplicius, an astronomer of the sixth century A.D., in a commentary on Aristotle's *De Caelo*, wrote that Sosigenes, Julius Caesar's astronomer, quoted Eudemus of Rhodes (fourth century B.C.) to the effect that Plato posed this problem.

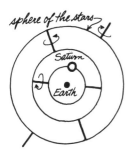

F I G. 5.2. Example of a system of concentric spheres proposed to account for the observed motion of Saturn. (After G. Holton and D. Roller, *Foundations of Modern Physical Science*, Addison-Wesley, Reading, Mass., 1952)

pied the attention of astronomers until the time of Kepler, almost 2000 years later, and led to some of the most fruitful creations of science.

Although Plato made no known effort to answer his question, a sometime student at his academy, Eudoxus,* who was probably far more skillful a geometer than Plato himself, proposed that 26 simultaneous uniform spherical motions† would duplicate the observed motion of the planets and one more would take care of the fixed stars. Aristotle later added 28 spheres, to remedy some of the most obvious failings of Eudoxus' system. There were problems within these systems, notably the problem that the planets appear brighter at some parts of the year than others. If, as the Greeks did, one assumes that the planets are immutable, the varying brightness is difficult to explain by any system of concentric spheres centered at the earth.‡

In spite of the overwhelming sentiment in favor of geocentric theories, Aristarchus (220–150? B.C.) proposed that the *sun* was the center of the universe, and that all the planets, including the earth turning once a day on its north-south axis in its proper place between Venus and Mars, revolved around it in great circles. This system, so strikingly modern, explained immediately the apparent change in brightness of the planets during the course of the year. However, it did substantial violence to the Greek conceptions about the place of the earth and to the dynamics developed, for example, by Aristotle. In addition, Aristarchus' proposal seemed to be in disagreement with at least one piece of evidence. If the earth moved in an orbit about the sun, the direction in which a fixed star appeared from the point of view of an observer on the earth would change during the course of the year—a phenomenon known as *parallax*. The observed angle between two stars (called star 1 and star 2 in the figure) would appear to change during the course of the year just as, for example, the angular separation of two trees on a road appears to change as we pass the trees in a car. No trace of such a parallax had ever been observed by the Greeks. Thus they had the alternative either of assuming that the stars were extraordinarily far away compared to the size of the earth's orbit, in which case the change in angle during the course of the year would become unobservably small, or of assuming that the earth remained in one place.

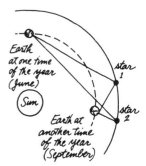

They chose the latter. Apollonius put the earth back in the center

* Sarton believes that it was Eudoxus who posed the problem.
† Three spheres each for sun and moon; four spheres each for the five planets.
‡ The measurements and observations on which these theories were based were real even if the instruments available were simpler than those in use today. Astronomers had been measuring the relative positions of the planets and the stars for centuries; by the end of the second century B.C. they had even observed the shift of the North Star due to the precession of the axis about which the earth rotates.

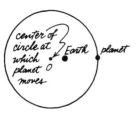

of the universe, with all other heavenly objects revolving about it in perfect circles, as was "required." But to explain the apparent change in brightness of the planets he displaced the center of the circles from the center of the earth, making the planetary orbits eccentric circles.

To explain the apparent halts and retrogressions of the planets, Hipparchus* of Nicaea (second century B.C.) added an ingenious new device, the *epicycle*. It was another circle centered on the original circle now "centered" off center. The planet moved uniformly on the epicycle while the center of the epicycle moved uniformly about O.

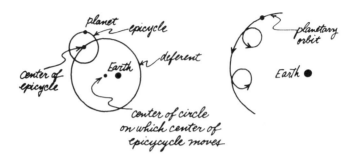

In the chaos that followed the decline of the Greek city-states, most intellectual life shifted to Alexandria. There, in the second century A.D., Claudius Ptolemy compiled his encyclopedic summary of the Greek effort to order the motion of the stars and planets. Called by the Arabs (to whom, as with many other ancient works, we owe its preservation) *Almagest*, meaning "The Greatest of Books," it was greatly influenced by Hipparchus and was replete with deferents, eccentrics, and epicycles. This massive codification of all the then-known astronomical theories and observations is one of the few Greek astronomical treatises that survived until medieval and modern times, thus exerting a permanent influence on all astronomical thought until Copernicus.

Ptolemy's system, remarkably complex, was the ultimate effort to order the motion of the planets within a geocentric system. He began by placing the earth at the center, with the fixed stars on a rotating spherical bowl:

---

* He also introduced several new instruments, and with these he was able to make more detailed and exact observations, on the basis of which he further refined Apollonius' theory, determining in much greater detail the exact eccentricity of the orbits of the sun and the moon.

The sun, the moon, and the planets rotated about the earth in an elaborate combination of various uniform circular motions—eccentric motion:

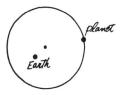

in which the center of the circular orbit was displaced from the center of the earth, epicyclic motion in which one circle was placed on another (the second circle rotating on the first),

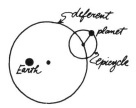

and uniform motion with respect to a displaced point (the equant):

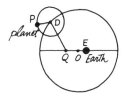

In this ultimate combination, the planet at *P* moves uniformly in a circle about the point *D*, which moves uniformly with respect to *Q* and appears as a circle with respect to *O*; the earth itself is placed at the point *E*.

With the help of all these it was possible to make orbits almost as complex as one wished.

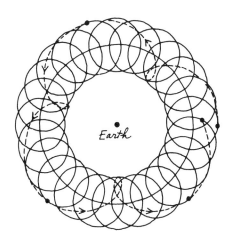

FIG. 5.3. Epicyclic motion that yields retrograde motions. (After G. Holton, *Introduction to the Concepts and Theories of Physical Science*, Addison-Wesley, Reading, Mass., 1952)

Using these various devices, he was able to produce a system of planetary orbits that was consistent with both his observations and the observations of Greek and other astronomers in centuries past. Thus, the Ptolemaic

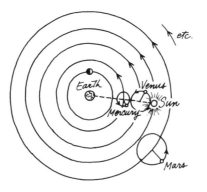

F I G. 5.4. A much simplified version of the Ptolemaic system. (After G. Holton, *Introduction to the Concepts and Theories of Physical Science*, Addison-Wesley, Reading, Mass., 1952)

system (Fig. 5.4), although possibly a little complex for our taste, represented an answer in his terms to the question posed by Plato. Ptolemy had created a system of orbits that reproduced the observed motions of the planets and of the stars. It was consistent with the general Greek tradition of dynamics. It satisfied one's intuitive feeling that the earth was still and fixed, and the most primitive observations—that the stars, the sun, the moon, and the planets moved. It was a theory consistent with both the presuppositions of his day and with the observations that had been made; its complexity was a result of the struggle to produce planetary orbits that would be consistent with these observations. But for all its complexity, it gave the planetary orbits as they had appeared in the centuries before and it predicted moderately well their future positions.

COPERNICUS' SUN-CENTERED SYSTEM

This planetary system of Ptolemy, corrected, amended, revised, with circles added to circles on circles, passed from generation to generation until the fifteenth century, was not incorrect or even inaccurate; but it lost any esthetic appeal it might once have claimed. In the thirteenth century, Alfonso X, King of Castile, confided that had he been consulted at the Creation, he would have made the world on a simpler and better plan. And later John Donne would complain:

> We thinke the heavens enjoy their Sphericall,
> Their round proportion embracing all.
> But yet their various and perplexed course,
> Observ'd in divers ages, doth enforce
> Men to finde out so many Eccentrique parts,
> Such divers downe-right lines, such overthwarts,
> As disproportion that pure forme: It teares
> The Firmament in eight and forty sheires, . . .[1]

In Nicolaus Copernicus (born in Polish Prussia in 1473), this dissatisfaction became so great that he was led finally to question in a strong and systematic way the hypothesis on which all ancient cosmology and physics was founded—the notion that the earth stood still and the heavens and planets revolved about it. He realized that the curves which described planetary motions would be remarkably simplified if the sun, rather than the earth, were considered the center of the solar system. This realization led him to conclude that the earth was neither at the center of the universe nor at rest, but was a planet, rotating about its own axis once a day and, along with the other planets, revolving around the sun.

Copernicus' deepest objection to the Ptolemaic theory was that he felt that in it circles had been so loosely combined with circles that the resulting motions, although consistent with the available numerical data, were not uniform in the sense originally desired. In particular he objected to the detailed assumptions (especially the equant) of the Ptolemaic theories, which in his mind destroyed the desired uniformity of motion. With reference to these he wrote:

> Having become aware of these defects, I often considered whether there could perhaps be found a more reasonable arrangement of circles, from which every apparent inequality would be derived and in which everything would move uniformly about its proper center, as the rule of absolute motion requires.[2]

He strengthened his proposal that the earth moves by referring to classical writers with similar ideas:

> . . . according to Cicero, Nicetas had thought the Earth moved, . . . according to Plutarch, others had held the same opinion . . . when from this, therefore, I had conceived of its possibility, I myself also began to meditate upon the mobility of the Earth.
>
> And although it seemed to me an absurd opinion, yet, because I knew that others before me had been granted the liberty of supposing whatever circles they chose in order to demonstrate the observations concerning the celestial bodies, I considered that I too might well be allowed to try whether sounder demonstrations of the revolutions of the heavenly bodies might be discovered by supposing some motion of the Earth . . . I found after much and long observation, that if the motions of the other planets were added to the motions of the Earth . . . not only did the apparent behavior of the others follow from this, but the system so connects the orders and sizes of the planets and their orbits, and of the whole heaven, that no single feature can be altered without confusion among the other parts in all the Universe. For this reason, therefore, . . . have I followed this system.[3]

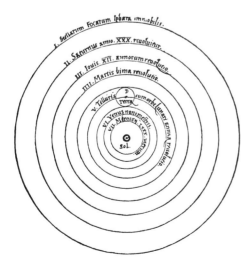

F I G. 5.5. Copernican orbits followed by the earth and the other planets as they move around the sun and the large, immobile sphere on which the fixed stars are located.

One qualitative simplification obtained within the Copernican system is the orbits of Mercury and Venus, the morning and evening stars—the bright objects that sometimes rise just before the sun or set just after it. In the Copernican system this phenomenon is easy to understand. Both Mercury and Venus circle close to the sun; when viewed from the earth, they follow close behind its rising or setting.

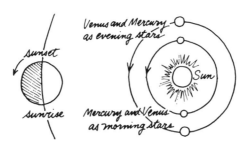

In the Ptolemaic system, if Mercury, Venus, and the sun travel independently on separate orbits, they should appear at some time widely separated in the sky. He thus must assume these planets are somehow attached to the sun and travel with it. Why are these two planets singled out to be attached to the sun, while the others are free to roam?

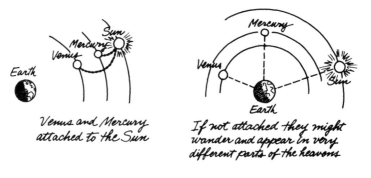

Venus and Mercury attached to the Sun

If not attached they might wander and appear in very different parts of the heavens

About his system Copernicus said:

> The first and highest of all the spheres is the sphere of the fixed stars. It encloses all the other spheres and is itself self-contained; it is immobile; it is certainly the portion of the universe with reference to which the movement and position of all the other heavenly bodies must be considered. If some people are yet of the opinion that this sphere moves, we are of a contrary mind. . . .[4]

Then, describing the planetary spheres and their periods of rotation in which the earth appears as one of six planets and the moon is clearly designated as a satellite of the earth, he concluded:

> In the midst of all, the sun reposes unmoving. Who, indeed, in this most beautiful temple, would place the light-giver in any other part than that whence it can illumine all the other parts?[5]

Frame of Reference

One can take the view that the difference between Copernicus and Ptolemy rested in their choice of a frame of reference. Ptolemy was standing on earth and preferred to believe that the earth was standing still. Copernicus found it simpler to view the motion of the planets as though he were standing on the sun. Since we now believe that the question: Does or does not the earth move? is not entirely answerable[*] we replace it by: Where is it most convenient to stand if we want to describe the motion of the solar system? Whether and how a body is moving depends, as Oresme said, upon the motion of the person observing the body. The motion of the cannonball, dropped from the mast, which appears to be a straight vertical line to the sailor on the ship, appears to be a parabola to the observer on shore. If we take into account the motion of the observers, we realize that the two are seeing the same event. But one sees it more simply than the other. The choice of a frame of reference plays a similar role in the description of celestial observations.

In Copernicus' own time, however, whether or not the earth moved was a serious and meaningful question. As he realized, the notion that the earth moves seemed absurd. All of medieval cosmology and physics was based on the idea that the earth was the center of the universe. Objects such as stones were thought to fall to the earth because they tended to fall toward the center of the universe. If the earth were in rapid motion, it could no longer easily be considered to occupy a preferred position in space. Why then should objects fall toward it? The fact that to us the entire question may seem so simple (there may not even seem to be a problem) is a demonstration of how thoroughly we have accepted modern ideas of inertia and the uniformity of space. If, for example, the heavy sluggish earth really had to be pushed in order to move (as people believed in Copernicus' time), would we be so willing to accept the statement that the earth is moving at about 18 miles/sec in its orbit around the sun?

Objections to the Copernican Theory

To achieve his simplified orbits it was necessary for Copernicus to discard the entire picture of the world that had been developed from the time of Aristotle. He expected much criticism—which was one of the reasons he delayed publication of his book so long that he did not see a printed copy of it until the day on which he died.

Anticipating many of the objections, he attempted to answer them beforehand. To the argument that the birds in flight should be left behind by the rapidly moving earth, Copernicus answered that the atmosphere is dragged along with the earth. To the argument that his earth, rotating so rapidly about its own axis, would surely burst like a flywheel driven too fast, he countered: "Why does the defender of the geocentric theory not fear the same fate for his rotating celestial sphere—so much faster because so much larger?" However, to this the proponents of the Ptolemaic theory would answer that the heavenly bodies were made of fine subtle stuff, and it was their nature to revolve in a circle. They argued that the earth, in

---

[*] Its rotation about its axis (with respect to the fixed stars) we can detect, but its uniform motion through space we cannot.

contrast, was sluggish, heavy, and immobile, therefore very difficult to move. (Only later, when Galileo saw details of the moon and the planets through his telescope, did the possibility that the heavenly bodies were also made of heavy earthlike material become a serious consideration.)

There were many other arguments, counterarguments, and violent invective. Luther branded Copernicus a fool and a heretic. The Copernican theory was denounced as "false and altogether opposed to the Holy Scriptures." For it had been written that the sun moved:

> Then spoke Joshua to the Lord in the day when the Lord gave the Amorites over to the men of Israel; and he said in the sight of Israel,
> "Sun, stand thou still at Gibeon,
> and thou Moon in the Valley of Aijalon."
> And the sun stood still, and the moon stayed, until the nation took vengeance on their enemies. Is this not written in the book of Jashar? The sun stayed in the midst of heaven, and did not hasten to go down for about a whole day.[6]

Controversy concerning this new conception of the universe raged for over a hundred years. About 1600 Hamlet listed the motion of the sun as an unquestionable verity, as true as his love for Ophelia:

> Doubt thou the stars are fire;
> Doubt that the sun doth move;
> Doubt truth to be a liar;
> But never doubt I love.[7]

But in 1611 Donne lamented the passing of the certitude of the old cosmology:

> And new Philosophy calls all in doubt,
> The Element of fire is quite put out;
> The Sun is lost, and th'earth, and no man's wit
> Can well direct him where to looke for it.[8]

Finally, even though the sun-centered solar system would require a new physics, the idea that the earth could move became accepted. And for the new physics yet to come, for the purpose of a dynamical analysis of the solar system, the sun-centered point of view (the point of view of Kepler and Newton) was overwhelmingly preferable.

Tycho Brahe (born in Denmark, 1546), an astronomer, did not question the simplicity of the Copernican system; but simplicity was not a sufficient reason for him to accept the notion that the earth moved. The earth, according to Brahe, was "too sluggish" to move. (Today we say too sluggish to stop.)

His refusal to accept the Copernican theory did not hinder him in his chosen work of mapping the positions of the stars and the planets over a long period of time. He began observing with an instrument consisting of a pair of joined sticks, one leg to be pointed at a fixed star, the other at a planet, so that he could measure their angular separation. Later he constructed very large sextants and compasses with which, over a period of more than 20 years, he made wonderfully careful observations of the positions of the planets in the heavens. He catalogued the positions of a thousand stars far more accurately than had ever been done, and his measurements of the planetary angular positions over the period of 20 years contains no error larger than $\frac{1}{15}$ of a degree, an angle about as small as that made by the width of the point of a fine needle held at arm's length.

His record of the actual positions of the planets over this long period of time composed the raw material from which a more accurate system of curves that described the planetary orbits could be drawn. It was soon clear from his observations, which were far more accurate than those available to Copernicus, that the Copernican orbits were only roughly correct. A search then began for a more accurate system of planetary orbits—carried out, after Brahe's death, by one of his students, the German astronomer, astrologer, and mystic, Kepler.

KEPLER'S PLANETARY SYSTEM

Johannes Kepler, born in 1571, was a striking contrast to Brahe. Brahe, the observer, gathered data and recorded what he saw. Kepler, the theoretician, was fascinated by the power of mathematics, akin to Pythagoras in his reverence for numbers and intrigued by puzzles concerning number and size. After he had learned the elements of astronomy he became possessed with the problem of finding a numerical scheme underlying the planetary system. He says: "I brooded with the whole energy of my mind on this subject." He devoted most of his life to the analysis of the tables of planetary positions that Brahe had left in order to construct a system of curves along which the planets moved.

Kepler began his long analysis of Brahe's tables with an exhaustive study of the motion of Mars. On what curve had Mars moved during the 20 years of Brahe's observations? Was it a simple curve that kept repeating itself so that, once found, one could forevermore predict where Mars would be? All Brahe's observations of the planets had been made, of necessity, from the earth. However: Did the planets move on the simplest curves if one imagined that the earth stood still, or was it simpler to imagine that the earth, too, moved, as Copernicus had proposed? Kepler believed that the Copernican idea was basically correct, and that the earth spun about its axis while moving in an orbit about the sun.

He first, as everyone before, tried using a system of circles moving on circles for possible orbits. Had it been easy to reproduce the observations in this way, Kepler would have succeeded very soon. However, matters turned out to be more complicated. He made innumerable attempts; each involved long and laborious calculation. He needed, in every case, to translate Brahe's measurements of the angle made by the position of the planets with respect to the stars at a certain hour of the night to a position of the planet in space with respect to a fixed sun about which the earth was moving.

After about 70 trials using the circle-on-circle orbits for Mars, placing the sun in various positions, Kepler finally produced one that agreed reasonably well with observations. Then, to his dismay, he found that this curve, when continued beyond the data he had matched, disagreed with observation by about $\frac{8}{60}$ of a degree—the angle that the second hand on a watch would move in about 0.02 sec. Could Brahe have been wrong? Could the cold of a winter's night have numbed his fingers or blurred his observations? Kepler, knowing Brahe's methods and the painstaking care he took, could judge the accuracy of the data; he decided that Brahe would not have been wrong by so large an amount and thus rejected the curves he had constructed. He could have paid no greater tribute to the memory of Brahe, the observer. And Kepler trusted well, for Brahe had been right.

Saying "Upon this eight minute . . . [discrepancy he] . . . would yet build a theory of the universe," Kepler began again. He now tried changing the speed of the planet as it moved in its orbit about the sun, thus discarding an ancient and cherished belief, the belief that led Copernicus to revise the Ptolemaic system. As an aid, he used an imaginary spoke connecting the sun to the planet, and in experimenting with this arrangement he made his first great discovery. He found that the spoke moves so that it sweeps out equal areas in equal times. This has come to be known as *Kepler's second law*. With this discovery, Kepler finally abandoned his attempts to build up the planetary motions out of combinations of circles and began to try various ovals as orbits. After more calculation, he achieved his most important result, the *first law*. The planets, he found, moved on elliptic orbits with the sun at one focus of the ellipse:

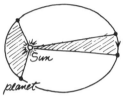

equal areas in equal times

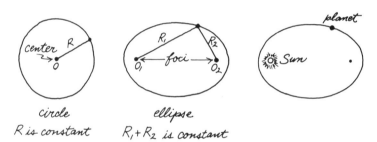

circle
R is constant

ellipse
$R_1 + R_2$ is constant

After years of effort, he had finally found simple curves that fitted the known motions of all the planets. Further, the planets moved on these curves in such a way as to sweep out equal areas in equal times.

Kepler then attempted to determine the connection between the size of the planetary orbits and the time it took them to complete a circuit, called the *period of the motion*. It had been thought since the Greeks that the planets with the largest orbits took the longest time to traverse them. After many additional trials, Kepler found that the regularity for which he was searching lay in the ratio of the cube of the radius to the square of the time. Once he happened upon this ratio, the regularity was most striking. In Table 5.1 this ratio is given using modern values for radius R and time T. The radius for an elliptic orbit is defined as one half the major axis.

He was ecstatic and did not conceal it: "What sixteen years ago I urged as a thing to be sought—that for which I joined Tycho Brahe . . . at last I have brought to light, and recognize its truth beyond my fondest expectations . . . The die is cast, the book is written, to be read either now or by posterity, I care not which—it may well wait a century for a reader, as God has waited six thousand years for an observer."[9]

TABLE *5.1*

Planet	R (in A.U.)	T (in sec)	R^3/T^2 (in (A.U.)^3/sec^2)
Mercury	0.387	7.60×10^6	1.003×10^{-15}
Venus	0.723	1.94×10^7	1.004×10^{-15}
Earth	1.000	3.16×10^7	$1.00 \ \times 10^{-15}$
Mars	1.523	5.94×10^7	$.997 \times 10^{-15}$
Jupiter	5.202	3.74×10^8	1.006×10^{-15}
Saturn	9.554	9.30×10^8	1.008×10^{-15}
Uranus	19.218	2.66×10^9	1.003×10^{-15}
Neptune	30.109	5.20×10^9	1.009×10^{-15}
Pluto	39.60	7.82×10^9	1.015×10^{-15}

Values of R^3/T^2 for the nine planets orbiting the sun.

NOTE: An astronomical unit of length (A.U.) is defined as one-half the sum of the longest and shortest distances of the earth from the sun (very nearly the radius of the earth's orbit). 1 A.U. = 1.495×10^{15} cm. For an elliptical orbit, R is defined as one-half of the major axis.

Here are his three laws, the solution of the problem of the planets.

I.   Each planet moves in an elliptical orbit about the sun with the sun at one focus.

II.   The line joining the sun and planet sweeps out equal areas in equal times.

III.   The ratio radius^3/time^2 is a constant for all the planets:

$$\frac{R^3}{T^2} = \text{constant} \tag{5.1}$$

It was a momentous result. Twenty years of observation and thousands of measurements had been condensed into a simple system of curves and rules. Those thereafter trying to construct a system of the heavens would struggle to reproduce these three laws in order that their theory give the motion of the planets. After Kepler (legislator of the heavens) the question would become: What theory will give us Kepler's rules?

But the planets—what of them? They no longer moved uniformly, no longer in circles, no longer in harmonic proportions . . . but perhaps

There is music even in the beauty, and the silent note which Cupid strikes, far sweeter than the sound of an instrument; for there is music wherever there is a harmony, order, or proportion; and thus far we may maintain the music of the spheres.[10]

6

NEWTON'S SYSTEM
OF THE WORLD

Between the time of Kepler and Newton, thinking in Europe evolved rapidly. Kepler had envisioned spokes that connected the sun and the planets, driving the planets in their orbits; Tycho Brahe had objected to the Copernican theory because the earth was too sluggish to move. It had been difficult to understand how the earth could move while its inhabitants did not feel its motion. Now a new point of view concerning the motion of bodies had developed. In Newton's time the discussions about motion were very often concerned with trying to find the law of force between the sun and planets which would produce the planetary motion that Kepler had described.

The question was no longer, How are the planets driven? Rather it was, Why do they remain in orbits? The earth's sluggishness prevented it from stopping rather than from moving; the uniform motion of Descartes and Galileo was accepted as the starting natural motion. The feeling was growing that there were laws governing the motion of bodies; these applied in the heavens as well as on earth. Could it be that these laws govern the motion of the planets? If so, what were the forces that would produce the orbits of Kepler?

There was a risk in the attempt to understand the motion of the heavenly bodies with rules that had been arrived at from observations on earth. The success of this attempt must be considered one of the great achievements of seventeenth-century science. In 1687 Newton published *Principia*, in which he wrought a system that described planetary motions, in fact, all motions in such detail that his name ever since has been associated with this monumental first, precise view of the world as a mechanical system, a machine whose rules could be known and whose behavior could be deduced, predicted, or controlled through knowledge of the rules.

Newton's first effort to organize the motion of the planets was directed toward an understanding of the orbit of the moon. This orbit is nearly circular, a fact that simplifies the calculations considerably. If there were no force on the moon, it would move in a straight line (first law); however, the moon appears to move in a circle about the earth. Therefore, if one accepts the first law, it is necessary to assume that there is some force acting on the moon—perhaps between the earth and the moon. Newton says:

> . . . nor could the moon without some such force be retained in its orbit. If this force was too small, it would not sufficiently turn the moon out of a rectilinear course; if it was too great it would turn it too much and draw down the moon from its orbit toward the earth.[1]

What then is the origin of the force which keeps the moon moving about the earth? Newton later told Henry Pemberton, who spent much time with him in his old age, that thinking about this problem while sitting in a garden, he saw an apple fall and was startled

> . . . into a speculation on the power of gravity: that as this power is not found sensibly diminished at the remotest distance from the center of the earth, to which we can rise, neither at the tops of the loftiest buildings, nor even on the summits of the highest mountains; it appeared to him reasonable to conclude, that this power must extend much farther than is usually thought; why not as high as the moon, said he to himself? and if so, her motion must be influenced by it; perhaps she is retained in her orbit thereby.[2]

The acceleration experienced by an apple or any other freely falling body near the surface of the earth due to the earth's attraction was by then known to be about 32 ft/sec². Could the moon be considered a freely falling body but one so far from the earth and moving so fast that it missed the earth in "falling" and continued to move in a stable orbit? In fact, could the moon's acceleration be the same as that of an object near the surface of the earth?

Calculation of the Acceleration of the Moon

The moon moving in a circle about the earth with a constant speed is accelerated toward the center about which it rotates (just as any object rotating with constant speed about a center) with an acceleration whose magnitude, as we have discussed before, is $a = v^2/R$:

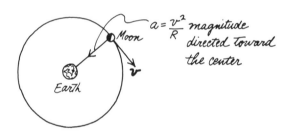

Since the moon is about 240,000 miles (1.27×10^9 ft) away from the earth,* and since it completes its circuit of the earth in about 27.3 days (2.36×10^6 sec), its acceleration is

$$a = \frac{v^2}{R} = \frac{1}{R}\left(\frac{2\pi R}{T}\right)^2 = \frac{4\pi^2 R}{T^2} = 0.00900 \text{ ft/sec}^2 \tag{6.1}$$

which is very much smaller than the 32 ft/sec² acceleration of bodies released near the surface of the earth. From this rough calculation it is obvious that the moon's acceleration is much smaller than it would be if it were close to the surface of the earth. This might suggest that the force with which the earth attracts a body grows smaller when the body is farther away; or it is possible that the proposed law of motion is not correct in general and cannot be extended to distances as large as the distance between the earth and the moon or to bodies as large as the moon. It was proposed, after all, on the basis of very local experience. On the other hand, one might assume that the second law is true in general and that

---

* Newton used 60 earth radii as the distance of the moon from the earth—a figure that had been determined by triangulation.

somehow a variation of the force is the explanation for the small acceleration of the moon.

Newton took the latter approach; he explained many years later that
the force law was suggested to him by working backward from Kepler's
third law. He took the point of view that it is the sun that attracts each of
the planets in the solar system in such a way as to keep them in their
orbits and that the way the sun attracts the planets can be related in some
simple way to the way that the earth attracts the moon. He then could use
Kepler's rule relating the square of the period (the time it takes a planet
to complete a revolution about the sun) and the cube of the radius,

$$\frac{T^2}{R^3} = \text{constant (the same for all the planets)} \qquad (6.2)$$

to determine how the force varies with the distance.

These assumptions were not Newton's alone. One might say they
were a part of the spirit of the time; Hooke, Wren, Halley, Huygens, and
very likely all literate scientists of the period were ready to do the same.

UNIVERSAL GRAVITATION

To reproduce Newton's argument we consider that each planet moves
roughly in a circular orbit (in spite of Kepler, this last statement is true;
the elliptical orbits are roughly circular) and that the acceleration

$$\frac{v^2}{R} \qquad (6.3)$$

is produced by a force between the sun and that planet directed toward
the sun (the center of the circle). The idea is that only a special force can
produce planetary motions such that

$$\frac{(\text{period})^2}{(\text{distance from sun})^3} = \frac{T^2}{R^3} = \text{constant} \qquad (6.4)$$

From the second law we have for the magnitude of the force between the
planet and the sun

$$F = M_{\text{planet}} \frac{v^2}{R} \qquad (6.5)$$

The planet's speed, v, can be related to the radius of its orbit, R, and the
period, T, by

$$v = \frac{\text{distance traveled}}{\text{time interval}} = \frac{2\pi R}{T} \qquad (6.6)$$

or

$$v^2 = \frac{4\pi^2 R^2}{T^2} \qquad (6.7)$$

Putting this in Eq. (6.5) gives

$$F = M_{\text{planet}} (4\pi^2 R) \frac{1}{T^2} \qquad (6.8)$$

(which says no more than $F = ma$ and that the planet moves in a circle
around the sun).

Now we can use Kepler's rule. For all the planets

$$\frac{T^2}{R^3} = \text{constant} \qquad (6.2)$$

or

$$\frac{1}{T^2} = \frac{1}{R^3(\text{constant})} \qquad (6.9)$$

Therefore,

$$F = \frac{4\pi^2}{\text{constant}} \frac{M_{\text{planet}}}{R^2} \qquad (6.10)$$

Thus the force between the sun and a planet is equal to a number (the same for all the planets) multiplied by the mass of the planet and divided by the square of the distance between the sun and the planet (the inverse-square law).

This is what could be concluded from Kepler's third rule. Newton, at this point, however, was interested in the force between the moon and the earth. So he made the now-famous assumption that the force with which the sun was apparently attracting the planets was really a special case of force between any two bodies whose magnitude was

$$F = G\frac{M_1 M_2}{R^2}$$

Here G is constant, the same for all systems, and M_1 and M_2 are the masses of the two bodies in question. From Kepler's rule, Newton could properly deduce only that the mass of the planet appeared in Eq. (6.10). But he made the force proportional also to the mass of the sun. For one thing, it looked more elegant that way. For the magnitude of the force between the sun and a planet he wrote

$$F = G\frac{M_{\text{sun}} M_{\text{planet}}}{R^2} \qquad (6.11)$$

Thus he came to the rather remarkable conclusion that the gravitational force was proportional to the mass of the planet (generalized to any body) on which it pulled. Now the mass is the measure of the body's inertia, its resistance to a change of motion. Why the gravitational force should be proportional to that mass (other forces—friction, pushes, pulls, electrical forces that we shall come to later—do not have this property) is something not clarified by Newton.* This very deep relation between the gravitational force and inertia is the starting point of Einstein's theory of gravitation.

Now Newton could proceed as follows: If the force between the sun and the planets is a special case of a force between all bodies, perhaps this force also acts between the earth and the bodies near the surface of the earth and between the earth and the moon. If the attraction of the earth falls off as $1/R^2$, so then does the resulting acceleration. The moon

* It is, however, something he worried about to the extent of doing experiments with pendula by which he was able to verify that the ratio of the gravitational force to the mass was the same for various different materials to 1 part in 1000.

is removed from the earth by approximately sixty earth radii, so its acceleration compared to that of a body near the surface of the earth should be approximately

$$\frac{32\ \text{ft/sec}^2}{60 \times 60} = 0.00889\ \frac{\text{ft}}{\text{sec}^2} \tag{6.12}$$

This Newton did when he was about 24 years old, at a time when he had retired from Cambridge due to the plague. He wrote

> . . . And the same year I began to think of gravity extending to y^e^ [the] orb of the Moon, and . . . from Kepler's Rule [Kepler's third law] . . . I deduced that the forces which keep the Planets in their Orbs must [be] reciprocally as the squares of their distances from the centers about which they revolve: and thereby compared the force requisite to keep the Moon in her Orb with the force of gravity at the surface of the earth, and found them answer pretty nearly. All of this was in the two plague years of 1665 and 1666, for in those days I was in the prime of my age for invention, and minded Mathematicks and Philosophy more than at any time since.[3]

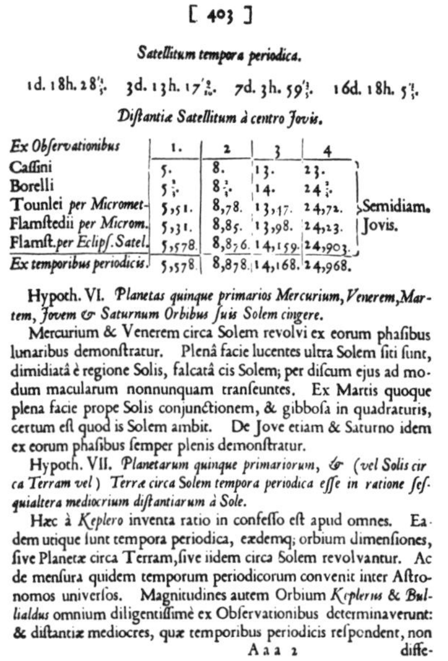

[403]

Satellitum tempora periodica.

1d. 18h. $28'\frac{1}{3}$. 3d. 13h. $17'\frac{2}{10}$. 7d. 3h. $59'\frac{1}{3}$. 16d. 18h. $5'\frac{1}{3}$.

Distantiæ Satellitum à centro Jovis.

Ex Observationibus	1.	2	3	4	
Cassini	5.	8.	13.	23.	Semidiam. Jovis.
Borelli	$5\frac{2}{3}$.	$8\frac{2}{3}$.	14.	$24\frac{2}{3}$.	
Tounlei *per Micromet.*	5,51.	8,78.	13,17.	24,72.	
Flamstedii *per Microm.*	5,31.	8,85.	13,98.	24,23.	
Flamst. *per Eclips. Satel.*	5,578.	8,876.	14,159.	24,903.	
Ex temporibus periodicis.	5,578.	8,878.	14,168.	24,968.	

Hypoth. VI. *Planetas quinque primarios Mercurium, Venerem, Martem, Jovem & Saturnum Orbibus suis Solem cingere.*

Mercurium & Venerem circa Solem revolvi ex eorum phasibus lunaribus demonstratur. Plenâ facie lucentes ultra Solem siti sunt, dimidiatâ è regione Solis, falcatâ cis Solem; per discum ejus ad modum macularum nonnunquam transeuntes. Ex Martis quoque plena facie prope Solis conjunctionem, & gibbosa in quadraturis, certum est quod is Solem ambit. De Jove etiam & Saturno idem ex eorum phasibus semper plenis demonstratur.

Hypoth. VII. *Planetarum quinque primariorum, & (vel Solis circa Terram vel) Terræ circa Solem tempora periodica esse in ratione sesquialtera mediocrium distantiarum à Sole.*

Hæc à *Keplero* inventa ratio in confesso est apud omnes. Eadem utique sunt tempora periodica, eædemq; orbium dimensiones, sive Planetæ circa Terram, sive iidem circa Solem revolvantur. Ac de mensura quidem temporum periodicorum convenit inter Astronomos universos. Magnitudines autem Orbium *Keplerus* & *Bullialdus* omnium diligentissimè ex Observationibus determinaverunt: & distantiæ mediocres, quæ temporibus periodicis respondent, non

Aaa 2 diffe-

FIG. *6.1.* Page from Newton's *Principia.*

The Magnitude of the Gravitational Force

Since the force of gravitation that Newton assumed to explain the motion of the planets and the moon is assumed to exist between any two bodies, why would any two bodies not attract one another with an easily observable force? Newton's answer was: "The gravitation toward them must be far less than to fall under the observation of our senses." To calculate this force one would have to know G. It was not possible for Newton to determine G from a measurement of the force of attraction of the earth's on a known mass because the earth's mass was not known. However, he knew that G must be small enough so that two normal-sized bodies on earth would not attract one another with an easily observable force.

In 1798, over 100 years later, Cavendish measured the gravitational interaction for laboratory-sized objects and calculated from his observations the value of the constant G. His apparatus consisted of two small spheres, 1 and 2, mounted on opposite ends of a light horizontal rod that was supported at its center by a fine vertical fiber. A small mirror was mounted to this fiber so that a beam of light, reflected from the mirror, would give a sensitive measure of any rotation of the thread. Finally, he used two large masses (A and B) and placed them as indicated in Fig. 6.2 with respect to the two spheres.

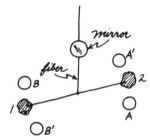

FIG. 6.2. View of the Cavendish balance.

After recording the equilibrium position of the reflected light beam, Cavendish moved the large masses to position A' and B'. The gravitational attraction between the spheres and the masses gave rise to a clockwise twist of the fiber when in the first position and counterclockwise when in the second. The resulting angular displacement of the system could thus be observed and measured. He had independently determined what force was required to produce a given twist of the fiber. From this he could determine the gravitational force between the masses and the spheres and thus obtain G. The modern value* for the constant G is

$$G = 6.67 \times 10^{-8} \text{ dyne-cm}^2/\text{g}^2 \qquad (6.13)$$

This G, together with the observed acceleration of objects at the earth's surface, is sufficient to permit calculation of the earth's mass. It gives†

* The experiment is quite difficult to do. Cavendish obtained $G = 6.71 \times 10^{-8}$ dyne-cm^2/g^2. Some later measurements (all in dyne-cm^2/g$^2 \times 10^{-8}$) are compared with his:

1. Cavendish, 1798	6.71
Phil. Trans. Roy. Soc. (_London_), 17	
2. Poynting, 1892	6.698
Phil. Trans. Roy. Soc. (_London_), A182, 565	
3. Boys, 1895	6.658 ± 0.006
Phil. Trans. Roy. Soc. (_London_), 17	
4. Heyl, 1930	6.670 ± 0.005
J. Res. Nat. Bur. Std., 5, 1243	
5. Heyl and Chzrnowski, 1942	6.673 ± 0.003
J. Res. Nat. Bur. Std., 29, 1	

Note: Boys favors 6.658, although his readings average about 6.663 ± 0.006 (± 0.018 extreme). He himself gives no error estimate.

† Given the radius of the earth, which can be determined in various ways, the most pedestrian by measuring the circumference and assuming it is roughly spherical, we can obtain

$$\frac{GM_e}{R^2} = g = 980 \text{ g/sec}^2$$

Therefore,

$$M_e = \frac{gR^2}{G} = \frac{980 \times (6.37 \times 10^8)^2}{6.67 \times 10^{-8}} = 5.96 \times 10^{27} \text{ g}$$

$$M_{\text{earth}} = 5.96 \times 10^{27} \text{ grams} \tag{6.14}$$

Using this value of G we can compute the force between two 1000-g masses 100 cm apart:

$$F = G \frac{(1000 \text{ g})^2}{(100 \text{ cm})^2} = 6.67 \times 10^{-6} \text{ dyne} \tag{6.15}$$

or a force of about a ten-trillionth of 1 lb.

Pisa Explained

It follows also that, neglecting air resistance, all bodies fall to the earth at the same rate, independent of their masses. Since a body's acceleration is related to the force exerted on it, by Newton's second law

$$a = \frac{F}{M_{\text{body}}} \tag{6.16}$$

while the force due to the earth's gravitational attraction is just

$$F = G \frac{M_{\text{body}} M_{\text{earth}}}{R^2} \tag{6.17}$$

which is proportional to the mass of the body. The acceleration a body experiences is

$$g = \frac{F}{M_{\text{body}}} = \frac{G M_{\text{body}} M_{\text{earth}}}{M_{\text{body}} R^2} = \frac{G M_{\text{earth}}}{R^2} \tag{6.18}$$

which is independent of that body's mass. Thus all bodies experience the same acceleration. The fact that the gravitational force is proportional to the mass of a body while the same mass determines the body's inertia is what leads to equal rates of fall near the surface of the earth.

The acceleration a body experiences near the surface of the earth, $g = 32 \text{ ft/sec}^2$, we have now identified with

$$g = G \frac{M_{\text{earth}}}{R^2} \tag{6.19}$$

The earth's mass and G are constant; however, R should vary as we move closer to or farther away from the center of the earth. The acceleration due to the earth's attraction should be different at the top of a mountain and in the valley below; even differences in g, due to the change in the distance from the center of the earth from one floor to the next in a building, can be measured, although this measurement requires an accuracy of about 1 part in 10,000,000.

TABLE 6.1

Various constants	Sym-bol	CGS	MKS	English
Average acceleration of gravity near the surface of the earth	g	980 cm/sec^2	9.8 m/sec^2	32 ft/sec^2
Gravitational constant	G	6.67×10^{-8} cm³/g-sec²	6.67×10^{-11} m³/kg-sec²	3.43×10^{-8} ft³/slug-sec²

Weight

The gravitational force between an object on the earth and the earth itself is what we call the weight, W, of the object. This can be written

$$W = \frac{GM_{earth}}{R^2} M_{body} \qquad (6.20)$$

and, using the definition of g,

$$g = \frac{GM_{earth}}{R^2} \qquad (6.21)$$

we can write the weight in the very common form

$$W = M_{body}g \qquad (6.22)$$

From our discussion of g, it is now clear that an object's weight will vary in different parts of the earth. This is so, in fact, and can easily be observed. Were the object on the moon, the force would be smaller, since the moon's mass is smaller than that of the earth. The mass of a body is a property intrinsically associated with the body. Its weight is associated with its mass and its environment: the large mass on which it is resting (earth, moon, etc.) and its distance from the center of this mass. On a rocket ship in space, objects have negligible weight, but their inertia (resistance to changes of motion) is still the same.

Up and Down

The gravitational force attracts a body toward the center of the earth. It is this force that makes apples fall to earth, whether in England, the United States, or China. This is the source of the oddity, faced at one time or another by every child, that the people on the other side of the earth are standing on their heads. Up and down are defined relative to the center of the earth. If a person could tunnel through the center of the earth, he would be going down until he reached the center, and at that point his direction of motion would change, by definition, to up.* This force has the property that it does not vary in magnitude as we travel over a surface, which is a constant distance from the center of attraction, but varies slowly in a direction always pointing toward the center, a variation too small to feel under normal circumstances.

With some methods more sophisticated (such as the calculus—invented for this purpose), and others very much like those we have used, Newton succeeded in showing, given the three laws of motion and the law of universal gravitation, that the planets would move in elliptical orbits with the sun at one focus of the ellipse and would obey Kepler's other laws: equal areas in equal times and T^2/R^3 = constant. He showed that there were in general three types of orbits a body would follow

---

* Dante remarks on this as he passes the bottommost point of hell (the center of the earth). From the Newtonian point of view, the force at the center of the earth goes to zero, and one would experience there a sensation of floating. In the Aristotelian system, the force remains constant in magnitude but suddenly reverses direction at the center. It is this last that Dante, a thirteenth-century figure, describes.

under a gravitational force—depending upon the initial velocity and position of the body.

One was the ellipse (planet initially not too far away or moving too fast):

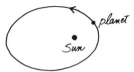

The others were open orbits—a parabola or hyperbola (planet initially moving fast and/or far away, so not captured):

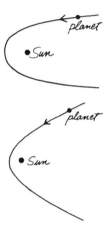

or a hyperbola (planet initially moving very fast, so just deflected a bit):

For a hyperbolic or parabolic orbit, the object would appear once and disappear forever. Thus, those transient visitors of the solar system, comets, might be bodies that trace orbits of the last two types; recurring comets (such as Halley's) could be thought of as following elongated elliptical orbits, as shown in Fig. 6.3.

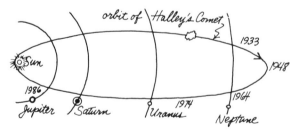

F I G. 6.3. Orbit of Halley's comet. The dates indicate approximate positions of the comet for those years.

Before Newton, Shakespeare might say, through Calpurnia in *Julius Caesar,*

When beggars die, there are no comets seen;
The heavens themselves blaze forth the death of princes.

Or as Bedford in Part I of *King Henry VI,*

Comets, importing change of times and states,
Brandish your crystal tresses in the sky.

The poets of today must search for another metaphor, for after Newton a comet becomes another respectable, if temporary, occupant of the solar system.

Newton also was able to analyze a large number of more complicated phenomena such as small irregularities (perturbations) in the planetary orbits. These deviations of the planets from their elliptical paths can be explained by the interactions between the planets themselves. Not only is the earth attracted by the sun but also, in varying degrees, by each of the planets. The latter attractions are relatively small because of the small masses of the planets in comparison to the mass of the sun. Their effects are referred to as *perturbations* (small irregularities) of the main orbital motion; these are easily observed and are predicted by taking into account attractions between planets.

Perturbation theory has resulted in the discovery on occasion of entirely new planets whose existence had not been suspected before. On March 13, 1781, Sir William Herschel, during a routine observation of a portion of the heavens, discovered a seventh planet (later named Uranus). No one was too astonished that there should be a seventh planet (Kepler was long since dead), but in computing the orbit of Uranus, it became apparent that this planet was not behaving as it should. Even when the disturbances that the nearby large planets, Jupiter and Saturn, might produce had been taken into account, the orbit of Uranus was not as calculated.

The Englishman J. C. Adams and the Frenchman Leverrier independently arrived at the conclusion that there must be, still farther from the sun, but close enough to influence the motion of Uranus, a yet undiscovered planet. On September 23, 1846, the German astronomer, Galle, found the new planet where Leverrier had told him to look. It was named Neptune.

Perturbation Theory

If the planets exerted no force upon each other, each would follow an elliptical orbit around the sun under the influence only of the sun's gravitational force. Suppose two masses P and C are sufficiently close so that the force between them becomes important. Mass P feels, in addition to the sun's attraction, a force toward mass C while, according to Newton's third law, the force on C due to P will be equal in magnitude but opposite in direction. As a result, both masses will depart from their unperturbed elliptical orbits and adopt new perturbed paths around the sun.

Imagine that P was a planet and C was a comet. (Typically, the mass of a comet is very much smaller than the mass of a planet.*) Then the

---

* The mass of a typical comet is estimated to be from 10^{15} to 10^{21} g. In comparison, the mass of the earth is about 6×10^{27} g.

effect of the force between P and C would be much greater on C. Thus no effect on the orbit of a planet due to a comet has ever been observed, but the orbits of comets are changed quite radically when they pass close to a planet.

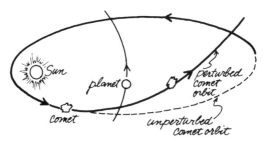

F I G.   6.4. Elementary illustration of the effect of a perturbation.

Often an elliptical (closed) comet orbit is changed to a hyperbolic (open) orbit. Thus, due to its interaction with the planet, a comet sometimes leaves the solar system.

Newton also explained the movements of the tides as being produced by the gravitational attraction of the moon and the sun; he explained the bulge of the earth at the equator as being produced by the rotation of the earth,* and he successfully explained a very complicated seeming motion of the earth's axis (a small precession that is completed every 26,000 years) all as consequences of the laws of motion and the law of gravitational attraction.

How early Newton obtained these results we do not know. For about a generation following those plague years 1665–1666 (when he was at the "prime" of his "age for invention") he published little or nothing on the subject. And during this time, others, both in England and on the Continent, were contemplating similar problems. Could it be proved, on the assumption that there is a force which falls off as $1/R^2$ between the sun and the planets, that the planets would move according to Kepler's laws? By 1684 several members of the Royal Society proved, as we have done, that if the orbits were circles, Kepler's third law would hold. Hooke claimed that he could deduce Kepler's elliptic orbits from an inverse-square force law, but never produced the details. At this point Halley decided to consult Newton at Cambridge. So, in August, 1684, he called on Newton, and, according to an account of this visit by John Conduitt (who later married Newton's niece):

> Without mentioning either his own speculations, or those of Hooke and Wren, he at once indicated the object of his visit by asking Newton what would be the curve described by the planets on the supposition that gravity diminished as the square of the

---

* When La Condamine measured the bulge at the equator by an actual trip around the world which took two years, Voltaire, with characteristic charity, wrote:

> *Vous avez trouvé par de longs ennuis*
> *Ce que Newton trouva sans sortir de chez lui.*

This might be translated:

> "What you have found by tedious route,
> I. Newton found not going out."

distance. Newton immediately answered, *an Ellipse.* Struck with joy and amazement, Halley asked him how he knew it? Why, replied he, I have calculated it; and being asked for the calculation, he could not find it, but promised to send it to him.[4]

To quote Professor Gillispie:

> While others were looking for the law of gravity, Newton had lost it. And yielding to Halley's urging, Newton sat down to rework his calculations and to relate them to certain propositions *On Motion* (actually Newton's laws) on which he was lecturing that term . . . Besides proving Halley's theorem for him, he wrote the *Mathematical Principles of Natural Philosophy.*[5]

In 1687, with the help of Halley (who paid for part of the costs of publication), the book was published. Our conception of the world has not been the same since.

Example 1. Earth satellites. If a satellite is in a circular orbit at a height R above the center of the earth (assumed spherical), how long does it take to complete an orbit? The moon, 240,000 miles away, circles the earth once every $27\frac{1}{4}$ days. We can apply Kepler's third rule, $\dfrac{T^2}{R^3} =$ constant. The constant may be obtained from our information about the moon:

$$\text{constant} = \frac{(27\frac{1}{4})^2\,(\text{days})^2}{(240{,}000)^3\,(\text{miles})^3}$$

$$= 5.35 \times 10^{-14}\,(\text{days})^2/(\text{miles})^3 \qquad (6.23)$$

Thus for the earth satellite:

$$\frac{T^2}{R^3} = 5.35 \times 10^{-14} \qquad (6.24)$$

Project Gemini orbited at 100 miles above the earth's surface ($R = 4100$ miles). Therefore,

$$T = 5.35 \times 10^{-14}\,(4100)^3$$

$$\simeq .061\ \text{days} \simeq 88\ \text{minutes} \qquad (6.25)$$

Example 2. Syncom. How high is Syncom, the satellite that always stays above the same point on earth?

The earth rotates once a day. If a satellite has a 1-day orbit in the direction of the earth's rotation, it should stay over the same spot all the time. Such an orbit is possible:

$$\frac{T_{\text{Syn}}}{T_{\text{moon}}} = \frac{R_{\text{Syn}}^{3/2}}{R_{\text{moon}}^{3/2}}$$

$$\frac{1}{27\frac{1}{4}} = \frac{R^{3/2}}{(240{,}000)^{3/2}}$$

$$R = \frac{240{,}000}{(27\frac{1}{4})^{2/3}}$$

$$R \simeq 26{,}700\ \text{miles} \qquad (6.26)$$

or about 22,700 miles above the earth's surface. As can be seen in the illustration, three Syncom satellites can effectively provide communication over all the inhabited earth.

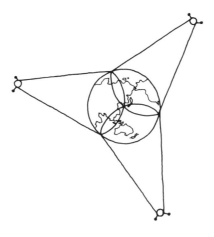

7

PARTICLES IN COLLISION

The world created by *Principia*, inhabited mentally by philosophers, scientists, and finally everyone for the next 200 years, begins with a void—empty, unchanging, the same from one place to another.

> Absolute space, in its own nature, without relation to anything external, remains always similar and immovable.[1]

And this void is peopled by particles: hard, massy, and indivisible, the idealization of something like a billiard ball. Like billiard balls, they possess a definite quantity of matter, occupy a definite position at a given time, and trace out a precise orbit. The subject of primary interest becomes to determine the orbits.

Descartes had proposed that the only forces would be those exerted when two particles collide: a force of contact, easily understood as a palpable thing. Newton found it necessary to add to this a force of gravitation (the subject of long argument between the Newtonians and the Cartesians) universally attracting any two particles at a distance. In the absence of forces, Galileo and Descartes had proposed that a particle continues in uniform motion or remains at rest; this was incorporated by Newton as the first law of motion. In the presence of forces, Newton, generalizing the suggestion of Galileo, proposed the second law: The force is equal to the mass times the acceleration. And further, he proposed that all forces whatever they are, gravitation or otherwise, obey the third law: Action equals reaction.

These are the raw materials of the Newtonian world. Is it really possible to construct from these few stones a building diverse enough to encompass the marvelous variety of our experience? Is it possible from these few definitions and postulates to construct a mathematical world that is a faithful representation of the real, or perhaps as Plato would put it, the mathematical essence of which the world is a true shadow?

By straightforward, but not necessarily easy, application of the second law to arbitrary systems of forces acting on particles, the trajectories of the particles can be obtained. For example, subjected to a uniform force (such as that near the surface of the earth), a particle follows a parabolic path; due to the gravitational force of the sun, a planet or comet follows an elliptic (or circular), parabolic, hyperbolic, orbit, etc. From the first law, the force necessary to balance any number of others at a point can be found; thus the equilibrium of forces acting at a point can be determined.

In the years after *Principia* was published a series of increasingly elegant developments of Newton's theory—variations on his theme—were introduced, some of which outlive and are more general than mechanics. It is convenient to develop such concepts as impulse, energy, and work. These are constructed from the materials we have—time, and mass, using the laws of motion (the original postulates) in much the same way that Euclid constructed triangles or circles from his original definitions and postulates. They are convenient tools, sometimes more convenient and general than the materials from which they were made.

THE THIRD LAW

What happens when two particles collide is a very old problem; and as with the problem of free fall and the problem of projectiles, the solution provides part of the foundation of the theory of the motion of bodies. Descartes successfully formulated the law of motion of a single particle unaffected by others in the void, proposing the principle of inertia, but never successfully answered the question of what happens when two particles collide. Galileo thought about the problem enough to realize that it could be very confusing. Does the force upon impact become infinite? This seemed to him unlikely:

> I have concluded that this question of impulsive forces is very obscure, and I think that, up to the present, none of those who have treated this subject have been able to clear up its dark corners which lie almost beyond the reach of human imagination; among the various views which I have heard expressed one, strangely fantastic, remains in my memory, namely, that impulsive forces are indeterminate, if not infinite.[2]

The difficulty is that in an impact, the force grows large over a very short time, so that it is difficult to follow the force as a function of time.

However, the product **F** Δt (as we shall see) remains nicely under control, for if the force grows very large the time it acts grows very small. (This is one of several different situations in which a product or a ratio remains under control, while the individual factors grow infinitely large or infinitesimally small: situations which have caused great confusion historically. Some of the paradoxes of Zeno [see the Appendices, p.486] rest on the fact that a time interval and a distance interval grow infinitesimally small, although their ratio remains constant. The tools we use today [limits, calculus] to resolve these problems are not very difficult technically. However, that their invention took several thousand years suggests they have a subtlety that was not easy to grasp.)

In 1668 the Royal Society of London proposed an investigation of the question of collisions; solutions were produced by the mathematician John Wallis; the architect of St. Paul's, Sir Christopher Wren; and the Dutch physicist Christian Huygens. Some of their conclusions and experiments are discussed by Newton in *Principia*. And Newton also provides a solution—the same solution as proposed by the others. What he achieved was to unite what seemed like a separate problem with the other problems of motion, and he was able to solve all using the three postulates of motion.

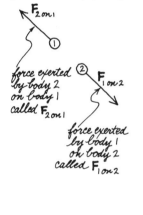

Newton's third law (on the nature of forces)

$\mathbf{F}_{2\,on\,1}$

①

force exerted by body 2 on body 1 called $\mathbf{F}_{2\,on\,1}$

② $\mathbf{F}_{1\,on\,2}$

force exerted by body 1 on body 2 called $\mathbf{F}_{1\,on\,2}$

The relevant postulate, Newton's third law of motion, is really a statement about the nature of the forces in the world. Newton's second law relates changes of motion to the impressed forces. He proposes one force —gravitation—between all bodies. About all the other forces that exist in the world—pushes, pulls, and so on—he says only what is contained in his third law: No matter what the nature of the forces between bodies, they always behave in such a way that, if at an instant of time a body, call it 1, exerts a force on another body, call it 2, body 2 exerts an equal and opposite force on body 1. Using the symbols defined we can write

THE THIRD LAW: *At all times*

$$\mathbf{F}_{2\,on\,1} + \mathbf{F}_{1\,on\,2} = 0$$

In the slapstick-comedy situation, the man stepping out of the rowboat gains an immediate experience of the third law. To get from the rowboat to the dock he needs a force to accelerate himself. He expects this to be provided by the boat. But at the same time (if the interaction between the man and the boat obeys Newton's third law) he must push the boat with a force that is equal and opposite. If the boat were infinitely heavy, this would be no problem—but then it would be no boat. A small rowboat begins to accelerate away from the man due to the force with which he pushes (second law). And just when he is perched between the boat and the dock, when he needs a push most, the boat is no longer where he expects it to be.

We may think of the third law as the postulate that permitted Newton to deduce the results already obtained by Wren, Huygens, and Wallis. His own reasons for the form in which he proposed it are given in the *Scholium*. As an immediate consequence of this law concerning the nature of forces, we have one of the most famous theorems of mechanics: conservation of momentum. The idea of momentum conservation is so deep that it pervades physics even when the Newtonian form of mechanics is no longer used. This is one of several instances in which a result obtained as a relatively special theorem is found later to be more important and more general than the postulates from which it came. One can just as well develop physics from a point of view in which conservation of momentum would be among the initial postulates of the subject, and such propositions as Newton's third law, with the proper restrictions, would be theorems.

CONSERVATION OF MOMENTUM

Momentum, the quantity of motion, is the fundamental entity of the Newtonian theory. For a particle of mass m the momentum is defined as:

DEFINITION: $p = mv$

Newton's second law relates the force to the change of momentum

$$F \, \Delta t = \Delta p \qquad (7.1)$$

or dividing by Δt:

$$F = \frac{\Delta p}{\Delta t} \qquad (7.2)$$

If the force on the particle is zero, the change in momentum is zero, Newton's first law.

Impulse

It is the product of the force multiplied by the time it acts that produces the change of momentum. Thus a large force acting for a short time can produce the same momentum change as a small force acting for a longer time. It becomes convenient, then, to define the product of the force acting on a body multiplied by the time it acts as the impulse:*

DEFINITION: **impulse** $= F \, \Delta t$

If the force is a constant in both magnitude and direction, then the magnitude of the total impulse after a time t is just

$$\text{total impulse} = Ft \qquad (7.3)$$

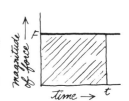

which is the shaded area under the curve. Usually, in situations of interest, the force will vary in time. If its direction is constant, the total impulse is defined as the area under the curve, force as a function of time. This can be approximated as before by dividing the time into n small intervals.†

* It is this quantity that was Newton's motive force.
† It is a matter of convention that physicists searching for a letter to represent a specific but arbitrary number often choose n.

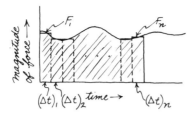

Then the magnitude of the total impulse is the sum over the individual impulses as Δt becomes very small:

$$\text{total impulse} = F_1(\Delta t)_1 + F_2(\Delta t)_2 + \cdots + F_n(\Delta t)_n \qquad (7.4)$$

As the time intervals become very small, this approaches as closely as we wish the area under the curve, force as a function of time. In the most general case—that in which the force changes both in magnitude and direction—the total **impulse** is defined as the vector sum of the impulses over the small time intervals $(\Delta t)_1 \cdots (\Delta t)_n$ into which the entire time interval has been divided.

DEFINITION: *total* **impulse** =

$$\mathbf{F}_1(\Delta t)_1 + \mathbf{F}_2(\Delta t)_2 + \cdots + \mathbf{F}_n(\Delta t)_n$$

From the definition, we see that the total impulse is a vector.

THEOREM 7.1: *The total impulse acting on a body is equal to the total change in momentum of the body.*

This follows directly from the second law:

$$\mathbf{F} \, \Delta t = \Delta \mathbf{p} \qquad (7.5)$$

The essential point is that the equation above is meant to be valid at every instant of time.[*] Therefore, if we call the force at time t_1, $\mathbf{F}_1$, the change of momentum in the time interval $(\Delta t)_1$ is

$$\mathbf{F}_1(\Delta t)_1 = (\Delta \mathbf{p})_1 \qquad (7.6)$$

The same is true for the time interval $(\Delta t)_2$

$$\mathbf{F}_2(\Delta t)_2 = (\Delta \mathbf{p})_2$$

$$\mathbf{F}_3(\Delta t)_3 = (\Delta \mathbf{p})_3$$

$$\cdot$$
$$\cdot$$
$$\cdot$$

$$\mathbf{F}_n(\Delta t)_n = (\Delta \mathbf{p})_n \qquad (7.7)$$

where we have again divided the time interval into n small parts $(\Delta t)_1 \cdots (\Delta t)_n$. If we add all these together, we get

$$\mathbf{F}_1(\Delta t)_1 + \mathbf{F}_2(\Delta t)_2 + \cdots + \mathbf{F}_n(\Delta t)_n = (\Delta \mathbf{p})_1 + (\Delta \mathbf{p})_2 + \cdots + (\Delta \mathbf{p})_n \qquad (7.8)$$

[*] This is something we might not have realized on first seeing the laws of motion. It is typical that proving theorems or exploring the consequences of a system of postulates reveals the precise meaning of the postulates.

The left side of the equation is by definition the total impulse, and the right side is the total change of momentum. Therefore, we have

$$\text{total } \mathbf{impulse} = \text{total change of } \mathbf{momentum} \qquad \text{Q.E.D.}^*$$

The total change of momentum means (final momentum minus initial momentum),

$$\text{total change of } \mathbf{momentum} = \mathbf{p}_f - \mathbf{p}_i \qquad (7.9)$$

If the mass of the body remains constant during the process, its initial and final momenta are

$$m\mathbf{v}_i \text{ and } m\mathbf{v}_f \qquad (7.10)$$

so that the theorem takes the form

$$\text{total } \mathbf{impulse} = m(\mathbf{v}_f - \mathbf{v}_i) \qquad (\text{mass constant}) \qquad (7.11)$$

The momentum of a single body is defined as its mass multiplied by its velocity. Suppose now that we have N bodies; the total momentum of the system is defined as the sum of their individual momenta:

$$\mathbf{P} = \mathbf{p}_1 + \mathbf{p}_2 + \mathbf{p}_3 + \cdots + \mathbf{p}_N \qquad (7.12)$$

For two bodies this is

$$\mathbf{P} = \mathbf{p}_1 + \mathbf{p}_2 \qquad (7.13)$$

As an example, consider two bodies:

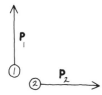

The total momentum of bodies 1 and 2 is the vector sum of $\mathbf{p}_1$ and $\mathbf{p}_2$:

Or, as in the next case, two bodies with equal and opposite momentum have a total momentum that adds to zero:

Now, using Newton's third law, we can prove the following theorem:

---

* Q.E.D. is the abbreviation for the Latin *quod erat demonstrandum,* "which was to be demonstrated," and terminates a proof.

THEOREM 7.2 (Conservation of Momentum): *In the absence of external forces, the total momentum of a system of particles remains constant.*

PROOF. We consider two bodies only, as this case contains the heart of the proof and can be generalized in a straightforward way to many bodies.

Let body 1 have momentum p_1 and body 2 have momentum p_2. The total momentum of this two-body system will be

$$\mathbf{P} = p_1 + p_2 \quad \text{(definition of total momentum)} \quad (7.14)$$

The essence of the argument is that any change in the momentum of body 1 is equal and opposite to the change in the momentum of body 2. This is shown as follows:

Consider body 1. Any change in its momentum is produced by a force from body 2, which we call $\mathbf{F}_{2 \text{ on } 1}$ (because they are the only two bodies in the vacuum). By Newton's second law, the momentum change of body 1 is

$$(\Delta p)_1 = \mathbf{F}_{2 \text{ on } 1} \, (\Delta t) \quad (7.15)$$

During the same time interval, body 1 exerts the force $\mathbf{F}_{1 \text{ on } 2}$ on body 2, so body 2 changes its momentum by

$$(\Delta p)_2 = \mathbf{F}_{1 \text{ on } 2} \, (\Delta t) \quad (7.16)$$

The total change in momentum is defined as

$$\Delta \mathbf{P} = (\Delta p)_1 + (\Delta p)_2 \quad (7.17)$$

which, using the above, is equal to

$$\Delta \mathbf{P} = (\mathbf{F}_{2 \text{ on } 1} + \mathbf{F}_{1 \text{ on } 2}) \, (\Delta t) \quad (7.18)$$

If the forces obey the third law,

$$\mathbf{F}_{2 \text{ on } 1} \text{ is equal and opposite to } \mathbf{F}_{1 \text{ on } 2} \quad (7.19)$$

so that

$$\mathbf{F}_{2 \text{ on } 1} + \mathbf{F}_{1 \text{ on } 2} = 0 \quad (7.20)$$

From this it follows that

$$\Delta \mathbf{P} = 0 \quad (7.21)$$

Since the change in total momentum is zero for any time interval, the momentum always retains its initial value. (If the momentum does not change, it remains what it was.) Q.E.D.

All the bodies that produce the forces must be included. For the cannonball falling toward the earth, gravity produces the force mg on the cannonball and the equal and opposite force on the earth. The earth, however, does not move very much.

The importance of this result is easy to underestimate. We have shown that there exists a quantity called the total momentum which remains a constant no matter what happens (and how complicated it is) to the system. If bodies collide, if an object explodes—no matter what happens—the total momentum remains the same. As we shall see, this result enables us to analyze the motion of bodies even when the details of forces are not known. This is one of the many general results that exist in the body of physics which state that even if all the details of the forces or the motions are not known, as long as the forces are of a certain type, the motion will obey certain general rules. In this case, no matter what the detailed nature of the forces, as long as they obey Newton's third law, the total momentum of the system will remain constant.

This theorem stating momentum conservation is more fundamental than the third law, on which it is based. From a twentieth-century point of view, the conservation of momentum is related immediately to the sameness of space from one point to another. It may be a valid statement even in cases when it is difficult to think of the forces as being strictly Newtonian. History often decides the original order in which propositions come to life, but it does not necessarily place them in the order of their final importance.

APPLICATION OF MOMENTUM CONSERVATION TO PARTICLE COLLISIONS

Using the theorem of momentum conservation, we can now analyze the motion of particles in collision in the absence of a detailed knowledge of the forces between them. As we proceed, we shall see that it is convenient to classify collisions into various categories. Some types of collisions can be analyzed completely using the principle of momentum conservation alone; others cannot. In the latter cases, it is possible, assuming some additional properties of the forces, to deduce much that we wish to know about the motion, but still without a detailed knowledge of the forces.

Let us begin with a simple case involving motion in one direction. Imagine two particles; to be specific, let us think of billiard balls. (Billiard balls on a table, although graphic, are not ideal because they roll. What we want to imagine are two particlelike objects—ice pucks on a frozen lake, or billiard balls in outer space—which have the translational motion (their velocity) but no internal motion such as rolling. However, collision theory without billiard balls, as probability theory without roulette, would be like a meal without wine or . . .) Imagine the two billiard balls (initially at rest) touching, and then being separated by a force between them. We do not have to know what the force is. It could be due to a small explosion, a spring, or any other interaction between the balls. After the explosion the two balls move away from each other. Without a detailed knowledge of the interaction, can we conclude anything about the motion of the balls after the explosion? If we were to analyze this problem, using Newton's laws directly, we would have to know exactly the force produced by the explosion on each of the billiard balls—something rather difficult to achieve, as the force would be large over a very short period of time and change in a complicated way. However, we can deal with the situation by using the principle of momentum conservation.

Analysis of the motion of two particles of equal mass which start at rest.

View from above, before the explosion:

$$v_1 = 0 \qquad p_1 = mv_1 = 0 \qquad\qquad (7.22)$$

$$v_2 = 0 \qquad p_2 = mv_2 = 0 \qquad\qquad (7.23)$$

Therefore, the total momentum of the two-particle system is

$$P = p_1 + p_2 = 0 \qquad\qquad (7.24)$$

This remains unchanged no matter what the two balls do to each other.
During the explosion:

After the explosion:

(We use the convention that the velocities after the collision are denoted by v'.)

After the explosion, ball 1 moves with velocity v'_1, ball 2 with velocity v'_2. There is a relation between v'_1 and v'_2 because the final momentum P', using Theorem 2, is

$$P' = p'_1 + p'_2 = mv'_1 + mv'_2 = 0 \qquad\qquad (7.25)$$

(The final momentum is equal to the original momentum, and that was zero.) Therefore, we can conclude that

$$mv'_1 + mv'_2 = 0 \qquad\qquad (7.26)$$

or

$$v'_1 = -v'_2 \qquad\qquad (7.27)$$

This means that the balls move with equal and opposite velocities independent of the detailed nature of the force that separated them.[*]

In general, the principle of momentum conservation gives us a single equation that is satisfied at all times:

[*] If the masses were unequal, the result would be $v'_2 = -\left(\dfrac{m_1}{m_2}\right) v'_1$.

$$p_1 + p_2 + p_3 + \cdots + p_n = \text{constant} \qquad (7.28)$$

That is, the total momentum of the system remains constant. In practice, in what one calls a collision problem, one usually speaks of the initial momentum, that is, the momentum before the collision, and the final momentum, the momentum after the collision. What is typical in these problems is that the particles interact for a short time only, and it is during this short time that their individual momenta are changing. Before the interaction the particles all move according to the first law, and after the interaction they move again according to the first law. It is only during the collision itself that the individual momenta change.

For a two-particle system, in general, the relation of momentum conservation gives us the following equation:

$$\mathbf{P}_1 + \mathbf{P}_2 = \mathbf{p}_1' + \mathbf{p}_2' \qquad (7.29)$$

What this means is that, given any of the three momenta, we can determine the fourth. For example, given p_1, p_2, and p_1', we can determine p_2', the final momentum of the second particle:

$$\mathbf{p}_2' = \mathbf{P}_1 + \mathbf{P}_2 - \mathbf{p}_1' \qquad (7.30)$$

Thus, for example, in nuclear collisions involving two incoming and two outgoing particles (a non-Newtonian system, but one which is presumed to conserve momentum) the measured momenta of three of the particles enables one to deduce the momentum of the fourth—a procedure in daily use.

8

MECHANICAL ENERGY

Momentum conservation provides one relation between the initial and the final momentum of particles in a collision. For two particles, this means that if any three momenta are known, the fourth can be determined. To obtain this relation, we use the fact that the forces are Newtonian (i.e., obey the third law). We now ask whether there are any other general properties forces might have which, regardless of their detailed nature, would enable us to determine other quantities that remained unchanged. An exploration of this question leads to the concepts of conserved forces, work and energy, which we now discuss.

Work

The work done on a particle is equal to the force acting on that particle times the distance it moves in the direction of the force, or

DEFINITION:

work = force × distance moved in direction of force

There is some relation between this technical definition and the common notion of work. We would normally say that a person worked harder if he pushed a heavier than a lighter load. Also, we would say, roughly, that one would work twice as hard pushing the same load a distance twice as far. However, there are situations in which a person might say that he was working, let us say standing and holding a suitcase, when there would be no work done in the sense defined above, since the suitcase would be moved through no distance. The definition we have given is one of many instances in which a word is taken from the body of normal usage and given a special technical significance. Its technical meaning bears only a limited relation to common usage. It would be possible, as an alternative, to coin another word. However, if we label such concepts by entirely different words, then their origins and their connections with common usage would be entirely lost, and they would evoke no associations, correct or incorrect.* When we say work, it is to be emphasized, we mean only what is defined above and in what follows.

The definition as we have given it is in its simplest form. It applies to those cases in which the force is constant and in the same direction as the motion:

Here work equals force × distance moved when the force is a constant and when the motion is in the same direction as the force.

If the force is in pounds and the distance is in feet, the work is given in foot-pounds. If the force is in $\left(\dfrac{\text{dynes}}{\text{newtons}}\right)$ and the distance is in $\left(\dfrac{\text{centimeters}}{\text{meters}}\right)$, the work is then given in units of $\left(\dfrac{\text{ergs}}{\text{joules}}\right)$.

If the force is a constant but not in the same direction as the motion, the work is defined as that component of the force parallel to the direction of the motion multiplied by the distance moved. This is obtained as shown:

Here the magnitude of $\mathbf{F}_{\|}$ is $F \cos \theta$ and the magnitude of $\mathbf{F}_{\perp}$ is $F \sin \theta$ (see the Appendices, pp. 490-493). The force $\mathbf{F}$ is resolved into two components whose vector sum is equal to $\mathbf{F}$. One of these, $\mathbf{F}_{\perp}$, is perpendicular to the direction of motion; the other, $\mathbf{F}_{\|}$, is parallel to the direction of motion:

* An example we shall come to later is the invented word "entropy."

$$\mathbf{F}_\perp + \mathbf{F}_\parallel = \mathbf{F} \qquad (8.1)$$

The work done in this case is defined to be

$$\text{work} = F_\parallel \times \text{distance moved} \qquad (8.2)$$

We can then generalize our original definition to say the following:

> DEFINITION: *For a constant force acting on a body, the work done equals the component of the force in the direction of the motion multiplied by the distance moved.*

We can see from this that for a given magnitude of force, the maximum work will be done if the force acts in the same direction as the motion of the body. If it is directed along the motion of the body, it will increase its speed. If it is directed opposite to the motion of the body, it will decrease its speed.

Three special and commonly encountered situations in which work is done by constant forces:

1. Force in the same direction as motion:

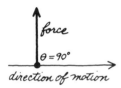

$$\text{work done} = Fd \qquad (8.3)$$

2. Force perpendicular to motion:

force

$\theta = 90°$

direction of motion

$$\text{work done} = F_\parallel d = 0 \qquad (8.4)$$

$(F_\parallel = F\cos 90° = 0$, since $\cos 90° = 0)$

3. Force opposed to the direction of motion:

$\theta = 180°$

force — direction of motion

$$\text{work done} = F_\parallel d = -Fd \qquad (8.5)$$

$(F_\parallel = F\cos 180° = -F$, because $\cos 180° = -1)$

From the definition we can construct a situation in which a force acts on a body and does no work. This will be true if the force is perpendicular to the direction of motion of the body. We know, however, that the force will change the direction of motion. In order, then, that there be no work done, the force, too, must change its direction to remain always perpendicular to the direction of motion. This, however,

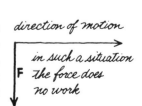

direction of motion

in such a situation F *the force does no work*

is not too difficult to achieve. Consider, for example, a ball at the end of a string of fixed length:

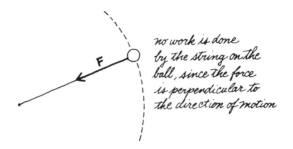

no work is done by the string on the ball, since the force is perpendicular to the direction of motion

We can complete our definition of work by considering the case in which the force does not remain constant throughout the motion. In this situation one takes the component of the force in the direction of the motion and plots it as a function of distance (Fig. 8.1) Then the work

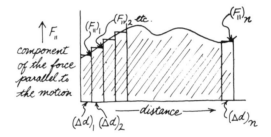

$\uparrow F_{\parallel}$ component of the force parallel to the motion

$(F_{\parallel})_1$ $(F_{\parallel})_2$ *etc.* $(F_{\parallel})_n$

— distance —

$(\Delta d)_1$ $(\Delta d)_2$ $(\Delta d)_n$

FIG. 8.1. Total work done, $W = (F_{\parallel})_1(\Delta d)_1 + (F_{\parallel})_2(\Delta d)_2 + \cdots + (F_{\parallel})_n(\Delta d)_n$, becomes equal to the area under the curve when the Δd becomes small enough.

done is the shaded area under the curve. We are saying, in effect, that if one moves the body through a very small distance with one force and then through the next small distance with another force, and so on, the total work done is the sum of the work done in the first displacement plus the work done in the second displacement, etc. In the special case that the force remains constant, the curve is just a rectangle, and the work, as stated previously, is the force times the distance.

In general, the definition of the work done by a force on a body is the following:

> DEFINITION: *For constant forces, it is the product of the component of the force in the direction of the motion times the distance moved; for forces that change in the course of motion, it is the area under the curve—the component of the force parallel to the motion as a function of the distance moved.*

Example 1. Sisyphus pushes his stone 10 ft along a rough level section of Hades, requiring a constant force of 100 lb. The work he does is $(100 \text{ lb}) \times (10 \text{ ft}) = 1000$ ft-lb. In the ninth circle, however, along the lake of ice the work he does is close to zero, since it requires practically no force to push. (He must, however, choose a path that avoids such obstacles as the partially embedded condemned.)

Example 2. A good spring has the property that the restoring force is proportional to the distance it is extended (Hooke's law*). If a spring exerts a force of 10^3 dynes per cm of extension, how much work is required to extend it 50 cm? The graph of force vs. extension is shown here.

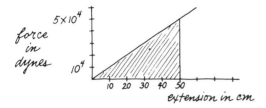

The work done is just the area of the shaded triangle bounded by the vertical line corresponding to 50 cm

$$W = \tfrac{1}{2} \times 50 \text{ cm} \times 5 \times 10^4 \text{ dynes} = 1.25 \times 10^6 \text{ ergs}$$

Kinetic Energy

The kinetic energy of a particle (descendant of what was known as the *vis viva*, the living force of the scholastic scholars), often denoted by the letter *T*, is defined as one half of mass times the speed squared, or

D E F I N I T I O N : *kinetic energy* $= \tfrac{1}{2}mv^2$

In early attempts to understand the motion of bodies, the concepts of vis viva and of impetus (now called momentum) were proposed, and both were called the force of motion. From our point of view, the impetus, or momentum, is related to the force necessary to change motion over a given time, whereas the kinetic energy, or vis viva, is related to the force necessary to change the motion over a given distance.

T H E O R E M 8.1: *The work done on a body is equal to its change of kinetic energy.*

This theorem is true for forces acting in the same or in a different direction as a motion—for forces that are constant, for forces that are varying in time, or under any circumstances. We shall prove it, however, only in the simplest case, the case in which we have a constant force in the same direction as the motion.†

---

* After years of conflict with a sometimes testy Newton about almost all scientific matters and orders of priority (who thought of what idea first), history has awarded almost all of the points (perhaps somewhat unjustly) to Newton. It is said, for example, that in order to determine experimentally the variation of the force of gravity with the distance from the center of the earth, Hooke measured *g* at the surface and at the top of a mountain. With the apparatus he had available he could find no variation. We are left, however, with Hooke's "law" for the force exerted by an "ideal" spring:

$$\text{force} = \begin{pmatrix} \text{number depending upon} \\ \text{the material of the} \\ \text{spring} \end{pmatrix} \times (\text{extension})$$

† As in all cases in which we give a simple form of a proof, we imply that all the ideas needed for the proof are contained in the simple form. The full proof requires a technical elaboration of these ideas, which is possible but is beyond either our present means or interest.

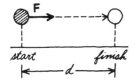

start   finish
|←——d——→|

PROOF. Consider a body that starts at rest, acted on by a constant force, **F**, over a distance, d. The force is in the direction of motion. Therefore all vectors can be treated as numbers. From the definition, the work done, W, is equal to the force, F, multiplied by the distance, d:

$$W = Fd \tag{8.6}$$

The change in kinetic energy is the difference between the final and initial kinetic energy. Since the initial kinetic energy is zero (the body starts from rest), we have

$$\text{change in kinetic energy} =$$
$$\text{final kinetic energy} - \text{initial kinetic energy} =$$
$$\tfrac{1}{2}mv^2 - 0 = \tfrac{1}{2}mv^2 \tag{8.7}$$

To calculate $\tfrac{1}{2}mv^2$ we need to find v (the speed after the body has moved the distance d). This we find by using Newton's second law. The force acting on the body is constant; therefore, from Newton's second law, the acceleration is constant also and has the magnitude

$$a = \frac{F}{m} \tag{8.8}$$

Under a constant acceleration, the relation between speed and time is

$$v = at \tag{8.9}$$

But what is the time required to go the distance d under a constant acceleration? This question was answered by Galileo, who gave the distance as a function of time:

$$d = \tfrac{1}{2}at^2 \tag{8.10}$$

This relation, inverted,† gives the time as a function of the distance,

$$t = \sqrt{\frac{2d}{a}} \tag{8.11}$$

Now we have all the pieces; we put them together as follows. The final speed is

$$v = at \tag{8.12}$$

but

$$t = \sqrt{\frac{2d}{a}} \tag{8.13}$$

Therefore,

$$v = a\sqrt{\frac{2d}{a}} = \sqrt{2ad} \tag{8.14}$$

† Inversion, why one takes $+\sqrt{}$ and not $-\sqrt{}$, is discussed in the Appendices, p. 481.

Further,

$$a = \frac{F}{m} \qquad (8.15)$$

so that

$$v = \sqrt{2 \, \frac{F}{m} \, d} \qquad (8.16)$$

Using this we have

change in kinetic energy

$$= \tfrac{1}{2}mv^2$$

$$= \tfrac{1}{2}m \left(\sqrt{2 \, \frac{F}{m} \, d} \right)^2 = \tfrac{1}{2}m \times 2 \, \frac{F}{m} \, d$$

$$= Fd = \text{work done} \qquad (8.17)$$

Q.E.D.

Example. As an application, consider a question that seems to appear on almost all written examinations for a driver's license. An automobile moving at 20 miles/hr can be stopped in, let us say, 30 ft. How many feet will it take to stop the same automobile if it is moving at 40 miles/hr? The answer that is marked correct is well known to be 120 ft. We are now in a position to list the suppositions that underlie this.

Presumably, if one applies the brakes, there is a maximum force that can be exerted by the road on the tires before the tires begin to skid. Call this force F. To stop the automobile, its kinetic energy must be changed from an initial value, $\tfrac{1}{2}mv^2$, to a final value, 0. According to the theorem, the change in kinetic energy is equal to the work done. Therefore, the maximum force times the distance in which the automobile can be stopped is equal to the initial kinetic energy, $\tfrac{1}{2}mv^2$. The distance required to stop the automobile, therefore, is given by the following expression:

$$-Fd = \text{final kinetic energy} - \text{initial kinetic energy}$$
$$= -\tfrac{1}{2}mv^2 \qquad (8.18)$$

(The work is negative since the force opposes the motion.) Therefore,

$$d = \frac{mv^2}{2F} \qquad (8.19)$$

The mass of the automobile and the maximum force enter into the equation as constants. The ratio of the distance required to stop the automobile for two initial speeds, call them v_1 and v_2, is

$$\frac{d_2}{d_1} = \frac{(m/2F)v_2^2}{(m/2F)v_1^2} = \left(\frac{v_2}{v_1} \right)^2 \qquad (8.20)$$

Or, the ratio of the distance goes as the square of the ratio of the speeds. Thus, if the speed is doubled, the distance required to stop goes up by a factor of four. The major assumption is that the maximum force the road can exert on the tires is independent of the initial speed of the automobile, which is reasonably close to the truth.

DISCUSSION OF THEOREM 8.1. There are several points we would like to make about Theorem 8.1.

1. There is nothing in the theorem that tells us about the rate at which the work is done or how long it took to do it. It asserts that independent of time, the change in kinetic energy is equal to the work done.

2. We see that work done on a body when the force is in the same direction as the motion increases the kinetic energy, whereas work done if the force is opposed to the direction of motion decreases the kinetic energy.

3. If the force is perpendicular to the motion, as in the case of a ball moving in a circle at the end of a string, the force does no work on the ball and there is no change in kinetic energy—the case of circular motion around a central force.

Potential Energy

Upon reflection, it seems curious that the kinetic energy, which is the same before and after, changes during the course of an elastic collision. For some collisions, for example that of two particles of equal mass moving with equal and opposite velocities, there exists an instant in which both particles have stopped moving entirely (in this case, the instant that separates the ingoing from the outgoing part of the collision), and the kinetic energy of the system is zero, seeming to have disappeared entirely.

This is contrasted with the momentum of the system which remains constant before, during, and after the collision. Of course, there is no reason that the kinetic energy must remain constant, but its disappearance and reappearance leads naturally to the question: Is there something else that increases as the kinetic energy decreases in such a way that the sum of the kinetic energy and the something else remains constant during the entire collision?

This something else exists and, in the case of forces whose magnitude and direction depend only on the position of the particles of the system, develops into a concept with usefulness far beyond collision problems, so powerful that it shapes the development of future branches of physics and remains even when the Newtonian form of mechanics is no longer employed. It is named the *potential energy*—a good word, because it evokes the right image. The kinetic energy or, as we might call it, the energy of motion, is visible, being associated with the speed of the particle. The potential energy, related to some internal configuration of the system, is less apparent, but it has the possibility of being reconverted into kinetic energy.

In common usage, the potential energy can be compared to a stored energy that can be converted into energy of motion. There are many examples. A spring that is coiled contains potential energy that can produce motion if it is released. Or, a weight lifted to some height contains potential energy in a sense that when released it falls and acquires a kinetic energy.

We define the potential-energy difference between two points as minus the work done on a particle by conservative forces (forces whose magnitude and direction depend only on the position of the bodies in the system) in moving from one point to the other.

DEFINITION: *Potential-energy difference between a and b =*
— work done on a body in moving it from a to b.

To illustrate, consider a particle subjected to the earth's gravitational force. The potential-energy difference between the points x_1 *and* x_2 is

— work done on particle in going from x_1 to x_2

$$= -(-mg)(x_2 - x_1) = mgx_2 - mgx_1 \qquad (8.21)$$

Putting this result in Eq. (8.18) and writing the kinetic energy explicitly, we have

$$\tfrac{1}{2}mv_1^2 + mgx_1 = \tfrac{1}{2}mv_2^2 + mgx_2 \qquad (8.22)$$

Suppose the particle is thrown upward from the surface of the earth with a speed v_0. In this case $x_1 = 0$, and we have

$$\tfrac{1}{2}mv_0^2 = \tfrac{1}{2}mv^2 + mgx \qquad (8.23)$$

Thus the speed v is given as a function of the position of the particle alone. For example, we might ask: How high does the particle go before coming to rest and falling again? At its maximum height the particle will have zero speed ($v = 0$). Thus

$$\tfrac{1}{2}mv_0^2 = mgx_{max} \qquad (8.24)$$

or

$$x_{max} = \frac{v_0{}^2}{2g} \qquad (8.25)$$

When it has risen to its maximum height, its speed and therefore its kinetic energy are zero. The potential energy has increased to its maximum and an inspection of the equations above reveals that the potential energy at this point is just equal to the original kinetic energy:

$$mgx_{max} = mg\left(\frac{v_0^2}{2g}\right) = \tfrac{1}{2}mv_0^2 \qquad (8.26)$$

Then the particle begins to fall again and eventually returns to the surface of the earth. Just before it strikes, the distance from the surface is zero, so that the speed is just equal to the original speed. However, the direction of motion has changed. The particle is now going down rather than up. This does not affect the kinetic energy, which is proportional to the square of the speed and is therefore independent of the direction of motion. From this point of view, we see that what happens in the motion is a transfer of kinetic energy into potential energy and back in such a way that the sum remains constant.

Among complicated systems of particles, things are not necessarily quite so transparent. Kinetic energy may be converted into potential energy and remain that way. For example, if we were to throw a particle onto a ledge in such a way that it remained there, the kinetic energy of the particle would be converted into potential energy, but the reconversion would not take place until the particle was pushed off the surface of the ledge.

Perhaps the most important property of the potential energy is contained in

THEOREM 8.2: *If the forces acting on a body are conservative, the work done on that body in going from one point to another depends only on the initial and final points—or—The potential-energy difference between two points depends only on those points.*

This powerful result—the dependence of the work done only on the initial and final points—can be stated more formally by saying that under conservative forces, minus the work done in going from a to b (or the potential-energy difference between a and b) is a function[*] only of a and b. To say that something is a function only of (depends only on) something else is, although not entirely specific because there are large numbers of functions, a very strong and limiting statement. In this case it is the not-obvious proposition that work done does not depend upon the path; it does not depend upon the initial speed; etc. It depends only on the initial and final points. This enables us to write

$$\begin{array}{c} \text{potential-energy difference between } a \text{ and } b \text{ defined} \\ \text{as minus the work done on body going from } a \text{ to } b \qquad (8.27) \\ = V(b) - V(a) \end{array}$$

where V designates the potential energy and is a function only of the point at which it is evaluated. (In this way we can assign to a point in space a property that influences the motion of bodies. This theme is developed later into the concept of a field.) And with this we can now write

$$T_a = T_b + V(b) - V(a) \qquad (8.28)$$

or

$$T_b + V(b) = T_a + V(a) \qquad (8.29)$$

To this point the natural quantity with which to work has been the potential-energy difference between two points. We now should like to assign some significance to the potential energy at one point, e.g., $V(P)$. We define the potential energy at the point P as the potential energy difference between some fixed (but arbitrary) point O and the point P. The point O can be chosen to make the form of the potential energy as simple as possible, as any shift of the fixed point shifts the potential energy by a constant, independent of the final points:

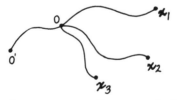

Suppose originally we took O as the fixed point; then the potential energy at x_1 would be by definition minus the work done between O and

[*] A general discussion of functions is included in the Appendices; see pp. 476-479.

x_1, the potential energy at x_2 would be minus the work done between O and x_2, and so on. If we decided on another fixed point—say O'—then the potential energy at x_1 would be minus the work done between O' and O in addition to minus the work done between O and x_1. The potential energy at x_2 would be minus the work done between O' and O in addition to minus the work done between O and x_2, and so on.

$$\begin{pmatrix} \text{potential energy at } x \\ \text{with the fixed point } O' \end{pmatrix} = \begin{pmatrix} \text{potential energy at } x \\ \text{with the fixed point } O \end{pmatrix}$$
$$- (\text{work done between } O' \text{ and } O)$$

But the work done in going from O' to O is independent of the point x. So the potential energy at all points would be shifted by a constant quantity: minus the work done between O' and O. This means that if we have calculated the potential energy using one fixed point, another choice of a fixed point would give us the original potential energy plus a constant. All these definitions turn out to be equivalent, because in any actual use of the potential energy we always subtract the potential energy at one point from that at another, and a constant drops out of the equation.[*]

With this we have completed our search for the quantity which, in spite of kinetic energy changes, remains constant during the motion of the system—it is the sum:

$$\text{kinetic energy} + \text{potential energy}$$

We are therefore led to define the mechanical energy, E, as follows:

D E F I N I T I O N : *The mechanical energy $E = T + V$.*

It is this mechanical energy of the system, E, that remains constant throughout the motion.

It follows from the definition of potential energy that it is measured in the same units as work (since the potential energy is equal to minus the work done in moving from one point to another). Thus potential energy is measured in ergs (CGS), joules (MKS), or foot-pounds (English). The potential energy of a particle of mass 10 g at rest 10 cm above the surface of the earth (the fixed point taken at the surface of the earth) would be

$$V = mgx = 10 \text{ g} \times 980 \, \frac{\text{cm}}{\text{sec}^2} \times 10 \text{ cm} =$$

$$98,000 \, \frac{\text{gm-cm}^2}{\text{sec}^2} = 98,000 \text{ ergs} \tag{8.30}$$

It follows from Theorem 8.1 (the work done on a body is equal to its change of kinetic energy) that the kinetic energy of a body is also measured in the same units as work (ergs, joules, or ft-lb). Thus the kinetic energy of a 10-g object moving with a speed of 10 cm/sec is

$$T = \tfrac{1}{2}mv^2 = \tfrac{1}{2} \, 10 \text{ g} \times \left(\frac{10 \text{ cm}}{\text{sec}} \right)^2 =$$

$$500 \, \frac{\text{gm-cm}^2}{\text{sec}^2} = 500 \text{ ergs} \tag{8.31}$$

[*] $V(b) - V(a) = [V(b) + \text{constant}] - [V(a) + \text{constant}]$

From this we see that E, the mechanical energy, which is the sum of T and V, must also be measured in the same units as work (ergs, joules, or ft-lb).

A particle of mass 10 g subjected to the earth's gravitational force has the mechanical energy:

$$E = \tfrac{1}{2}mv_x^2 + mgx, \tag{8.32}$$

where x is measured up from the surface and v_x is its speed at the point x. For a particle of mass 10 g at rest 10 cm above the surface of the earth,

$$E = 0 + 10 \text{ g } 980 \frac{\text{cm}}{\text{sec}^2} \, 10 \text{ cm} = 98,000 \text{ ergs} \tag{8.33}$$

If it is released, its kinetic energy when it reaches the surface is just the original potential energy, since the potential energy at the surface is zero, and the total energy remains constant. Thus

$$E = \tfrac{1}{2}mv_0^2 + 0 = 98,000 \text{ ergs} \tag{8.34}$$

or

$$v_0 = \sqrt{\frac{2E}{m}} = \sqrt{\frac{2 \times 98,000 \text{ g } \frac{\text{cm}^2}{\text{sec}^2}}{10 \text{ g}}}$$

$$= \sqrt{19,600 \frac{\text{cm}^2}{\text{sec}^2}} = 140 \text{ cm/sec} \tag{8.35}$$

We are now in a position to state, for these mechanical systems, the

PRINCIPLE OF ENERGY CONSERVATION: *For a system acted on by conservative forces, the energy, defined as the kinetic plus the potential energy, remains a constant throughout the motion.*

As in the case of momentum, the energy as thus defined for mechanical systems turns out to be a more general concept than its origins might indicate. The concept of energy and energy conservation has been preserved and generalized for systems that are non-Newtonian and for forces that are far more complex than those we have treated. As with the momentum, a concept that originated in a theory of the motion of Newtonian or Cartesian particles has proved to be deeper than the theories themselves.

The energy of a many-particle system is defined as

$$E = T_1 + T_2 + \cdots + T_N + V(x_1, \ldots, x_N) \tag{8.36}$$

That is to say, the energy is equal to the sum of the kinetic energies of all the particles plus the potential energy, which is a function of the positions of all the particles. And by methods similar to those we used above, one can show that the energy so defined remains a constant during the motion of a many-particle system.

POTENTIAL ENERGY FOR SEVERAL
SIMPLE FORCE SYSTEMS

We now consider several commonly encountered and useful conservative force systems. The first is a particle subjected to a constant force, as, for example, a particle near the surface of the earth in the uniform gravitational force of the earth. The work done on the particle

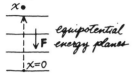

moved from the point zero to the point x is $-mgx$ and the potential energy of the particle is

$$V(x) = mgx \qquad (8.37)$$

If we consider a series of planes parallel to the surface of the earth, on any one of these planes the potential energy is a constant, because it depends only on the distance of the particle from the earth. These planes may thus be called equipotential energy planes. It is useful to remember that a particle can be moved along such a plane with no work done on it.

Now consider the potential energy of a particle in the actual gravitational force surrounding the earth. For a particle of mass m, this force is

$$\mathbf{F} = \frac{GMm}{R^2} \qquad \text{magnitude}$$

toward center
of earth
$\qquad$ direction $\qquad (8.38)$

as shown:

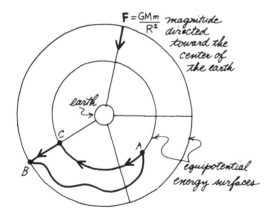

In analyzing this situation, the most convenient thing to do first is to define the equipotential energy surfaces. The surfaces again are those on which the potential energy is constant, or on which no work is done to go from one point to another. For a force, such as the gravitational force, always directed toward the center of the earth, these surfaces are very easy to determine. Any spherical surface centered on the center of the earth will be a surface on which the potential energy is the same. This can be seen as follows. To move a particle from one point on the sphere to another involves a motion in which the force is always perpendicular to the motion of the particle, because a radius is always perpendicular to the surface of a sphere. Therefore, no work will be done by the gravitational force in such a motion and the potential energy will be the same at any point on the surface. From this we see that the potential energy due to the gravitational force can depend only on the distance of the particle from the center and can have nothing to do with the direction. (This results in a symmetry which will be of great interest later.)

To calculate the potential energy of a particle subject to the gravitational force, then, we have to calculate the work done in moving a particle along the simplest possible path from one circle to another. For example, as illustrated above, if the particle were to go from point A to point B via curved path AB, the potential-energy difference could be computed by going first from A to C and then going radically outward from C to B. The potential-energy difference between A and C is zero, because it is an equipotential energy surface. Therefore the potential-energy difference between A and B is just that between C and B.

Thus we have only to compute the potential difference as a particle moves outward along the radial line.

The force on the particle with such motion is

$$\mathbf{F} = \frac{GMm}{R^2} \qquad \text{magnitude (directed toward the center)} \qquad (8.39)$$

Let us say the direction of motion is toward the center.

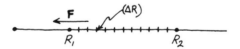

The work done in going from R_2 to R_1 is given by breaking the path up into many small intervals and adding:

$$F_1 (\Delta R)_1 + F_2 (\Delta R)_2 + \cdots + F_n (\Delta R)_n \qquad (8.40)$$

This is given by the shaded area under the curve.

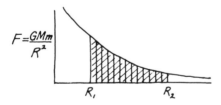

The summation can be carried out exactly, and the result is

minus the work done going from R_2 to R_1

$$= V(R_1) - V(R_2) = \frac{GMm}{R_2} - \frac{GMm}{R_1} \qquad (8.41)$$

For this system the fixed point is often taken at infinity ($R_2 = \infty$). This has the meaning that $GMm/R_2 = 0$ when $R_2 = \infty$. Thus the gravitational potential energy (with this fixed point) is

$$V(R) = -\frac{GMm}{R} \qquad (8.42)$$

Example 1. The High-Altitude Rocket. Imagine that there were some reason to fire a rocket to an altitude of 4000 miles (one earth radius) above the surface of the earth before it stopped and returned. What blast-off speed should it have? At the blast-off, when the rocket is given its maximum speed and its fuel is consumed, say that it has the speed v_0. Its energy is then

$$E = T + V = \tfrac{1}{2}mv_0^2 - \frac{GMm}{R_e}$$

and remains constant during the motion.

At its maximum height, $2R_e$, its speed is zero. Therefore,

$$E = \tfrac{1}{2}mv_0^2 - \frac{GMm}{R_e} = -\frac{GMm}{2R_e}$$

and

$$\tfrac{1}{2}mv_0^2 = \frac{GMm}{2R_e}$$

so that

$$v_0 = \sqrt{\frac{GM}{R_e}} = \sqrt{gR_e} = \sqrt{32 \text{ ft/sec}^2 \times 2.11 \times 10^7 \text{ ft}}$$

$$= 2.6 \times 10^4 \text{ ft/sec} = 4.9 \text{ mi/sec}$$

(We recall that g was defined as $g = \dfrac{GM}{R_e^2}$; thus the arithmetic can be simplified by using $\dfrac{GM}{R_e} = gR_e$).

Example 2. Escape Speed. What speed would be required so that the rocket would never return to earth? The minimum speed (escape speed) is that which takes it "infinitely" far away ($R = \infty$) and leaves it there with zero speed. The total mechanical energy is

$$E = \tfrac{1}{2}mv^2 - \frac{GMm}{R} \qquad (8.43)$$

At $R = \infty$, the potential energy, $-\dfrac{GMm}{R}$, is equal to zero. Since the speed is to be zero or greater at infinity, the kinetic energy at $R = \infty$ (and hence the total mechanical energy) is equal to or greater than zero. But the total mechanical energy is constant. Thus in order that the rocket escape, E must be greater than zero.

At the earth's surface, this requires that v^2 be at least as large as

$$v^2 = \frac{2GM}{R_e}$$

R_e

$R_e = 4000 \text{ miles} \approx 2.11 \times 10^7 \text{ ft}$

or

$$v = \sqrt{\frac{2GM}{R_e}},$$

but

$$\frac{GM}{R_e^2} = g = 32 \text{ ft/sec}^2$$

Therefore,

$$v = \sqrt{2(32 \text{ ft/sec}^2) R_e} = \sqrt{\frac{2(32 \text{ ft/sec}^2)(4000 \text{ mi})}{5280 \text{ ft/mi}}}$$

$$v \simeq 7 \text{ mi/sec}$$

(We observe that with this definition of the gravitational potential energy, an object with energy smaller than zero is bound to the earth, while one with energy larger than zero escapes. In a similar manner, the closed planetary elliptical orbits are characterized by energies smaller than zero, and the open hyperbolic orbits are associated with energies larger than zero.)

9

CONSERVATION OF ENERGY
(The First Law of Thermodynamics)

The idea of energy predates, is implicit in, and outlives Newtonian mechanics. Leibnitz argued with Newton about the relative importance of force and *vis viva* (the living force that eventually was to be equated with kinetic energy). D'Alembert showed that the argument was semantic, and that one could obtain the conservation of kinetic energy under the proper circumstances as a consequence of Newtonian mechanics. But if we confine our definition of energy to potential plus kinetic energy, then it is easy to find circumstances in which energy is not conserved. As Julius Robert von Mayer said, "In numberless cases we see motion cease without having caused another motion or the lifting of a weight."

This motivates us to generalize the definition of energy so that it is always conserved (if necessary creating new forms of energy or particles). We ask: Is there some macroscopic quantity that has changed when motion ceases "without having caused another motion"? The answer could be that the objects become warmer. When a hammer strikes an iron bar, it is common experience that the bar becomes hot. Two sticks rubbed together become warm. The kinetic and potential energy that have seemingly disappeared from the system may have become heat. May we then say that heat is a form of energy that under appropriate circumstances can do work?

In 1824 Sadi Carnot in a short memoir, *Reflections on the Motive Power of Heat*, analyzed first steam engines, and then all engines, considering the problem of what is the best way to build an engine so that the most useful work can be obtained from a given quantity of heat. And he raised what he called an interesting and important question:

> Is the motive power of heat invariable in quantity, or does it vary with the agent which one uses to obtain it—that is, with the intermediate body chosen as the subject of the action of heat?

With such questions begin the merger of the concepts of heat and temperature with the body of mechanics and molecular theory.

DEFINITION OF A TEMPERATURE SCALE

According to current conventions, it is convenient to measure what we might like to call the heat content of bodies by measuring their temperature. However, the relation between these two concepts is somewhat subtle. If we have two bodies, one large and the other small, both at the same temperature, we might imagine that the larger body in some sense has a greater heat content than the smaller one. Two containers might have the same level of liquid and yet, depending on their size or their shape, could contain different quantities of liquid. We first define a temperature scale, and later attempt to construct a relation between heat and temperature.

We all have an intuitive idea of what temperature is. The hand can distinguish roughly between hot and cold; however, we know that the hand can be fooled. There is a familiar trick in which one hand is placed in cold and the other in hot water. When the two hands are simultaneously placed in a bowl of lukewarm water, the hand that was in hot water registers cold, while the hand that was in cold water registers hot—an illustration of how our senses, usually reliable, can mislead us and a reason to construct an independent measure of temperature that will not be influenced by its previous history or psychology.

Such a measure of temperature is provided by the common thermometer, a glass tube with a bulb beneath, partially filled with mercury or another liquid. Because the mercury expands more rapidly than does the glass when heated, the mercury column is seen to rise or fall in the glass tube when the bulb is immersed in a warm or cool substance. We are all familiar with this phenomenon, and with the method of reading the temperature off the scale of a thermometer by observing the level to which the mercury column has risen.

The relation, however, between the level of the mercury and what will finally be convenient to call temperature is not entirely clear. For example, if we construct a thermometer of other materials so that both it and the mercury thermometer stand at the same levels at the boiling point and the freezing point of water, there is no guarantee that

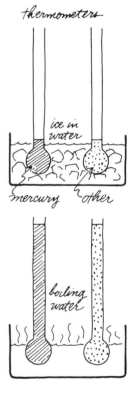

thermometers

ice in water

mercury   *other*

boiling water

lukewarm water

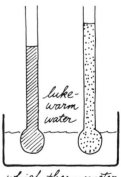

which thermometer measures the temperature?

they will be at the same level in between, because there is no guarantee that the other materials will expand at the same rate as the mercury and glass. In this situation, which thermometer would we say read the temperature?* We cannot answer the question at the moment. Now we shall content ourselves with constructing a crude definition of temperature. Later, in the course of our investigation of heat, gases, and the kinetic theory, we shall see how the definition of temperature may be refined.

After they realized that how hot a person felt was an indication of his well-being, doctors attempted to measure the temperature of patients when examining them. Glass tubes filled to some level with water, mercury, or a colored liquid were used. The doctor would presume that the higher the liquid rose, the higher the temperature. In the absence of uniform markings, a doctor would compare the reading given by his own body with that of the patient and no doubt come to some learned conclusion about the patient's health. Devices of this kind had been used in meteorology and for various other purposes, but since they did not give the same readings (that is to say, two such devices would not necessarily show the same level in a given bowl of water), it was difficult to make comparisons, and it was not even clear that, for example, water would always boil at the same "degree of heat."

When Fahrenheit read of the discovery of Amontons† "that water boils at a fixed degree of heat," he was at once inflamed with a great desire to make for himself a thermometer ". . . so that I might with my own eyes perceive this beautiful phenomenon of nature. . . ." He realized "that the height of the column of mercury in a barometer was a little (though sensibly enough) altered by the varying temperature of the mercury," and he constructed a thermometer based on the expansion of mercury in a glass column; he was the first to succeed in making thermometers that were comparable with each other (gave the same reading) throughout the whole length of the scale.

He chose for the two fixed points a level corresponding to what possibly was his wife's body temperature (100°F on that day, if we are still using the same thermometer) and another, 0°F, which is said to be the lowest level to which the mercury fell during the course of one winter in Northern Ireland. (Possibly he was trying to avoid negative temperatures and held the not uncommon opinion that Northern Ireland in the midst of winter was as cold as a place can be on earth.) The distance between these two points he divided into 100 equal intervals, each one of which he called a degree—now known as 1°Fahrenheit. On this thermometer, in which the boiling and freezing points of water occur at 212° and 32°, he was able to observe that various liquids boiled at "fixed degrees of heat."

Anders Celsius (1701–1744) introduced the convention of associating the two fixed points with a property of matter. He called the level of the mercury zero in a bath of melting ice and 100 in a bath of boiling water. Dividing the interval into 100 equal spaces, he arrived at what is now called the centigrade scale.

* It is not trivial, even using the same materials, to ensure that the thermometers will give the same readings throughout the scale. (The cylindrical hole has to be uniform, for example.) Fahrenheit was the first to succeed in doing so.

† He constructed the first thermometer that measured temperature using the pressure of air.

Liquids	Specific Gravity[a] of Liquids at 48° of heat	Degree Attained by Boiling
Spirits of Wine or Alcohol	8260	176
Rain Water	10000	212
Spirits of Niter	12935	242
Lye prepared from wine lees	15634	240
Oil of Vitriol	18775	546

[a] The specific gravity of a substance is now defined as the weight of a given volume of that substance divided by the weight of the same volume of water. Therefore, the specific gravity of water is 1.0000. For some reason, Fahrenheit doesn't seem to have included the decimal point.

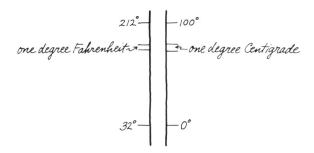

To convert between the centigrade and the Fahrenheit scale, it is necessary to take into account the fact that the divisions on the Fahrenheit scale are smaller than those on the centigrade scale ($\frac{5}{9}$ of 1 centigrade degree = 1 Fahrenheit degree) and that the centigrade 0° is equal to the Fahrenheit 32°.

$$\tfrac{5}{9}\left(t_{\text{fahrenheit}} - 32\right) = t_{\text{centigrade}} \qquad (9.1)$$

The centigrade scale is no less arbitrary than the Fahrenheit scale; however, it is used more commonly in scientific work. Later, when we have discussed gases and the kinetic theory, we will define what is called an absolute temperature scale.

Example: Comfortable room temperature for some is 72°F. What is this on the centigrade scale?

$$t_c = \tfrac{5}{9}\left(t_f - 32\right) = \tfrac{5}{9}\left(72 - 32\right) = \tfrac{5}{9}\left(40\right) = 22.2°C \qquad (9.2)$$

HEAT

Once a temperature scale has been decided upon, we can define a quantity of heat in the following way. We call that amount of heat that will increase the temperature of 1 g of water from 14° to 15° centigrade a calorie. It does not matter how the heat is produced—by pounding, rubbing, or flame. When 1 calorie of heat is added, the water increases in temperature 1 degree. Or when 1 calorie of heat is taken away, one reduces the temperature of 1 gram of water 1 degree centigrade. The standard quantity of heat is thus defined in terms of a unit of tempera·ture change and the mass of a standard material (water).*

When we speak this way, we are treating heat almost as though it were a substance. We come close to saying: Pour 1 calorie of heat into 1 gram of water and the temperature increases 1 degree. The language is what remains of a theory that did treat heat as a fluid substance (caloric). Although an object is very cold, it may still contain heat. For example, an ice cube can increase the temperature of a block of dry ice while itself being cooled. A block of dry ice can increase the temperature of liquid helium. (Is there a limit to how far this can go on?)

One calorie of heat will change the temperature of 1 gram of a substance other than water by an amount not necessarily equal to 1 degree. For example, on the addition of 1 calorie of heat 1 gram of copper is increased in temperature 10.9 degrees centigrade. This relative capacity of different substances to absorb different amounts of heat for the same change in temperature is called the *specific heat capacity* (one has a greater capacity than another to absorb heat for a given change of temperature) of a substance. This is defined as the number of calories required to increase the temperature of 1 gram of the substance 1 degree centigrade. As an example, the specific heat capacity of copper is 0.092 cal/g-°C.

It is an observed property of many substances that the heat capacity remains approximately constant over wide variations of temperature. Thus 1 calorie of heat increases the temperature of 1 gram of water approximately 1 degree centigrade independent of the initial temperature of the water. At melting or boiling points, however, substances can frequently absorb relatively large quantities of heat without a change in temperature. For example, to melt ice, at 0°C requires 80 calories per gram of ice. To vaporize 1 gram of boiling water at atmospheric pressure requires 540 calories. Thus ice cubes floating † in water hold a constant temperature of 0°C, since any heat added or taken away from the water-plus-ice system serves to freeze or melt the ice rather than changing the temperature (Fig. 9.1).

It is for systems insulated in such a way that heat can neither enter nor escape that heat appears to be a conserved quantity—a substance that can neither be created nor destroyed. If one puts a warm spoon into an insulated bowl of cool water, the spoon will cool and the water warm until they reach the same temperature. For given quantities of spoon and water and given initial temperatures, the final temperature will always

---

* It is understood that we shall have to be careful about various things that we do not necessarily entirely appreciate at the moment (for example, pressure at which the gram of water is held; will the same calorie increase the temperature of the same gram of water from 18° to 19°C or from 78° to 79°C?) in order to construct the most useful definition.

† Almost all substances contract on freezing and become more dense. Water, an exception, expands anomalously on freezing so that ice at 0°C is less dense than water at that temperature. It thus floats at the surface of lake or ocean, helping to retain the heat of the body of water below and to keep it liquid, rather than sinking to the bottom, letting the heat escape and the rest of the water freeze. The consequences of this exceptional behavior, for fish, our climate, and our way of life, are considerable.

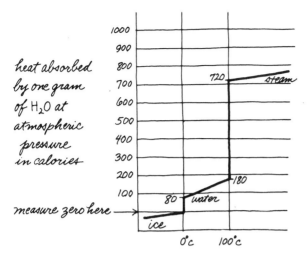

FIG. 9.1. In melting, the gram of ice absorbs 80 cal; 100 cal is absorbed by the gram of water in going from zero to 100°C. To convert the gram of water at 100°C to steam requires 540 cal. Since the specific heats of both ice and steam are about one half that of water, less heat is absorbed by these for a given temperature increase.

be the same—as though the total heat is distributed among all the parts of the system in such a way as to bring them all to the same temperature (thermal equilibrium).

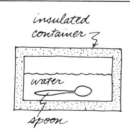

$$\text{heat lost by spoon} = (\text{mass of spoon}) \times \left(\begin{array}{l}\text{specific heat capacity:}\\ \text{calories to increase}\\ \text{temperature of 1}\\ \text{gram of spoon 1 degree}\end{array}\right)$$
$$\times (\text{change in temperature})$$

$$\text{heat gained by water} = (\text{mass of water}) \times \left(\begin{array}{l}\text{specific heat capacity:}\\ \text{calories to increase}\\ \text{temperature of 1}\\ \text{gram of water 1 degree}\end{array}\right)$$
$$\times (\text{change in temperature})$$

If the spoon were made of copper (specific heat capacity = 0.092 cal/g-°C), had a mass of 10 g, and were initially at a temperature of 100°C, and the water a mass of 100 g initially at the temperature 20°C:

$$\text{heat lost by spoon} = 10 \times 0.092 \times (100 - T)$$
$$\text{heat gained by water} = 100 \times 1 \times (T - 20)$$

Therefore,

$$0.92(100 - T) = 100(T - 20) \qquad 100.92T = 2092 \qquad T \simeq 20.8°C \tag{9.3}$$

The fact that bodies become warmer when they collide or are rubbed together is possibly the origin of the extremely old idea that heat is in some sense motion. In 1620, Francis Bacon stated that "Heat itself . . . is motion and nothing else."[1] He was led to this conclusion by observing that heat was produced whenever friction or collisions took place between solid bodies. Later in the seventeenth century, both Robert Boyle and Robert Hooke expressed similar ideas. However, the theory had little support at the time, mainly because no one was able to explain why, if heat were motion, this motion would be conserved in such experiments as those above in which materials at different temperatures are mixed together in a thermally insulated container.

During the eighteenth century, the theory came into being in which heat was assumed to be a subtle, elastic fluid whose particles repelled each other but were attracted to the particles of ordinary matter and whose presence within a body was the heat. This matter of heat came to be known as caloric (the word was coined by Lavoisier later on in 1787) and the treatment of heat as a material substance was called the *caloric theory*.

At the foundation of the caloric theory was the idea that heat is conserved. The majority of the observations and experiments of the calorists were performed under limited conditions which allowed the total quantity of heat as they defined it to remain constant, suggesting that heat is a conserved quantity. It was thus convenient to think of heat as a substance that could neither be created nor destroyed but which could flow from one material to another. Assuming, as was customary, that the basic units of matter were impenetrable, then despite the mutual attractions of the particles of the substance, these particles could not be in actual contact. (Otherwise, materials could not be compressed.) Therefore, there must be some counterbalancing repulsive force, and this force was ascribed to caloric. Because of the self-repulsion of the caloric substance, heat would flow from a hot material to a cold one. Whether a substance existed in the solid, liquid, or gaseous state depended upon the quantity of caloric that entered into its composition. When it contained large amounts of caloric, a substance would assume the form of a gas. By virtue of the self-repelling property of caloric, a large quantity of caloric would overcome the mutual gravitational attractions among the particles of the substance and cause these particles to remain separate from each other. As a substance was cooled, it was thought to be drained of caloric, which was consistent with the fact that most substances contract upon cooling. Solids and liquids would contain less caloric and thus occupy less volume.

Although the caloric theory is no longer in fashion, some of its language has been carried over into the modern organization of heat, particularly when one discusses flow and transfer of heat. We still speak of heat flowing or of an object "soaking up heat." This leads to the mild confusion that heat is sometimes spoken of as though it were a substance, even when we are told that it is not.

Benjamin Thompson, Count Rumford (1753–1814), an early expatriate, the author of the almost Cartesian dictum,

> It is certain, that there is nothing more dangerous, in philosophical investigations, than to take anything for granted, however unquestionable it may appear, till it has been proved by direct and decisive experiment,[2]

and a man of legendary energy, chose as one of his occupations the "science of Heat:—a science, assuredly, of the utmost importance to mankind!"[3] His interest in "this curious subject" was aroused, as he says:

> When dining, I had often observed that some particular dishes retained their Heat much longer than others; and that apple pies, and apples and almonds mixed (a dish in great repute in England) remained hot a surprising length of time. Much struck with this extraordinary quality of retaining Heat, which apples appeared to possess, it frequently occurred to my recollection; and I never burnt my mouth with them, or saw others meet with the same misfortune, without endeavouring, but in vain, to find out some way of accounting, in a satisfactory manner, for this surprising phenomenon.[4]

Burning his mouth with rice soup and later his hand at the hot baths near Naples did not discourage his interest. He was, on the contrary, inspired to do an almost classic series of experiments challenging the caloric theory.*

One might argue that if caloric were matter, then it ought to possess the fundamental property of all matter and have mass. In a series of experiments, Rumford showed to his own satisfaction that: "All attempts to discover any effect of heat upon the apparent weights of bodies will be fruitless." His demonstration that caloric was weightless, however, offered no serious challenge to the calorists. They could counter that caloric, as celestial material had been at an earlier time, was not ordinary matter and was therefore not necessarily subject to gravitational forces.

An accident ("It was by accident that I was led to make the Experiments of which I am about to give an account") led to Rumford's inquiries on the production of heat by friction.

> Being engaged, lately, in superintending the boring of cannon, in the workshops of the military arsenal at Munich, I was struck with the very considerable degree of Heat which a brass gun acquires, in a short time, in being bored; and with the still more intense Heat (much greater than that of boiling water, as I found by experiment) of the metallic chips separated from it by the borer.[5]

From the point of view of the caloric theory, one is pressed to explain where all the caloric came from. But with a little imagination, one introduces the hypothesis that the attractive force that was supposed to exist between the molecules of the metal and caloric might have been diminished when the metal was broken into chips, releasing caloric which appeared as heat.

---

* We are told that historians take this work more seriously than did his contemporaries.

However, the heat released in the course of the boring seemed to be inexhaustible. And this was enough to convince Rumford:

> . . . it appears to me to be extremely difficult, if not quite impossible, to form any distinct idea of any thing, capable of being excited and communicated in these Experiments, except it be MOTION.[6]

MECHANICAL EQUIVALENT OF HEAT; HEAT AS ENERGY

Among the first to realize the importance of the identification of heat as energy was a doctor (not of philosophy but of medicine): Julius Robert von Mayer (1814–1878). "In numberless cases," he writes,

> we see motion cease without having caused another motion or the lifting of a weight. . . .

And then he proceeds to propose a convention that had become almost a commonplace at the beginning of the nineteenth century:

> but an [energy]* once in existence cannot be annihilated, it can only change its form; and the question therefore arises, What other forms is [energy]* . . . capable of assuming?

Then arguing that as work can be converted to heat (e.g., rubbing two sticks together warms them up), heat must be a form of energy:

> If potential energy and kinetic energy are equivalent to heat, heat must also naturally be equivalent to kinetic energy and potential energy.

Now he makes his most penetrating observation. If heat is a converted form of kinetic or potential energy and energy as a whole is conserved, then a given quantity of heat must be the result of a given amount of mechanical energy. Or, a given amount of work must produce a given amount of heat. From experiments that already had been done on gases, Mayer was able to infer a quantitative relation between mechanical work and heat that is in fairly good agreement with the best values that have been obtained since.

A direct measurement of the mechanical equivalent of heat was made by the Englishman, James Prescott Joule (1818–1889). Throughout his life Joule carried out a long series of experiments in which various forms of energy were converted to heat. He first compared the mechanical work required to run an electric generator with the heat produced by the current generated.

Later he measured the heat resulting from the friction due to water flowing through thin pipes and the work needed to produce the flow. He measured the work done to compress a gas and the heat so created. Then he performed the famous experiment in which a paddle wheel was rotated in an insulated bucket so that the mechanical work done in rotating

*Mayer used "force" where we use "energy" today.

the paddle wheel could be compared with the temperature rise of the water due to the friction between the paddle wheels and the water. It

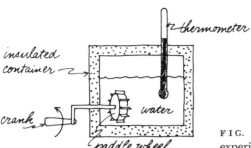

F IG. 9.2. Joule's paddle-wheel experiment.

is said that even on his honeymoon he could not be diverted from his principal preoccupation and that he measured the temperature of the water above and below a waterfall in Chamonix so that the temperature rise due to the collision of the water with the rocks below the falls (its potential energy being converted to kinetic energy and then to heat) could be determined.

The result of this devotion was as follows: He could conclude that a given amount of work produced a given amount of heat. The numerical relationship as it is used today is

$$1 \text{ cal} = 4.18 \text{ joules} = 4.18 \times 10^7 \text{ erg} \qquad (9.4)$$

This means that if we stir, rub, paddle, or otherwise do mechanical work on a system that is 1 gram of water thermally insulated from its surroundings, then for every 4.18 joules of work we do, the temperature of the water will rise 1°C. If the system were 2 g of water, it would take 8.36 joules of work to raise its temperature 1°C. If the system were 1 g of copper, then $0.092 \times 4.18 = 0.385$ joules of work would be required to raise the temperature 1°C. By now it was clear that not only was caloric not conserved but it was produced in definite quantities when work was done on a system, and it was this realization, no doubt, that signaled the end of the caloric theory.

The Conservation Principle That Failed

The caloric theory was founded on the idea that heat was a conserved quantity. If this were so, it was then easiest to visualize heat as a substance that could neither be created nor destroyed but could flow from one material to another. It was just at this point that the most effective attacks on the theory were made. One could easily show that in friction experiments (such as Rumford's cannon) heat could be produced in seemingly endless quantities.

The history of physics is strewn with the debris of conservation laws: laws which state that in isolated systems certain quantities cannot be created or destroyed. The feeling that such laws exist is old and runs deep. Lucretius echoes an ancient intuition when he says, ". . . things cannot be created out of nothing, nor, once born, be summoned back to nothing. . . ."[7] And there has been no lack of them since: conservation of matter, energy, charge, baryons, and so on. Often such laws organize fruitfully only a limited range of phenomena. In chemical reactions, matter can be thought to be conserved, whereas in nuclear reactions it is difficult to do so since, for example, the end products of uranium fission have less mass than the original uranium. It was not too different in the case of heat. Heat or caloric we now say is not conserved, but energy we say is. As has become clear, heat could be identified as a form of energy, and the restricted principle of conservation of heat became the more general principle of energy conservation.

HELMHOLTZ' CONSERVATION OF ENERGY

Possibly the earliest comprehensive statement of the principle of energy conservation was written in a memoir by Helmholtz in 1847. Energy as a term had not yet come into common use, and Helmholtz used the word "force." In mechanics, a particle could be assigned what we have called a kinetic energy (*vis viva*) and potential energy in such a way that the sum of kinetic plus potential energy, for the class of conservative forces, remains constant. The essential observation of Mayer, Joule, and others is that there is a large class of systems—living and otherwise—that seems to possess something which has the power to do work and yet cannot very easily be classified as either kinetic or mechanical potential energy. In addition, kinetic energy can seemingly disappear (be converted) into heat—a given amount producing a given amount of heat.

With this in mind, Helmholtz proposed that one should generalize the definition of energy* from that which comes from mechanics:

$$\text{energy} = \text{kinetic energy} + \text{potential energy} \tag{9.5}$$

to

$$\begin{aligned} \text{energy} = &\; \text{kinetic energy} + \text{potential energy} + \text{heat} \\ &+ \text{electrical energy} + \text{other forms to be} \\ &\text{discovered or invented} \end{aligned} \tag{9.6}$$

The dominating idea was that the work done on a system should be equal to the increase in energy of that system. But this energy might go into either mechanical energy, kinetic or potential, heat, electrical energy, or some other form not yet discovered.

This, of course, gives one a great deal of freedom. If we insist on energy conservation in any physical process, then we may have, some day, to invent a new form of energy to balance both sides of the ledger. (Sooner than give up energy conservation, Henri Poincaré once remarked, we would imagine new forms of energy to save it.) It has happened in the realm of particle physics, that new particles, such as the neutrino, have been proposed to save the principle of energy conservation. For if we observe a process in which the energy seems to disappear, at present we are inclined to ask where it disappeared: what particle carried it off, or what other manifestation it has assumed. If one is always willing to invent a new particle or a new form of energy, the principle of energy conservation can always be saved. It is a manner of viewing

---

* When we use the word energy with no qualification we mean what might be called the "total energy"—that is, the sum of all the individual forms.

the world. If, however, the neutrino is proposed to balance energy in nuclear reactions, then we might reasonably expect to find other traces of this object. When these are in fact found, when the neutrino is finally "discovered," we agree that the principle of energy conservation has been fruitful once again.

Each form of energy (old or new) must bear a specific relationship to the others, and all should be equivalent to a given amount of work. Thus, 4.18 joules of work (the work done pushing with a force of 4.18 newtons through a distance of 1 m—or about 1 lb through 3 ft) can produce 1 calorie of heat. It can also move 2.6×10^{19} electrons through a potential difference of 1 volt, raise a mass of 100 g 426 cm. When the weight falls, its maximum kinetic energy is 4.18 joules, and when it strikes the ground and comes to rest, 1 calorie of heat has been produced, warming both the weight and the earth to which it has returned.

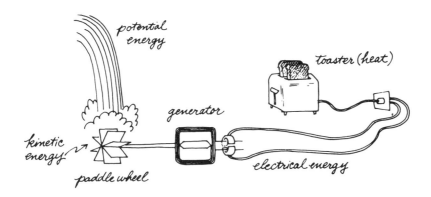

This has come to be known as the *first law of thermodynamics** or the *principle of energy conservation;* its meaning is that in any physical process we can define a quantity called "the energy" [the sum of various terms as in Eq. (9.6)] whose value remains constant no matter what changes take place in the system. Thus the concept of energy that originates in mechanics has been generalized to include all processes: electrical, chemical, and even living. As Helmholtz was to say:

> From a similar investigation of all the other known physical and chemical processes, we arrive at the conclusion that nature as a whole possesses a store of [energy]† which cannot in any way be either increased or diminished; and that, therefore, the quantity of [energy] in nature is just as eternal and unalterable as the quantity of matter. Expressed in this form, I have named the general law "The principle of the conservation of [energy]."[8]

And he continues:

> We cannot create mechanical [energy], but we may help ourselves from the general store-house of nature . . . The brook and the wind, which drive our mills, the forest and the coal-bed, which supply our steam engines and warm our rooms, are to us the

---

* Thermodynamics: defined in at least one dictionary (*Webster's*) as the physics that deals with the mechanical action or relations of heat.

† His word is "force."

bearers of a small portion of the great natural supply which we draw upon for our purposes, and the actions of which we can apply as we think fit. The possessor of a mill claims the gravity of the descending rivulet, or the living [energy] of the moving wind, as his possession. These portions of the store of nature are what give his property its chief value.[9]

10

HEAT DEATH (The Second Law of Thermodynamics)

WHEN CAN HEAT DO WORK?

In mechanics the concept of energy is intimately related to that of work. Work done on a particle gives the particle energy and a particle with energy can do work. A pendulum at the peak of its swing pauses an instant—it has zero kinetic and maximum potential energy. At the nadir of its swing it has minimum potential energy and maximum speed. If the pendulum collides with a billiard ball it can do work on the billiard ball and start it moving toward the side pocket. Conservation of energy in mechanics seems to involve a simple ledger book. The energy that goes in is equal to the energy that comes out. In generalizing the mechanical definition to electrical and other forms of energy (although one adds a bit in complexity), there is the same property. Electrical energy can be converted to mechanical energy, and so on; energy can be freely converted to work. But with the introduction of heat, there is something curious and new.

Mechanical energy, electrical energy, or what is equivalent, work, can be freely converted into heat—there is no problem. We know precisely, after Joule, how much work has to be done to produce a given quantity of heat—4.18 joules produce 1 calorie. And it is easy to do: no harder than rubbing our palms together, or rubbing two dry sticks in a forest. It is the question in reverse that leads to a concept that has excited as much popular response as almost any other in physics—a concept apparently charged with mystery and romance, blamed by some for the *fin de siècle* weariness of the nineteenth century, expressed in particular by the American, Henry Adams. A concept that has been associated with the disillusionment of a society no longer reaching toward perfection but running downhill toward mediocrity: a culture with television and without genius, an architecture with suburbia and without Notre Dame.

A concept that has produced gloomy visions of the heat death of the universe. Where, as Swinburne said:

> Then star nor sun shall waken,
> > Nor any change of light:
> Nor sound of waters shaken,
> > Nor any sound or sight:
> Nor wintry leaves nor vernal,
> > Nor days nor things diurnal
> Only the sleep eternal
> > In an eternal night.

The question which led to the idea that excited this fervor and pessimism seems relatively straightforward: Under what circumstances can that form of energy we have called heat be converted into work?

CARNOT'S STEAM ENGINE

> The steam engine [wrote Carnot] works our mines, impels our ships, excavates our ports and our rivers, forges iron. . . . To take away today from England her steam engines would be to take away at the same time her coal and iron. It would be to dry up all her sources of wealth. . . . Notwithstanding the work of all kinds done by steam engines, notwithstanding the satisfactory condition to which they have brought today, their theory is very little understood.[1]

In practical situations one has a source of heat (say coal) and a job to be done (say moving a train). There is a definite amount of heat contained in the coal (1 pound of coal when burned will raise the temperature of a fixed large number of grams of water 1 degree centigrade). The question to which Carnot addressed himself was: What is the way to build an engine that gets the most work from a given quantity of heat? One can easily enough build an engine that gets no work from the coal. (Burn the coal in a fireplace.) In the steam engine the coal is burned to produce heat; the heat boils water and converts it to steam; the steam pushes the piston; the piston pushes something else, and finally wheels are turned, trains propelled, or mines kept dry. How can this process be made most efficient? Given a pound of coal what is the most efficient way its heat can be used to do work?

Carnot generalized, in the fashion of France, the particular steam engine of England to heat engines of any kind. He constructed, mentally, the ultimate heat engine—more efficient than any real engine. And he showed that the amount of work that can be done by a given quantity of heat put into this ideal engine depends only on the temperature differential that is available.

The Ideal Engine

Carnot's engine is ideal in two senses: (1) there is no internal friction, and (2) it operates between only two temperatures. He thus imagines a frictionless engine which is so arranged that, for example, the gas that pushes the piston and does the work takes in all the heat from the fuel at one temperature, T_1 (high), and expels heat to its surroundings at the temperature T_2 (low). With this he proved:

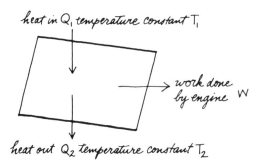

CARNOT'S THEOREM: Any heat engine working between the two temperatures T_1 (hot) and T_2 (cool) is no more efficient than the ideal engine.

Thus the answer to the question: What is the maximum efficiency of an engine working between two temperatures? is: It is no more efficient than the ideal engine. The efficiency of the ideal engine was simple enough so that he could calculate it.

This means, for example, that if an engine takes its heat from a furnace which heats the gas that pushes the piston, let us say, to a high temperature, then the amount of work that can be obtained depends upon the low temperature to which this gas can be reduced before it is reheated again by the furnace. For the gas when it is hot might do work on a piston and be cooled in the process. But at the low temperature it gives its heat to the surroundings and thus (as far as the engine is concerned) wastes it. The engine would be most efficient if all the heat energy could be taken out of the gas and none wasted on the surroundings. But can this be done? If yes, why not take ordinary air or water from the ocean and allow it to cool itself, convert its heat energy to work and then expel it colder.

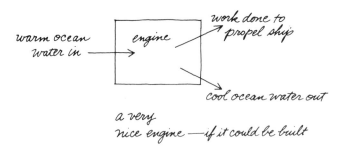

There is a huge amount of energy in the ocean due to the fact that the ocean is not as cold as it can be. From the point of view of energy conservation, it would be perfectly possible for a ship to propel itself by extracting energy from the ocean, leaving the ocean a little cooler; there would be no reason not to make ice cubes in the ship's refrigerator at the same time. The reverse process is perfectly feasible; there is no trouble whatsoever in converting an arbitrary amount of work into heat to warm up the ocean. Xerxes, in his anger, had the Hellespont flailed, converting the cheap labor of his slaves into warmth for this ancient sea. But the reverse, as a matter of fact, does not happen. We cannot get the seas to do work for us and to cool themselves in the process.

What is called the *second law of thermodynamics* can be thought of as a formal statement of this empirical fact: There is something about how the world is made that prevents us from converting heat into work such that the sole result is to extract heat from a reservoir (say the ocean) and convert it to an equivalent amount of work, cooling the reservoir in the process. Using this (what we now call the second law) Carnot showed that his ideal engine would be more efficient than any real engine and that its efficiency would depend only on the initial (high) and final (low) temperatures. Thus any real engine working between those two temperatures would be less efficient than the ideal engine (usually quite a bit less). The greater the temperature difference (in particular, the lower the final temperature), the more efficient the engine could be made.

A steam engine, any heat engine, takes in heat (provided by the fuel) and expels heat in the exhaust or to the surrounding atmosphere. The heat expelled represents energy, not easily available perhaps, but energy. And the more energy that is carried away, the more is lost. Therefore, to produce an engine that extracts all the energy possible from what heat is available, the final temperature at which the engine operates (the temperature at which the exhaust is expelled) must be a temperature at which a material in some sense contains no heat. This temperature we later identify with the absolute zero.

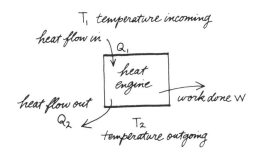

THE ARGUMENT REFORMULATED

Carnot phrased his arguments in the language of the caloric theory. We are not sure he understood the first law before he formulated the second; when he wrote, it was still not generally agreed that heat was equivalent to energy. Because of this there were apparent difficulties in Carnot's argument. Did it really rely on the caloric theory of heat and did the demise of the caloric theory invalidate his conclusions? Or was he trying to avoid the hypothesis that heat was equivalent to energy and were his arguments essentially independent of the first law?

Rudolf Clausius reanalyzed the operation of that ideal heat engine,

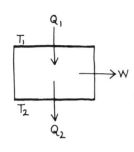

more efficient than any existing engine, making no assumptions about the conservation of caloric, using instead the first law—which Carnot may have been trying to avoid. The analysis, in a modern form, follows. We call Q_1 the heat that flows into the engine (the heat supplied from the burning coal), Q_2 the heat that flows out of the engine (the heat carried away when the air cools the outside of the furnace or the cyclinders), and W the work that the engine does. We assume that the engine goes through cycles (say a complete stroke of a piston), so its internal state is the same at the end of each cycle.*

Since heat is equivalent to energy, the total energy input is equal to the heat coming in, Q_1. The total energy output is the heat and the work going out, $Q_2 + W$. And from the first law of thermodynamics (conservation of energy) one can write

$$\begin{pmatrix}\text{heat going into the engine}\\\text{from the burning fuel}\end{pmatrix} = \begin{pmatrix}\text{heat expelled}\\\text{to surroundings}\end{pmatrix} + \begin{pmatrix}\text{work done}\\\text{by engine}\end{pmatrix}$$

$$Q_1 \qquad = \qquad Q_2 \qquad + \qquad W \qquad (10.1)$$

If one wants to obtain the maximum amount of work from the quantity of heat Q_1, the quantity of wasted heat Q_2 must be zero. (To convert all the heat energy into work, one should not waste it by heating up the outside air, or a river.) What is the best one can do? Under what circumstances can one make Q_2 equal to zero?

For the ideal (Carnot) engine it is shown that the ratio of the heat supplied, Q_1, to the heat expelled, Q_2, is a function only of the initial and final temperatures:

$$\frac{Q_1}{Q_2} \qquad \text{depends only on } t_1 \text{ and } t_2 \qquad (10.2)$$

It does not depend on details such as whether steam or air is the working substance. And if the temperature scale is properly arranged the ratio has the startlingly simple form

$$\frac{Q_1}{Q_2} = \frac{T_1}{T_2} \qquad (10.3)$$

One therefore arranges the temperature scale so that this is so—thus defining what is known as the absolute, or Kelvin, scale; this is related to the centigrade scale simply by a shift in the zero.

temperature in centigrade = temperature in absolute − 273.16°

100°C ┼┼   ┼┼ 373.16°K
0°C ┼┼   ┼┼ 273.16°K

−273.16°C ┴┴   ┴┴ 0°K

---

* We want to ensure that no energy is left inside the engine. For a real engine a little energy might be left for the first few cycles—when the engine starts it might be cold, and it becomes warmer; but this process comes to an end when the engine has reached its operating temperature. From that time one has cycles in which the energy internal to the engine remains the same.

If we insist on measuring temperature in the centigrade scale, the relation Eq. (10.3) becomes

$$\frac{Q_1}{Q_2} = \frac{t_1 \, (\text{in} \, ^\circ\text{C}) + 273.16}{t_2 \, (\text{in} \, ^\circ\text{C}) + 273.16} \qquad (10.4)$$

We can now rewrite Eq. (10.3) to obtain

$$Q_2 = \left(\frac{T_2}{T_1}\right) Q_1 \qquad (10.5)$$

The wasted heat, Q_2, is equal to the total heat input multiplied by the ratio of final to initial temperatures (in the Kelvin scale). The definition of the absolute zero in this way permits us to make the following plausible-sounding statement: If an ideal engine operates between high temperature T_1 and low temperature $T_2 = 0$ (the absolute zero), then the heat exhausted, Q_2, is equal to zero, as

$$Q_2 = \left(\frac{0}{T_1}\right) Q_1 = 0 \qquad (10.6)$$

Since no heat is exhausted no energy is lost—thus all the heat can be converted to work if and only if the final temperature is the absolute zero.

This then, written in the modern form, is the answer to the question posed by Carnot. The maximum work that can be obtained from a source of heat at the temperature T_1 by an engine operating between the temperature T_1 (high) and T_2 (low) is obtained by the ideal (Carnot) engine and this is

$$W = Q_1 - Q_2 = Q_1 - \frac{T_2}{T_1} Q_1 = Q_1 \left(1 - \frac{T_2}{T_1}\right) \qquad (10.7)$$

A real engine can do no better.

Example. What is the maximum possible efficiency of a steamship engine whose boiler is at 100°C (T_1) and whose condenser is at sea temperature, 15°C? For the ideal engine

$$\text{efficiency} = \frac{\text{useful work}}{\text{energy input}} = \frac{W}{Q_1} = \frac{Q_1 - Q_2}{Q_1}$$

$$= 1 - \frac{Q_2}{Q_1} = 1 - \frac{T_2}{T_1} = 1 - \frac{288}{373} \qquad (10.8)$$

$$= 1 - 0.77 = 23 \text{ percent}$$

Thus the steamship engine has an efficiency of less than 23 percent.

ENTROPY

Clausius then showed, by an argument as abstract and difficult to understand as any in physics, that it was possible to define a function, S, which depended only on the variables of the system (for a gas engine, for example, the pressure and temperature of the gas and *not on how the engine got to that pressure and temperature*) and that the change of this function is related to the quantity of heat put into the system. (It is an argument not unlike the argument by which one shows that for conservative forces a potential energy exists that is a function only of the initial and final points—not of the path along which the particle is moved.)

To find a name for this function Clausius said:

> I prefer going to the ancient languages for the names of important scientific quantities, so that they may mean the same thing in all living tongues. I propose, accordingly, to call S the *entropy* of a body, after the Greek word "transformation." I have designedly coined the word *entropy* to be similar to energy, for these two quantities are so analogous in their physical significance, that an analogy of denominations seems to me helpful.[5]

By doing this, rather than extracting a name from the body of the current language (say: lost heat), he succeeded in coining a word that meant the same thing to everybody: nothing.

When we interpret heat and temperature in terms of mechanical concepts—total energy, average speed—we shall find a natural and very simple interpretation for entropy. For the moment any formula would only make it more mysterious. Because more important than its explicit form is the fact that such a function exists and that with it we can formulate a sweeping principle.

Just as energy can be defined for all systems and for the universe in general, so can entropy. The function introduced by Clausius can be defined not only for the steam engine, but also for the fuel that drives the engine, the furnace that holds the fuel, the platform on which the furnace stands, the earth on which the platform stands, the solar system, and the universe which contains the solar system. By the principle of conservation of energy the total energy of the universe remains conserved, whether the universe is young or old, before or after the explosion of a supernova, before or after a comet enters the solar system, and so on. The principle of entropy (which is equivalent to the second law of thermodynamics) is that entropy always increases. In every physical process, although the entropy of parts of a system can decrease, the entropy of the entire system goes up. Just as the statements that for an isolated system (1) $E = \frac{1}{2}mv^2 + V + $ electrical energy $+ \cdots$ and (2) E is a constant number are a mathematical formulation of the principle that energy is neither created nor destroyed, so Clausius' statement that (1) the function S (entropy) exists and can be defined for an isolated system and (2) that this function increases in all physical processes,

S increases

is a mathematical formulation of the principle that one cannot get a warm thing to do work and by so doing cool itself to a temperature below that of its environment.*

Applied to the universe as a whole, it gives a direction to all physical processes. The increase in entropy means that hot things tend to be-

---

* One can cool a warm thing to a temperature below that of its environment (a refrigerator, for example) but that requires work to be put in; that is, one must have some kind of a motor.

come cool, cool things to become warm (all objects tend to attain the temperature of their surroundings) until the entire universe approaches its lukewarm end—where entropy is a maximum and nothing further happens.

Was this the reason for Henry Adams' pessimism? It is hard to believe.

11
THE WORLD AS A MACHINE

The idea did not originate with Newton. Democritus, Epicurus, Lucretius, and Gassendi had argued that change could be regarded as the rearrangement of atoms in a void—that atoms and their arrangements were all there were. Descartes, not a believer in the void, applied mechanical laws to the motion of corpuscles to attempt to deduce the observed properties of the world. But the Newtonian system put in a concrete, explicit form, and, with remarkable success, that ancient and often disreputable conjecture that all matter, all phenomena, all experience, could be considered the rearrangement of primordial atoms, following mechanical laws. If so, then why would not economics, history, and finally human behavior all be governed by (natural) laws, consequences of the laws of motion governing the behavior of the fundamental corpuscles. The question of determinism and free will and the problems of knowledge (often meaning Newton's system) were rephrased to take a Newtonian form.

But perhaps the most important result was psychological. Medieval man could live without embarrassment in a world of which he was the center, a world whose reason, awareness, and concern were centered about man, just as the motion of the stars and that of matter, heavy or light, was centered about the earth; a universe built around the drama of salvation, where all things had a purpose, an almost magical luminous world where all of creation from the angels to the beasts, even the inanimate stones, knew their place, their purpose, and their relation to everything else.

It had been conjectured that the universe might be material, an assortment of atoms with no intrinsic purpose or direction and with no particular relation to man. Depending on the time and on how obstreperous they were, the advocates of such views could be burned or ignored. It was an interesting (if perhaps cold and inhuman) possibility—it was a view motivated possibly more by one's moral preference (Epicurus) than by its economy or efficacy as an ordering of the phenomena.

Possibly it is too much to say that Newton forced the abandonment of the medieval world; but his success made one less comfortable living there. After Newton, justifiably or not, it was no longer easy to believe that the world had been constructed about man; that all of creation, no longer centered about the earth, the result of the motion of particles subject to mechanical laws, was yet directed and ordered with man as the principal character in a grand drama. It was only a step to the materialism of Holbach:

> The universe is a vast assemblage of everything that exists, presents only matter and motion. The whole offers to our contemplation nothing but an immense and interrupted succession of causes and effects.

British universities were teaching Newtonism by the end of the seventeenth century. Newton was accorded a royal funeral when he died in 1727. And after attending the funeral, Voltaire wrote, "Not long ago a distinguished company were discussing the trite and frivolous question: 'Who was the greatest man, Caesar, Alexander, Tamerlane, or Cromwell?' Some one answered that without doubt it was Isaac Newton. And rightly: for it is to him who masters our minds by the force of truth, not to those who enslave them by violence, that we owe our reverence." Then Newtonism swept the continent. Before 1789, some 18 editions of the difficult and technical printings of *Principia* were called for; there had appeared about 40 popular books in English and 17 in French on the subject. And there were lectures for the ladies: *"Le Newtonisme pour les dames."*

It became so fashionable to read about the new science that it is said young women hesitated to give their hands to men who had not achieved some scientific status. Benjamin Franklin's great success in Paris was not hindered by his renown for his discoveries in electricity. Perhaps the lady who carried a cadaver in her carriage so that she could study anatomy during her promenades through the Bois de Boulogne was somewhat extreme. And Cotton Mather could write: "Gravity leads us to God and brings us very near to Him."

We do not have to believe that the ladies or the gentlemen who attended the lectures had read *Principia* or understood the details of Newton's arguments. Newton, Newtonism, Newton's system had become one of the uncriticized preconceptions of European thinking. From that time any other system of the world would be measured and challenged by the one created by Newton; from that time he would dominate the thinking of the West almost as Aristotle had dominated it for the centuries before.

Newton had suggested that the planetary system did not run precisely according to his laws but would on occasion have to be adjusted, as though each millennium the hand of God would correct his watch, which ran a little slow. By the beginning of the nineteenth century LaPlace had shown that rather than running slow or fast, the watch ran precisely on time. Those minor adjustments that troubled Newton could be taken into account exactly if one calculated the influence of one planet on another. When Napoleon asked LaPlace (and one must be impressed by the aptness of the question) where God was in his system,

LaPlace we are told answered, *"Je n'ai pas eu besoin de cette hypothèse"* (I did not need that hypothesis).

But not everyone greeted this universe in which man walked as a stranger, in which atoms and planets moved in orbits irrelevant to his destiny. It was a universe that had to be reckoned with but not necessarily welcomed. It was a universe influencing all—philosophers, economists, political scientists, theologians, and moralists; welcomed by some, angrily rejected by others, and finally accepted with resignation and perhaps a certain bitter joy.

> I opened my heart to the tender indifference of the universe.
> To feel it so like myself, indeed so brotherly, I realized that
> I had been and was still happy.

—CAMUS

EXPERIENCE, ORDER, AND STRUCTURE

12

EXPERIENCE AND ORDER

Man comes into the world with a cry: a burst of light, a slap, initiate him into the universe of sensation. The material of science is this: our experience of the natural world, the world that is—not the ones that might be. Somehow in the mind this raw experience is ordered, and this order is the substance of science. What happens there is the application to a great diversity of the phenomena of the world of many of those elements called common sense, used every day and resting on certain suppositions we make concerning the world about us. Some of these are probably universal to man and beast. Others are more particular. We tend to accept them without special awareness, and some are so well hidden we are scarcely conscious of their existence.

The foremost supposition is the belief that the world outside ourselves, outside our own mind, exists. This belief is so primitive that it is very likely shared by all, except animals lowest on the evolutionary scale and some philosophers (whose position on the evolutionary scale we cannot guess). It may be that a newly born child is not aware that the patterns of light, sound, touch, smell, and taste to which he is exposed have their origin in objects outside his mind. He may not know where he ends and something else begins. The first realization that an often-repeated pattern of sensation is another person—mother—is then a discovery whose magnitude is never equalled. Yet, it is a discovery all of us who grow up and function make.

What is called a chair is a concept created to unify what we see from one side, and then from another; what we feel with our hand, and what supports us when we sit on it. We believe without analysis that all these sensations proceed from a single object, the chair. There is no additional evidence for its existence. We might experience the same sensations in a dream or hallucination in the absence of an actual chair. When we propose the existence of any object (electron, chair, or neutrino) it is done to unify such a variety of experience. This is probably the most primitive, yet the most important, theory we create.

The identification of a single object as the agent for various sensations is not always easy. For a long time men thought of the morning star and the evening star as two separate celestial objects. It was probably an astronomer of Babylon who first identified them as one and the same—the planet Venus. Western mountaineers report the surprise of their Sherpa guides when they are told that peaks they have seen in different profiles for their entire lives are many faces of the same mountain.

As we climbed into the valley we saw at its head the line of the main watershed. I recognized immediately the peaks and saddles so familiar to us from the Rongbuk (the north) side: Pumori, Lingtren, the Lho La, the North Peak and the west shoulder of Everest. It is curious that Angtarkay, who knew these features as well as I did from the other side and had spent many years of his boyhood grazing yaks in this valley, had never recognized them as the same; nor did he do so now until I pointed them out to him.[1]

An infant rapidly develops expectations about the sensations to which it is exposed (to the sorrow of the overindulgent parent). A child usually does not have to be burned twice to learn to keep his fingers away from fire. Without analysis, he believes that there is order in the world; it is not difficult to understand why. In a world where similar things repeat, the animal that is so made that it can grasp this fact survives more easily than its fellow who cannot or will not. The animal philosopher, whose mind is sophisticated enough to argue—"The tiger ate my brother but that does not permit me to conclude that he will (given the chance) eat me"—increases the likelihood that he will provide a meal for the next hungry tiger. One's view of the world thus affects one's chances of survival. Those who can adapt themselves to the order (just as do those who can adapt themselves to the temperature, moisture, and other environmental conditions) in the world survive best.

We all share this—let us call it instinctive—belief in the order of the world. But our response to the belief is quite varied, and it is in this response that men differ among themselves, and in which animals differ so much from men. Thinking—according to Pierce, an activity that begins when the mind is uncomfortable and ceases when it is comfortable again—has led man in many diverse directions, sometimes enlightening, other times not. To the animal mind—at least as one can imagine it— simple associations occur: a paw placed on a hot coal is burned; thereafter, hot coals are avoided. A bell rings and food comes; and salivation after a time begins when the bell rings. Whether animals venture further we do not know, but without prejudice to our fellow creatures we may imagine a state of mind in which no questions are ever asked.

Men, however, being featherless bipeds, seem characteristically unwilling to remain in this passive condition. Demons, spirits, purpose, or the machinelike laws of nature have been proposed to "explain" the association of two events. Once explanations begin, frequently what has actually been observed becomes confused with the explanation. A belief, strongly implanted, has a life of its own. Man, alas, is mortal, but ideas and superstitions seem to live forever; and when they become involved with our emotions, it takes a rare ability to distinguish what we see from what we believe should have been seen.

It seems simple to see what the world is, yet it requires the perception that hopefully comes with age, together with an innocence sentimentally attributed to children. A child, for example, pushes a switch on a television set and sees moving images appear on the screen. To him, this is as plausible as anything else; he is surprised only when the switch is pushed and the screen does not light up. To the young mind, any relation or correlation is as believable as any other, because it has no predisposition. It is no more remarkable that a button should be pushed and a picture appear than that a mouth should open and a voice be heard.

If there is such a thing as the much-mentioned scientific method, one element certainly would be just this openness and honesty—the interest, the skill, and the commitment to look at the world as it is, a commitment expressed by Descartes when he said that his library was the calf he was dissecting, by Galileo when he looked at the heavens through his telescope, or by Aristotle when he watched the bees:

> There is a kind of humble-bee that builds a cone-shaped nest of clay against a stone or in some similar situation, besmearing the clay with something like spittle. And this nest or hive is exceedingly thick and hard; in point of fact, one can hardly break it open with a spike. Here the insects lay their eggs, and white grubs are produced wrapped in a black membrane. Apart from the membrane there is found some wax in the honeycomb; and this wax is much sallower in hue than the wax in the honeycomb of the bee.[2]

It would include, further, a commitment to compare one's theoretical conceptions with what is in fact observed. For, as Aristotle said more generally, in a critique perhaps intended for one of his better-known instructors:

> Lack of experience diminishes our power of taking a comprehensive view of the admitted facts. Hence, those who dwell in intimate association with nature and its phenomena grow more and more able to formulate, as the foundation of their theories, principles such as to admit of a wide and coherent development; while those whom devotion to abstract discussions has rendered unobservant of the facts are too ready to dogmatize on the basis of a few observations.[3]

However, seeing is not easy when belief is strong (and we live in a world with no lack of strong belief). The so-called man of science must, therefore, develop an objectivity that can separate him from his contemporaries and, like the artist, make him seem an observer. Jenner's discovery of a vaccine against smallpox, tearing from a morass of opinion and superstition the fact that the girls who milked the cows did not catch smallpox, required an eye like a painter's, an eye able to see what is and to separate that morsel from all that is believed. It is said that when Thales told the natives of Miletus that the sun and stars were fire, they looked at him in astonishment, for they worshipped them as gods. And when Galileo said that the sun had dark spots and that Jupiter had satellites, some of his contemporaries would not even look; he wrote to Kepler: "Shall we laugh or shall we cry?" In this aspect, the face of science is hard. For it must, if it is to be at all, look at the world without illusion, just as in the greatest works of art the world of human emotion and experience is viewed as it is, without sentimentality.

This need for trustworthy observation, the raw material from which we construct our view of the world, has driven the scientist into the laboratory and has created the public image of the man in the white coat. The laboratory is a place where certain kinds of observations can be carried out more accurately and perhaps with less distraction than on a city street. The readings on a meter may seem impersonal and objective, but they are so only if the eye that watches and the mind that interprets are themselves objective. For the man with a theory, a career, a reputation at stake, the flickering of a needle on a meter generates enough emotion so that objectivity has occasionally been strained.

If a man attempts to measure the temperature of some water with his hand, he may or may not estimate it accurately. If he measures the temperature with an instrument, he may gauge it more precisely. In either case he might be right or wrong; but we have learned as a matter of experience that an instrument, properly built and maintained, will give a measurement of temperature that is usually more accurate than that of a hand placed in water. We can measure time roughly by our pulse beat, as Galileo once did. But there is a limit to the accuracy of such a measurement, a limit than can be improved by using a pendulum, a watch, or a very modern clock stabilized by hydrogen-atom vibrations, losing only about 1 sec in 100,000 years.

The necessity for measurement pervades science for several reasons. One is the recognized need to separate our observations from our beliefs; the use of machines, impersonal machines, is one way to do this. But we must always remember that a skillful eye often can achieve this separation as well as any machine. A second reason is the obvious desire for more accurate measurements than our senses can produce unaided. And a third reason is the usefulness of measurements that can be duplicated in different places and by different people; machines that can be exactly reproduced facilitate this duplication. Thus, when we say that science weighs and measures, what we mean is that science attempts to achieve a knowledge of the world as it is, a knowledge that is at once accurate and public, so that men all over the world can duplicate it if they create the proper circumstances.

But the gathering of facts without organization would yield a filing cabinet in disarray, a random dictionary, that dull and useless catalogue sometimes confused with science. Yet what is there in experience itself to indicate that order can be found? What is it that produces a conviction strong enough to have kept Kepler working for years calculating and recalculating orbits, or Galileo working for a lifetime to understand the motion of bodies? We have no guarantee—in fact, it is a little surprising—that we can find any relations such as those between the orbit of the moon and the path of a projectile near the surface of the earth. What is there that leads us to believe that an order we might create would be any less complex than the events themselves, that the symbols we write down on paper will somehow permit us not only to know but also to manipulate the world?

There is a mystery in writing, and in times when writing was not common, the act itself was considered magic. Woton, to learn the mystery of the runes (writing), suffered heroic agonies, as does many a student today. *Ode*, among the Greeks, originally meant a magic spell, as did the English *Rune* and the German *Lied*. A comic-strip superman tames his enemies by uttering the right formula; another unlocks his powers by saying "SHAZAM." In some of the most primitive magic we can find that effort to attribute mystical qualities to numbers, the conviction that relations between symbols are like relations between objects in the real world, and that somehow the manipulation of the symbols gives us a power over the stubborn material of the world.

From Egypt, in a passage of the *Book of the Dead,* based on a spell of the "Pyramid Texts," comes the "Spell for Obtaining a Ferry-Boat." Professor Neugebauer writes:

> The deceased king tries to convince the Ferryman to let him cross a canal of the nether world over to the Eastern side. But the Ferryman objects with the words: "This august god (on the other side) will say, 'Did you bring me a man who cannot number his fingers?'" However, the deceased king is a great "magician" and is able to recite a rhyme which numbers his ten fingers and thus satisfies the requirements of the Ferryman.
>
> It seems obvious to me that we are reaching back into a level of civilization where counting on the fingers was considered a difficult bit of knowledge of magical significance, similar to being able to know and to write the name of a god. This relation between numbers (and number words) and magic remained alive throughout the ages and is visible in Pythagorean and Platonic philosophy, the Kabbala, and various other forms of religious mysticism.[4]

And in Pythagoras, schooled in the mysteries of the Orient, we find possibly the earliest suggestion of a theme that has become the leitmotif of physics: that the order of the world is somehow to be found in orders and relations among numbers.

In the sixth century B.C., when the Pythagorean school flourished, numbers as we know them were not in use. Instead, numbers, or relations between numbers, were discovered by placing pebbles or making dots in the sand. These pebbles or dots could be grouped in various arrangements. Two of the arrangements studied were triangles or squares;

square array   *triangular array*

it was apparently in this fashion that the notion of triangular and square numbers arose and that the relations among such numbers were deduced. For example, an inspection of two square arrays 4×4 and 3×3 leads eventually to the relation

$$4 \times 4 - 3 \times 3 = 2 \times 4 - 1$$

and finally to the abstract relation (probably written down later)

$$n \times n - (n-1) \times (n-1) = 2n - 1$$

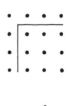

A particular configuration of pebbles that seemed singularly perfect, the monad, fascinated the Pythagoreans, and they made much of it (to the amusement of Aristotle, among others, who came later). This array of pebbles that add up to ten, having survived Aristotle's sarcasm, reappeared recently in a letter by 33 authors called "Observation of a Hyperon with Strangeness Minus Three" (see Fig. 12.1).

Among the discoveries of Pythagoras was the relation between the hypotenuse and sides of a right triangle, which today we write in the form:

$$a^2 + b^2 = c^2$$

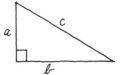

We are told that he sacrificed one hundred oxen to the muses in thanks for this, the most famous theorem of antiquity.

Once, it is said, when he passed a blacksmith's shop, he heard different sounds made by the blacksmith's hammer striking metal rods of varying lengths. He then experienced an astonishing revelation: Uniform

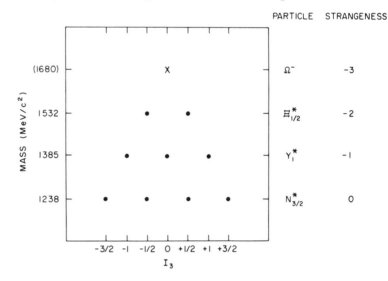

F I G. *12.1.* Decuplet of ³⁄₂+ particles plotted as a function of mass versus third component of isotopic spin. (V. E. Barnes, et al., "Observation of a Hyperon with Strangeness Minus Three," *Physical Review Letters,* Vol. 12, No. 8, February 24, 1964; illustration courtesy of Brookhaven National Laboratory)

rods whose lengths are in simple ratios produce harmonic sounds. If one can say that physics began at all, it was at a moment such as that. For Pythagoras realized that the relation between two properties of the real world—the sound emitted when a rod is struck and the length of that rod—is mirrored in a relation between whole numbers. Uniform rods whose lengths are in simple ratios produce harmonic sounds: 12 to 6, the octave; 12 to 8, the fifth; 8 to 6 the fourth; etc.

The tradition that music possesses powers is very old: Orpheus charmed man and beast when he played his lyre. The Pythagorean school, imbued with mystery, having seen the harmony of music reflected in relations between numbers, was captured. They attempted to order much of what they saw in the world, using the idea of harmonic proportions. Applying this to astronomy, they proposed that the spheres on which planets and stars were fixed rotated so that their speeds of motion were in integer ratios. Thus the planets as they sped through the heavens would emit different notes in harmony. According to Hippolytus, "Pythagoras maintained that the universe sings and is constructed in accordance with harmony, and he was the first to reduce the motion of the seven heavenly bodies to rhythmn and song."

We know now, to our sorrow, that the planets sing no song, sacred or profane. They do not move in the harmonies of Pythagoras, on the epicycles of Hipparchus, the circles of Copernicus, or even on the ellipses of Kepler. They move instead on orbits, almost elliptical, but of such complexity that they have no special name and no easy description, according to two rules whose simplicity and elegance are such that Newton could state them in two lines. Because they do the latter rather than the former, we may, if we wish, conclude that the Pythagoreans were incorrect; but that they do either is no small wonder. And the Pythagoreans were correct in their belief that the entire world could be ordered in some stunningly simple way at its very roots, and that this order could somehow be related to the structure of numbers.

It is a belief that is strong in Plato—of ideal forms behind the flux of phenomena—of rules according to which the planets move, of elementary atomic entities that have the form of solid geometric figures, ultimately triangles, to which they owe their existence. Whether the order is there and we find it (discover the laws of nature) as Plato might have said, or whether the world is so made that we can impose an order on it, is perhaps not relevant. But this belief, daring, naïve, and, considering the almost endless variety of our experience, not at all obvious, that order can be found, has moved scientists from Thales to Kepler to authors in the latest issues of *Physical Review Letters* to create science. For what we call modern science is as much an attitude of mind—a belief in the possibility of a certain kind of understanding—as a set of principles or methods; an attitude of mind such as Albert Einstein expressed after seeing a compass needle at the age of 4 or 5: "Something deeply hidden had to be behind things."

13

THE LANGUAGE OF PHYSICS

IS PHYSICS QUANTITATIVE?

In science, the saying goes, one weighs and measures, one deals in numbers; there is perhaps a suggestion that those elsewhere deal in a currency of less substance. At this point eyes close and minds go to sleep; the decimals unfortunately awaken memories of sultry afternoons in algebra class. It is, of course, true that physics deals with numbers and occasionally with very accurate measurement. One is impressed when an extraordinarily complex calculation of the quantity of magnetism the electron possesses, referred to a certain standard magnet μ_0, gives $\mu_e/\mu_0 = 1.0011596 \ldots$ and is measured to be $\mu_e/\mu_0 = 1.001165 \pm 0.000011$, where the $\pm$ means that the accuracy of the measurement is not sufficient to distinguish any value from 1.001176 to 1.001154. Or when the period of Mercury, calculated according to the laws of Newton, is found to differ from the observed period by ¾ sec out of 8,000,000.[*] The test of a theory, it is thought, is its ability to be in exact numerical agreement with the observations. However, the fact that numbers appear wantonly throughout the body of physics tends to conceal rather than to reveal their significance.

We have celebrated Kepler for taking so seriously the discrepancy between one of his proposed orbits and the observations of Tycho Brahe of the planet Mars: a discrepancy of only 8 min of arc. This discrepancy finally led him to his famous law: The planets describe elliptic orbits with the sun at one focus of the ellipse. Had Tycho Brahe measured even more accurately, however, it would have been apparent to Kepler that the orbit of Mars was not exactly fitted by an ellipse. Would he then have discarded the elliptic orbit? If he had done so, he would have abandoned one of the most important results in the history of science.

The fact that the planetary orbits are very nearly elliptical is as important a result in this case as if they were exactly elliptical. It is a remarkable regularity; such regularities can be clues to a more precise order or, it is quite possible, could be the only regularities that exist and therefore in some circumstances the limit of our knowledge. We have no a priori guarantee that all the relationships we can discover in the world can be stated with unlimited numerical precision.

Yet, we all know that mathematics is used throughout physics. It is the language of physics. What this means Galileo said very well:

> The method which we shall follow in this treatise will be always to make what is said depend on what was said before. . . . My teachers of mathematics taught me this method.

"To make what is said depend on what was said before." The result is a structure in which each element has a well-defined meaning and con-

---

[*] The extra ¾ sec is given by Einstein's theory of gravitation.

nection to every other element. Pythagoras saw the relation between notes emitted when a metal rod was struck and the length of the rod mirrored in relations between integer numbers. In the years since, physicists have introduced a variety of other mathematical structures in which they have attempted to mirror the world of real objects.

Mathematics, distinct from arithmetic, is a study of the structures that arise out of relations between various well-defined but possibly abstract objects. Much like a game, the rules are defined and all the situations follow as a consequence. In the game of chess, for example, one has a domain of activity—the chess board; one has the pieces, each of which is allowed certain moves. Every situation that can occur on a chess board is a consequence. In the game of mathematics, the mathematician begins with certain pieces and a set of rules and then explores the situations or the structure that arises as a consequence of the rules. The rules themselves can be what the mathematician wishes; they need not have correspondence with anything in or out of the world; they need only be consistent. It is the mark of a mathematician, as Bertrand Russell once said, "that he does not know what he is talking about." With inferior rules, however, the structure of a mathematical system becomes trivial, just as when a game is too simple it becomes uninteresting (as in tic-tac-toe, which can always be drawn with the proper moves).

When we think of mathematics, however, we most often think of numbers and perhaps geometry. For it was to the structure of numbers and geometry that mathematicians first turned their attention. The reason is apparent; counting is so primitive that we possibly possessed numbers before we possessed words, or before we could have been identified as human. Animals, more easily taught to count than to understand words, have progressed further than some of our human neighbors. The Watchandies of Australia count to two, *co-ote-on* (one), *u-tay-re* (two), *booltha* (many), and *bool-tha-bat* (very many). The Guaranis of Brazil adventure further, saying one, two, three, four, innumerable. Was it the ability to count that first separated us from nature? "An honest man," lamented Thoreau, "has hardly need to count more than ten fingers, or, in extreme cases he may add his toes, and lump the rest. I say, let our affairs be as two or three, and not as a hundred or a thousand; instead of a million, count half a dozen, and keep your accounts on your thumbnail."[1]

Possibly one of the earliest motivations for counting was the desire to keep track of a group of animals. At the end of a day of grazing were there as many sheep as there had been at the beginning? A pebble (*calculus* is Latin for little stone) representing each sheep put in a pile at the beginning of the day could be used in the evening to determine whether the same number of sheep returned. This method, both simple and successful, is even more primitive than counting. It involves only the ability to determine that there are the same number of pebbles as there are sheep, regardless of how many sheep and pebbles are involved. Its success depends also on a fundamental property of the world. Neither pebbles nor sheep disappear into thin air. We know intuitively that the system would not work if we matched the sheep to soap bubbles. Today we might call this the conservation of sheep. With a special provision for antimatter, similar conservation laws exist for nucleons and other particles. If we lived in a world where there were no objects to count, the mathematics of the number system would probably not be among the first to appear; we would not have so primitive and intuitive a knowledge of numbers—and their discovery, rather than being an ordinary jaunt for every child, would require a difficult adventure into a realm of abstract thought.

Geometry arose out of attempts to measure such things as boundaries and areas of land. (The name itself, geo-metry, means earth measure; much of what we know as geometrical theorems was first observed empirically.) It is easy to see how many times a yardstick goes along the side of a field. Of course, the ability to do this presupposes that the world is favorable. The yardstick must not shrink as it is being moved and, for the measurement to make sense, the field itself must maintain its shape to some reasonable degree. Although such suppositions are not always conscious, we should recall that we can measure because we live in a world where we believe we can find yardsticks that do not shrink and fields that do not change their shape. In a liquid world, in which no rigid bodies such as yardsticks exist, it might be much more difficult to measure in a meaningful way.

However, such questions probably did not trouble our ancestors, nor do they usually trouble us. We assume that the lengths and shapes of bodies remain the same when they are moved or turned in space. This seems like an obvious or necessary assumption, but it is not, for we can imagine a world in which such things would not be true. Furthermore (although we do not have to assume this at the moment), we are inclined to believe that the length of the body should remain the same whether it is in motion or not. We shall see later that this assumption will be abandoned.

We thus imagine that we have rods of which we can make triangles, squares, parallel lines, etc.; that we can follow the paths of light rays to produce straight lines of any length; and that we can construct the various geometric figures required. We are willing to assume that rigid bodies exist, that a triangle remains a triangle when we displace it in space, and that when moved it will remain the same triangle. Such properties, when imposed on the world, are not true of necessity but rather something we observe can be true. If we lived in a world without rigid bodies, it would be possible (but not easy) to choose and utilize the same conventions.

Consider the following, possibly outrageous, but perhaps of interest when we consider regions of space as small as those in the interior of a nucleus. Suppose that the space in which we live did not have that quality of rigidity which allows us to mark off distances that remain as marked. Imagine instead that we lived in a world that had some of the qualities of a rubber sheet, continually stretching, twisting, and being otherwise contorted in a completely unpredictable manner. Imagine that living creatures such as ourselves were in this world and that some of them, having nothing better to do, attempted to understand its nature.

Now, clearly it would not be fruitful for them to construct rulers, straight lines, triangles, or to measure distances in their particular world, since the distances between any two points would be continually changing and a measurement at one moment would bear absolutely no relation to a measurement at any other moment. It would be no use to tell a person that it was 2 miles farther from city A to city B than from city B to city C since, having begun the journey, he might find at this particular moment that *B* was closer to *A* than to *C;* it would be better under the circumstances not to mention the matter at all.

In such a world, distance, straight lines, etc., would be meaningless concepts and the mark of a good physicist would be not to introduce them—which does not mean, however, that this world would be without order. Some of the concepts that occur in Euclid's geometry, such as inside and outside, could be defined. Here we see illustrated a dot inside a curve; no matter how the rubber sheet is contorted, the dot will always remain inside the curve.

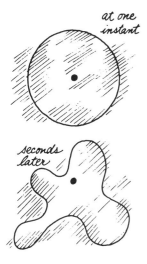

Therefore one could mark off private property or could keep men in prison although one could never say how many acres one owned or whether the prison cell was smaller or larger than the world outside.

From the study of the structures of numbers and geometry, mathematicians have gradually turned their attention to the study of many others. Which structure the mathematician will study is not entirely a different problem from that a painter faces when he asks himself what painting he will paint next. In making such a decision, both the mathematician and the painter have in mind, among other things, the previous history and development of the subject—those areas of investigation which have become overexploited, those which are sterile, and those which seem likely to lead to new and fruitful results.

The underlying program of theoretical physics, both as it is defined by tradition and as it is practiced today, is to find a structure (whose elements are explicitly displayed and logically connected to one another—almost by definition a mathematical structure) some part of which can be put into a correspondence with that domain of the world under analysis. The ultimate aim is to find a single structure which is rich enough so that it can be put into correspondence with all the phenomena of the world. Thus the relations between material things in the real world will be mirrored in the relations between the abstract objects in the mathematical world. Every theory we shall discuss is an attempt of this kind.

Imagine that we are at a baseball game with no knowledge whatsoever of the rules. As we watch, we see many situations, complex or ludicrous. We might begin to see that certain sequences repeat. A batter hits the ball; a fielder catches it, and quite often, but depending on the circumstances, throws the ball toward a position we could call first base. We might after a while have in our minds a certain abstract team in the field consisting of right fielder, center fielder, left fielder, and so on. Any actual team we see will have players with widely varying personalities occupying these positions. In spite of the fact that the size, looks, personalities of the players will vary from team to team, in spite of the fact that the Giant and Dodger shortstops are different people, as a baseball team these players in their positions will bear certain relations to each other that are independent of their individual characteristics. In that sense, any actual team (one that is in the world) will be a realization of the abstract team we have envisioned in our mind and pictured here. If

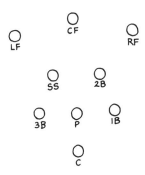

we watch long enough, we may actually be able to divine the rules according to which the game is being played. But in the game of life there is no set of rules written somewhere against which we can compare our guesses. What systems we invent can be judged by only one criterion—the degree to which they can successfully organize our experience.

"There seems to be a pattern of some kind, but what I can't figure out is *why.*"

(Drawing by O'Brian; © 1967 The New Yorker Magazine, Inc.)

At first it would seem that the problem which theoretical physics sets itself—that of finding a mathematical structure, at least a part of which can be put into a correspondence with the phenomena of the world—would result in nothing more illuminating or less complicated than the world itself, a system governed by rules as numerous as the events observed. If this were so, the enterprise would not be particularly rewarding. However, it has turned out that we can find mathematical structures which can be put into a correspondence with the events of the world, such that the rules underlying the working of the structures are simple and few in number. This could hardly have been anticipated and is perhaps the most remarkable fact about our world.

And if so inclined, one can identify these few rules according to which everything might be thought to move as being similar in spirit to those ubiquitous ideas of Plato, those ideas which to Plato meant what numbers meant to Pythagoras: those essences which lay behind the working of reality.

ON THE
NATURE OF
LIGHT

14

THE "PHAENOMENA OF COLOURS"

Isaac Newton again:

> I procured me a Triangular glass-Prisme to try therewith the celebrated *Phænomena of Colours*. And in order thereto having darkened my chamber, and made a small hole in my windowshuts, to let in a convenient quantity of the Sun's light, I placed my Prisme at his entrance, that it might thereby be refracted to the opposite wall. It was at first a very pleasing divertisement, to view the vivid and intense colours produced thereby; but after a while applying myself to consider them more circumspectly, I became surprised to see them in an *oblong* form, which, according to the received laws of Refraction, I expected should have been *circular*.[1]

PLATE *14.1*. ". . . in an *oblong* form . . ." (Courtesy of J. A. Lubrano and D. Scales)

. . . the true cause of the length of that image was detected to be no other, than that light is not similar or homogenial, but consists of *Difform Rays, some of which are more Refrangible than others;* so that without any difference in their incidence on the same medium, some shall be more Refracted than others; and therefore that, according to their *particular Degrees of Refrangibility* they were transmitted through the prism to divers parts of the opposite wall.[2]

Thus Newton concluded that white light, which hitherto had been considered a pure and uniform substance, was in fact composed of rays of different colors.

It was Newton's first paper, simple, straightforward, and full of delight at the discoveries he had made. But older men objected. The traditional view stated by Robert Hooke, for example:

Light is nothing but a simple and uniform motion, or pulse of a homogeneous and adopted [that is, transparent] medium, propagated from the luminous body in orbum, to all imaginable distances in a moment of time. . . .[3]

and then in Newton's direction:

I believe Mr. Newton will think it no difficult matter, by my hypothesis, to solve all the phenomena, not only of the prism, tinged liquors, and solid bodies, but of the colors of plated bodies, which seem to have the greatest difficulty.[4]

But Newton had seen the composite nature of white light:

. . . the most surprising, and wonderful composition was that of whiteness. There is no one sort of rays which alone can exhibit this. . . . I have often with admiration beheld that all the colors of the prism being made to converge, and thereby to be again mixed, as they were in the light before it was incident upon the prism, reproduced light, entirely and perfectly white, and not at all sensibly differing from the direct light of the sun. . . .[5]

Then he hesitates:

. . . unless when the glasses, I used, were not sufficiently clear; for then they would a little incline it to their color.[6]

The nature of light, omnipresent, third in the order of creation, source of much of our knowledge of the world, has been an object of speculation for so long no one can date its origins. And the pursuit of the question seems to lead continually to depths of increasing profundity.

How do we see it? The view held now that sight is produced by something external entering the eye and exciting vision makes light an objective quantity, independent of our senses and of our mind. As was said long ago by Diogenes, speaking about what Democritus believed, "We see by virtue of the impact of images upon our eyes."[7] (A view, however, that raises the question of how we can be sure our interpretation of what we see is correct.)

The world contains objects that respond to light in various ways: some opaque, some transparent, some translucent. Some objects are luminous, themselves sources of light. Others absorb and appear black. The sun is a luminous body; the moon, a reflecting body. Whatever

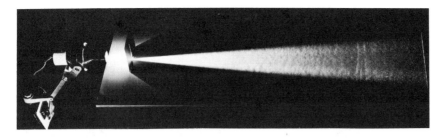

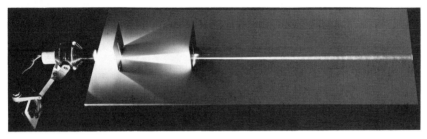

PLATE *14.2. Top:* Light coming from the source through a pinhole forms a cone-shaped beam. *Below:* A second pinhole in line with the first produces a narrow pencil of light. (Physical Science Study Committee, *Physics,* D. C. Heath, Boston, 1967)

light is, it must travel very quickly. When we strike a match the room is illuminated in an instant; and it disappears just as quickly. When the match goes out, all is darkness. Whatever it is, it travels in straight lines. We do not see around corners. Under the proper circumstances a sharp object has a sharp shadow. Properly prepared, a beam of light visibly travels in a straight line.

Two beams of light, apparently, will pass through each other without any effect (Plate 14.3). This is obvious in our normal experience. The image of the stranger across the room will not be affected by someone else who enters to one side.

Does light always travel in straight lines? Euclid hesitated to follow Plato and define a straight line essentially as that path which light took. He was aware that at the boundary of two media, for example, water and air, the path of a light ray is bent.

Even within a single medium, light does not necessarily travel in a straight line. The road ahead on a hot day seems wet because the light, moving through the hotter air near the surface of the earth and then through the cooler air higher up, travels in a curve; we interpret what we see as the reflection from a wet road. So convinced are we that light travels in straight lines that we would rather interpret the road as wet or the oar as bent than agree that the path of light is curved.

Light travels through some materials such as glass but not through others such as metal. It travels through air and it travels through the vacuum of outer space. We see the stars and the distant galaxies after intervals of time enormous by the standards of a human life; the light emitted by these objects has traversed cosmic distances without any apparent fatigue. And when it reaches our eye, to record events long dead, we witness events—supernova, the creation of galaxies—that took place perhaps before our planet came into existence.

The messenger that carries this information, whatever it is, is itself

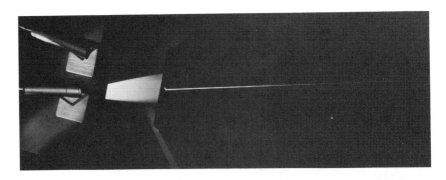

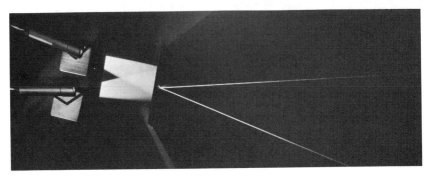

PLATE *14.3*. Two beams of light can pass through each other. (Physical Science Study Committee, *Physics*, D. C. Heath, Boston, 1967)

invisible. Its source is visible; the object it illuminates is visible. But the beam itself is not (Plate 14.4). However, if the space between the source and the object is contaminated with particles which themselves may reflect the light, so that they become nonluminous sources, we "see" the beam, as shown in Plate 14.5.

This messenger light is swift in its course. Does it move instantaneously from one body to another "to all imaginable distances in a moment of time"? Galileo attempted to measure its speed using lanterns on top of

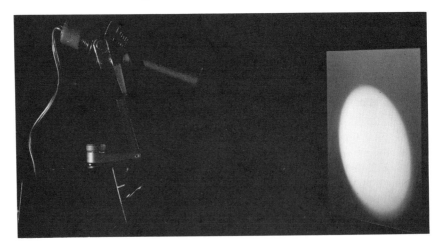

PLATE *14.4*. Light beams are invisible. Here we see the light source and the reflection from the white cardboard, but nothing between source and target. (Physical Science Study Committee, *Physics*, D. C. Heath, Boston, 1967)

P L A T E *14.5.* If particles are introduced into the air, the light beam shows clearly all the way from the source to the cardboard target (compare Plate 14.4). (Physical Science Study Committee, *Physics*, D. C. Heath, Boston, 1967)

two mountains, but the moment of time is too small for him to have seen it. The first observation of the finite speed of light was made by Römer in 1676. He found that the eclipses of the moons of the planet Jupiter occurred about 11 minutes early when the earth was nearest Jupiter and 11 minutes late when it was farthest away. He associated this curious variation in the eclipse time with the position of the earth in its orbit with respect to Jupiter. The times of the eclipses of the satellites of Jupiter presumably should not depend on the position of the earth in its orbit. Römer proposed that what was at issue was the fact that it took a finite time for the light to travel from Jupiter to the observer on the earth, thus taking longer when the earth was farther away from Jupiter. From this he could estimate the speed of light.

We now know that light moves through the vacuum of space with a finite but enormous speed, approximately 3×10^{10} cm/sec (about 186,000 miles/sec), a number so violently out of proportion with common experience that we shrug our shoulders. However, this finite speed may become a matter of immediate experience in the near future, when telephone conversations are held via communications satellites 25,000 miles from the earth. The time for the beam to move from a sender on the earth to the satellite and back to a receiver—about 50,000 miles—is approximately $\frac{1}{4}$ sec, which means a delay of this amount between the time a speaker begins and the listener hears. If the listener begins to speak at the same time, there results a confusion which has been simulated at World Fairs and will probably necessitate a new politesse on the part of international conversations.

REFLECTION AND REFRACTION

We think of light as a messenger that carries information from an object to our eye. A luminous object emits light in all directions. The light itself is invisible, but when it strikes our eye, vision is excited and we see.

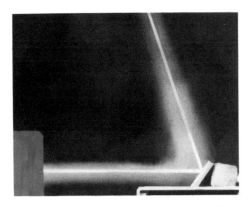

PLATE *14.6.* Specular reflection. A laser beam (made visible by blowing steam into it) striking a polished metal surface produces a sharply defined reflected beam. (R. Stevenson and R. B. Moore, *Theory of Physics*, W. B. Saunders, Philadelphia, 1967)

It is our brain that interprets what we see as an external object. For an infant the patterns registered on the eye are not necessarily so interpreted. It takes time and learning before a child comes to interpret, *we* would say properly, the sensations and the patterns that appear to him through his eyes. If we take this point of view, we ask next what is it that goes between the object and the eye—what is it and how does it behave? The first question leads to mystery after mystery. But the second can be answered in a relatively straightforward way. And the answers have been accumulated through centuries of observation.

Reflection

When a beam of light strikes a polished reflecting surface such as a mirror, it is reflected in a remarkably simple way (Plate 14.6). The single beam is reflected from the mirror as a single beam. What is the direction of the reflected beam? Observation yields a simple result that can be summarized: The angle of reflection is equal to the angle of incidence (see Fig. 14.1). If the surface is not flat, it might be visualized as many individual flat and polished surfaces. The beam of light reflected from each

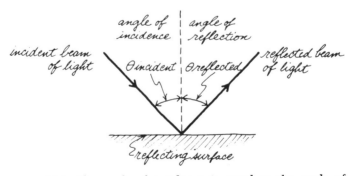

FIG. *14.1.* The angle of incidence is equal to the angle of reflection.

PLATE *14.7.* Diffuse reflection. A laser beam is shown hitting a scatter plate from an infrared spectrometer. The scatter plate is made of roughened glass (about the same roughness as coarse sandpaper) coated with aluminum. (R. Stevenson and R. B. Moore, *Theory of Physics,* W. B. Saunders, Philadelphia, 1967)

of these small surfaces so that the angle of incidence is equal to the angle of reflection appears from an over-all point of view to be reflected in all directions, or diffusely. Such a reflector would be called *diffuse* (see Plate 14.7), as opposed to the previous highly polished surface, called *specular.*

Refraction

Light traveling through a vacuum, or through a uniform medium in a straight line and with finite speed, has what seems to be an inertial property. It is hard to avoid the comparison with a Newtonian particle moving with uniform velocity. What is it that changes the motion of a particle from uniform motion in a straight line? "Forces," answers Newton. Will we be able to say the same for light? We have seen that when it strikes a hard, polished surface it bounces back, such that the angle of incidence equals the angle of reflection.

We now consider the behavior of light as it passes from one homogeneous medium to another, for example as it passes from air to water. The water, so to speak, is not as hard as a polished mirror. Some of the light is reflected, and for that portion the angle of incidence is equal to the angle of reflection. But some light enters the water, too. This is shown schematically in Fig. 14.2, which defines the symbols that are used.

We note that refraction is in a sense a generalization of reflection. There is an incident beam, a reflected beam, an angle of incidence, and an angle of reflection. But there is also the refracted beam. In the absence of a theory, one does not know, of course, what the relations between the incident, reflected, and refracted beams will be. However, one can observe what in fact happens very easily. And such observations date at least from the time of the Greeks. Measurements, such as those illustrated in Plate 14.8 for a light beam entering glass from air, can determine for any two media the relations both in direction and intensity of the incident, reflected, and refracted beams.

Once we have seen the phenomenon of refraction (and it was seen by

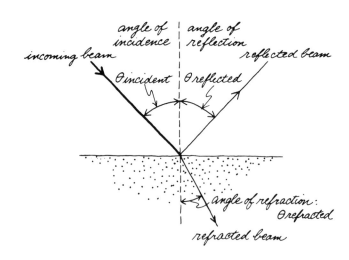

F I G. *14.2.* Incident, reflected, and refracted beams.

Aristotle, who described the apparent bending of an oar dipped into water, and by Ptolemy, who published a table listing various different angles of incidence and refraction for a beam of light refracted from air into water), the question that naturally arises is: what is the relation

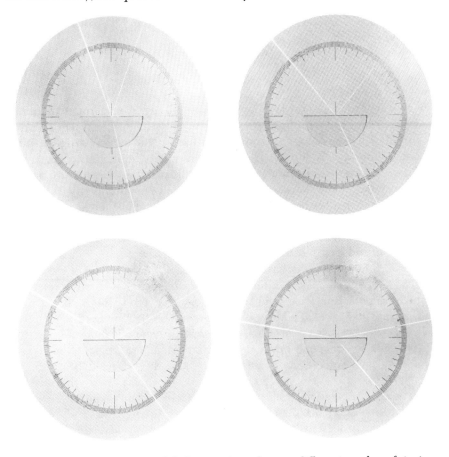

P L A T E *14.8.* Refraction of light entering glass at different angles of inci-dence. (Physical Science Study Committee, *Physics*, D. C. Heath, Boston, 1967)

TABLE *14.1 Angles of refraction for an air-to-water boundary according to Claudius Ptolemy (140 A.D.).*

Angle in air	Angle in water
10°	8°
20°	15.5°
30°	22.5°
40°	29°
50°	35°
60°	40.5°
70°	45.5°
80°	50°

between the incident ray and the refracted ray? In the case of reflection, the relation (the angle of incidence equal to the angle of reflection) is so simple it is hard to miss once one is directed to it.

In refraction there are three angles—the angle of incidence, the angle of reflection, and the angle of refraction. In most simple glass-air, or glass-water surfaces, the first observation one makes is that all three rays lie in the same plane. Further, for the incident and reflected rays, the angle of incidence is equal to the angle of reflection. Therefore, simple reflection becomes a special case of refraction in which a refracted ray has zero intensity.

The relation between the incident angle and the refracted angle is a trifle more complex. It had been proposed that the ratio of the angle of incidence to the angle of refraction was constant:

$$\frac{\theta_{incident}}{\theta_{refracted}} = \text{constant} \qquad (14.1)$$

This turns out to be true only for very small angles. Kepler was later to try to improve this but not to succeed. In 1621 Snell proposed the following, now accepted, ratio between the incident and refracted rays[*] (see Table 14.2):

$$\frac{\sin(\theta_{incident})}{\sin(\theta_{refracted})} = \text{constant} \qquad (14.2)$$

TABLE *14.2 Calculated angles of refraction according to* $\sin \theta_i = n \sin \theta_r$ *for an air-to-water boundary* $(n = 1.33)$.

Angle in air (incident)	Angle in water (refracted)
10°	7.5°
20°	14.9°
30°	22.0°
40°	28.8°
50°	35.0°
60°	40.5°
70°	44.8°
80°	47.6°

[*] Sine and cosine are defined in the Appendices, p. 488.

That is, the sine of the incident angle divided by the sine of the refracted angle is a constant. This constant, known as the *index of refraction,* is a property of the two materials and differs for different materials. For example, there is one index of refraction for an air-water surface, another for an air-glass surface, and a third for a glass-water surface.

For any given combination of materials, say glass and air, one measurement of the incident and refracted ray suffices to determine the index of refraction. Once we have it, we know what the angle of refraction will be for any given angle of incidence. For the moment this relation simply *is*. And we want at once to ask: Does it follow from some fundamental view of the nature of light? We shall consider several possible answers.

The intensity of the various beams reflected and refracted also varies with the incoming angle. In Plate 14.9 we see several pencils of light passing through a liquid. In this case the light passes from the liquid into the air, and for each pencil we see the incident, refracted, and reflected beams. As the angle of incidence increases, the refracted beam is bent more and becomes less intense. This happens continuously, until at a maximum angle there is no refracted beam at all. Thus, at this angle, the surface reflects completely—a phenomenon known as *total internal reflection.*

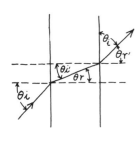

The path of a ray of light from air and through a glass plate with parallel faces and out into air again is shown here. If we follow the beam carefully, marking the angles of incidence and angles of refraction and observing that the angle of incidence at the glass-air surface is equal to the angle of refraction at the air-glass surface, we find, as shown above, that the final ray emerges parallel to the initial ray, but displaced somewhat.

If the faces of the glass plate are not parallel, as for a prism, then the light beam leaves at a different angle than that with which it entered, producing what Newton called "the celebrated *Phænomena of Colours.*" A beam of white light entering the prism on one side exits as a beam of light of many colors. Less celebrated today, known without fanfare as the *phenomenon of dispersion,* it is explained by assuming that

PLATE *14.9.* Pencils of light passing through a liquid-air surface. Total reflection occurs for the last pencil going clockwise. (Courtesy of J. A. Lubrano and D. Scales)

the different colors of which white light is composed are refracted differently or, that the index of refraction differs for the different colors. The amount of bending depends upon the material, but it is always true that cool colors—violet or blue—are bent more than warm colors—yellow or red (Fig. 14.3).

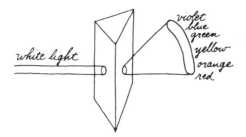

F I G. *14.3*. ". . . the celebrated *Phænomena of Colours* . . ."

These phenomena convinced Newton that white light was not pure but was in fact composed of the different colors. Newton tried then to break up a single color, such as yellow, by bending it again through a prism. He found that although the rays spread apart more, a yellow would remain yellow and a red would remain red. He then recombined the colors of the spectrum by bending them through another prism and again found the white light (Fig. 14.4). Dispersion, suggested Descartes, is what pro-

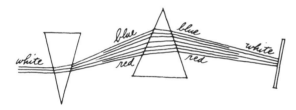

F I G. *14.4*. "I have often with admiration beheld that all the colors of the prism being made to converge . . . reproduced light, entirely and perfectly white. . . ."

duces the rainbow. Sunlight reflected from droplets of water is dispersed due to the varying indices of refraction of the colors that compose the white light. This dispersion, plus the reflection from the droplets, separates the white light of sunlight into the bands of color we see in a rainbow (Fig. 14.5).

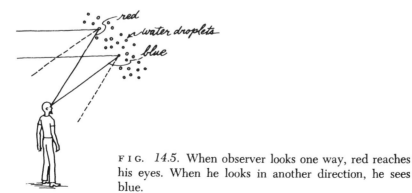

F I G. *14.5*. When observer looks one way, red reaches his eyes. When he looks in another direction, he sees blue.

Because of his enormous prestige, Newton's belief, or his apparent belief, that light was a stream of particles tended to suppress examination of other points of view, especially in England. Newton did feel that it was unlikely that light could be a wave "because waves bend as they pass corners, while light apparently does not." And it is likely that the corpuscular theory of light appealed to Newton because it seemed consistent with the general corpuscular interpretation of nature. Matter could be thought of as particles in the void, and it seemed consistent to think of light in the same way.

However, he was relatively ascetic in the introduction of hypotheses. He says, for example,

> It is to be observed that the doctrine which I explained concerning refraction and colors, consists only in certain properties of light, without regarding any hypotheses, by which those properties might be explained . . . for hypotheses should be subservient only in explaining the properties of things, but not assumed in determining them, unless so far as they may furnish experiments.[8]

and further:

> Tis true, that from my theory I argue the corporeity of light; but I do it without any absolute positiveness . . . I knew, that the *properties,* which I declar'd of *light,* were in some measure capable of being explicated not only by that, but by many other mechanical hypotheses.[9]

He provided no explanation for the source of gravity:

> You sometimes speak of gravity as essential and inherent to matter [this is in a letter to Bentley]. Pray do not ascribe that notion to me, for the source of gravity is what I do not pretend to know, and therefore would take more time to consider of it.

And again in the Scholium at the end of *Principia:*

> . . . to us it is enough that gravity does really exist, and act according to the laws which we have explained, and abundantly serves to account for all the motions of the celestial bodies, and of our sea.[10]

About the nature of gravity, about the nature of light, Newton tended to retain an open mind. His colleagues and his contemporaries pressed him to give an explanation. Yet he preferred not to. The assumption of a gravitational force, the hypotheses that light consisted of a stream of particles, led to consequences in agreement with experience. This was as far as Newton wanted to go. Was it with exasperation that he wrote near the end of *Principia, "hypotheses non fingo"* (I don't make hypotheses)?

15

WAVES

WHAT IS A WAVE?

In 1678 Christian Huygens published his *Traité de la Lumière,* in which he proposed that light was a pulse of some kind in a medium that pervaded all space—as we would say now—that light was a wave. The disturbance in the medium moved from one point to another as the medium jostled itself and fell back into place, somewhat as a disturbance moves along a line of dominos. Huygens developed this idea, discussed how the pulse would move through empty space and what it would do when it encountered a barrier or a surface between two media. Thus was born, or reborn, the idea that light, contrary to the majority of prevailing ideas of the time, was a wave. So powerful was Newton's influence that not even a review of Huygens' book occurs in the *Proceedings* of the Royal Society.

If we regard light as a particle, we have at least a clear, if misleading, image of what light is supposed to be. But what is a wave? The word itself has two meanings: something that undulates, occurs over and over, or a sudden billow that overwhelms—a human wave or a heat wave. The concept of a wave dominates nineteenth- and twentieth-century physics in much the same way as the concept of a particle dominated the physics of the seventeenth and eighteenth centuries. Waves are to light, the theory of electricity and magnetism, and quantum physics what corpuscles were to the Cartesian, Galilean, and Newtonian systems. Perhaps the most striking result of twentieth-century physics has been the subtle blending of these two traditions—the properties of corpuscles and waves into a single entity—which perhaps could be called "the quantum."

When talking about or explaining waves, a physicist will often point to ripples on the surface of water, the disturbance that moves along a stretched spring, or the disturbance that moves through air—sound. These represent attempts to make visual the abstract notion of a wave in the same way that speaking of particles as tennis balls or billiard balls is an attempt to make visual the abstract idea of a particle. All these disturbances, to a certain extent, have the properties that we associate with waves, just as tennis balls have some of the properties we associate with particles.

But in the end, a wave, like a particle, a vector, or a straight line, is a mathematical construction, given certain properties, obeying certain rules. These particular constructions are suggested by the properties of the world; in another world other constructions would be more natural. When we say force is a vector, we understand that this means that those objects in the physical world that we identify as forces can be associated in a well-defined way with vectors in the mathematical world. We developed the concept of a vector in such a way that this could be done. But once developed, the vector has a life of its own. It can be associated with, but is not, a force. Waves are mathematical objects in the same sense. In all the illustrations we shall give below—water waves, waves on springs, and so on—it is to be understood that these are given because in some respects the properties of what we see on the surface of water or on a spring are the properties that we wish to give to our mathematical waves. The physical objects we are able to build in the macroscopic world do not necessarily have to have all the properties of our mathematical waves or vice versa.

Several physical ideas underlie the concept of a wave. The first is that a wave, in some sense, is a disturbance that moves through a medium without any part of the medium itself moving very much. Consider, for example, the standing row of dominos, famous as a child's game and political metaphor. When the domino on one end tumbles, it pushes its neighbor, which pushes its neighbor, and so on until the entire row has fallen. Each domino, it is perfectly true, has moved a bit. But the disturbance itself can move indefinitely far, depending on the number of dominos in the row, as compared with the very small motion of each domino. Another example is obtained by stretching a spring between two supports. A clothesline would do as well. In Plate 15.1 we see a pulse moving from the right side of the spring to the left. The pulse, or the vertical displacement of the spring, can move as far as the spring is long, whereas the parts of the spring themselves move relatively little; the pulse moves past the ribbon, while the ribbon remains in place.

These illustrations are two of many that could be given to realize the idea of a disturbance that moves through a medium without any substantial motion of matter accompanying it. In contrast, a particle, which also moves through space, has associated with it a definite mass. When a particle moves from one point to another, it carries with it energy and momentum. A wave also can carry energy and momentum. Clearly, the domino at the end of the line can do work on or transfer momentum just as can the pulse in the spring. The first property, then, that we associate with the wave is that a wave is an entity which propagates through space, which can carry energy and momentum, but which does not necessarily carry any matter or any mass with it.

Now we might ask, in analogy with the motion of a particle, if a wave has some inertial property—a motion it pursues in the absence of external forces. Can one consider the motion of waves in some sense analogous to the motion of particles, decide on an inertial motion, and then consider all deviations from this inertial motion as produced by external forces? We have not yet said what an external force would be in the case of a wave, nor is it clear that we can successfully proceed this way.

But there is a sense in which one can. One can write rules or equations for the motion of waves (with reasonable justice called *wave equations*) which play the same role as Newton's first two laws of motion play for the motion of particles. There, from the postulates—Newton's laws of motion—we deduced the motion of particles under various systems of forces: no force, a uniform force, or a central gravitational force. The steps of the deduction were technically within our grasp. For the motion of waves, we can again write equations that would form the postulates whose consequences would give the behavior of waves under various circumstances. We do not, however, have the technical apparatus to follow the steps. But we can state what the solutions are—what the properties of waves are under the various circumstances in which we shall be interested, those

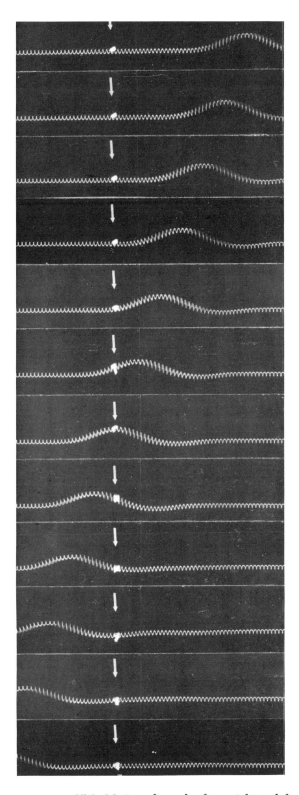

PLATE *15.1*. Motion of a pulse from right to left along a spring with a ribbon at the midpoint. The ribbon moves up and down as the pulse goes by but does not move in the direction of motion of the pulse. (Physical Science Study Committee, *Physics*, D. C. Heath, Boston, 1967)

which correspond to uniform motion, to uniform forces, to no forces, and so on. It is these properties of waves that are needed to describe the physical situations to be encountered; there is really more concern with the properties of the solutions of the wave equations than with the equations themselves. The only statement we have to accept without evidence is that there exist a consistent set of postulates whose consequences give us the properties we require.

SOME PROPERTIES OF WAVES IN ONE DIMENSION

To exhibit the properties we require of a wave we begin with what are called *one-dimensional systems*—a stretched spring as opposed, for example, to the surface of a pond. In the absence of displacement, the spring is more or less a straight line. If part of the spring is pulled upward, or disturbed in some way, a pattern is formed:

We can display the pulse (displacement as a function of distance) as

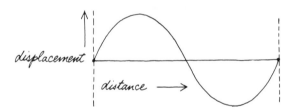

Thus a certain displacement of the spring is shown as a function of the position along the spring. Such pulses propagate along the length of the spring, as in Plate 15.1.

We now find the beginning of what we might call an *inertial property:* A pulse moving along a uniform spring, moves with constant speed and maintains its shape. For actual springs, the speed with which the pulse moves down the length of the spring depends on the density of the material of the spring and on the tension under which it is held; if the shape of the pulse is maintained, the spring is called nondispersive.

The most fundamental property postulated for waves can be arrived at by asking the question: What happens when two pulses collide? In the case of particle motion, this question could be analyzed by considering either the forces between the particles and by using Newton's second law, or by postulating conservation of momentum and energy in the collision. It is usually presumed for particles that some forces will act between them when they collide, so that they will not go through one another. However, it is always a possible solution to the equations of motion that two particles do just go through one another. If no forces act between them, clearly momentum and energy will be conserved, if they continue in their original paths after *collision:*

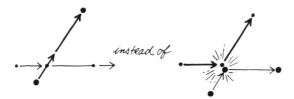

However, that is not what one usually thinks happens when particles collide.

For wave motion this is precisely what we postulate in what is known as the *principle of superposition,* certainly the property of deepest significance assigned to waves. As opposed to the collision of two particles, two pulses moving toward each other from opposite ends of the spring do not rebound but essentially pass right through one another (Plate 15.2).

What happens when the two pulses are crossing one another contains the essence of the superposition principle: The total displacement of the spring is equal to the sum of the individual displacements due to each of the pulses. Thus, if a pulse displaced upward were moving from

left to right at the same time that a pulse displaced downward were mov

ing from right to left on the same spring, the pattern formed by the two pulses would be

When the pulses are widely separated, one sees them individually. As the pulses approach—this time in the middle of the figure—we see that because one is displaced downward, the other upward, they cancel, and the spring looks almost straight. Finally, we see the two pulses emerging as they would have individually.

This cancellation of the two pulses that occurs when one is a displacement upward and the other is a displacement downward is the single most important property assigned to waves. It results in the curious circumstance that two waves can serve to annul as well as to reinforce each

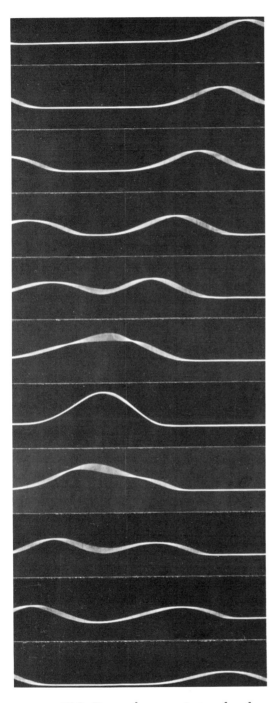

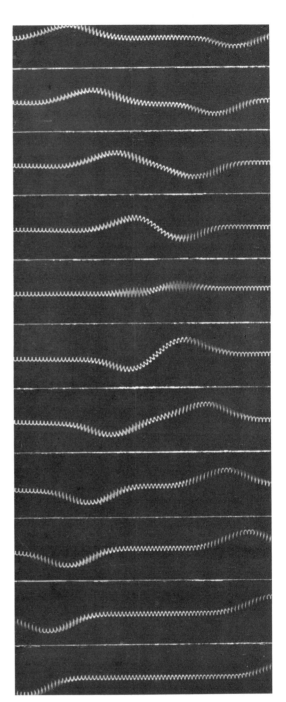

PLATE 15.2. Two pulses crossing each other. Because the two pulses have different shapes we can see that the one that was on the left at the beginning is on the right after the crossing, and vice versa. (Physical Science Study Committee, *Physics*, D. C. Heath, Boston, 1967)

PLATE 15.3. The superposition of two equal and opposite pulses on a white coil spring. In the fifth picture, when there is a net displacement of zero, they almost cancel each other. (Physical Science Study Committee, *Physics*, D. C. Heath, Boston, 1967)

other, a phenomenon known as *interference* (Plate 15.3). If the displacements are in the same direction, the interference is known as constructive; if in the opposite direction, it is called destructive. In Figs. 15.1 to 15.3 we show several other examples of the superposition principle in the crossing of two pulses.

What we mean by a one-dimensional wave, then, can be summarized as follows. It can be thought of as a displacement that is a function of position along a line—a displacement that moves in time, sometimes with a constant speed and sometimes maintaining its shape. Most important, we can define an operation of addition between two separate waves on the same line. At any time the total displacement is the sum of the individual displacements so that the wave that results when two waves are

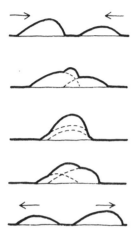

FIG. *15.1.* The superposition of two pulses. The displacement of the combined pulse is the sum of the separate displacements.

placed on the same line is the wave whose displacement is the sum of the displacements of the two original waves.

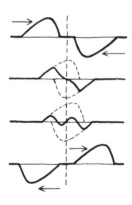

FIG. *15.2.* The superposition of two asymmetric but similar pulses.

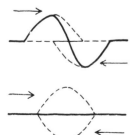

FIG. *15.3.* The superposition of two equal and opposite pulses: *above*, before complete cancellation; *below*, at complete cancellation.

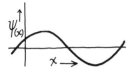

Possibly the concept of a wave will be clearer if we become more abstract. Consider the function $\psi(x)$. For every value of x there is assigned a number $\psi(x)$. (We use the Greek letter ψ rather than f because we understand by ψ a special class of functions—known as *wave functions*—which we define in what follows. By doing this we phrase in the symbols of the twentieth century an idea dating at least from the seventeenth century.) What we shall do is select from the class of all possible functions a subclass that has the special properties we want. The functions that have these properties are the solutions of what are known as *wave equations.* These are the governing equations for such functions just as Newton's laws of motion are the governing equations for the possible orbits that satisfy Newton's laws of motion. Since we do not have the technical equipment to analyze wave equations, what we shall do is display the properties of the solutions without writing down the governing equation. But the properties of the solutions can be considered more important, or more fundamental, than the equations themselves.

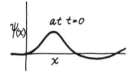

Imagine that $\psi(x)$ changes in time. If at the time $t = 0$, $\psi(x)$ looks as shown, what will the function $\psi(x)$ look like for t later than $t = 0$, say $t = 1$ sec, or $t = 2$ sec? If we graph $\psi(x)$ at the times $t = 0$, $t = 1$, $t = 2$, and so forth, we get the result shown in Fig. 17.4. This graph reveals how

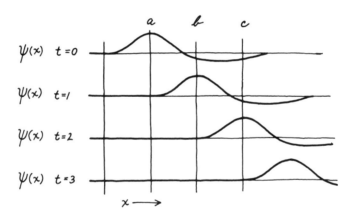

F I G. *15.4.* Time development of the wave function.

the function $\psi(x)$ changes in time or, as is conventionally said, the sequence shows the time development of the wave function. The pulse maintains its shape and moves at a constant speed from left to right.

We can view this wave function as a function not only of x but also of time, and write it $\psi(x, t)$. The meaning of this is precisely the same as before, except that now the value of the function depends not only on the distance and the spatial point x, but also on the time, t. In Fig. 15.4, for example, the value of the function at the point $x = a$ is quite different at $t = 0$ and at $t = 1$. There is a crest at $t = 0$, but at $t = 1$ the crest has passed and the value of the function is almost zero. If one follows the crest, one sees that it occurs at $x = a$ for $t = 0$, at $x = b$ for $t = 1$, and at $x = c$ for $t = 2$. If the wave moves with uniform speed, the distance between a and b is the same as the distance between b and c if the time interval between zero and 1 is the same as the time interval between 1 and 2. This is somewhat like the property of the particle, which, in the absence of forces, moves at a constant velocity.

The property of a superposition can be expressed as follows: If $\psi_1(x, t)$ is a solution of a wave equation (is a wave function) under some given set of circumstances and if $\psi_2(x, t)$ is another solution of this wave equation under the same circumstances, then the sum

$$\psi_1(x, t) + \psi_2(x, t) \qquad (15.1)$$

will also be a solution of the wave equation under the same circumstances. This is the most fundamental property of waves.

For the problem often posed is: Given two waves (whose behavior we understand individually), what will be the resulting wave? The resulting wave, using the property of superposition, is just the sum. Recall, for example, the two waves $\psi_1(x, t)$ and $\psi_2(x, t)$, both traveling down the same line. The resulting wave is the sum $\psi_1 + \psi_2$:

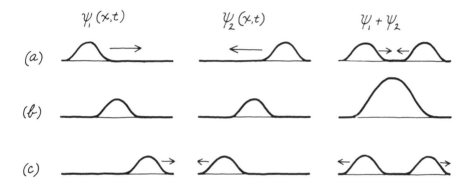

The wave which began as two pulses moving toward each other (a) finished as two pulses moving away from each other (c).

Superposition is a natural property, which we understand completely under some circumstances. The sum of two numbers is a number. The sum of two vectors is a vector, and now, as we have constructed them, the sum of two waves gives a wave. If we ask what this wave is, the answer tends to be given in terms of specific examples. For the spring the wave is the displacement of the spring as a function of position and time. For ripples on the surface of a pond, the wave is the displacement of the water as a function of position and time. The abstract notion of a wave arises, of course, from the observation of what are called actual waves in such things as ponds and in springs. However, we shall talk finally about waves that are waves in nothing—where, if we think of ψ as a displacement, it is a displacement of nothing. A wave, like a vector or a number, is a well-defined mathematical object which, developed according to the rules, results in a structure in the same sense as geometry or Newtonian mechanics. The criteria for the success or failure of the identification of certain aspects of the real world with this structure depend on whether or not the correspondence is accurate.

The waves we have considered have been pulses (like little bumps) that move uniformly along a line. A nondispersive medium is defined as one in which a wave of arbitrary shape at a time $t = 0$ will move at a constant speed without being altered in shape. In a dispersive medium the shape will change in time. Thus we could have a wave that developed in time as:

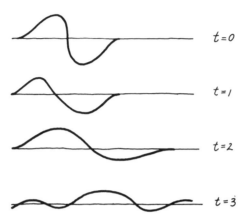

Now we might ask: Is there a special shape that always maintains itself whether the uniform medium is dispersive or not? or, Is there a certain function of position that maintains its form as time passes, whether or not the medium is dispersive? The answer is "Yes." And the special wave forms that result have properties that are so important and far-reaching that we shall study them for a while.

We consider the wave—that is to say, the wave function—of the form

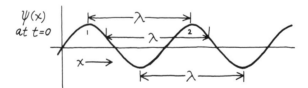

a series of crests and troughs that fall at regular intervals such that the distance from crest to crest is always the same, as is the distance from trough to trough or the distance between any point of a given height and slope and the following point of the same height and slope. This characteristic distance in which the wave form repeats itself is known as the *wavelength*, λ. To have the necessary smoothness that precisely characterizes such a periodic wave, we could write the functional form as:

$$\text{at the time } t = 0: \quad \psi(x) = \sin\left(\frac{2\pi}{\lambda} x\right) \tag{15.2}$$

Understanding the properties of sines and cosines is not essential, but if we do we note that the sine function has the same value every time the argument in the parentheses is some multiple of 2π, which occurs whenever x is some multiple of λ.

Among the properties of such a periodic wave is that if we view it at a later time, just that time required for the wave to move through a distance λ, then the crest we have labeled 1 will occupy the position of the crest we have labeled 2. But another crest, preceding 1 and not drawn on the graph, will by then occupy the position we labeled 1. And so the wave form at that particular time will be precisely what it was at the original time, $t = 0$. We call that special time τ, or the period of the wave. We now have a natural definition for the speed of the wave: the distance it moves, λ, divided by the period, τ:

$$v = \frac{\lambda}{\tau} \tag{15.3}$$

We can define a nondispersive medium as one for which the speed of all such waves is the same, independent of their wavelength.

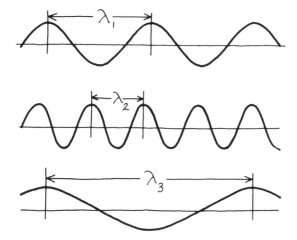

F I G. 15.5. In a nondispersive medium, the speed of all such periodic waves (wavelengths: λ_1, λ_2, and λ_3) is the same.

The periodic wave function can be written

$$\psi(x, t) = \sin\left(\frac{2\pi}{\lambda} x + \frac{2\pi}{\tau} t\right) \tag{15.4}$$

For some given time, regarding ψ as a function of x, we see that when x increases by λ, the value of $\frac{2\pi}{\lambda} x + \frac{2\pi}{\tau} t$ increases by 2π. Thus, for example,

$$\left(\frac{2\pi}{\lambda} x + \frac{2\pi}{\tau} t\right) = \begin{cases} 0 & x = 0, \quad t = 0 \\ 2\pi & x = \lambda, \quad t = 0 \end{cases} \tag{15.5}$$

An increase of the argument, the quantity $\left(\frac{2\pi}{\lambda} x + \frac{2\pi}{\tau} t\right)$, by 2π brings the sine back to its original value:

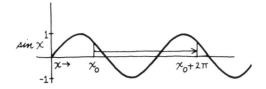

In the same way, for some given position in space (x fixed), an increase in t by τ also brings the function back to its original value.

We now introduce one additional term which, although not necessary, is often used and therefore worth defining. This is the *frequency* of the wave, which we define as below.

DEFINITION: $frequency = \dfrac{1}{\text{period}}$

$$\nu = \frac{1}{\tau} \tag{15.6}$$

The meaning of frequency is as follows: If we place ourselves at a given position, the number of crests that pass in 1 sec will be the frequency. Using this we can rewrite

$$v = \frac{\lambda}{\tau} \tag{15.7}$$

in a form that is very commonly employed:

$$v = \lambda \nu$$
$$\text{speed} = \text{wavelength} \times \text{frequency} \tag{15.8}$$

Our special interest in periodic waves arises because all such sine and cosine waves maintain their shape in any medium. In those media in which the periodic waves of different wavelength travel with different speeds (dispersive media), another wave form will change its shape in time. In those special media (such as the vacuum) in which periodic waves of all wavelengths travel with the same speed (nondispersive media), an arbitrary wave form will maintain its shape.

One reason for the importance of periodic sine or cosine waves is a theorem which states that an almost arbitrary shape can be constructed out of the sum of such periodic shapes, as shown.

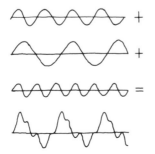

(After F. A. Jenkins and H. E. White, *Fundamentals of Physical Optics*, McGraw-Hill, New York, 1937)

This fact, plus the superposition principle, often enables us to construct any wave-function solution we want out of a sum of periodic functions. We know first that an arbitrary function can be constructed of the sum of periodic functions, using the theorem mentioned above. Second, all the periodic functions often are possible solutions of the wave equation. Thus the mathematical analysis of waves usually proceeds using periodic functions, because the properties of periodic functions are particularly simple and very well understood. Then at the end it is always possible to construct an arbitrary form as a sum over the simple periodic functions.

ONE-DIMENSIONAL WAVES AT BOUNDARIES

The properties of waves, when they pass through a region in which we might say forces were acting, we treat by returning to the realization of a wave as a displacement along a spring and observing what happens at boundaries. We assert that these are among the properties that are to be possessed by typical wave functions.

First we consider a wave pulse approaching an impenetrable boundary. Imagine that the spring is attached to a wall, and a pulse approaches the fixed end. In Plate 15.4 we see a pulse approaching the fixed end from the right, moving toward the left. When the pulse reaches the end there is a curious moment, in the sixth frame, when the spring seems to be flat and still; from the blur at the end, however, we can see that the spring is moving a bit. Then the pulse comes back—it is reflected. And, strangely, what was an upward bump comes back reflected as a downward bump. This is one of the properties of wave functions, and it is one of the consequences of the wave equation at such a boundary. In the same sense that Newton's laws of motion applied to a particle striking an impenetrable wall and rebounding elastically give a final speed equal to the initial speed, so the wave equation plus this particular boundary gives the above reflection property to the solution.

Next, we consider a wave pulse passing from a light into a heavy spring—a special case of a wave passing from one uniform medium into another. We expect that the force (or what corresponds to a force for a wave) will be exerted only at the boundary between the two media—in this case the boundary between the two springs. (What we have examined previously is a kind of inertial motion of waves-the motion through a uniform medium. The force systems we are considering now are special cases in which a wave in a uniform medium rebounds from a rigid boundary or passes from one uniform medium into another. The force is exerted only at the boundary.)

At the rigid boundary the wave is reflected, and its amplitude is reversed (an upward pulse becomes a downward pulse going back). A pulse passing from a light spring to a heavy spring shows an effect charged with significance for what will come later. We imagine first the limit in which the heavy spring is infinitely heavy—very much like a rigid wall—and we expect that the pulse would be reflected as from a rigid wall. In the other limit, the heavy spring becomes the same as the light spring; the two springs would then be continuous and the pulse would

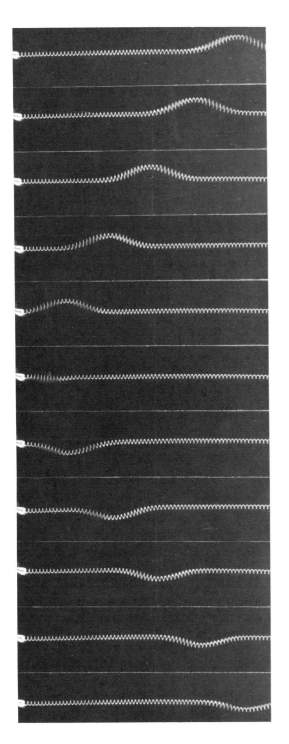

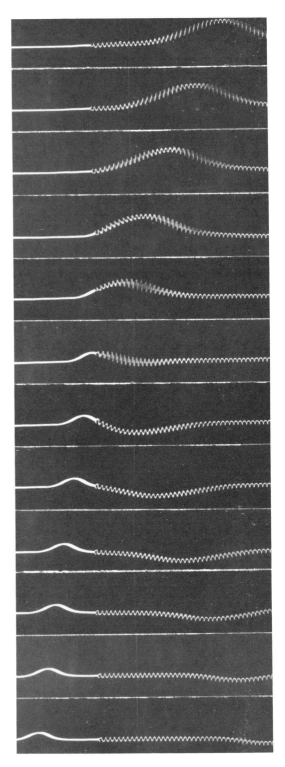

PLATE 15.4. Reflection of a pulse from a fixed end. The reflected pulse is upside down. (Physical Science Study Committee, *Physics*, D. C. Heath, Boston, 1967)

PLATE 15.5. A pulse passing from a light spring (right) to a heavy spring. At the junction the pulse is partially transmitted and partially reflected; the reflected pulse is upside down. (Physical Science Study Committee, *Physics*, D. C. Heath, Boston, 1967)

move as it does in a uniform medium. What happens at the boundary when the heavy spring is neither infinitely heavy nor exactly the same as the light spring is something between the two limits. We see in Plate 15.5 a pulse coming from the right toward the left, from the light spring toward the heavy spring. When the pulse reaches the heavy spring, one part of it is transmitted into the heavy spring and one part of it is reflected back into the light spring. The part reflected back reverses its amplitude, while the part that goes forward in the heavy spring does not—it is still an upward bump. A pulse passing from a heavy spring into a light spring also splits. However, in this case, the reflected pulse comes back without its amplitude reversed. If there is anything that distinguishes a wave pulse from a particle it is this striking property: its ability to split itself into parts at a boundary. As we shall see later, it is just this property that leads to rather deep mysteries in the interpretation of quantum mechanics.

STANDING WAVES

To conclude this discussion of one-dimensional waves, we treat the very important *standing waves*. In our previous analysis we have dealt with periodic waves or with pulses that moved at a given speed. It is possible under the proper circumstances to superimpose two waves—one traveling in one direction and the other traveling in the opposite direction —in such a way that, although the amplitude oscillates back and forth from positive to negative—upward bumps to downward bumps—the entire pattern seems to remain fixed. In particular, that place where there is no displacement, where the amplitude is zero (what is known as a *node*), seems to stand still. There are several situations in which one can produce standing waves. The easiest to illustrate is when one end of a spring or a tube or a rope is fixed at a rigid support. If a traveling wave of just the right wavelength is produced (and we shall discuss what just the right wavelength is), then the superposition of the original traveling wave, plus the reflected wave, which has the opposite amplitude, gives the pattern known as the standing wave (Plate 15.6).

Besides the fact that the pattern produced in this way is rather striking, a critical property of the standing wave is that it is only possible to produce it when a specific relation is satisfied between the wavelength and the distance between the two supports. In the first photograph in Plate 15.6 we see precisely one half of the wavelength, $\lambda/2$, between the two supports. If we call the distance between the two supports ℓ, then we have

$$\tfrac{1}{2}\lambda = \ell \tag{15.9}$$

In the second photograph, we see just one wavelength fitted into the distance ℓ, so that

$$\lambda = \ell \tag{15.10}$$

In the third photograph, we see one and one-half wavelengths, and in the fourth, two. The general condition that must be satisfied in order that a standing wave be possible is

$$\text{(integral number)} \times \text{(wavelength)} = 2\ell$$
$$n\lambda = 2\ell \qquad n = 1, 2, 3, \ldots \tag{15.11}$$

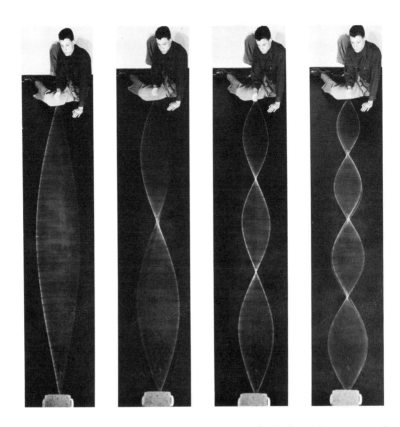

P L A T E *15.6.* Standing waves. As one end of the tube is moved from side to side with increasing frequency, patterns with more and more loops are formed. There are, however, only certain definite frequencies that will produce fixed patterns. (Physical Science Study Committee, *Physics*, D. C. Heath, Boston, 1967)

This restricts the possible wavelengths to

$$\lambda = \frac{2\ell}{n}, \tag{15.12}$$

which implies that for a given distance between two rigid supports only standing waves of certain wavelengths are possible.

The special importance of standing waves is in part due to the fact that they form a fixed wave pattern. For example, if one wants a specific sound from a violin string, it must be produced by a fixed wave pattern. A wave pattern that changes its shape in time too rapidly produces a mixture of sounds not identifiable as a note. Thus the sound that comes from an organ pipe or from a violin string (in general, many of the persistent phenomena of nature that we shall attempt to describe by waves) is the result of standing waves. But standing waves can only come in given wavelengths or frequencies for a given length between supports; this implies that for a given length of a violin string, or for a given length of organ pipe, one can produce only certain notes: those corresponding to the longest wavelength, plus those other shorter wavelengths (15.12) called overtones. A properly built organ pipe will yield not only the fundamental tone, the longest wavelength, but also overtones that enrich the sound—the ratio of the overtones to the fundamental tone depending on the details of construction. The waves on violin strings and organ pipes are quite different from the ones we have photographed. Thought of as realizations of wave functions, however, the abstract relationships are the same.

If we attempt to construct such a standing-wave pattern using other wavelengths than those allowed by the relations above, we find that we cannot get a pattern that persists. In particular, the nodes, the points of zero displacement, will move back and forth to produce a rather random-looking pattern. Thus the sound emitted over any reasonable time interval (long enough for our ear to detect a pitch) will not consist of a single pitch, say A above middle C, but a mixture—what we might call a noise.

WAVES IN TWO, THREE, AND N DIMENSIONS (INERTIAL PROPERTIES)

The mathematician has no problem generalizing the concept of a one-dimensional wave to a wave in two, three, or N dimensions.

Instead of $\psi(x,t)$, he writes for his wave function

$$
\begin{aligned}
&\psi(x,y,t) && \text{for a two-dimensional space} \\
&\psi(x,y,z,t) && \text{for a three-dimensional space} \\
&\psi(x_1, x_2, \ldots, x_N, t) && \text{for an } N\text{-dimensional space} && (15.13)
\end{aligned}
$$

where x_1 through x_N are the N-dimensions of the space. To begin with the last, what the wave function means is that given a point in the N-dimensional space (that is, given x_1, x_2, x_3, up to x_N, and given a time t), the function ψ has a certain value. Just as in the case of the one-dimensional wave, given the point x and the time t, the function ψ has a certain value that sometimes is interpreted as a displacement at that point of space and time. It is difficult for us to visualize more than three dimensions or to draw pictures of models in even three dimensions. So it is fortunate that most of the useful qualitative properties associated with waves appear in two dimensions.

All the general concepts we have developed for one-dimensional waves are equally valid in the case of two, three, or N-dimensions. Again, we can define periodic waves of a given wavelength or wave pulses that propagate at a given speed through a nondispersive medium. The superposition principle is valid in general. The only real difference is that the study of two- or three-dimensional waves yields patterns that do not appear in one dimension.

Even a two-dimensional wave is somewhat difficult to represent. At a given time, what we have to do is to associate a number (which could be interpreted as a displacement) with every point in a plane. A wave pattern on the surface of the pond, for example, would be what we call a two-dimensional wave. For every point on the surface of the pond at a given time there is a certain displacement of the surface. This can be represented for a given time as shown.

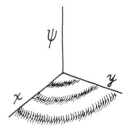

For the most part we have to deal only with wave shapes of the sinusoidal or periodic types we considered in one dimension. Thus a cross section of a wave that is of interest to us taken in a plane perpendicular to the xy plane might yield a one-dimensional pattern of the following kind—a series of crests and troughs in a periodic pattern.

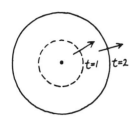

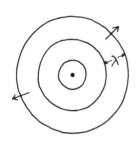

What concerns us most in the two-dimensional wave problem is the geometric pattern of the given sets of crests or troughs. Consider in particular a circular wave pulse, or a circular periodic wave. Looked at from above, a single nonperiodic circular pulse might move with a given speed, the radius of the pulse increasing uniformly. The archetype of a periodic circular wave is the pattern produced on the surface when a stone or pebble is thrown into the middle of a quiet pond. The stone creates a disturbance at the point in the pond where it strikes. The disturbance oscillates back and forth for a while, producing a wave pattern that looks from above as shown. The dark curves we have drawn (circles in this case) represent the crests of a wave moving outward from the point where the pebble struck the pond. If it is a periodic wave, the distance between the crests is the wavelength λ; the time required for one wave crest to move to the position of the next wave crest is τ. And again the wavelength divided by this time is equal to v, the speed with which the crests move outward:

$$v = \frac{\lambda}{\tau} \tag{15.14}$$

The problems that arise for one-dimensional waves occur also in two dimensions, resulting again in a particular interest in periodic two-dimensional waves.

Line Wave

A second two-dimensional wave pattern of very great interest could be called a *line wave*. The archetype of this wave would be produced by dropping a long stick or ruler onto the surface of a quiet pond. The pulse generated in that case, rather than originating at a point and going outward in circles, would originate along the line and would go outward in lines. In this case the crest of the pulse seen from above is a line, and this crest moves at a given speed. Again, and for the same reasons as before, this wave becomes of special interest when it is periodic, when the crests seen from above form the pattern shown. As usual, the distance between two crests is called λ, the time for one crest to reach another τ, and the ratio λ/τ is the speed with which the wave travels.

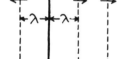

A ripple tank, essentially a shallow pond with a glass bottom, allows

PLATE *15.7.* Periodic line wave moving across a ripple tank. (Physical Science Study Committee, *Physics,* D. C. Heath, Boston, 1967)

us to photograph various two-dimensional wave patterns. A periodic line wave moving across a ripple tank is shown in Plate 15.7.

light source

The inertial properties of these two different surface waves, circular and line, are precisely those of the one-dimensional waves studied previously. We can paraphrase Newton's first law as follows: The wave front (that is, the leading crest) moves with a uniform speed and a uniform direction through the medium.

Behavior of Two-Dimensional (Surface)
Waves at Various Boundaries

In principle the behavior of two-dimensional waves at boundaries is no different from that of the one-dimensional waves considered previously. However, various new patterns occur that must be understood.

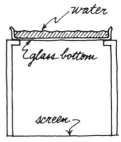
ripple tank

Reflection

Consider an incident line crest that approaches a solid barrier; if the crest is parallel to the barrier, it is reflected right back, its amplitude reversed (as for a pulse on a spring in one dimension). If the incoming line crest is at an angle to the barrier, however, it is reflected at an angle, as shown next. The direction of motion of such a crest is defined to be along the line perpendicular to the crest itself. The angle of incidence is defined as the angle between the direction of motion and the perpendicular to the barrier (above), and the angle of reflection is defined as the angle between the direction of motion of the outgoing crest and the perpendicular to the barrier. Perhaps it is no surprise that the properties of waves in the presence of a barrier of this kind require that the angle of incidence should be equal to the angle of reflection.

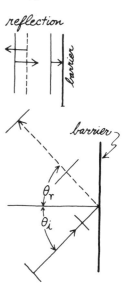
reflection

Refraction

Next consider the properties of such line waves passing from one medium to another. Again, there is a barrier, but now the barrier is not impenetrable. As in the case of the one-dimensional wave, part of the pulse is transmitted and part is reflected:

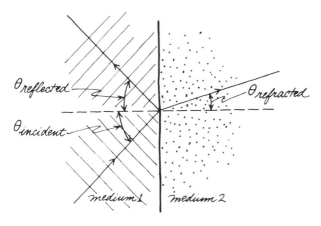

This we might expect using the following argument. Assume the wave is traveling from medium 1 to medium 2. If medium 2 becomes infinitely dense, it behaves as an impenetrable barrier and one will have reflection. If, however, medium 2 is the same as medium 1, then there is no barrier and the wave should move with its inertial motion. In any intermediate situation, one expects something between inertial motion and reflection: part of the wave reflected and part transmitted. For the reflected wave the angle of incidence is equal to the angle of reflection.

The wave that passes into the second medium—the refracted wave— follows a path that can be described by Snell's law:

$$\frac{\sin\left(\theta_{\text{incident}}\right)}{\sin\left(\theta_{\text{refracted}}\right)} = \text{constant} \qquad (15.15)$$

where the constant, or index of refraction, is again a property of the two media.

Diffraction

Consider now the properties of periodic surface waves when they pass through boundaries in which there are openings. The patterns that result can be deduced as consequences of the various properties, in particular the superposition principle, that we have assigned to the waves and can be observed among the surface waves produced in a ripple tank.

We first recall that a periodic point disturbance in the center of a pond, such as that produced for example by a pencil point going in and out of the water—let us say one time per second—will produce outgoing circular wave crests:

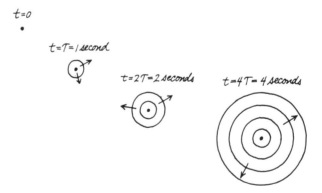

Now imagine a periodic line wave approaching a barrier that has a very small opening, an opening small enough to be considered a point. The crests of the approaching line wave produce a periodic disturbance at the opening of the barrier. This generates a circular wave in very much the same manner as the pencil point above. The part of the wave going backward interferes with the incoming line wave to produce a complicated pattern. But the part going forward takes on a relatively simple shape:

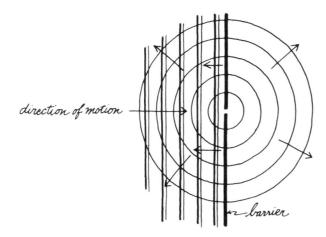

direction of motion

barrier

The wave pattern formed on the right of the barrier consists of a series of semicircles moving outward—a semicircular wave, with the same wavelength and the same frequency as the incoming line wave on the left (assuming that the speed of propagation of the waves is the same on both sides of the barrier).

The direction of motion of the wave crests is along a line perpendicular to the wave front—the leading crest. Therefore, the direction of motion of the incoming plane wave is horizontally toward the right (above), while the circular wave on the right spreads out, moving radially from a central point—the hole from which it originates.

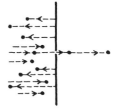

This we can contrast with the behavior of a particle that goes through a small hole in a similar barrier, as shown. We see a hail of particles striking a barrier moving horizontally from the left. If one assumes there are no forces between the hole in the barrier and the particles (this, of course, is a major assumption), then the direction of motion of a particle that manages to get through is unaffected by the barrier, and it continues to move horizontally to the right. However, if one assumes that some kind of force exists between the edges of the barrier and the particles, then one can have another situation, as might occur, for example, if the particles bounced off the edge of the hole.

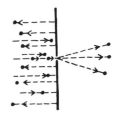

Grimaldi discovered that light was diffracted as it passed through a small hole. The image cast was larger than would have been expected assuming that light traveled in straight lines. From this one could conclude that light bends a bit passing through a small opening in a barrier. Whether this bending is better described by a wave bending as it passes through a hole in a barrier, or by particles with forces acting between them and the barrier, depends finally on how the details of the theory can be constructed.

To discuss the situation of particles in more detail, one would have to make some statement about the forces between the particles and the opening in the barrier. For waves, what turns out to be relevant are two lengths: the wavelength of the incoming wave and the size of the opening in the barrier. The situation can be summarized in a few sentences. If one considers the ratio λ/d, where λ is the wavelength of the wave and d is the size of the opening in the barrier, then when λ/d is large (when the wavelength is larger than the opening of the barrier),

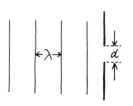

there is a large amount of spreading. The limiting case occurs when the opening is just a point and the spreading takes place uniformly in all directions. If, however, λ/d is much smaller than 1 (the wavelength is much smaller than the opening), then the spreading diminishes until finally there is almost no spreading at all.

This can be beautifully illustrated for surface waves in a ripple tank. In Plate 15.8 are shown waves of decreasing wavelengths from top to bottom passing through a barrier with an opening of constant size d. The decrease of bending as the wavelength becomes shorter, and therefore the ratio λ/d becomes smaller, is evident as we descend from the upper photograph to the lower one.

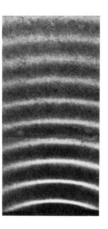

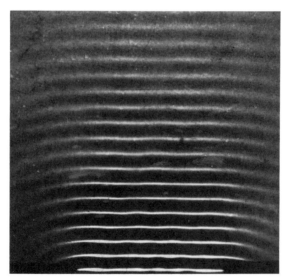

PLATE 15.8. Two views of waves passing through the same opening. The bending decreases at the shorter wavelength. (Physical Science Study Committee, *Physics,* D. C. Heath, Boston, 1967)

Whether or not waves bend or spread out as they pass a barrier becomes, then, a quantitative question. The answer in principle is "Yes, they always do." But the amount by which they spread depends on the ratio λ/d. It is not hard to imagine then that as λ/d, the wavelength divided by the opening, becomes extremely small, it might be very difficult to observe any bending of the waves. If, however, λ/d is large, the bending should become so apparent that it would hardly escape notice.

What is important as well as elegant in the association of waves with light is that the bending, for any normal barrier, depends only on the two distances λ and d. One does not have to construct a very complicated theory of what happens at the opening. And as we will see, this corresponds very nicely to what is observed in the bending or the diffraction of light that passes through a barrier. The phenomenon of diffraction depends only on the ratio of the wavelength of the light to the size of the opening in the barrier.

The rectilinear propagation of light through a vacuum, its inertial property, seems to be one of the most compelling arguments for a particle interpretation. Yet waves (for example, the line waves discussed) also propagate in straight lines through the vacuum. On passing barriers, the bending of waves is more apparent than the deflection of a particle. But the amount of bending depends on the ratio of the wavelength to the size of the opening. If we can in a consistent way assign such ratios that are small enough so that the bending is not easily apparent (as is the case for light), then we might manage to associate light with a wave and be consistent with the observed phenomenon.

Interference

Now, the *pièce de résistance:* interference. There are no new principles involved, but the phenomenon is striking.

Instead of a single disturbance on the surface of a pond producing a periodic circular wave, imagine two such point disturbances separated by a distance d. Imagine also that the periods of the disturbances are the same, so the wavelengths of the two separate waves produced are the same. When the two wave fronts are separated, we have what we

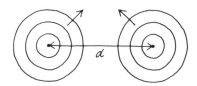

expect—two outgoing circular waves. We can obtain the wave pattern that results when the crests of the two circular waves cross one another by applying the superposition principle, just as was done for one-dimensional waves on a spring. The resulting amplitude pattern (in this case the resulting displacement of the surface of the water) is equal to the sum of the displacements produced by the individual waves.

The pattern predicted at any instant of time will be more complicated, of course, than the one-dimensional pattern, because the pattern now covers the surface of a plane. Where two wave crests meet, the amplitude (displacement) is large and positive (up). Where two troughs

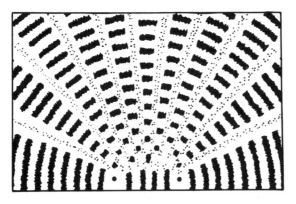

PLATE *15.9.* Pattern predicted by applying the superposition principle to the waves from two point sources. (Physical Science Study Committee, *Physics*, D. C. Heath, Boston, 1967)

meet, the amplitude is large and negative (down). But where a crest meets a trough, the amplitudes cancel, and the surface is relatively undisturbed. The pattern that results after a time is shown in Plate 15.9. Here the darkened areas indicate those places of very large amplitude where the crest meets a crest; the blank areas show those places (of very large negative amplitude) where trough meets trough. Those places where crests meet troughs and where the surface of the water is relatively undisturbed (the amplitude is small) are indicated by the dotted regions. An actual photograph on the surface of a ripple tank of two such interfering circular waves is shown in Plate 15.10. Here the white parts indicate the meeting of the crests, the dark parts the meeting of the troughs, and the gray intermediate parts those places where the water is relatively undisturbed.

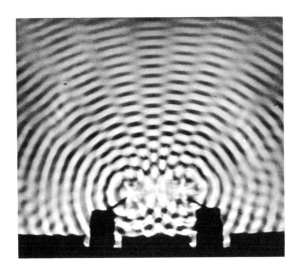

PLATE *15.10.* Photographed interference pattern due to two point sources. (Physical Science Study Committee, *Physics,* D. C. Heath, Boston, 1967)

It is of interest to analyze this pattern. Probably the most important observation is that there exist curves radiating outward from the sources of the two disturbances—curves along which the water is relatively undisturbed. These are the dotted areas in Plate 15.9. It is along these areas (called nodal areas) radiating outward that crest meets trough. The result, from the point of view of an observer some distance from the two sources—let us say an observer watching the progression of the pattern along a line—is:

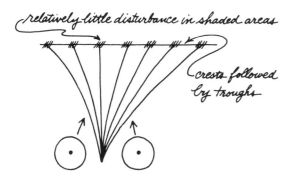

At those places where the nodal areas meet the line of observation, he sees relatively little disturbance of the water surface. In the other places he sees crests followed by troughs.

If only one disturbance existed on the surface of the water, he would see, along the line of observation, a relatively uniform crest or trough going across the entire line:

No part of the line of observation would be discriminated against. What is most startling is that by adding another disturbance, certain areas along his line of observation are produced at which the disturbance is diminished: the *addition* of a source of disturbance *diminishing* the disturbance.

This is a direct consequence of the superposition principle—a phenomenon so important, and so characteristic of waves, that it is given the special name *interference*. As on the one-dimensional spring, where crest meets crest, or trough meets trough, the amplitude, positive or negative, is enhanced; where crest meets trough the amplitude is diminished. Those gray areas in Plate 15.10 (places where the amplitude is small—where crest meet trough), which form more or less outward-going curves from the points of the disturbance, determine the *nodal lines*.

We are often interested in observing the nodal lines at a distance that is relatively large compared to the separation of the two sources. Under these circumstances we obtain a very simple result for the angle between the nth nodal curve and the perpendicular to the line between source 1 and source 2. It is

$$d \sin \theta_n = (n - \tfrac{1}{2})\lambda \qquad n = 1, 2, \ldots \qquad (15.16)$$

The pattern that results when circular crests and troughs cross one another is a particular, special, and very important case. But we can easily imagine more complicated situations in which, for example, the wavelength of the two circular waves is not the same, or in which there is superposition of a circular and a plane wave and so on. The number of such patterns is innumerable and their detailed analysis is a problem of straightforward but possibly complicated geometry. There is, however, a qualitative point of great significance that we emphasize again. In these wave patterns one gets continuous areas—lines or curves—in which the displacement of the surface is less than it would have been if only a single source existed. Thus by adding an extra disturbance the wave amplitude in some region may be diminished.

Coherence

There is another point of great practical significance. To observe a wave pattern, such as one on a water surface, it must have a certain kind of permanence. If the nodal lines were shifting from one place to another relatively rapidly, soon the water and perhaps even the eye would not be able to respond. We would see an agitated surface showing no clearly defined pattern. If the disturbances stop and start more or less at random, then the locations at which crests and troughs cancel will shift about over the surface of the water. Under these circumstances, we can expect that the pattern will not have any persistence or permanence. Two sources that behave in such a way would be called *incoherent*—leading characteristically to an agitated surface—a pattern that is all gray. In order that a sharp pattern be observed with well-defined highs and lows, the two sources have to beat in rhythm with one another, or, as is said, coherently.

16

LIGHT AS A WAVE

A century after Newton, Thomas Young made another small hole in a window shutter and covered it with a piece of thick paper which he perforated with a fine needle. He wrote in 1803:

> I brought into the sunbeam a slip of card, about one thirtieth
> of an inch in breadth, and observed its shadow, either on the wall,

or on other cards held at different distances. Besides the fringes of colors on each side of the shadow, the shadow itself was divided by similar parallel fringes, of smaller dimensions, differing in number, according to the distance at which the shadow was observed, but leaving the middle of the shadow always white. Now these fringes were the joint effects of the portions of light passing on each side of the slip of card, and inflected or rather diffracted, into the shadow. For, a little screen being placed either before the card, or a few inches behind it, so as either to throw the edge of its shadow on the margin of the card, or to receive on its own margin the extremity of the shadow of the card, all the fringes which had before been observed in the shadow on the wall, immediately disappeared, although the light inflected on the other side was allowed to retain its course. . . .[1]

But Isaac Newton had written:

Are not all hypotheses erroneous, in which light is supposed to consist in pression or motion, propagated through a fluid medium? . . . [for if light] consisted in pression or motion, propagated either in an instant or in time, it would bend into the shadow. For pression or motion cannot be propagated in a fluid in right lines beyond an obstacle which stops part of the motion, but will bend and spread every way into the quiescent medium which lies beyond the obstacle.[2]

and further:

The waves on the surface of stagnating water, passing by the sides of a broad obstacle which stops part of them, bend afterwards and dilate themselves gradually into the quiet water behind the obstacle. The wave, pulses or vibrations of the air, wherein sounds consist, bend manifestly, though not so much as the waves of water. For a bell or a cannon may be heard beyond a hill which intercepts the sight of the sounding body, and sounds are propagated as readily through crooked pipes as through straight ones. But light is never known to follow crooked passages, nor to bend into the shadow. For the fixed stars by the interposition of any of the planets cease to be seen.[3]*

In the experiments Thomas Young did with light, he attempted to demonstrate that light exhibits just those properties of bending around obstacles and of interference that one would expect when dealing with waves. He wrote:

In making some experiments on the fringes of colors accompanying shadows, I have found so simple and so demonstrative a proof of the general law of the interference of two portions of light, which I have already endeavored to establish, that I think it right to lay before the Royal Society a short statement of the facts which appear to me to be thus decisive. The proposition on which I mean to insist at present, is simply this, that fringes of colors are

---

* However, he also wrote: "The rays which pass very near to the edges of any body are bent a little by the action of the body . . . [But] as soon as the ray is past the body it goes right on."[4]

produced by the interference of two portions of light; and I think it will not be denied by the most prejudiced, that the assertion is proved by the experiments I am about to relate, which may be repeated with great ease, whenever the sun shines, and without any other apparatus than is at hand to everyone.[5]

Working independently in France, Augustin Jean Fresnel published in a paper given before the French Academy of Science in 1816 his own account of a wave theory of light, which predicted such effects as diffraction and interference. He was understandably disappointed to be told that Thomas Young had observed similar things before. In a letter to Young in 1816, he wrote:

When one thinks one has made a discovery, one does not learn without regret that one has been anticipated; and I shall admit to you frankly, Monsieur, that regret was my sentiment when M. Arago made me see that there were only a few really new observations in the Memoir which I had presented to the Institute. But if anything could console me for not having the advantage of priority, it would be to have encountered a scientist who has enriched physics with so great a number of important discoveries. At the same time, that experience has contributed not a little to increasing my confidence in the theory which I had adopted.[6]

The beginning of the nineteenth century saw a resurgence of interest in the wave theory of light. The phenomena were not entirely new. Diffraction had been observed by Grimaldi; some of the effects of interference had been observed by Newton; and a wave theory had been proposed by Huygens. But somehow the issue hung in abeyance until Young and Fresnel focused their attention on just that point which most decisively separates the two interpretations, the phenomenon of interference.

The essence of the argument made by Fresnel and Young is that, under the proper circumstances, adding one beam of light to another produces darkness. It is not easy to imagine particles that cancel one another; if one adds particles to other particles, one should get more particles. It is, however, a well-known property of waves that under the proper circumstances the addition of one wave to another produces regions where the disturbance is diminished.

Young wrote, in a paper presented to the Royal Society in 1802,

Wherever two portions of the same light arrive at the eye by different routes, either exactly or very nearly in the same direction, the light becomes most intense when the difference of the routes is any multiple of a certain length, and the least intense in the intermediate state of the interfering portions; and this length is different for light of different colors.

Fresnel was to say later:

The theory of luminous vibrations has that character and affords these precious advantages. We owe to it the discovery of the most complicated laws of optics, and the most difficult to divine. . . .

There was one dramatic instance that he may have had in mind. The French Academy in 1818 had proposed in its annual competition questions concerning the nature of the various diffraction and interference effects.* Fresnel presented his theories and experiments to a commission composed of LaPlaee, Poisson, and Biot, who were skeptical; Arago, who was enthusiastic, and Gay-Lussac, who was neutral. There Poisson pointed out that among the consequences of Fresnel's theory was one that was most remarkable. At the proper distance from a circular obstacle placed in the path of a beam of light, because of the diffraction of the waves around the circular obstacle, it was implied by Fresnel's theory that there should appear a bright spot in the shadow:

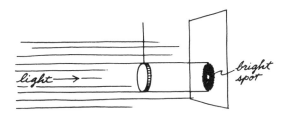

This seemed highly implausible, and Fresnel was challenged. We do not know what his expectations were when he performed the experiment,† but there must have been some satisfaction in finding a hitherto unsuspected bright spot in the middle of the shadow—a spot never before consciously seen, but a consequence of the equations he had proposed to describe the behavior of light.

INTERFERENCE

Interference is the essence. A line-wave crest striking a barrier with two small openings produces the pattern of interfering circular waves analyzed previously. For distances far enough away from the barrier, the minima (those places where the disturbance is weakest) are related to the wavelength of the incoming wave by

$$d \sin \theta = (n + \tfrac{1}{2})\lambda \qquad n = 0, 1, 2, \ldots \qquad (16.1)$$

and the maxima (those places where the disturbance is largest, where crest meets crest or trough meets trough) are given by

$$d \sin \theta = n\lambda \qquad n = 0, 1, 2, \ldots \qquad (16.2)$$

Just a half a wavelength produces the difference between crest-crest or crest-trough.

Young says:

Supposing the light of any given color to consist of undulations,

---

* The competition seems to have been posed to obtain convincing evidence for the corpuscular theory.

† There seems to be some disagreement about whether it was Fresnel or Arago who performed the experiment. Under the circumstances, both may have done it.

of a given breadth, or of a given frequency, it follows that these undulations must be liable to those effects which we have already examined in the case of the waves of water, and the pulses of sound. It has been shown that two equal series of waves proceeding from centers near each other, may be seen to destroy each other's effects at certain points, and at other points to redouble them.[7]

But

In order that the effects of two portions of light may be thus combined, it is necessary that they be derived from the same origin, and that they arrive at the same point by different paths, in directions not much deviating from each other.[8]

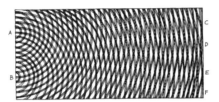

That is, the light must be coherent. If the crests arrived at the two openings at random, the pattern would shift. A region where trough meets crest in the ensuing wave pattern would become a region where crest meets crest. Maxima would become minima. Rather than a distinct pattern, one would see only some average gray.

The easiest way to achieve the needed coherence is to take the light from a single source, to split it by some means,

by diffraction, by reflection, by refraction, or by any of these effects combined. . . . But the simplest case appears to be when a beam of homogeneous light falls on a screen in which there are two very small holes or slits which may be considered as centers of divergence, from whence the light is diffracted in every direction.[9]

And then

In this case, when the two newly formed beams are received on a surface placed so as to intercept them, their light is divided by dark stripes, into portions nearly equal . . . The middle of the two portions is always light, and the bright stripes on each side are at such distances, that the light, coming to them from one of the apertures, must have passed through a longer space than that which comes from the other, by an interval which is equal to the breadth of one, two, three, or more of the supposed undulations, while the intervening dark spaces correspond to a difference of half a supposed undulation, of one and a half, or two and a half, or more.[10]

One can simplify the analysis and make the patterns more intense by replacing the holes by lines. The resulting pattern is almost identical

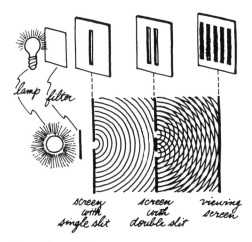

Young's interference experiment; holes re-placed by slits.

(After K. R. Atkins, *Physics*, Wiley, New York, 1965)

with what one expects from holes. A photograph of the interference pattern produced by white light passing through two narrow slits is shown in Plate 16.1. Besides the dark and light lines, we notice that the light has been separated into various colors. The same, repeated, using instead of white light as a source, pure red light or pure blue light, gives the pattern shown in Plate 16.2.

If we interpret what we see as a pattern due to the interference of waves (and it is becoming difficult not to), then a measurement of the spacing of the dark lines should tell us what the wavelength is. We have only to measure the length, d, the distance between the two openings, and θ_1, the angle to the first minimum. Then, using the relations we have already derived, the angle to the first minimum is given by

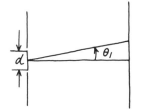

$$d \sin \theta_1 = \frac{\lambda}{2} \qquad (16.3)$$

And if one can measure this angle, one immediately can determine the wavelength of the radiation,

$$\lambda = 2d \sin \theta_1 \qquad (16.4)$$

This is not quite the most elegant way. For technical reasons it is better to measure the distance between the various dark bands; the result then is more accurate. But the principle is the same. If it is a wave property we are witnessing, and if the dark bands are due to the interference of the waves coming through the openings in the barrier, then the wavelength of the undulation is given from a knowledge of d (the separation between the two openings) and the sine of the angle from the central maximum to the first minimum.

We see from the photographs that the separations differ for red light and blue light and that, for white light, one has a mélange. This might lead one to suggest that the different colors of light correspond to different wavelengths, and therefore the distance to the first dark line differs slightly for each different color. This was proposed by Young:

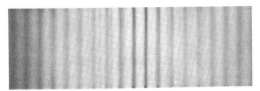

PLATE *16.1*. Interference pattern produced by white light passing through two narrow slits.

PLATE *16.2*. Interference patterns of red light and blue-violet light made with exactly the same arrangement used to make the white light interference pattern above.

PLATE *16.3*. White light passing through a single slit produced this pattern.

PLATE *16.4*. With the same arrangement as in Plate *16.3*, red light produced this pattern. Both patterns are called diffraction patterns.

PLATE *16.5*. Formation of a white light diffraction pattern from the diffraction patterns of the spectral colors.

(Plates *16.1–16.5* from Physical Science Study Committee, *Physics*, D. C. Heath, Boston, 1967)

For color versions of the above plates, see Insert *1*.

From a comparison of various experiments, it appears that the breadth of the undulations constituting the extreme red light must be supposed to be, in air, about one thirty-six thousandth of an inch, and those of the extreme violet about one sixty thousandth . . . From these dimensions it follows, calculating upon the known velocity of light, that almost five hundred millions of millions of the slowest of such undulations must enter the eye in a single second.[11]

In Table 16.1 we list the wavelengths of some of the common colors obtained from actual measurement of the diffraction patterns of these colors. They are not exact because what we see as a color usually includes a small range of wavelengths. What is known is that the visible spectrum ranges from about 4 to 7.2×10^{-5} cm in wavelength, the range of sensitivity of the typical human eye.

TABLE *16.1*

Color	*Wavelength deduced from measurement of diffraction pattern,* $\times 10^{-5}$ *cm*
Ultraviolet[a]	smaller than 4.0
Violet	about 4.0–4.5
Blue	4.5–5.0
Green	5.0–5.7
Yellow	5.7–5.9
Orange	5.9–6.1
Red	6.1–about 7.2
Infrared[a]	larger than 7.2

[a] Beyond the visible range.

DIFFRACTION

The observations and theory of Young and Fresnel were conclusive. In the nineteenth century light became a wave—leading to a search for other phenomena characteristic of waves. One of these, diffraction, had been seen by Grimaldi; light travels through an opening in a barrier, not quite in straight lines; it bends a bit. This could be explained, as by Newton, as due to an attraction between particles and the edges of the boundary. It can also be explained as a bending of waves when they pass through a barrier. Plate 16.3 shows white light passing through a single slit producing a typical diffraction pattern. In Plate 16.4 red light passes through the same single slit. The first dark line of a diffraction pattern occurs at that point in which the light from one end of the slit has to travel a half a wavelength farther than the light from the middle.Without going through the details, adding up the contributions of the light from every part of the slit, we state the result:

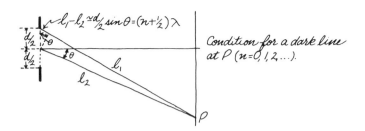

Condition for a dark line at P ($n = 0, 1, 2, \ldots$).

When $\ell_1 - \ell_2 \simeq d/2 \sin\theta = \frac{1}{2}\lambda$, we obtain the first minimum. When $\ell_1 - \ell_2$ differ by a wavelength, the waves reinforce each other and we obtain a maximum. And when $\ell_1 - \ell_2$ are equal to $\frac{3}{2}\lambda$, we obtain another minimum. The successive maxima are diminished in intensity, and the entire pattern looks as shown in Fig. 16.1.

Since the minima and maxima occur at slightly different angles for the different colors, the cooler colors being bent least, the warmer colors bent most, white light is broken into a spectrum, as seen in Plate 16.3, where the minima for different colors occur in slightly different places. Plate 16.5 shows the diffraction patterns of red, green, violet, and white

FIG. *16.1.* Single-slit diffraction pattern for light of one definite wavelength. The intensity of the light is plotted vertically as a function of the distance from the center line.

light through the same slit. The minima, the dark lines, occur at slightly different angles for the different colors. All the colors show a central maximum where the angle is zero; where these central maxima are combined, as in the bottom figure, one obtains white light. In a typical diffraction pattern of white light, the central maximum is white, but the lines on the wings are split into the colors of the rainbow due to the different angles of diffraction for the different colors.

Diffraction Grating

The diffraction grating provides the basic practical device for resolving spectra into their various components. It is essentially an obstacle with many (often thousands) of fine slits spaced equally (a distance d apart). On the distant screen, as for the double slit, maxima and minima are observed for the various colors. However, because more light is allowed through, the pattern is much more intense than for the double slit and so more useful.

The maxima occur at angles that satisfy

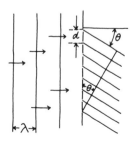

$$d \sin\theta = n\lambda \qquad n = 0, 1, 2, \ldots \qquad (16.5)$$

Therefore, with a knowledge of the spacing between slits (the basic parameter of the grating) from a measurement of θ, one can determine the wavelength or wavelengths of the incoming light.

WHAT IS LIGHT?

What, then, is light? Is it a wave? Is it particle? Is there any reason it should be either? Clearly light is no more a wave or a particle than force is a vector or pebbles are numbers. What we are led to believe from the mathematical structure of waves and from our observations of the world is that one can associate with the physical thing known as light the mathematical thing known as a wave and that the structure and relations of mathematical waves in their world somehow is a shadow or a mirror of the structure and relations of light in its world.

By the middle of the nineteenth century, after the work of Young and Fresnel, one would be inclined to associate with every beam of light a disturbance or a displacement perpendicular to the direction of motion that propagates in the vacuum with a constant speed (the speed of light). With a pure color one would associate a periodic displacement whose crests and troughs repeat with regularity at the given wavelength. The product of the wavelength and the frequency as usual give the speed, which in a vacuum is the same for all the different colors or all the different wavelengths. Thus the vacuum is a nondispersive medium. With this we can understand in a very straightforward way the origin of the different colors. We can also understand why light travels with the same speed, independent of its source, whether its source is the sun, a match, or a light bulb. For waves, characteristically, propagate through a medium with a speed independent of the source, independent of the origin of the disturbance that produced them. Further, we can understand dispersion. We postulate that certain materials exist in which the speed for the different colors (different wavelengths) is different. The phenomena of reflection and refraction can be put into correspondence with the behavior of waves at the boundary of two surfaces. In order that the refracted light in the denser medium be turned toward the perpendicular, it is required that light travel more slowly in the denser medium (as it does) opposed to the particle picture, where light must travel more rapidly.

A wide variety of diverse phenomena—interference, diffraction, and so on (one could make a long list)—fall into place from this point of view. The wavelengths determined from interference patterns are consistent with those determined from diffraction. The entire fabric makes a consistent whole. The old problem of rectilinear propagation is no longer a problem when we recall that a line wave maintains its direction of propagation until it meets a surface or a barrier. When light passes through a small opening in a barrier, it *does* bend. The bending is small (it depends on the ratio of the wavelength divided by the size of the opening), and it is observed.

The waves themselves are apparently invisible. But when they strike the retina, they excite it. When the waves strike materials, they sometimes go through; they sometimes are absorbed; they sometimes heat the material. Why or how, we do not know, for we know nothing about these waves. We know nothing about the nature of the displacement, how it interacts with matter, or what it is that is being displaced.

Why some materials are dispersive, some transparent, some opaque, why light travels more slowly in denser materials and more quickly in rare, how light excites vision in the retina, or chemical processes in a photographic plate, are questions we can answer only when we have proposed a theory that contains a statement of how light interacts with matter. If, for example, we take particles seriously, we might say that the particles collide with the atoms of matter and propose some force that acts in the collision. If we take waves seriously, we might propose that the atoms of matter bob on the surface of the waves the way corks bob on the surface of a lake. We shall have to make some such commitment before we obtain any consequences.

If light is thought of as a wave, then we might ask: How does the wave transfer its energy to matter? Suppose we did attempt to construct the transfer in the same way that a surface wave transfers its energy to a cork making the cork bob up and down. The amount of bobbing depends on the amplitude of the wave: the height of the crests and troughs. Now, as a circular wave spreads out from the source, the amplitude becomes smaller and smaller, and there seems to be no limit to the smallness of the amount of bobbing that could be transferred to a cork. Does light transfer energy to material substances in this way? Can as small an amount of energy as one wishes be transferred? From the point of view of the wave theory, nothing would be more reasonable, and it is what everyone believed. Yet it will turn out that a theory which allows light to transfer its energy continuously to matter is not in agreement with observation.

There are some questions, the answers to which lead to additional questions in such a way that finally one is at a depth not imagined at the beginning. The problem of the nature of light is one of these. Every attempt to understand what this luminous entity is has led deeper and deeper. Somehow the pursuit of this has always been fruitful. When light was accepted as a wave in the nineteenth century, and finally when light was interpreted as an electromagnetic wave by Maxwell, one asked with increasing persistence, What is the wave in? What is the medium that carries this wave? All the waves that form the prototype of the mathematical waves we have discussed are disturbances in a medium—sound, a disturbance in air; water waves, a disturbance on the surface of the water; or a pulse disturbance propagating in a spring. What was this medium that carried the disturbance called light? To speak about it, it was given a name: "the luminiferous ether." And, to parse a sentence, it was always the subject of the verb "to undulate." As Maxwell was to write:

> Ethers were invented for the planets to swim in, to constitute electrical atmospheres and magnetic affluvia, to convey sensations from one part of our bodies to another, and so on, until all space had been filled three or four times over with ethers.[12]

And he continues:

> The only ether which has survived is that which was invented by Huygens to explain the propagation of light.[13]

It must exist in the near-vacuum of outer space, for light propagates through vacuum. It must exist in transparent material. Yet its properties were extremely curious. To support an oscillation that moved as rapidly as light moves, in spite of its rarity it would have to have the elastic properties of a solid material. It would have to snap back to its original position somewhat the way steel does.

The properties of this ether were bizarre indeed. But what was worse is its seeming conspiracy to make itself totally unobservable. In trial after trial all attempts to observe the ether or any of its effects would fail, until finally all that was left was the statement that light moved with respect to the ether. And when finally not even this could be asserted, the stage was set for one of the great revolutions in physics.

ELECTRO-MAGNETIC FORCES AND FIELDS

17
ELECTROSTATIC FORCES: CHARGES AT REST

ELECTRIC CHARGES

"In amber," Plutarch wrote, "there is a flammeous and spirituous nature, and this by rubbing on the surface is emitted by hidden passages, and does the same that the loadstone does."

We know now that electrical forces, by any reasonable method of comparison, are many orders of magnitude stronger than gravitational forces and that there is no lack of electrically charged particles in matter. Thus it seems almost a surprise that gravitational forces should have been understood first. Electrical forces, as we shall see, are in some respects very similar to gravitational forces; point charges repel or attract one another with a force that is similar in form to the force of gravitation (constant/R^2) a force that falls off inversely as the square of the distance between the charges. The magnitude of this force, however, is enormous compared to the magnitude of the gravitational force. Why then is it so invisible that it was necessary to rub amber and watch it attract chaff in order to discover its presence?

In the answer to this we have the first major difference between gravitational and electrical forces. The gravitational force of Newton is one such that all matter attracts all other matter, the magnitude of the attraction being proportional to the product of the two masses. Electric charges, however, do not attract all other electric charges. They appear to divide themselves into two classes. This was first proposed by Charles François de Cisternay Du Fay (1698–1739).

> This principle is that there are two distinct electricities, very different from each other: one of these I call *vitreous electricity*; the other *resinous electricity*. The first is that of [rubbed] glass, rock crystal, precious stones, hair of animals, wool, and many other bodies. The second is that of [rubbed] amber, copal, gum lac, silk, thread, paper, and a vast number of other substances.
>
> The characteristic of these two electricities is that a body of, say, the *vitreous electricity* repels all such as are of the same electricity; and on the contrary, attracts all those of the *resinous electricity*.[1]

We no longer use the expressions "vitreous" or "resinous" electricity. Today we would say positive or negative electricity. But the point is the same. Positive charges repel one another; negative charges repel one another; but positive and negative charges attract:

charged bodies of the same class repel each other while oppositely charged bodies attract

Nothing intrinsic distinguishes positive from negative charges. It is at present regarded as a deep symmetry principle that the label positive or negative has no absolute significance and that all positives could be interchanged with negatives without altering any of our observations. But it is the existence of these two classes of charge that produces the first remarkable difference from the gravitational force. For, in spite of the magnitude of electrical forces and in spite of the fact that all matter supposedly contains huge numbers of electrically charged particles, in any ordinary piece of matter the positive and negative charges are so nearly evenly balanced that it is difficult to observe any electrical force. It is for this reason that rubbing or stroking was important in the early observations of electrical effects; because on the proper material, the rubbing or the stroking separates negative from positive charges and alters the precise balance of charge, thus revealing its presence.

Some idea of the enormous size of the electrostatic forces and of the almost perfect balance between negative and positive charges in ordinary matter can be obtained by considering, let us say, two ordinary-sized objects about 1 yard apart. Imagine that there were an excess of about one out of every 1000 charges. The force between the two bodies, either attractive or repulsive depending on whether the charges were the same or the opposite, would then be of the order of 10^{12} pounds (1000 billion pounds). If they were oppositely charged, sparks would fly between them until they had more evenly divided their charge. The size of this force, the fact that sparks can fly, that the charge leaks off, is of course the reason that most of the objects we observe are almost electrically neutral. Any serious charge imbalance produces forces which are so enormous that neutrality is restored one way or another.

CONDUCTORS AND INSULATORS

Among the common materials of our experience, there are two extremes of behavior with regard to electric charges that are worth mentioning. Indoors on a dry winter day we accumulate charge on our bodies as we walk across a rug, producing a sometimes unpleasant shock when we greet a friend. The same thing does not happen on a hot humid day. On the winter day, the air is dry and behaves as what is called an *insulator,* so the charge we accumulate by rubbing the rug stays put. On a damp day, the air is less of an insulator (more of a *conductor*), and the charge we accumulate leaks off our bodies.

"Insulator" or "conductor" are words that describe materials: insulators—such as glass—which do not permit charges to move very freely through them, conductors—such as metals—which do permit charges to move through them freely. This is an old distinction; materials do not all neatly fall into the two classes. We know now of an almost continuous range of materials beginning with such crystals as diamond, almost perfect insulators, to superconductors, metals at very low temperatures which one might call perfect conductors. Most ordinary metals are good conductors— the reason they are used in electrical wires. Glass, fabrics, and plastics are very good insulators—the reason plastics or fabrics are used to insulate copper wires carrying current.

An insulator is characterized by the fact that the charge placed on it remains roughly where it is placed. A conductor, on the other hand, allows the charge to move freely throughout its volume, so that the charge distributes itself wherever the forces acting direct. The reason early physicists such as Gray had trouble, sometimes keeping the charge and sometimes losing it, is because they were not aware at the time (it was Gray who discovered the distinction) that some of the materials they used to support their charged objects were conductors and allowed the charge to leak away.

Once one understands them a bit, electric forces are easier to study than those of gravity. When we have learned to accumulate and keep charges on a conductor or on an insulating surface, because the forces between charges are so strong, we can observe their effects on other bodies perceptibly in the laboratory. For the gravitational force one has little choice other than to use large massive objects such as the earth.

COULOMB'S FORCE

Experiments to determine the variations of electrical forces with the distance of the separation of the charges were carried out by Daniel Bernoulli around 1760, and by Joseph Priestley and Henry Cavendish about 10 years later. But the definitive work which determined that the electric force between two charges that are stationary falls off as the square of the distance was done by Charles Augustin de Coulomb in 1785, and it is his name that is associated with the law of attraction or repulsion of electrical charges. Coulomb constructed a torsion balance sensitive enough so that he could accurately measure the change in the force between two charges as the distance between them varied. The results of his experiments he summarized as follows:

> . . . the mutual attraction of the electric fluid which is called positive on the electric fluid which is ordinarily called negative is in the inverse ratio of the square of the distances; just as we have found . . . that the mutual action of the electric fluid of the same sort is in the inverse ratio of the square of the distance.[2]

The analogy with Newton's law of gravitation is striking. Coulomb said, in fact, that the electrical force varies directly "as the product of the electrical masses of the two balls." The direction of the force is along a line between the two charges. If the charges are of the same class, that is, both positive or negative, the force tends to separate them; if the charges

are of different classes, the force tends to bring them together. If we call a positive force "one that tends to separate," and a negative force "one that tends to pull together," then if q_1 and q_2 are the electrical masses of the two bodies the entire situation can be summarized as: *

$$\mathbf{F} = \text{constant } \frac{q_1 q_2}{R^2} \qquad \text{(magnitude and sense)} \dagger$$

(directed along the line joining the two charges)   (17.1)

The vector is directed along the line joining two charges; its magnitude is the magnitude of constant $q_1 q_2 / R^2$; its sense (+ or −) is the sign of $q_1 q_2$ and is shown above.

Example. If two point charges separated by 100 cm repel each other with a force of 100 dynes, with what force will they repel if separated by 1000 cm?

at 100 cm:

$$F_{100} = 100 \text{ dynes} = \text{constant } \frac{q_1 q_2}{(100 \text{ cm})^2} = \frac{\text{constant } q_1 q_2}{10^4 \text{ cm}^2} \qquad \text{(magnitude)}$$

---

* The most accurate experiment performed to date to test this relation on a macroscopic scale was done by Plympton and Lawton [*Phys. Rev.*, **50**, 1066 (1936)], who found that the exponent is 2 to 1 part in 10^9.

† Constant $q_1 q_2 / R^2$ has a magnitude (say 7 dynes) and a sign (+ or −). The magnitude of +3 is 3; the magnitude of −3 is also 3. The magnitude is by definition the size or largeness of a number, independent of its sign. The sign gives what is called the sense: attractive (−) or repulsive (+).

PLATE *16.1*. Interference pattern produced by white light passing through two narrow slits.

PLATE *16.2*. Interference patterns of red light and blue-violet light made with exactly the same arrangement used to make the white light interference pattern above.

PLATE *16.3*. White light passing through a single slit produced this pattern.

PLATE *16.4*. With the same arrangement as in Plate *16.3*, red light produced this pattern. Both patterns are called diffraction patterns.

PLATE *16.5*. Formation of a white light diffraction pattern from the diffraction patterns of the spectral colors.

(Plates *16.1–16.5* from Physical Science Study Committee, *Physics*, D.C. Heath, Boston, 1967)

at 1000 cm: $\qquad\qquad\qquad\qquad\qquad\qquad\qquad\qquad\qquad$ (17.2)

$$F_{1000} = \frac{\text{constant } q_1 q_2}{10^6 \text{ cm}^2} = (100 \text{ dynes}) \frac{10^4 \text{ cm}^2}{10^6 \text{ cm}^2}$$

Therefore, $F_{1000} = 1$ dyne.

NATURE OF THE ELECTRICAL MASS

An inspection of Coulomb's force immediately poses the problem of the nature of electrical masses or the quantity of electricity that resides on the charged particles—labeled q_1 and q_2 above. The similarity between the Coulomb force and the gravitational force is so striking that we are led to ask whether the properties of the electrical mass (charge) are in any way like the properties of the gravitational mass. There are immediate differences. The charge is divided into two classes, positive and negative, whereas all gravitational masses are in one class, making gravitational forces always attractive.[*]

Mass can be thought of as being infinitely divisible. (From the point of view of primitive atomic theories, one might think that masses should come in multiples of some unit. It turns out for reasons that will become clear later that in spite of the atomic nature of matter, it is thought now that the masses that occur in the gravitational equations can take any value.) In contrast to this, the charges that can occur in Coulomb's equation are not continuous. From a macroscopic view, regarding objects that are rubbed or otherwise electrified, the charge appears continuous. But it is our present view—supported by a large body of experimental evidence—that charge cannot be continuously divided and comes ultimately in multiples of a fundamental unit: the charge on an object, called the *electron* (from the Greek for amber).

According to present conventions the electron is taken to be negatively charged. The letter e is used to denote the magnitude of this fundamental unit of charge.[†] (Nowhere does the word "fundamental" seem more appropriate.) The charge $(+e)$ on the particle called the *proton* is as far as we know equal in magnitude but opposite in sign to the charge of the electron. Very sensitive experiments done to measure possible differences in the charge of these two particles enable us to say that the two charges are equal in magnitude—to an accuracy of about 1 part in 10^{20}.

All the currently identified elementary particles, all assemblies of particles, all atoms and known matter, are either neutral or are charged in integral multiples of the charge on the electron or the proton. No one has any deeper understanding of this. (Particles with charge, for example one third of the electron charge recently suggested, so far have not been observed.)

Perhaps the deepest property of the electric charge is that it is conserved. This means that if a certain amount of charge is contained in an

[*] Alternative hypotheses have, from time to time, been suggested, but there never has been found any experimental support for the conjecture that in some situations the gravitational force might be repulsive.

[†] We will use the convention that the charge e is a positive quantity. Thus the charge on the electron $= -e$.

isolated system (no charged matter passes in or out, in just the same sense that when sheep are contained in a fenced-in field no sheep pass in or out), then the total quantity of charge remains unaltered. The total quantity of charge is the sum of the positive and negative charges. If the isolated system is a closed container, in which there are no charges, we might, after a while, find two charges there. But one would be positive and the other negative, so that their sum would be zero. It is possible, for example, for a photon (a particle of light) to, as is said, create a positive and a negative charge (electron and anti-electron or positron), but the sum of the final charges adds up to what it was initially:

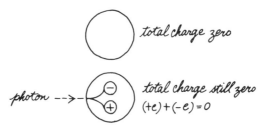

This fact has been elevated to the level of a principle: *charge conservation.* In a system in which no charge is brought in or taken away, the total amount of charge (the sum of the positive and the negative charges) remains a constant. This, along with a few others such as conservation of energy or momentum, seems to be among the deepest concepts that currently exist in the body of physics.

SYSTEMS OF UNITS

It is necessary now to define some unit system (unfortunately none of the existing systems are convenient) in which charge can be measured. We recall that from Newton's law of motion, with length measured in centimeters and time in seconds, one could define a unit of force, given a unit of mass. One procedure traditionally followed in electrical theory is to begin with Coulomb's law

$$F_{\text{electric}} = \text{constant} \; \frac{q_1 q_2}{R^2} \qquad \text{(magnitude and sense)} \qquad (17.3)$$

and to define the charge and the constant in terms of the unit of force and the unit of length, a somewhat awkward procedure but one that came into being because units of force and length already existed at the time that Coulomb did his work. In the centimeter, gram, second (CGS) system, the force is measured in dynes, distance in centimeters, and time in seconds. If the constant in Coulomb's law is arbitrarily set equal to 1, this defines a unit of charge (the electrostatic unit, abbreviated esu). Coulomb's law then takes on the convenient form

$$F_{\text{electric}} = \frac{q_1 q_2}{R^2} \qquad \text{(magnitude and sense)} \qquad \begin{array}{l} \text{CGS system:} \\ \text{charge in esu} \\ \text{distance in cm} \\ \text{force in dynes} \end{array} \qquad (17.4)$$

This system of units defines an electrostatic unit of charge, such that two point charges, each of 1 esu separated by 1 cm, exert on each other a force of 1 dyne, and eliminates the constant in Coulomb's equation:

CGS

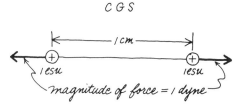

This is contrasted with what is called the "practical," or MKS, system, in which the distance is measured in meters, mass in kilograms, and time in seconds and which uses the coulomb (defined independent of Coulomb's equation) as the unit of charge. Since the coulomb of charge is not obtained from Coulomb's law, the constant is not 1—there is no reason why it should be; it then turns out that in the MKS system Coulomb's law reads:[*]

$$F_{\text{electric}} = 9 \times 10^9 \frac{q_1 q_2}{R^2} \quad \text{(magnitude and sense)}$$

MKS system:
charge in coulombs
distance in meters
force in newtons

(17.5)

This form of Coulomb's law is somewhat less convenient because of the constant that must be dragged around and also because of the size of the charges.

From the point of view of elementary particle physics, the electrostatic unit already is too large. One would prefer to, and often does, use the charge on the electron as the unit. The coulomb, even by macroscopic standards, is enormous. For example, two objects separated by 1 meter, each with 1 coulomb of charge, exert a force on each other of 9×10^9 newtons, about 2 billion pounds. On the other hand, by macroscopic standards the electrostatic unit is rather small. One dyne of force is about 1 millionth of a pound.

The advantage of the CGS system is that the fundamental equations are cleaner. Since we are interested primarily in applying these rules and equations to various fundamental situations, this system seems preferable most of the time. In most current research some variation of the CGS system is used. On the other hand, the MKS system, which measures charge in coulombs, measures potential difference in volts and currents in amperes; since we are used to household voltages stated in volts or currents in amperes, the MKS system gives a quicker intuitive feeling for what these quantities mean. A possible comfort is that there does not seem to be a corresponding English unit of charge with the force defined in pounds and the distance in feet or in yards. We shall use the CGS system primarily and translate to volts and amperes when we want to give some intuitive feeling for the magnitudes involved.

What is often done in practice is to change units as the situation changes, to keep them reasonably convenient. It does not seem worth

[*] The exact number now is taken to be 8.9875 . . . $\times 10^9$.

TABLE *17.1*

	CGS	*MKS*	*Relation*
Distance	centimeter	meter	There are 100 cm in 1 meter
Time	second	second	There is 1 second in 1 second
Mass	gram	kilogram	There are 1000 grams in a kilogram
Force	dyne	newton	There are 10^5 dynes in 1 newton
Charge	esu (also called the statcoulomb)	coulomb	There are about 3×10^9 esu in 1 coulomb

trying to use a single unit system exclusively, because a unit system convenient in one situation will often be extraordinarily awkward in another. In our daily lives we measure medicine in ounces and loads of coal in tons.

THE ELECTRIC FIELD

Additive Character of Electrostatic Forces

Coulomb's law and the agreement that the electrostatic attraction or repulsion is a force (and therefore can be associated with a vector) imply what is to be done in the presence of three or more charges. Suppose we consider the force on a third charge due to the first two. According to Coulomb's law one can calculate the force due to the first charge on the third and the force due to the second charge on the third (called in the diagram $\mathbf{F}_{1\,on\,3}$ and $\mathbf{F}_{2\,on\,3}$). The total force on the third charge is given

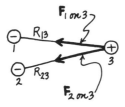

by the vector sum of $\mathbf{F}_{1\,on\,3}$ and $\mathbf{F}_{2\,on\,3}$. It is a very important property of electrical forces, just as for gravitational forces that the force with which two charges interact is not taken to be changed by the presence of a third, fourth, or fifth charge. This enables us, in calculating the effect of a large number of charges on any other charge, to superimpose the effects of the individual charges in a relatively straightforward fashion.

Example. What is the magnitude of the force between a charge of 5 esu and a charge of 25 esu if the charges are 10 cm apart?

$$F = \frac{q_1 q_2}{R^2} = \frac{(5\ \text{esu})\ (25\ \text{esu})}{(10\ \text{cm})^2} = 1.25\ \text{dynes} \qquad (17.6)$$

If the charges are both positive, the force is repulsive. If the charges had both been negative, the answer would have been the same (which points out the conventionality of positive and negative), but if one of the charges was negative while the other was positive (for example, −5 esu and 25 esu or 5 esu and −25 esu), there would have been an attractive force of 1.25 dynes on each of the charges.

A third charge of 10 esu is placed as below:

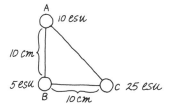

What are the horizontal and vertical components of the force on charge B?

Charge C exerts a force (to the left) on charge B. The 10 esu charge (A) also exerts a repulsive force, but this force is downward. Its magnitude is

$$F = \frac{(10 \text{ esu}) \, (5 \text{ esu})}{(10 \text{ cm})^2} = 0.5 \text{ dyne} \qquad \text{(magnitude)} \qquad (17.7)$$

Thus

$$\mathbf{F}_{\text{horizontal}} = 1.25 \text{ dynes to the left}$$
$$\mathbf{F}_{\text{vertical}} = 0.5 \text{ dyne downward} \qquad\qquad (17.8)$$

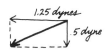

Introduction of Electric Field

The concept of a field that we introduce now could also have been introduced in the case of gravitational forces. It grows out of an attempt to describe what the force will be on a charge at a given point in space, due to the presence of many other charges, and assigns a property to a point in space that influences the motion of particles. We have agreed that the force on a point charge due to other charges will be the vector sum of the individual forces produced by the other charges. For the system of three charges above, the force on charge 3 would be written:

$$\mathbf{F}_3 = \mathbf{F}_{1 \text{ on } 3} + \mathbf{F}_{2 \text{ on } 3} \qquad\qquad (17.9)$$

The magnitude and sense of the force $\mathbf{F}_{1 \text{ on } 3}$ is

$$F_{1 \text{ on } 3} = \frac{q_3 q_1}{R_{13}^2} \qquad \text{(magnitude and sense)} \qquad (17.10)$$

while that of $\mathbf{F}_{2 \text{ on } 3}$ is

$$F_{2 \text{ on } 3} = \frac{q_3 q_2}{R_{23}^2} \qquad \text{(magnitude and sense)} \qquad (17.11)$$

Each of these is multiplied by the charge q_3; this follows immediately from Coulomb's law. If there were N charges acting on the particle 3, there would be N terms in the sum and the magnitude and sense of each of these would be proportional to the charge at the point 3.

This circumstance enables us to do a seemingly trivial thing and thus to arrive at the concept of the electric field. We factor the charge q_3 out of the expressions above and introduce a new entity, $\mathbf{E}$:

$$\mathbf{F}_{1\,on\,3} = q_3\mathbf{E}_{1\,on\,3} \qquad (17.12)$$

This new entity, the vector $\mathbf{E}_{1\,on\,3}$, we call the electric field produced by the charge at 1 at point 3. Note now that the electric field at the point three is independent of whatever charge exists at that point. And note further that the resultant electric field at point 3 will be the vector sum of the electric fields produced by all the other charges in the system; this enables us to define a total electric field at point 3 as

$$\mathbf{E}_3 = \mathbf{E}_{1\,on\,3} + \mathbf{E}_{2\,on\,3} + \cdots + \mathbf{E}_{N\,on\,3} \qquad (17.13)$$

In the CGS system of units the electric field strength is expressed in dynes per esu.

If it is not clear why we have defined this additional concept, it is not surprising. It was introduced originally by Faraday to aid him in visualizing the effects of charges on other charges. Maxwell attempted to visualize the electric field as a mechanical stress in the medium of space, the ether. Since that time, the electric field has acquired a significance deeper than any such mechanical interpretation and, like some other concepts, such as momentum or energy, finally becomes more important than the specific theories out of which it came. This we shall see later when we discuss electromagnetic radiation.

The electric field can be called a *vector function*. To every point in space it associates a vector and this vector multiplied by a charge will give the force on the charge at that point due to the charges that produced the field. If we know the electric field at all points in space, then, without further inquiry, we know what force will act on a charge at any point without having to know what the system of charges was that produced the electric field.[*] This is one of the great conveniences of the concept; in some cases it is easier to specify the electric field that exists than the charged particles that produced it.

It is difficult to draw a picture of an electric field because it involves placing a vector at every point in space. However, we can draw some of the electric field vectors due to a positive point charge as in Fig. 17.1.

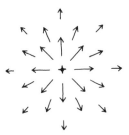

F I G. *17.1.* The electric field around a fixed positive charge is a set of vectors defined at every point in space. $\mathbf{E} = q/r^2$ directed away from the charge. Some of them are shown.

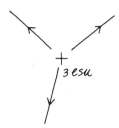

Their magnitude diminishes as $1/r^2$, while their direction is radially away from the charge. Sometimes, to better visualize the electric field, rather than drawing many vectors, one draws a continuous line that always lies parallel to the field at the point through which it is drawn. In such a construction the intensity is associated with the density of the lines. We might agree, for example, that one line is to flow out of every positive point charge of 1 esu. Thus a positive point charge of 3 esu would produce

[*] It is assumed that the introduction of the new charge does not change the distribution of the original charges from which the field was calculated.

field lines as shown, while a point positive charge of 6 esu would produce the field lines also shown. Several other field patterns are shown in Fig. 17.2.

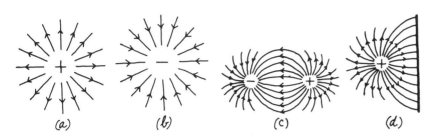

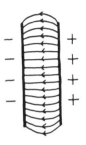

FIG. 17.2. Electric field lines *(a)* around a single isolated positive point charge, *(b)* around a single isolated negative point charge, *(c)* of two point charges of opposite sign. If the two point charges are very close together, compared to the distance at which one is sampling the field, one has what is called an *electric dipole*. Its total charge is zero; the field is produced due to the separation of the positive and negative charges. *(d)* The electric field lines of a point charge near a large plane conductor.

Electric field lines are shown here for a positively and negatively charged plate (called a *capacitor* or *condenser*). Except at the edges, the density of the electric field lines between the two plates is constant, as is the magnitude and direction of the electric field.

It sometimes occurs that small bodies will line up parallel to the electric field, if properly arranged. This happens, peculiarly enough, for grass seed suspended in an insulating liquid. (Why might that happen?) If we produce an electric field within such a liquid by putting charged objects there, the seeds line up in the direction of the field, and we can see from the patterns they form what the pattern of field lines is (Plate 17.1). In spite of their convenience, however, such pictures deceive us into attributing a physical reality to the field that probably is misleading.

We can summarize the above by stating that the electric field is a vector function of space (at every point of space a vector is defined) determined by a distribution of charges. If a charge of magnitude q is placed, without disturbing the original charges, at some point in the space, the force on that charge is then given by

$$\mathbf{F} = q\mathbf{E} \tag{17.14}$$

Thus the electric field enables us to determine, without knowledge of the particular charge distributions that exist, what the force will be on a charge at any point in space* (Fig. 17.3).

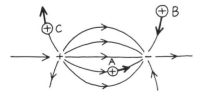

FIG. 17.3. The small positive charge placed at the point *A, B,* or *C* will have exerted on it the forces shown.

* Whether it is possible or desirable to interpret the electric field further is somewhat a matter of fashion. Like potential energy its meaning rests in its definition and its relation to the other elements of the structure of electromagnetic theory.

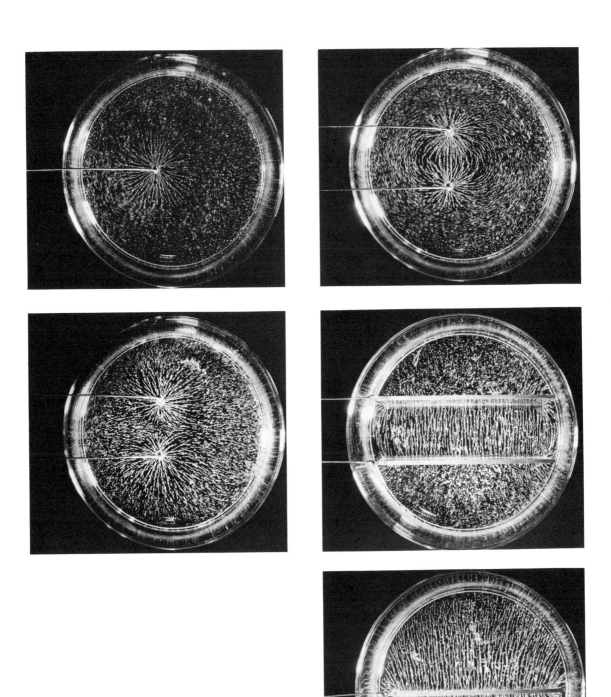

PLATE *17.1*. Electric force field patterns formed by grass seeds in an insulating liquid. *Upper left:* a single charged rod; *upper right:* two rods with equal and opposite charges; *center left:* two rods with the same charge; *center right:* two parallel plates with opposite charges; *bottom:* a single charged metal plate. (Physical Science Study Committee, *Physics*, D. C. Heath, Boston, 1967)

An important question that arises is how to determine the electric field from some given distribution of charges. We can always do this in simple situations by using Coulomb's law and adding the contribution of each charge. For complicated cases, involving elaborate charge distributions, very elegant methods have been devised. Beginning with Coulomb's law, several powerful theorems can be deduced that enable one to determine easily and quickly the fields due to relatively complicated but highly symmetric charge distributions. We do not need these technical devices, however, and so will not develop them.

ELECTRIC POTENTIAL

Although we have introduced new terms such as electric field and charge, so far we have done no more than to postulate an entirely Newtonian force between charged particles. Since the electrostatic force depends only on the distance between two particles, it is conservative in the sense discussed in Chapter 8. This enables us to define the extremely important concept of the electric potential energy.

We recall that the potential energy difference between points b and a is defined as minus the work done in bringing a particle from a to b (see Fig. 17.4):

$$-W_{a \to b} = V(b) - V(a) \tag{17.15}$$

FIG. *17.4.* The potential-energy difference between b and a is equal to minus the work done bringing the particle from a to b.

For conservative forces, the work done in bringing the particle from point a to b is independent of the path. (This is an alternative definition of a conservative force.) The gravitational force was one example of a conservative force; in the same way, Coulomb's force, which is identical in form to the gravitational force, is conservative. It is therefore possible to define the potential energy of a charge subject to the forces produced by a system of charges.

Suppose, as a simple illustration, we have a uniform and constant electric field, **E**. When multiplied by a charge this gives a constant force and so corresponds to the simple case treated in Chapter 8. Now consider the work done on a charged particle going from point a to point b in Fig. 17.5. If the charge on the particle is positive and of magnitude q,

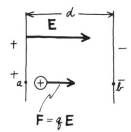

FIG. *17.5.* The work done on the particle when it goes from a to b is $Fd = qEd$.

the force on that particle will be

$$\mathbf{F} = q\mathbf{E} \qquad \text{(magnitude)} \tag{17.16}$$
in direction of the field

And if the distance between points a and b is d, the work done going from point a to b will be

$$W_{a \to b} = qEd \tag{17.17}$$

(The work is positive because the particle is moved with the force.) The potential-energy difference between points b and a, from the definition, is thus

$$V_b - V_a = -qEd \tag{17.18}$$

If by convention we set the potential energy of the point b equal to zero (that is, we choose b as the fixed point from which all potential-energy differences are measured), then the potential energy at point a would be $V_a = qEd$. Setting $V_b = 0$ is often indicated by saying we "ground" point b (that is, connect it with a conductor to ground—earth— so that it has the same potential energy as earth, which by convention is zero); a "ground" is designated on electrical diagrams by the symbol shown here.

The electric potential energy of a charge in complicated situations is not always easy to determine (it can become a very elaborate calculation), but the principle is always the same. One calculates the work done in bringing the charge from one point to another, in the given distribution.[*]

Because of the similarity between the gravitational and Coulomb forces the electric potential energy of a charge subject to the force of another charge is very similar to the gravitational potential energy of a mass subject to the force of another mass. The gravitational potential energy of a point of mass m a distance R from a point of mass M (we recall) is

$$V(R) = \frac{-GMm}{R} \qquad \text{(gravitational potential energy)} \tag{17.19}$$

In the same way the electric potential energy of a negative point charge $-q$ a distance R from the point charge Q is

$$V(R) = \frac{-Q}{R} q \qquad \text{(electric potential energy)} \tag{17.20}$$

(For convenience the potential energy is defined with the fixed point infinitely far away.)

A negative charge brought from infinity to a distance R from a positive charge experiences an attractive force, just as two masses subject to the gravitational attraction do. Thus the potential energy is negative, as is the gravitational potential energy. A positive charge $+q$ brought from infinity to the same point, R, experiences a repulsive force; the potential

---

[*] One can also calculate the energy necessary to assemble the system of N charges at the points 1, 2, ..., N. It is minus the work done by the forces within the system on the particles as the system is being assembled and is a function of the points 1, 2, ..., N: $V(1, 2, ..., N)$.

energy of the positive charge is then of the same form but of opposite sign:

$$V(R) = \frac{+Q}{R}q \qquad (17.21)$$

It is therefore convenient to define an electric potential, distinct from the electric potential energy, as minus the work done by the forces of the system in bringing a *unit positive charge* from infinity to the given point. It is just the potential energy given above, divided by the charge of the particle being brought in. The electric potential of a point positive charge Q is then

$$\text{electric potential of a point positive charge } Q \text{ (called } \phi) = \frac{Q}{R} \qquad (17.22)$$

This has some of the same convenience that the field has: The field at a point multiplied by a charge gives the force on that charge. The electric potential at a point multiplied by the charge at that point gives the potential energy of the charge at that point.

It is electric-potential differences that we commonly encounter in household circuits and appliances. In the CGS unit system, the unit for electric potential is ergs per esu,[*] which is given the name statvolt:

$$\frac{\text{ergs}}{\text{esu}} = \frac{\text{esu}}{\text{cm}} = \text{statvolt} \qquad (17.23)$$

In the MKS system the unit for the electric potential is joules per coulomb, which bears the familiar name volt:

$$\frac{\text{joules}}{\text{coulomb}} = \frac{\text{coulomb}}{\text{meter}} = \text{volt} \qquad (17.24)$$

If an electron falls through a potential difference[†] of one volt, then the work done on it by the electric forces is 1.6×10^{-12} erg, which by definition is an electron volt of energy (work). The origin of this as a unit is due to the fact that one puts electric potential differences, related to household voltages (measured in volts), across accelerators. In such machines, one often accelerates particles like electrons; it becomes convenient to characterize the energy that these machines can impart to the particles in terms of the potential difference across the plates of the machine, multiplied by the charge that is transported. In spite of its mixed origins, this unit is a very useful one; the reason is that it is a combination of a practical unit (voltage) and the electronic charge, and at the same time its magnitude is just right for atomic energies. As we shall see later, the kind of energies involved in atomic reactions are of the order of

[*] The esu of charge is sometimes called the statcoulomb; thus ergs/statcoulomb = statvolt. (See also Appendices, pp. 493-495.)

[†] Electric-potential differences are what are commonly designated as voltage. Thus a 12-volt battery provides an electric potential difference of 12 volts between one terminal and the other. (The red terminal is usually the one at the higher potential.) It thus has become conventional to designate electric potential difference (in volts) by the letter V, which produces a certain confusion with the symbol for potential energy, a related but different concept. (The same kind of confusion occurs elsewhere. When I read a book, am I doing something in the present or did I do it in the past?)

10^{-12} erg or 10^{-19} joule. It is awkward to keep repeating numbers such as 3.2×10^{-19} joule; 2 electron volts is easier to say.

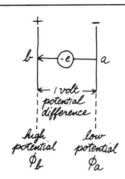

$$-e = \text{charge on electron}$$
$$= -1.6 \times 10^{-19} \text{ coulomb}$$

potential energy of the electron at b is: $V_b = -e\phi_b$ (17.25)

potential energy of the electron at a is: $V_a = -e\phi_a$ (17.26)

minus work done on electron going from $a \to b = V_b - V_a$
$$= -e(\phi_b - \phi_a) = -e \times (1 \text{ volt})$$
$$(17.27)$$

or

$$W_{a \to b} = e \times (1 \text{ volt}) = 1.6 \times 10^{-19} \text{ coulomb} \times 1 \text{ joule/coulomb}$$
$$= 1.6 \times 10^{-19} \text{ joule} = 1.6 \times 10^{-12} \text{ erg} = 1 \text{ electron volt}$$
$$(17.28)$$

Example 1. There is a potential difference of 12 volts between two charged plates that are 0.03 meter apart. What is the strength of the field between them? What force will be exerted on a proton in this field?

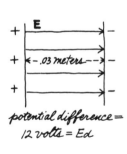

potential difference =
12 volts = Ed

$$E = \frac{12 \text{ volts}}{0.03 \text{ m}} = 400 \frac{\text{volts}}{\text{m}} \quad \text{(magnitude)} \quad (17.29)$$

The force on a proton in a 400 volt/m field is given by

$$F = qE \quad \text{(magnitude)}$$

$$= (1.6 \times 10^{-19} \text{ coulomb}) \times 400 \frac{\text{volt}}{\text{m}} = 6.4 \times 10^{-17} \frac{\text{coulomb joules}}{\text{coulomb-m}}$$

$$= 6.40 \times 10^{-17} \text{ newton} \quad (17.30)$$

Example 2. A machine accelerates protons by releasing them at rest from a plate that has an electric potential of 1,000,000 volts toward a second plate that is at zero potential. What are the energy and speed of a proton when it reaches the second plate?

The proton travels through a potential difference of 1,000,000 volts. The work done on it therefore is $e \times 10^6$ volts $= 10^6$ electron volts, abbreviated 1 MeV.

$$1 \text{ MeV} = 10^6 \text{ electron volts} = 1.6 \times 10^{-6} \text{ erg} \quad (17.31)$$

To find its speed,

$$E = \tfrac{1}{2}mv^2 = 1.6 \times 10^{-6} \text{ erg} \qquad (17.32)$$

$$v = \sqrt{\frac{2(1.6 \times 10^{-6} \text{ erg})}{m}} \qquad (17.33)$$

The mass of the proton is 1.67×10^{-24} g. Therefore,

$$v = \sqrt{\frac{2(1.6 \times 10^{-6} \text{ erg})}{1.67 \times 10^{-24} \text{ g}}} = 1.4 \times 10^9 \frac{\text{cm}}{\text{sec}} \qquad (17.34)$$

Example 3. There is a potential difference of 120 volts across the terminals of a wire 10 m long. If we assume that the electric field is constant in the wire and directed always in the direction of the wire,* what is its magnitude? What force does an electron feel due to this field; what would be its acceleration?

electric potential difference $= Ed$

$$E = \frac{120 \text{ volts}}{10 \text{ m}} = 12 \frac{\text{volts}}{\text{m}} \qquad (17.35)$$

The force on an electron in magnitude is

$$F = eE = 1.6 \times 10^{-19} \text{ coulomb} \times 12 \frac{\text{volts}}{\text{m}} \simeq$$

$$1.9 \times 10^{-18} \text{ newton} = 1.9 \times 10^{-13} \text{ dyne} \qquad (17.36)$$

Its acceleration is then†

$$a = \frac{F}{m} \simeq \frac{1.9 \times 10^{-13} \text{ dyne}}{0.9 \times 10^{-27} \text{ g}} \simeq 2.1 \times 10^{14} \text{ cm/sec}^2$$

$$\text{or about } 2 \times 10^{11} \text{ g} \qquad (17.37)$$

---

* The ordinary wires used in household circuits have just this property. The electric field produced within them by the potential difference is very close to constant and is directed along the wire.
† However, due to collisions, the electrons do not reach very high drift velocities.

18

MAGNETIC FORCES: CHARGES IN MOTION

ELECTRIC CURRENTS

Any motion of an unbalanced charge by definition constitutes an electric current—a charged ball moved at the end of a stick, a falling charged water droplet, a moving charged belt, or, as is the case in an ordinary conductor, the motion of the electrons in the metal. If we move an ordinary neutral piece of matter, we move a large number of charged particles (as many positive as negative charges), which, from the definition to come, results in no total current.

The current flowing in a wire is defined as the amount of charge passing a given area of the wire in a unit time: in electrostatic units, esu per second:

$$\text{current} \sim \frac{\text{esu}}{\text{sec}} \sim \text{statamperes} \qquad (18.1)$$

For a wire with a constant cross section, a given number of electrons pass through the cross section per second:

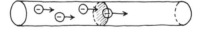

The current then is defined as the quantity of charge that passes the cross-sectional area per unit of time:

$$\text{current (denoted by } I) = \frac{\text{charge crossing area}}{\text{time}} \sim \left(\frac{\text{esu}}{\text{sec}}\right) \quad (18.2)$$

This is not the familiar unit. (The usual household appliance draws current measured in amperes, which is the MKS system.) In the MKS system —the system that measures charge in coulombs, potential in volts, and current in amperes—the unit of current is given in terms of coulombs per second:

$$1 \frac{\text{coulomb}}{\text{sec}} = 1 \text{ amp} \qquad (18.3)$$

The relation between the CGS current unit and the ampere is

$$3 \times 10^9 \text{ statamp} = 3 \times 10^9 \frac{\text{esu}}{\text{sec}} = \frac{1 \text{ coulomb}}{\text{sec}} = 1 \text{ amp} \qquad (18.4)$$

What is of significance is that when we say that an electrical appliance draws 1 amp of current, we mean that through any cross section of the wires that lead into and out of the appliance there flows 1 coulomb of charge, every second (or if we wish: 3×10^9 esu of charge every second). Fortunately, the second remains invariant.

We say now that what actually flows in a metal are the electrons, so that when we speak of 1 amp of current, what we mean is that 1 coulomb of negative charge (0.62×10^{19} electrons) passes a certain area in 1 sec. It was Benjamin Franklin who established the convention that current flow in a wire is due to the motion of positive electricity. Thus it is conventional to show the direction of the current as the direction of positive charge motion, opposite to the direction in which the actual flow of the electrons occurs; this is sometimes a little confusing but is not known to have affected his popularity in Paris:

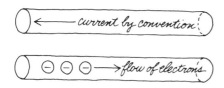

If we apply a potential difference across a material, the electrons feel a force. In those materials, called insulators, the electrons are held firmly to the atoms of the substance; there are no "free" electrons that can move in the direction of the force on them and no current flows. In other materials, such as metals, there are electrons not held tightly by the atoms, and these electrons can move under the influence of the field imposed in the material. Whenever there is a difference in potential between two points on a conductor, whether that conductor is a long copper wire or just a small block of steel, a current will flow between those two points.

In addition to household electrical outlets, which did not exist in Franklin's time, one can use lightning discharges, discharges from static accumulators of electricity, or batteries to produce potential differences. A battery is a device that by "chemical action" maintains an electric potential difference across two outlets:

In the early part of the nineteenth century, Georg Simon Ohm investigated the relation between the potential difference across a conductor and the amount of current that flows. He proposed that the current which flows in a given conductor at a given temperature is directly proportional to the potential difference applied across it. The proportionality constant is called the *resistance, R*; this is usually written

$$\text{potential difference} = \text{current} \times \text{resistance}$$
$$V = I \times R \tag{18.5}$$

(V is the conventional symbol to denote what in our notation should be $\phi_+ - \phi_-$.) In the MKS system the unit of resistance is called the ohm— the resistance of a wire that requires 1 volt of potential difference across it for each ampere of current that flows. This relation, which is an expression for the frictional force felt by electrons, is remarkably accurate for many materials.

Example. An electric toaster draws 10 amp of current. The household voltage is 110 volts. What is the resistance of the toaster?

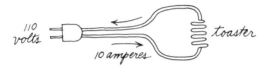

$$\text{household voltage} = \text{potential difference} = \text{current} \times \text{resistance}$$
$$\text{resistance} = \frac{110 \text{ volts}}{10 \text{ amp}} = 11 \text{ ohms} \tag{18.6}$$

If this toaster is taken to Europe and accidentally plugged into a 220-volt outlet, how much current will flow?

$$\text{current} = \frac{\text{voltage}}{\text{resistance}} = \frac{220}{11 \text{ ohms}} = 20 \text{ amp} \tag{18.7}$$

which will result in very rapid toast.

MAGNETIC FORCES

> The first experiments on the subject which I undertake to illustrate were set on foot in the classes for electricity, galvanism, and magnetism, which were held by me in the winter just past.[1]

In that winter just past, the winter of 1819–1820, electricity meant the forces due to stationary charges—Coulomb's law. Galvanism referred to those phenomena produced by moving charges or currents and magnetism dealt with those mysterious objects: loadstones and compass needles, in the earth's magnetic field. All three were thought to be separate subjects; although it seemed that there must somehow be a connection, the connection had never been found. In that winter, Oersted allowed a galvanic current to flow through a wire parallel to a small magnetic needle and as he says:

> The magnetic needle will be moved, and indeed, under that part of the joining wire which receives electricity most immediately from the negative end of the galvanic apparatus,* will decline toward the west.[2]

The forces between electrically charged particles could not have been more Newtonian to this point. Not only does Coulomb's force obey the third law, but its form is identical to that of the gravitational force. If the matter ended here, it might be hardly worth putting more than a footnote after the gravitational force to say that in some cases a similar force is exerted between what are called *charged particles*. The magnitude differs; there is the possibility of repulsion and attraction, but everything else is identical. However, the matter does not end here. Upon further exploration, the electrical force begins to display a subtlety and a variety that will stretch the Newtonian picture to its limits and finally force us to go beyond it.

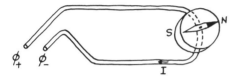

Oersted's discovery signaled the beginning of frenzied activity; within a decade Ampère and Faraday would have worked out the theory of the magnetic effects of currents. Not only had Oersted established that a moving charge or a current had an effect on a magnetic needle, but the effect had curious directional properties: the magnetic needle lined up in a direction perpendicular to the motion of the current:

And in such a way as to form a circular pattern in a plane perpendicular to the wire. This can be illustrated using a simple technique—a child's pastime for rainy afternoons. If we sprinkle small iron filings (each of which behaves, for these purposes, as a small magnetic needle) on a piece of paper, the patterns for various current arrangements become visibly displayed (Plate 18.1).

Among the more intriguing aspects of this discovery, and one reason it took so long to come upon, is the fact that a stationary charge will have no effect on a magnetic needle. For the effect that Oersted observed to occur, it is necessary that the charge be moving. Thus for the first time we have a force that seems to depend on the motion of the bodies that produce it.

Less than a year later (October 2, 1820) Ampère published, in the *Annals of Chemistry and Physics,* the work that established that two current-carrying wires exert a force on each other. He found that if currents flow in two wires in the same direction, the wires attract one another, while if the currents flow in opposite directions, they repel. These new forces seem to be fundamentally different from electrical forces, since they do not seem to depend on any unbalanced charge that is on the wires. If one has a very long wire carrying a current, as in Fig. 18.2, then if

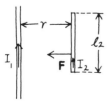

F I G. *18.2.* A very long wire carrying the current I_1 attracts the wire of length ℓ_2 carrying the current I_2.

---

* A device that produces electric potential differences through chemical action, for example, a battery.

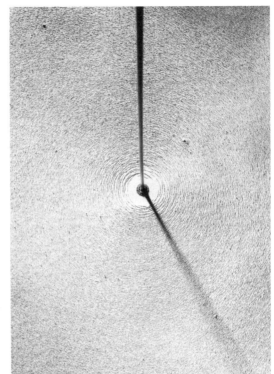
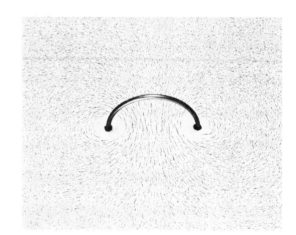

PLATE *18.1.* Magnetic field patterns due to: *upper left:* a bar magnet; *upper right:* current flowing in a long straight wire; *center:* current flowing in a loop of wire; *bottom:* current flowing in a solenoid. (Three photographs: Physical Science Study Committee, *Physics,* D. C. Heath, Boston, 1967. Bottom figure: after R. Kronig, ed., *Textbook of Physics,* Pergamon Press, New York, 1959)

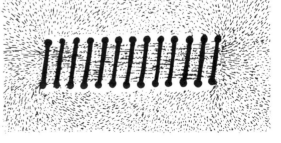

the nearby wire is parallel to the first wire, the force will be such as to attract it if the current is running in the same direction and will be such as to repel it if the current is running in the opposite direction. The magnitude of the force depends on the distance by which the two wires are separated, the amount of current and the length of the second wire; in the CGS system it is given by

$$F = \frac{2}{c^2} \frac{I_1 I_2 \ell_2}{r} \qquad \text{(magnitude)} \qquad (18.8)$$

I_1 is the current in the first wire, I_2 the current in the second wire, ℓ_2 the length of the second wire, and r the distance between them. The letter c which occurs in the denominator is a constant equal to

$$c = 2.998 \cdots \times 10^{10} \frac{\text{cm}}{\text{sec}} \qquad (18.9)$$

It has the dimensions of speed, and we now know its magnitude is identical with that of the speed of light.[*]

To give an idea of the order of magnitude of the force between such wires, imagine that the second wire is 1 cm long and 1 cm away from the first, and that both wires carry currents of 10 amp. (To convert the amperes into CGS current units refer to Table 18.1 (10 amp = c statamp). Putting these numbers into Eq. (18.8) we obtain the force

$$F = \frac{2 \cdot c \cdot c \cdot 1}{c^2 \cdot 1} = 2 \text{ dynes} \qquad (18.10)$$

Two dynes is not a very large force (about four millionths of 1 pound), but it can easily be measured. [In contrast, a charge imbalance of 1 electron for every 10^6 atoms would produce a force (for wires 0.1 cm in diameter) of about 10^8 dynes (about 0.1 ton) per centimeter.]

We might expect from this that a current should exert a force on a moving charge. This is the case. It is the moving charges that constitute the current in the wire on which the force is being exerted. This communicates itself as a force to the entire wire. The force a current-carrying wire exerts on a beam of moving charged particles, such as electrons, can be very directly demonstrated by a device such as an electron gun, producing rapidly moving electrons. Such a beam of electrons can be seen visibly to be deflected by a current-carrying wire nearby:

---

[*] The answer to the question "What is the speed of light doing in the denominator of this equation?" (and this is not the only place that it appears; it occurs, as we shall see, in many of the equations that relate electric and magnetic phenomena) comes later, with the electromagnetic theory of light. As the relation was originally written, the speed of light did not appear. One measured the current in the two wires and the distance between them and found that the magnitude of the force satisfied the relation

$$F = (\text{constant}) \frac{I_1 I_2 \ell_2}{r}$$

The constant depends on what system of units is being used. In the CGS system (also not originally used in this connection) the constant would be about

$$\frac{2}{9 \times 10^{20}} \frac{\text{sec}^2}{\text{cm}^2}$$

It is later (as we shall see) that this constant would be identified with 2/(speed of light).

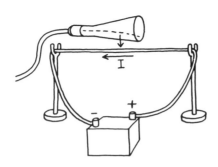

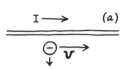

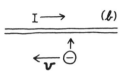

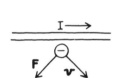

The qualitative properties of this force seem complicated and quite extraordinary. Consider the current-carrying wire shown. An electron (a) moving in the same direction as the current is pushed away from the wire, while an electron (b) moving in the direction opposite to the current is pulled toward the wire. For electrons moving in an arbitrary direction, with respect to the wire, the force changes its direction; we observe, however, that it is always perpendicular to the velocity of the electron—that the magnitude of the force increases as the electron moves more rapidly and decreases when the electron is farther away from the wire.

We therefore are dealing with a force that not only depends on the position of the electron, but also on its speed and its direction of motion. Compared to those considered before, this is a force of great complexity; to study it further we find it convenient to introduce the concept of the magnetic field.

THE FORCE ON A MOVING CHARGE

Just as the electric field, **E**, the magnetic field, which we denote by **B**, is a vector defined at every point in space. Once the electric field is specified, the magnitude and direction of a force on an electric charge at any point in space is immediately given by multiplying the electric field at that point by the charge placed there. The magnetic field, although somewhat more complex, serves the same purpose. Given a large number of currents, moving charges, and so on, each of which produces its own force on any other current or moving charge that is introduced into the environment, we are able to replace all these by a single vector magnetic field that is sufficient to give the force on any charged, moving particle. In doing this we assume that the force produced by one current-carrying wire can be added as a vector to the force produced by another current-carrying wire and that the total force is the sum of the two forces, in the same way as for electrostatic or gravitational forces. In the end it is possible to state what the total force on a moving charge will be in terms of two fields: an electric and a magnetic field.

This is done as follows: Consider first a charged particle with a charge, q, and suppose that it is at rest. Then, from our knowledge of electrostatics, we know that any electric force acting on this particle can be expressed as

$$\mathbf{F} = q\mathbf{E} \tag{18.11}$$

The meaning of this equation is that the force on the charged particle is equal in magnitude and sense to the charge times the magnitude of the electric field and is in the direction of the electric field. The force on a moving charge is in the direction opposite to **E**. If the charge is at rest, whether or not there are any currents flowing nearby will not matter, because, according to Oersted, Ampère, and to all of our experimental observations, currents or magnets produce no effect on charges that are not in motion.

Now imagine that the charged particle begins to move. If there are currents or magnets nearby, we find that the force on it is no longer q multiplied by the electric field; further, as the particle moves more and more rapidly, the force begins to differ more and more from the force on the static charge. We assert that there is an additional force acting on the

charged particle which is proportional to its speed and which is due to the existence of magnets or currents in the environment. This additional force can be added to the electric force to produce the total force; the entire situation is summarized in

$$\mathbf{F} = q\mathbf{E} + \frac{q}{c}\,\mathbf{v} \times \mathbf{B} \qquad (\text{the Lorentz force}) \qquad (18.12)$$

The total force on the moving charge at some point is that force which is produced by the electric field, plus another force due to currents and magnets which can be characterized completely if one knows the magnetic field at that point. The magnetic force is proportional to the charge on the particle, to the speed with which the particle moves, and is related in a somewhat intricate matter to the directions of motion and the magnetic field. Again and for no reason we can see at the moment, the speed of light has mysteriously appeared in relating electric and magnetic forces. For magnetic fields, the more commonly used unit is the CGS unit known as the *gauss* $\left(\text{gauss} \sim \dfrac{\text{esu}}{\text{cm}^2}\right)$;[*] in the MKS system the unit is the *tesla*. The following relation holds between the two:

$$10^4\ \text{gauss} = 1\ \text{tesla} \qquad (18.13)$$

The total force on a moving charged particle due to both electric and magnetic fields (known as the Lorentz force) is given in dynes if the speed is in cm/sec, the magnetic field in gauss, the charge in esu, and the electric field in esu/cm². Where a relation between currents and magnetic fields is given, the current must be given in statamps (the CGS current unit).

The cross (**X**) represents a new kind of product, a product of two vectors (see Appendices, p. 492). It is called by various names, the cross product, the outer product, or the vector product of the two vectors. It is

[*] From Eq. (18.12) we can conclude that the units of magnetic field must be related to those of electric field by

$$q\mathbf{E} \sim \frac{q}{c}\,\mathbf{v} \times \mathbf{B}$$

which implies that E has the same units as B or $B \sim \text{esu/cm}^2$ (CGS).

by definition another vector, and therefore its magnitude, sense, and direction must be specified. The new vector is perpendicular to the plane formed from the original two vectors. The magnitude of the new vector is related in a fairly simple way to the two original vectors when they are perpendicular to one another. Since this is as much as we need for the moment, we shall restrict our definition to this case.

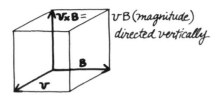

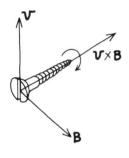

The vector $\mathbf{v} \times \mathbf{B}$ is defined as the vector perpendicular to $\mathbf{v}$ and $\mathbf{B}$ as above whose magnitude is vB.

The sense of the vector is by convention (in a right-handed system) in the direction of advance of a right-handed screw when turned, $\mathbf{v}$ going into $\mathbf{B}$ (see Appendices, p. 492), as shown.

All of this may be summarized as follows: It is possible to define an electric field $\mathbf{E}$ and a magnetic field $\mathbf{B}$, both vectors defined at every point of space such that the force on a moving charged particle is given by the Lorentz equation. The electromagnetic force is far more subtle and complicated than the forces with which we have dealt before, since the magnitude and the direction depend not only on the position of the particle but also on its velocity. But from the point of view of dynamics, since we need to know only the force acting on a particle in order to determine its acceleration, we are stating that the motion of the particle can be found given $\mathbf{E}$ and $\mathbf{B}$.

THE MAGNETIC FIELD

The Lorentz equation implies that to determine the electromagnetic force on a charged particle it is necessary to determine both the electric and the magnetic field at all points in space. The subject of electrostatics concerns itself with the determination of the electric field given the distribution of charges. The actual construction of an electric field from given charges, except in the simplest cases, might involve very elaborate calculation; we have sampled only the simplest rules. However, the principle is straightforward. One calculates the force or field due to each charge using Coulomb's law. In magnetostatics one determines magnetic fields from any given distribution of magnets or currents. The fundamental relation, which is the equivalent of Coulomb's law in electrostatics, is Ampère's law, which relates the magnetic field to the current distributions.

We give this relation between the magnetic field and a small element of current, in a form due to Biot and Savart: The magnetic field produced by a small wire of length ℓ carrying the current $\mathbf{I}$ is

$$\mathbf{B} = \frac{\ell}{c} \frac{\mathbf{I} \times \mathbf{r}}{r^3} \qquad \text{CGS system} \qquad (18.14)$$

where the symbols ℓ, **r**, and **I** are as shown:

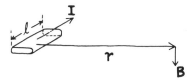

The magnetic field vector is also given as the cross product of two vectors. It is perpendicular to the direction of the current element, as well as to the direction of the position vector from the current element to the point at which the field is being evaluated. The relation is somewhat more complex than we will need. The current element has a direction that must be related to the direction of the magnetic field. If one has many such current elements, made up to form a wire, for example, the total field at a point is the sum of the fields produced by each of the current elements.

In the development of the subject of magnetostatics, the Biot-Savart law, or some equivalent relation, is used to compute the magnetic fields produced by any distribution of currents in the same way that Coulomb's law is used in electrostatics to compute the electric field given any charge distribution. The logic is straightforward, but the actual application of the Biot-Savart law to current distributions is not easy. We shall write down the magnetic fields that are produced by some normal current distributions. Some of the qualitative features, such as the general directions of the magnetic fields, can be guessed easily; but the magnitudes are not too easy to get, and we shall not go through the steps that are necessary to deduce them.

One of the simplest current distributions is that produced by the current flowing in an infinitely long straight wire. The field close to the middle of a long wire is not very different from the field of an infinite wire. Iron filings placed on a sheet around a long wire reveal the pattern of the magnetic field. The magnetic field circles around the wire in planes perpendicular to the direction of the wire.[*] In magnitude the field at a distance r from the wire is equal to

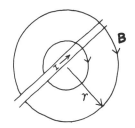

$$B = \frac{2}{c}\frac{I}{r} \qquad (18.15)$$

where I is the current flowing in the wire, c the speed of light, and r the distance from the wire to the point at which the field is being evaluated. Looking down from above, the field pattern looks as shown here, if the current is flowing out of the page. The direction of the magnetic field is perpendicular to the direction of the current and to the direction of the vector from the current element, to the point at which the magnetic field is being evaluated. Therefore, it has no choice other than to circle around the wire, because a curve that is always perpendicular to a radius must describe a circle.

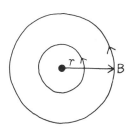

[*] The direction of the field, with the conventions we are using, is such that the current moves in the direction of motion of a right-hand screw turned with the direction of the field.

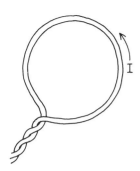

Plate 18.1 showed the magnetic field produced by a loop of wire such as this one. We can get an idea of how this field is produced by considering the field produced by each segment of the loop and by regarding each such segment as part of a long wire:

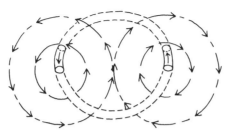

In the center of the loop both fields have components upward, but their components to the side cancel each other (the superposition principle is satisfied by magnetic fields as well as electric fields, because they are both vectors). Thus the field at the center of the coil is straight up; all together one gets the following pattern:

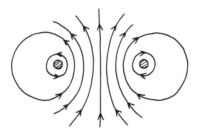

A *solenoid* consists of a large number of current loops that are some-

times wound on a spool. By adding vectorially the fields produced by each current loop, the field in the solenoid can be found:

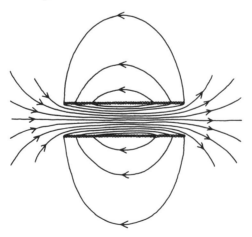

The magnetic field produced by a solenoid, relatively uniform in the center, is often used in experimental work, where a constant, controlled magnetic field is desired. The magnitude of the field at the center of an infinitely long solenoid is given by

$$B = \frac{4\pi IN}{c} \qquad (18.19)$$

where I is the current in esu/sec, N the number of loops per centimeter, and c the ubiquitous speed of light. The field at the center of a solenoid coil which is 10 cm long and 2 cm in diameter differs by only 2 percent from the infinite solenoid value. If the same solenoid were 50 cm long, the error would be less than one tenth of 1 percent.

A MOMENTARY SUMMARY

The sum total of what has been said is that such things as charges exist, and that charges produce forces on other charges, which depend on the position of the charges as well as, in a somewhat complicated way, on their relative motions. We have introduced two auxiliary concepts to make the calculations of the forces simpler, the electric and magnetic fields, and have stated that we can calculate electric fields from charge distributions and magnetic fields from current distributions. Then the force on any charged particle is given by the expression

$$\mathbf{F} = q\mathbf{E} + \frac{q}{c}\,\mathbf{v}\,\mathbf{X}\,\mathbf{B} \qquad (18.20)$$

Although the Lorentz force law is more complicated than the Newtonian forces that we considered previously, depending on the velocity of a particle as well as its position, the entire development has been thoroughly Newtonian so far in character. This is most clearly illustrated perhaps by the methods by which the forces on charged particles are actually determined. The presence of electric forces in the first place is revealed by the fact that small objects that are properly rubbed show a tendency to move toward other objects, properly rubbed. We see motion, and from the existence of the motion we infer the force.

When Coulomb measured the magnitude of the force between two charged particles he did so by using a balance, balancing an electric force against a mechanical force whose magnitude he already knew. He then used the law of inertia: When a body is in uniform motion or at rest, the sum of the forces acting on it is equal to zero. Thus the presumption is that the electric force is a force in the same sense as a mechanical force and can be combined with mechanical forces as vectors.

The total force on a body is then considered to be the electromagnetic plus the mechanical force and is equal to the mass times the acceleration. The classification of electrical forces, just as that of other forces, results from the decision to view the uniform motion of inertia as the natural motion of the theory. The velocity dependence is secondary. When we consider the motion of charged bodies in the presence of other charged bodies or currents, what we do in the end is to calculate the orbits of these charged bodies, using the second law of motion, after having calculated the Lorentz force from the electric and magnetic fields. This has been done before for a charged particle moving through a uniform electric field.

We now consider a charged particle moving through a uniform magnetic field. This is extremely useful in itself and is an illustration of how one deals with the velocity-dependent force, a force whose direction and magnitude depend on the velocity and the magnetic field.

MOTION OF CHARGED PARTICLES IN A UNIFORM MAGNETIC FIELD

Consider a particle of mass m, positive charge q, moving in a uniform magnetic field B. By a uniform magnetic field we mean a field whose direction and magnitude are constant throughout the volume in which the particle moves. There is a subtle simplification here, which we might mention. We do not inquire what is the current distribution that produces this magnetic field. (The field might to a reasonable approximation be produced in the interior of a solenoid.) We assert merely that by some arrangement of currents we have managed to produce it. This is one way in which the field simplifies the treatment of problems. It essentially divides them into two steps:

The first involves the calculation of the field from some distribution of currents.

The second involves the calculation of the force and then the motion of the particle given the fields. We assume that by some means or another, we have produced a region in space where the magnetic field is uniform. We assume further that no electric field exists in this region (there are no unbalanced charges about). The force that the charged particle feels in this uniform magnetic field is therefore given by the second term of the Lorentz force law:

$$\mathbf{F} = \frac{q}{c} \mathbf{v} \times \mathbf{B} \qquad (18.21)$$

The magnitude of the vector $\mathbf{v} \times \mathbf{B}$ if $\mathbf{v}$ is not perpendicular to $\mathbf{B}$ is $vB \sin \theta$, where θ is the angle between $\mathbf{v}$ and $\mathbf{B}$:

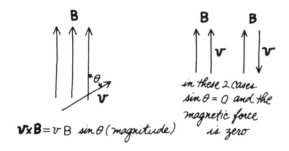

The Lorentz force is zero if the sine of the angle between the particle's velocity and the magnetic field is zero. This occurs when the particle is moving parallel or antiparallel to the direction of the field. For a given speed and a given magnetic field, the force reaches its maximum when $\sin \theta$ is equal to 1 or when $\mathbf{v}$ is perpendicular to $\mathbf{B}$ as here.

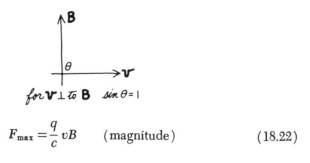

for $\mathbf{v} \perp \text{to } \mathbf{B}$ $\sin \theta = 1$

$$F_{\text{max}} = \frac{q}{c} vB \qquad (\text{magnitude}) \qquad\qquad (18.22)$$

We lose nothing and gain in simplicity by considering the case in which the motion of the charged particle is perpendicular to the field

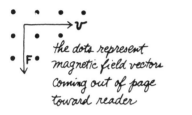

the dots represent magnetic field vectors coming out of page toward reader

FIG. *18.3.* Because of the three dimensions involved, matters are simplified a bit by arranging things so that the magnetic field comes up toward us out of the paper, and so that the paper itself is in the plane of the velocity and the force.

(Fig. 18.3). For this situation the force due to the magnetic field on the moving positively° charged particle is

$$\mathbf{F} = \frac{q}{c} vB \qquad (\text{magnitude}) \qquad\qquad (18.23)$$

and in direction always perpendicular to the velocity as above. Without doing any algebra, we can immediately draw some remarkable qualitative conclusions about the effect of this force on the charged particle.

First consider a consequence of something we have repeated several times: The force is perpendicular to the velocity. And it remains perpendicular to the velocity, even as the velocity continually changes direction. From the definition of work we can infer that such a force does no work on the particle; thus the kinetic energy of the particle does not change. And, therefore, we can conclude that in the absence of other forces the speed of the particle remains constant in a magnetic field: remarkable but true. However, a force does act and does produce an acceleration. The acceleration changes the direction of the motion of the charged particle but not the speed. Under these circumstances the charged particle moves in a circle. We know that the acceleration for motion of a particle in a circle takes a particularly simple form, $a = v^2/R$. We then obtain

$$\mathbf{F} = m\mathbf{a}$$

$$F = \frac{q}{c} vB \qquad (\text{magnitude}) \qquad\qquad (18.24)$$

° If the charge is negative, the magnitude of the force is the same, but it points in the opposite direction.

$$a = \frac{v^2}{R} \qquad \text{(magnitude)}$$

Therefore,

$$\frac{q}{c} vB = m \frac{v^2}{R} \tag{18.25}$$

or

$$mv = \frac{q}{c} BR \tag{18.26}$$

From this we see that if the magnetic field is known (it can either be calculated from a knowledge of the external current distribution producing the field or measured by various devices—usually, both are done), if the charge of the incoming particle is known, and the radius of curvature of the circular trajectory can be measured, we can determine the momentum of the particle. If the speed of the particle is also known, then we can determine the mass of the incoming particle.

PLATE *18.2.* Electron track in liquid-hydrogen bubble chamber. Track is curved due to uniform magnetic field pointing out of the page. Radius of curvature decreases as electron is slowed down by liquid hydrogen. (Why?) (Courtesy of the Lawrence Radiation Laboratory, University of California, Berkeley)

Example. A particle moving through a uniform magnetic field of magnitude $B = 10$ gauss with the known speed $v = 10^9$ cm/sec is seen to move in a circle of radius $R = 5.65$ cm as below. What is the ratio of its charge to its mass q/m?

From Eq. (18.26),

$$\frac{q}{m} = \frac{cv}{BR} = \frac{3 \times 10^{10} \times 10^{9}}{10 \times 5.65}$$

$$= 5.3 \times 10^{17} \frac{\text{esu}}{\text{g}}$$

(18.27)

a number of some interest, because, as we shall see later, this is the ratio, first determined by J. J. Thomson, of charge to mass of the electron.

Plate 18.2 is a photograph of a charged particle in a magnetic field. The visible track is produced by the bubbles made in the liquid hydrogen due to the passage of the charged particle. If the particle had the opposite charge, and everything else remained the same, the curvature of the track would be reversed, because the force would be in the opposite direction. Such tracks are seen over and over again in photographs of elementary particles taken in bubble or cloud chambers (see Plates 18.3 and 18.4).

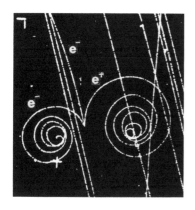

PLATE *18.3.* Curved tracks of electrons in a magnetic field that is perpendicular to the page and points toward the reader. A certain process suddenly produces two negatively charged electrons and a positron (which is identical with an electron except that it is positively charged). A white track is a trail of bubbles left behind after the passage of a charged particle. The electron e^- which moves off to the left follows a circular path in the magnetic field, but slows down as it loses energy in collisions with the atoms of the liquid through which it is passing. As its velocity decreases, the radius of its circular path decreases, and so it spirals inward. The positron e^+ which moves off to the right has the opposite charge and its path is curved in the opposite direction. The electron e^- in the center has a very high initial velocity, and so the radius of its circular path is very large. (Courtesy of the Lawrence Radiation Laboratory, University of California, Berkeley)

MAGNETS

Great [said Gilbert] has ever been the fame of the loadstone . . . in the writings of the learned: many philosophers cite the loadstone and also amber whenever, in explaining mysteries, their minds become obfuscated and reason can no farther go. Over-inquisitive theologians, too, seek to light up God's mysteries and things beyond man's understanding by means of the loadstone as a sort of Delphic sword. . . .[3]

Should we, having spoken of magnetic forces, not attempt to light up the mystery of the loadstone? A magnetized sliver of metal or a compass needle produces a magnetic field pattern similar to that produced by a bar magnet (displayed using iron filings), as shown in Plate 18.5.

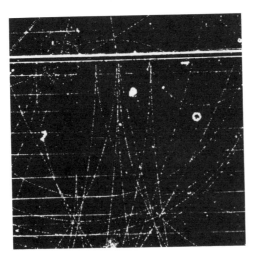

PLATE *18.4.* Cloud-chamber tracks of three electron-positron pairs in a magnetic field. Three gamma-ray photons entering at the top materialize into pairs within a lead sheet. The coiled tracks are due to low-energy photoelectrons ejected from the lead. (Courtesy of the Lawrence Radiation Laboratory, University of California, Berkeley)

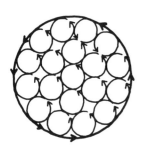

It was first proposed by Ampère that the fields of such permanent magnets were produced by currents, which, although undetectable, continually flow inside the magnetized object, to produce the effect of a solenoid. Thus, from Ampère's point of view, the behavior of a loadstone or a compass needle, the palpable action at a distance which one feels when one holds a magnet, is produced by an interaction between two currents, precisely the same interaction that Ampère himself had observed in his laboratory. And it is the viewpoint today that all permanent magnets*

TABLE *18.1*

	CGS	MKS	Relation
Electric potential difference	statvolt $\left(\dfrac{\text{esu}}{\text{cm}}\right)$	volt	There are 1/299.8 statvolts in 1 volt
Current	statamp $\left(\dfrac{\text{esu}}{\text{sec}}\right)$	ampere	There are 2.998×10^9 statamps in 1 ampere
Magnetic field	gauss	tesla	There are 10^4 gauss in 1 tesla

are produced by currents that flow within the material. These currents may be atomic or molecular in their origins, possibly due to orbital motions of electrons, but they are fundamentally currents.

The magnetic fields of a solenoid and of a magnetized rod are com-

---

* There exist materials (in particular, iron alloys, steel, or alloys such as aluminum-nickel-cobalt: alnico) that become permanently magnetized when placed in a magnetic field. Thus such a needle, initially unmagnetized, might be used to detect the presence of a magnetic field in a region by testing to see if the needle had become magnetized when left in that region.

PLATE *18.5*. Magnetic field pattern produced by bar magnet. (Physical Science Study Committee, *Physics,* D. C. Heath, Boston, 1967)

pared below in Fig. 18.4. The north and south poles of a magnet are by convention placed as shown. Thus if the earth's magnetic field was interpreted as due to a current flow around a solenoidal core, the direction of flow would be as shown in Fig. 18.5. Unfortunately, the *south pole* of the earth's magnet would be in the arctic (north), because the north (seeking) pole of a compass needle must be attracted to it.

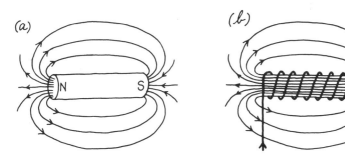

FIG. *18.4.* Field lines due to (*a*) magnetized rod; (*b*) solenoid of identical shape. (After J. Orear, *Fundamental Physics,* Wiley, New York, 1961)

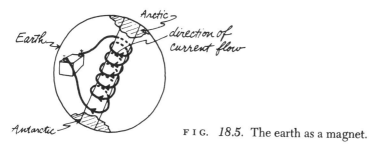

FIG. *18.5.* The earth as a magnet.

19

INDUCTION FORCES: CHARGES AND CHANGING CURRENTS

MICHAEL FARADAY: INDUCTION OF ELECTRIC CURRENTS

"He smells the truth," said Kuhlrausch. Michael Faraday, son of a black-smith, trained as a bookbinder, by one of those happy chances of history had himself discovered by Sir Humphrey Davy; some say Davy's greatest discovery. (Faraday attended a series of Davy's lectures in London, took careful notes, amplified and bound them himself into a book, then sent the book to Davy along with a request for a job in Davy's laboratory—any job. When he would not be discouraged, Davy hired him. Thus began one of the most illustrious careers in the history of science.)

The bookbinder, not having studied much mathematics, by necessity thought in pictures, which made mathematicians extremely uneasy. As matters developed, he outdid with his poor pictures all the mathematicians of that generation. He wrote, "It is quite comfortable to me to find that experiment need not quail before mathematics, but is quite competent to rival it in discovery." One can say of him that he looked nature in the face, with a mind so extraordinarily alert and open that those things he observed first in his laboratory, which may have been accidental, never had to happen more than once.

By the end of the 1820s, the work done by Coulomb, Oersted, and Ampère had established that stationary charges produced electric fields, therefore forces, on other stationary charges, and that moving charges produced magnetic fields, thus forces, on other moving charges. But, as Faraday put it:

> . . . it appeared very extraordinary, that as every electric current was accompanied by a corresponding intensity of magnetic action at right angles to the current, good conductors of electricity, when placed within the sphere of this action, should not have any current induced through them, or some sensible effect produced equivalent in force to such a current.[1]

It is his feeling that there is a symmetry lacking that prompts his surprise. Since electricity produces magnetism he expects that somehow magnetism should produce electricity. He continues:

> These considerations, with their consequence, the hope of

obtaining electricity from ordinary magnetism, have stimulated me at various times to investigate experimentally the inductive effect of electric currents. I lately arrived at positive results . . . which may probably have great influence in some of the most important effects of electric currents.[2]

The possible practical implications of a connection between magnetism and electric fields, which underlies much of our present technology, perhaps can be appreciated better if we recall that at that time the only way one could produce an electric current was either to accumulate charge mechanically by a frictional method or to use some variation of a battery. To produce a steady current the only available method at that time was a battery. It was with batteries that Ampère, Faraday, and others produced their galvanic currents. But batteries then, if anything, were even less efficient than those of today. Some new way to produce an electric current, it was clear, would be of the greatest possible practical significance.

Faraday tried at first to see if a current in a wire would produce a current in a nearby wire. He reasoned that a current produces a magnetic field, so if one put another wire in the magnetic field why should that magnetic field not produce another current? He constructed coils of wire, winding them one on the other so that they would lie very close together, yet separated electrically by some insulator such as paper. The idea was to let a current flow through one of the coils; the other coil, separated electrically from the first, was then tested to see if a current flowed through it at the same time. (Such a test could be made, for example, by seeing if the second wire would deflect a magnetic needle placed far enough away from the first wire so that the first current had a negligible effect on it.)

Faraday was disappointed to find that a current in one wire would produce no observable current in the second nearby wire. He connected one of his coils with a sensitive galvanometer (a device to detect currents) and the other with, as he describes it, "a voltaic battery of ten pairs of plates four inches square, with double coppers and well charged"; "yet," he says, "not the slightest sensible deflection of the galvanometer needle could be observed."[3]

But then, working with 203 ft of copper wire coiled around a large block of wood, and another coil of 203 ft of similar wire, separated electrically by an insulator (metallic contact everywhere prevented by twine) and connecting one of these coils with a galvanometer to detect current and the other to a battery with 100 pairs of plates, 4 inches square, with double coppers and well charged, there was a surprise. He wrote: "When the contact was made, there was a sudden and very slight effect at the galvanometer, and there was also a similar slight effect when the contact with the battery was broken."[4] But while the current flowed steadily, he could detect no effect.

The galvanometer is basically a device with a magnetic needle and one or several loops of wire through which a current might flow. If current flows, a magnetic field is produced and the needle is deflected:

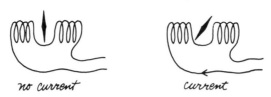

no current   current

Since the current flowed in Faraday's second circuit only during the opening or closing of the switch, it was observable only by what Faraday calls "a sudden and very slight effect at the galvanometer."

Now Faraday replaces the galvanometer by an unmagnetized iron needle inside a solenoid connected into the second circuit. If a current flows through the solenoid it produces a magnetic field that leaves the iron needle magnetized. He thus has a permanent record of the current flow rather than the momentary flicker of the needle. Further, he can easily verify that the current in the second circuit flows in the opposite direction when the switch is opened because the needle becomes magnetized with opposite polarity (north-south reversed).

How much this was in the air is perhaps indicated by the fact that Joseph Henry, an American physicist, made the same discovery at almost the same time. Working with coils not too different from those of Faraday, he found, too, that his magnetic needle was deflected just when the switch was closed and then resumed its former undisturbed position,

> . . . although the galvanic action of the battery, and consequently the magnetic power, was still continued. I was, however, much surprised to see the needle suddenly deflected from a state of rest . . . when the battery was withdrawn from the acid and again deflected to the west when it was reimmersed. The operation was repeated many times in succession and uniformly with the same results. . . .

And at almost the same time in Russia, H. F. E. Lenz (1804–1865), studying the same phenomenon, found and formulated a principle, now known as Lenz's law, which related the direction of the current induced in the second coil to the direction of the rate of change of the current in the first coil, which we shall state more precisely later.

And this was the magical fact. It was the changing current, the changing magnetic field, which produced a current in the second circuit. When the current in the first circuit was increasing (when the switch was first closed) or when it was decreasing (when the switch was opened) an effect was observed in the second circuit. In the interim period, when a steady current flowed in the first circuit so that the magnetic field produced was steady, no current was observed in the second circuit. It was thus a changing magnetic field that produced a force on a stationary charge. Faraday then proceeded to track down the details of this new phenomenon, to explore its every facet, and to reveal every feature by which magnetic fields could produce electrical effects.

As often happens when something new is discovered, it manifests itself in diverse and seemingly different ways. Electromagnetic induction, which Faraday first discovered as a current induced in a second circuit by a changing current in nearby wires, appears also as a current produced in a circuit when the magnetic field through that circuit is changed, as an electric field produced in a wire when the magnetic field is changed around the wire, as a field produced in a wire as the wire is swept through the magnetic field, as a current produced in a circuit if it is turned through a stationary magnetic field, in a variety of seemingly different ways—all of which can be summarized qualitatively by: Changing magnetic fields produce electric fields.

This is known as Faraday's law. We present it in the following form. If one has a loop of wire, such as

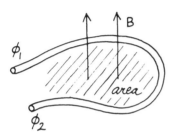

then the average electric potential difference across the ends of that loop multiplied by the time interval Δt is equal to $1/c \times$ the change of the flux of the magnetic field through the loop in that time interval:[*]

$$\text{electric potential difference} \times \text{time interval} =$$
$$-\frac{1}{c}\,(\text{change of magnetic flux through loop}) \qquad (19.1)$$

This is a little tricky because of the definition of the magnetic flux through the loop and the sign. If the magnetic field is uniform (constant in magnitude and direction) and is perpendicular to the plane of the loop, the simplest case, then the magnetic flux is defined as the field times the area of the loop:

$$\text{magnetic flux (denoted by } \Phi) = B \times (\text{area of the loop}) \qquad (19.2)$$

If the magnetic field is uniform and the area of the loop is not perpendicular to the magnetic field, then the flux is defined as that component of the magnetic field perpendicular to the area, multiplied by the area. If the magnetic field varies over the area of the loop, then the flux is defined by dividing the area into pieces small enough so that the field can be considered constant over these smaller areas, multiplying each "constant" field by the smaller area (giving the flux through that small area) and adding the result.

The flux through a loop can change for several reasons. In one simple situation, a magnetic field uniform over the area of the loop and

---

[*] As the time interval Δt becomes very small, "average," as usual, becomes "instantaneous."

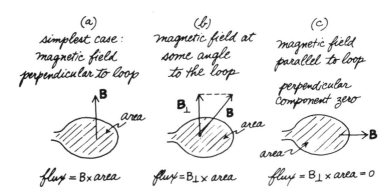

simplest case:
magnetic field
perpendicular to loop

(b)
magnetic field at
some angle
to the loop

(c)
magnetic field
parallel to loop

perpendicular
component zero

B

area

$B_\perp$ **B**

area

area

→**B**

flux = B × area

flux = $B_\perp$ × area

flux = $B_\perp$ × area = 0

FIG. 19.1. For a given magnitude of magnetic field the flux through a given loop is a maximum when the field is perpendicular to the loop (a) and a minimum (zero) when the field is parallel to the loop (c).

perpendicular to the plane of the loop can change in magnitude. In another, the loop can turn while the magnetic field remains constant. In either case the flux through the loop changes, producing an electric potential difference.

Lenz's Rule

If the circuit is completed, the direction of the current produced by a changing magnetic field is contained in a rule stated in 1834 by H. F. E. Lenz: The current induced in a loop by the changing flux through the loop flows in the direction that will produce a field that tends to maintain the previous flux through the loop. That is, the current induced will tend to retard the change of flux. For instance, consider a single loop in a magnetic field toward the right which is decreasing. Since the field

B decreasing

I

→**B**

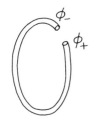

ϕ_-
ϕ_+

is decreasing, the current induced in the loop will flow in such a manner as to produce an additional magnetic field to the right, as above. Thus a magnetic field toward the right decreasing will induce a current flow in the same direction as a field to the left increasing. The high potential (if the circuit were open) would be as shown.

Example 1. Imagine that one has a loop of wire of area 30 cm² and that the magnetic field is uniform throughout space, perpendicular to the plane of the loop, and increasing uniformly at the rate

$$\frac{\Delta B}{\Delta t} = 1000 \, \frac{\text{gauss}}{\text{sec}} \qquad (19.3)$$

Then the electric potential difference across the two ends of the wire would be

electric potential difference $= \frac{1}{c} 1000 \, \frac{\text{gauss}}{\text{sec}} \times 30 \, \text{cm}^2$

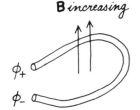

B increasing

ϕ_+
ϕ_-

$$= \frac{1}{3 \times 10^{10} \text{ cm/sec}} \, 1000 \, \frac{\text{esu}}{\text{cm}^2/\text{sec}} \times 30 \text{ cm}^2$$

$$= 10^{-6} \frac{\text{esu}}{\text{cm}} \simeq 3 \times 10^{-4} \text{ volt} \qquad (19.4)$$

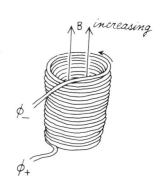

A coil is constructed of many loops, and each loop can be thought of as contributing its separate component to the total area. Thus if a uniform magnetic field B is perpendicular to the area of a coil with N loops, the flux is

$$\Phi = BN \times (\text{area of loop}) \qquad (19.5)$$

or N times what it would be for a single loop. If the area of a single loop is 30 cm², there are 1000 turns in the coil and the magnetic field is increasing as previously at the rate $\Delta B/\Delta t = 1000$ gauss/sec, then the potential difference across the ends of the wires would be about 0.3 volt.

Example 2. In the system below,

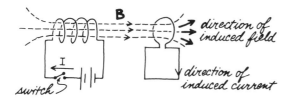

the switch in the left circuit is suddenly opened. What direction does the current flow in the right circuit? With the switch closed, the left circuit produces a field toward the right. As the switch is opened, this field decreases. Therefore, the current in the right circuit flows so as to produce a field toward the right.

ELECTRIC GENERATORS

In 1831 Michael Faraday rotated a copper disk edgewise between the poles of a horseshoe magnet and obtained a constant (direct) voltage between two rubbing contacts, one on the edge and the other on the shaft of the disk. The voltage output was low and the efficiency was poor. One year later Hippolyte Pixii generated an alternating current by rotating a wire loop in the magnetic field of a permanent magnet. Such a generator consists of a loop rotating in a constant magnetic field. As the loop turns, the area of the loop perpendicular to the field changes—going from the full area of the loop down to zero area (when the plane of the loop is parallel to the field). Since the flux through the loop changes, a voltage is induced across its ends. All modern generators of electricity are based on some variation of this technique.[*] We outline it below in a form crude enough to embarrass any apprentice engineer.

Imagine a rectangular coil of 1000 turns each of whose loops has an area of 30 cm² in a magnetic field of 1000 gauss:

---

[*] A possible exception in the future might be a plasma device which converts heat directly into electricity.

$$\Phi = 30 \frac{\text{cm}^2}{\text{turn}} \times 1000 \text{ gauss} \times 1000 \text{ turns} \tag{19.6}$$

$$= 3 \times 10^7 \text{ gauss-cm}^2$$

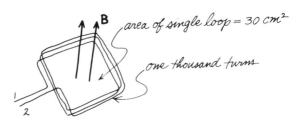

If the coil rotates 3600 turns/min (60 turns/sec), in ¼ turn beginning with the plane of the coil perpendicular to **B** as above, the flux goes from 3×10^7 gauss-cm² to zero (not exactly uniformly, but for an estimate we allow the change of flux to be uniform).

As this ¼ turn takes ½₄₀ sec,

$$\frac{\Delta\Phi}{\Delta t} = \frac{3 \times 10^7}{1/240} = 7.20 \times 10^9 \text{ gauss-cm}^2/\text{sec} \tag{19.7}$$

This produces an average potential difference across the ends of the wires (1 and 2) of magnitude

$$\text{average potential difference} = \frac{1}{c}\frac{\Delta\Phi}{\Delta t} = 0.24 \frac{\text{esu}}{\text{cm}} \simeq 72 \text{ volts} \tag{19.8}$$

In the second ¼ turn the flux through the coil goes from 0 to 3×10^7 gauss-cm² (in the opposite direction). The average potential difference is again 72 volts. In the third and fourth quarters the average voltage is the same, but high and low are reversed, as shown in Fig. 19.2.

If for convenience we call the potential at end 1 zero (ground), then with respect to this the potential at end 2 as a function of angle appears below:

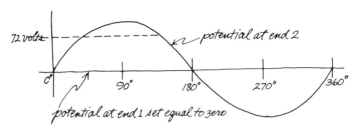

Therefore, the high potential continually shifts from one end to the other, producing what is called an *alternating potential* or *alternating current*. To convert such an alternating potential to a more direct potential, various devices might be used, for example:

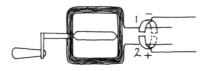

If we turn a hand generator without connecting anything across its open wires, we feel a certain resistance to the turning motion due to friction. As soon as we connect something across the wires the generator becomes more difficult to turn, because work must be done to force the electrons through a circuit, such as through a light bulb. The energy put into the system by the force° moving the wire through the magnetic field or turning the crank of the generator is regained (partially for real systems, completely only for an ideal frictionless system) as electrical energy. It is through this phenomenon that we are able to convert mechanical energy, the energy of turbines moved by steam or by falling water, into the energy of electrons moving through circuits far away.

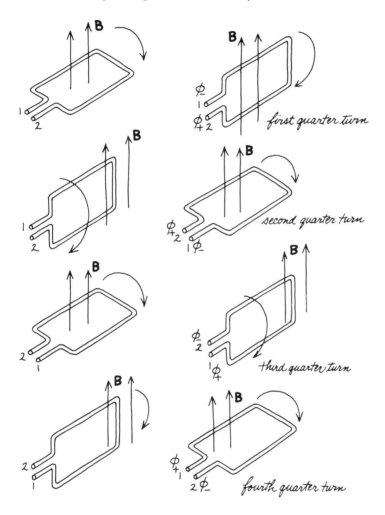

F I G. 19.2. Electric potential produced across the ends of a turning loop of wire in a uniform magnetic field.

° It is this force that does the work; the magnetic force—always perpendicular to the velocity—does no work.

20

ELECTROMAGNETIC THEORY

JAMES CLERK MAXWELL
TO WILLIAM THOMSON
(LORD KELVIN)

Trin. Coll., *Feb. 20, 1854.*

Dear Thomson

Now that I have entered the unholy estate of bachelorhood I have begun to think of reading. This is very pleasant for some time among books of acknowledged merit wh one has not read but ought to. But we have a strong tendency to return to Physical Subjects and several of us here wish to attack Electricity.

Suppose a man to have a popular knowledge of electrical show experiments and a little antipathy to Murphy's Electricity, how ought he to proceed in reading & working so as to get a little insight into the subject wh may be of use in further reading?

If he wished to read Ampère Faraday &c how should they be arranged, and at what stage & in what order might he read your articles in the Cambridge Journal?

If you have in your mind any answer to the above questions, three of us here would be content to look upon an embodiment of it in writing as advice. . . .[1]

Regarding electricity and magnetism, the situation in 1830 after Faraday's work was as follows:

1. Electric charges produce forces on each other that can be described by Coulomb's law and by the introduction of an electric field.

2. Wires carrying currents produce forces on each other that can be described using Ampère's law and by the introduction of magnetic fields.

3. No magnetic charges exist.

4. Changing magnetic fields produce electric fields—Faraday's law.

5. Electric charge is conserved: The total amount of charge in any region of space remains constant unless charges cross in or out of that region.

It took about two generations to realize that these five statements are logically contradictory.

The episode is one of the more interesting in the interplay between observation and theory. Each of the five statements can now be put into a highly compact mathematical form that reveals its meaning immediately to a person familiar with the symbolism. It is clear that one of the reasons the contradiction did not become apparent was that such a condensed form for the equations was not available in Faraday's time. Maxwell, who first wrote the equations of magnetism and electricity, in a compact form, found, considering the five statements above as postulates, that they implied a contradiction. The situation logically was as though Euclid had found, via one route, that the sum of the angles of a triangle is larger than 180° and, via another, that the sum of the angles is equal to 180°. This would imply that there was a self-contradiction among the postulates, and, according to the rules, one of them would then have to be modified.

We might object, however, that all five statements about electricity and magnetism have been carefully culled from experimental observation. In what sense, then, can they contradict each other or be modified? The answer comes, as it has come before and no doubt will come again. Any observation, experimental or otherwise, samples only a finite portion of experience. The equations or the rules one writes down have a generality that goes beyond the portion of experience sampled. They imply statements about experience we have not had and about phenomena we have not seen. And the modification of such postulates, if it is to be consistent with experience, must be made in such a way that the implications in agreement with experience already observed are not altered; those concerning phenomena not yet observed may, however, be different.

Maxwell, always regarding Faraday as the source of wisdom in electricity in contrast to the mathematicians of the Continent, who tended to think of him as a rather unsophisticated experimentalist, began his work in this field with an attempt to formulate mathematically Faraday's ideas —an attempt to write down in a precise language, simple enough so the mathematicians could understand it, what he felt Faraday had discovered.

> I have endeavored to make it plain that I am not attempting to establish any physical theory of a science in which I have not made a single experiment worthy of the name, and that the limit of my design is to show, how, by a strict application of the ideas and methods of Faraday to the motion of an imaginary fluid, everything relating to that motion may be distinctly represented, and thence to deduce the theory of attractions of electric and magnetic bodies, and of the conduction of electric currents. The theory of electromagnetism including the induction of electric currents, which I have deduced mathematically from certain ideas due to Faraday I reserve for a future communication.[2]

He attempted to formulate Faraday's concept of lines of force in terms of stresses and strains in a fluid filling up all space—one of the origins of the famous ether. About this Maxwell wrote:

> It is not even a hypothetical fluid which is introduced to explain

actual phenomena. It is merely a collection of imaginary properties which may be employed for establishing certain theorems in pure mathematics in a way more intelligible to many minds and more applicable to physical problems than that in which algebraic symbols alone are used.[3]

Further:

My aim has been to present the mathematical ideas to the mind in an embodied form, as a system of lines or surfaces, and not as mere symbols, which neither convey the same ideas nor readily adapt themselves to the phenomena to be explained.[4]

When he read what Maxwell had done, Faraday wrote:

I was at first almost frightened . . . when I saw such mathematical force màde to bear upon the subject, and then wondered to see that the subject stood it so well.

Maxwell tracked the contradiction among the postulates of electromagnetism to Ampère's law. If Ampère's law was exact in the form it had been stated, then it was not consistent with the conservation of charge. This law relates the magnetic field produced solely to the current that flows, which in itself, if we were directed to it in the right way, might seem peculiar; electric fields can be produced both by charges and, according to Faraday's law, by changing magnetic fields. With a passion for symmetry we might then believe that magnetic fields could be produced both by currents and by changing electric fields. It was just by adding this possibility to Ampère's law that Maxwell was able to make it consistent with the conservation of electricity.

Why had it not been observed? Under the circumstances that existed in the laboratories at that time, the term that Maxwell added, called a *displacement current term,* was too small to have been observed and so had escaped notice. With the experimental arrangements then available, the effect of electric currents in producing magnetic fields completely obscured any possible effect due to changing electric fields.

Briefly, the situation was as follows: Ampère had considered the magnetic field produced by a steady continuous current as might flow through an infinitely long wire or through a closed circuit. In such a wire charge is

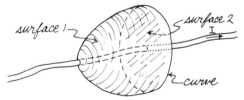

always conserved. That is, the quantity of charge entering through surface 1 always equals the quantity of charge leaving through surface 2.

Suppose, however, that we did not have an infinitely long wire. Imagine instead that we have an accumulation of charge on a ball and a wire from that ball. The current flowing through surface 2 will be just the current flowing through the wire. Yet no current flows through surface 1. (We have chosen surface 1, of course, so that it has just this property.) The reason is clear. The current results from the flow of charge from the

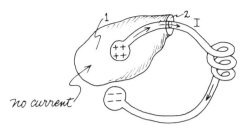

positive ball to the negative ball, and in the volume enclosed by surfaces 1 and 2, the total amount of charge is decreasing. Here we have a situation in which charge is conserved and yet the current flow in the circuit is not continuous. However, Ampère's law demands that the current flow should be continuous—that there should be current flowing through the surface 1.

It was on this that Maxwell focused his attention. He wrote:

> I am trying to form an exact mathematical expression for all that is known about electromagnetism without the aid of hypotheses, and also what variations of Ampère's formula are possible without contradicting his expressions.[5]

The situation discussed above is quite different from that with which Ampère worked in his laboratory. Ampère measured the force one wire exerted on another when continuous steady currents flowed. In the example above the current will not be steady. The charge will rush from one ball to the other and then back, very much as a pendulum bobs first to one side and then to the other. Maxwell was quite conscious of this, writing:

> We must recollect, however, that no experiments have been made on these elements of currents except under the form of closed currents either in rigid conductors or in fluids, and that the law of closed currents can only be deduced from such experiments.[6]

He realized that Ampère's law had been attained for closed currents and posed the question of what will happen if the circuit is open. In Faraday's "electro-atomic state," he uses the "equation of continuity for closed currents," and says, "Our investigations are therefore for the present limited to closed currents; in fact we know little of the magnetic effects of any currents which are not closed."[7]

Maxwell proposed that to Ampère's law should be added another term which becomes significant only when the currents change very rapidly. This new term, called a *displacement current,* not detectable under the circumstances of Ampère's experiments, removed the contradiction between Ampère's law and the law of conservation of charge and restored the symmetry of the equations of electricity and magnetism. For the new term related a magnetic field to a changing electric field.

With Maxwell's modification of Ampère's law, consistency and symmetry are restored to the equations of electromagnetism: Faraday's rule says that a changing magnetic field can produce an electric field, and now, through Maxwell's displacement current, a changing electric field can produce a magnetic field.

For various technical reasons, it turns out that the effects of this displacement current are really rather hard to see unless the rate of change of the fields becomes very rapid; it took 20 years, not until after

Maxwell's death, before the first direct confirmation of his theory was obtained by Hertz.*

With the work of Maxwell, the equations for the electric and magnetic fields could be summarized in four statements:

1. The electric field produced by a charge distribution is given by Coulomb's law.

2. No magnetic charges exist.

3. Faraday's law: A changing magnetic field produces an electric field.

4. Maxwell's modification of Ampère's law: The magnetic field is related to the currents, and to changing electric fields.

0. Charge is conserved.

5. Electric and magnetic fields produce forces on charges according to the Lorentz equation.

The statement labeled (0), conservation of charge, is no longer logically independent of the others. It follows as a consequence of (1) and (4). Therefore, five statements are the postulates for electricity and magnetism: (1), (2), (3), and (4) state how the electric and magnetic fields arise from charges and currents, and (5) tells us how an electric or magnetic field affects a moving or a fixed charge.

In his earlier papers Maxwell developed the electromagnetic theory in terms of explicit mechanical models interpreting the various electrical concepts as stresses, strains, and vortices, in an elastic medium. He says:

> I propose now to examine magnetic phenomena from a mechanical point of view, and to determine what tensions in, or motions of, a medium are capable of producing the mechanical phenomena observed. If, by the same hypotheses, we can connect the phenomena of magnetic attraction with electromagnetic phenomena and with those of induced currents, we shall have found a theory which, if not true, can only be proved to be erroneous by experiments which will greatly enlarge our knowledge of this part of physics.[8]

The vortices by themselves were inadequate. The question arose: How can the vortices be side by side, yet turning in the same direction? To answer this Maxwell put idler wheels between the vortices, writing:

> The vortices being separated by a layer of particles revolving each on its own axis in the opposite direction to that of the vortices so that the contiguous surfaces of the particles in the vortices have the same motion.[9]

It was probably one of the most complicated models ever proposed.

Later Maxwell made it clear that his theory was independent of any mechanical interpretation:

> I have on former occasion attempted to describe a particular kind of motion and a particular kind of strain, so arranged as to

---

* D. E. Hughes, seven years before Hertz, demonstrated the propagation of electromagnetic waves in air to an audience that included the well-known physicist Stokes. For some reason, it was believed that only electromagnetic induction (Faraday's law) was involved, and Hughes, apparently discouraged, did not publish his results until 1899, 12 years after Hertz. He who hesitates . . .

account for the phenomena. In the present paper I avoid any hypothesis of this kind; and in using such words as "electric momentum" and "electric elasticity" in reference to the known phenomena of induction of currents and the polarization of dielectrics, I wish merely to direct the mind of the reader to mechanical phenomena which will assist him in understanding the electrical ones. All such phrases in the present paper are to be considered as illustrative, not as explanatory.[10]

As Hertz was to say later: "Maxwell's theory is Maxwell's equations."

21

ELECTROMAGNETIC RADIATION

LIGHT AS AN ELECTROMAGNETIC WAVE

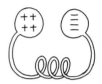

For steady currents, or for charge distributions that vary slowly in time, the implications of Maxwell's equations are almost identical with those of the equations of electricity and magnetism before his addition of the displacement current. However, for currents or charge distributions that vary, and especially those that vary rapidly in time, such as might occur, for example, if the charge were allowed to rush back and forth from ball to ball, new solutions of the equations of electromagnetism exist that did not exist previously.

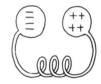

Consider a magnetic field produced by a current (let us say flowing through a wire). Now imagine that the current is shut off. As the current decreases, the magnetic field surrounding the wire decreases; but as the magnetic field decreases an electric field is produced (by Faraday's law —changing magnetic fields produce electric fields). When the magnetic field stops changing as rapidly, the electric field begins to decrease. In the pre-Maxwell form nothing more happens—the electric and magnetic fields vanish as the current goes to zero. The changing electric field does nothing.

However, with Maxwell's modification, the collapsing electric field produces a magnetic field as well as the collapsing magnetic field producing an electric field, and the two can combine in such a way that, as the field collapses, a new field is produced somewhat farther on, so that the entire pulse moves through space. Maxwell's equations imply that if the magnitude of **B** is equal to the magnitude of **E**, and if the two are perpendicular, then the pulse propagates itself through space with a definite speed.

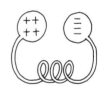

This pulse has all the properties that we have required previously of wave motion. If instead of a single pulse we had a large number, as might be produced by electric charges rushing back and forth, as between the two balls above, then we could generate a series of pulses and could associate with these repeated pulses a wavelength—the distance between the crests of the pulses. The propagation from one point to another is of the kind that we expect of a wave. And, most important, from the additive properties of the electric and the magnetic field, we obtain the basic property of superposition. The electric and magnetic pulses generated display the properties we have required of waves.

When he realized that such solutions were possible, Maxwell calculated the speed with which the waves would be propagated through space. He writes:

> The velocity of transverse undulations in our hypothetical medium, calculated from the electromagnetic experiments of MM. Kohlrausch and Weber, agrees so exactly with the velocity of light calculated from the optical experiments of M. Fizeau, that we can scarcely avoid the inference that *light consists in the transverse undulations of the same medium which is the cause of electric and magnetic phenomena.*[1]*

And in a letter to William Thomson (Lord Kelvin):

> I made out the equations in the country before I had any suspicion of the nearness between the two values of the velocity of propagation of magnetic effects and that of light, so that I think I have reason to believe that the magnetic and luminiferous media are identical. . . .[2]

(As this famous result first appeared to Maxwell, it was even more dramatic than we might guess. We have for convenience used c, the speed of light, to relate the change of magnetic field to the electric field it creates, substituting $4\pi/c$ for arbitrary-looking numbers such as 4.18×10^{-9} sec/cm. And we have used it again to relate the magnetic field to currents and changing electric fields. What Maxwell had before him were the measured amounts of electric field produced by a given rate of change of magnetic field.)

When Maxwell put these equations together, to find the solution that corresponded to the propagation of the pulse of electromagnetic radiation, he found from these measured numbers another number that corresponded to the speed with which the pulse was propagated. And this number turned out to be about 3×10^{10} cm/sec. But 3×10^{10} cm/sec was the measured speed of light. Therefore, Maxwell proceeded to identify this pulse of radiation with light itself. He writes:

> . . . it seems we have strong reason to conclude that light itself (including radiant heat, and other radiations if any) is an electromagnetic disturbance in the form of waves propagated through the electromagnetic field according to electromagnetic laws.[3]

Amazement was almost universal, but there were some doubts:

> The coincidence between the observed velocity of light and your calculated velocity of a transverse vibration in your medium seems a brilliant result. But I must say I think a few such results are

* Italics in the original.

wanted before you can get people to think that every time an electric current is produced a little file of particles is squeezed along between rows of wheels.[4]

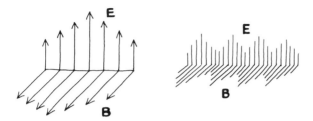

F I G. *21.1.* The solution of Maxwell's equations which propagates through the vacuum with the speed of light has the form shown above. **E** is perpendicular to **B**, and the two are equal in magnitude. Either pulses or periodic solutions which correspond to waves of a given wavelength are possible. The vacuum behaves as a nondispersive medium, which means that all periodic waves propagate with the same speed. (After E. M. Purcell, *Electricity and Magnetism,* Berkeley Physics Course, II, McGraw-Hill, New York, 1963)

If one could identify light as an electromagnetic wave—the different colors of light corresponding to different frequencies or wavelengths of the radiation, visible light a very small proportion of the total spectrum of electromagnetic radiation—then, since the interaction of electric and magnetic fields with charged particles was known (the Lorentz force), for the first time a theory of the interaction of light with matter (possibly composed of charged particles) became possible. After Maxwell published his treatise, Lorentz and Fitzgerald, for example, calculated what would be the behavior of an electromagnetic wave passing a boundary from one substance to another, attempting to obtain behavior similar to the observed reflection and refraction of light; the results corresponded to what was then known of the behavior of light.

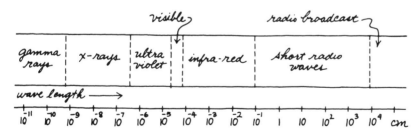

F I G. *21.2.* The electromagnetic spectrum. What are known as x-rays, visible light, radio waves, etc., are identified as electromagnetic waves of different wavelengths. What distinguishes visible from "invisible" light is the range of sensitivity of the human eye.

Even if Maxwell had not succeeded in identifying electromagnetic radiation with light, the magnitude of the discovery would have been startling. To appreciate it we might recall that an electric field can move and do work on a charge. Thus a charge, rushing back and forth at one point in space, would produce an electromagnetic pulse, which could propagate through space however far one wished from the original moving charge and whose electric field would be capable of doing work on another charge far away.

It had not been long since electrical energy had been sent through wires to do work in regions removed from the generators that were producing the current. Maxwell was now proposing that energy could be sent without wires—by wireless—through space to do work on distant charged bodies. In addition, information could be sent, contained in controlled variations of such an electromagnetic wave, and this information could be retranslated at some distant point.

In 1887 Edward Bellamy wrote[5] with a certain optimism

> . . . If we could have devised an arrangement for providing everybody with music in their homes, perfect in quality, unlimited in quantity, suited to every mood, and beginning and ceasing at will, we should have considered the limit of human felicity already attained, and ceased to strive for further improvements.

If Her Majesty's government had become convinced that its survival depended upon the introduction of music into every home and had made some organized effort to achieve this, our museums might now display the resulting million-dollar mechanical pianos. Maxwell, to justify his research, would not be likely to have pointed out its relevance to the national effort to bring music into every home. Nor would a Whitehall executive with divine foresight be likely to have defended Maxwell against the charge of irrelevance. One of the probable effects of an effort to redirect nineteenth-century English research into areas relevant to music in homes would have been to eliminate the theoretical basis for wireless, vacuum tubes, and transistors. The mechanical piano perhaps would now be in our home.

HERTZ OBSERVES MAXWELL'S RADIATION

Maxwell published the first extensive account of his theory in 1867. But it was not until about 1887, eight years after Maxwell's death, that Heinrich Hertz was able to convince his colleagues that he had produced and detected electromagnetic radiation of the type Maxwell had predicted.

Hertz's apparatus consisted of two polished metal spheres separated by a small air gap. The two spheres were attached to an induction coil (a specially designed transformer that enables one to convert low constant voltages to very large voltage), which built up a huge potential difference between them—the same way that an induction coil builds up a large voltage and produces a spark across the gap of the spark plug in a car—so large that a spark finally rushed between one sphere and the other (Fig. 21.3). When the spark discharges across the air gap, it carries charge from one ball to the other, building up charge on the opposite ball. There follows another discharge in reverse direction and charge rushes back. And so back and forth until equilibrium has been reestablished, at which point the induction coil builds up the potential on the two balls, repeating the process over again.

According to Maxwell's theory, the charge rushing back and forth between one ball and the other produces an electromagnetic wave that

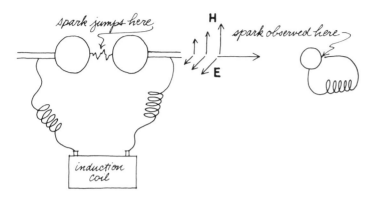

F I G. *21.3*. Schematic diagram of Hertz's apparatus.

then propagates out through all space; this means, for example, that at some distance from the spark strong electric fields are produced. Now, if one had a little circuit, such as the one shown in Fig. 21.3, which would act as a kind of receiver (this Hertz made of a piece of circular wire, as shown, with a ball on one end), the propagating electric fields produced by the first discharge would produce currents in the receiver and potential differences across the gap. If these are large enough, a spark might be seen to jump across. And this was Hertz's observation—although there was no physical connection between the sender and the receiver, Hertz observed sparks jumping across the receiver when sparks were made to jump across the sender, even though the distance between them was as much as several meters—perhaps no surprise today when transistor radios yield their message in all the quiet corners of the earth.

By measuring the wavelength and the frequency of the radiation, Hertz was able to determine its speed, which turned out to be about 3×10^{10} cm/sec—the speed of light. He then reflected it from polished surfaces, refracted it between one medium and another, made it interfere with itself, polarized it, in short did everything to this radiation that had been done before with light. In every case it behaved just as light did, except that it was not visible.

From this Hertz concluded:

> The described experiments appear, at least to me, in a high degree suited to remove doubt in the identity of light, heat radiation, and electrodynamic wave motion.

SPACE AND TIME REEXAMINED

22

ABSOLUTE MOTION, ABSOLUTE REST

"All nature, then," Lucretius wrote, "as it exists, by itself, is founded on two things: there are bodies and there is a void in which these bodies are placed and through which they move about." Is there a limit to this void?

> . . . if for the moment all existing space be held to be bounded, supposing a man runs forward to its outside borders and stands on the utmost verge and then throws a winged javelin, do you choose that when hurled with vigorous force it shall advance to the point to which it has been sent and fly a distance, or do you decide that something can get in its way and stop it? for you must admit and adopt one of the two suppositions; either of which shuts you out from all escape and compels you to grant that the universe stretches without end.[1]

Later Giordano Bruno was to say in one way or another until silenced by fire: * "Thus let this surface be what it will, I must always put the question, what is beyond?"[3] The argument is easy to understand and perplexing to deny. If space ends, what is beyond?†

Whether space is infinite or finite; whether it is merely a relationship between material bodies, or exists independently in its own right; whether it is a receptacle in which matter is placed and can be visualized in the absence of material bodies; whether space is the same from one point to another, or whether it has special properties of direction; whether it is neutral, or whether it guides bodies placed in it; whether its properties are given intuitively and are known without external evidence to the mind, or whether they are deduced from experience—all of these are questions that have been asked at one time or another about the nature of the thing called "space."

---

* He also wrote—with some prophetic vision:
> Henceforth I spread confident wings to space;
> I fear no barrier of crystal or of glass;
> I cleave the heavens and soar to the infinite.
> And while I rise from my own globe to others
> And penetrate ever further through the eternal field,
> That which others saw from afar, I leave far behind me.[2]

† Suppose, however, space were like the surface of a sphere. The javelin would fly its winged course; nothing would get in its way to stop it; yet the universe would not stretch without end.

The space of Galileo and Newton is Euclidian, infinite, homogeneous (the same from one point to another), isotropic (the same in all directions), partly filled, partly empty; it possesses no preferred points, no preferred direction; it is an empty receptacle, a void* in which matter is contained.

"Absolute space," wrote Newton, "in its own nature, without relation to anything external, remains always similar and immovable."[4] It is this view of absolute space that leads so naturally to the first law of motion. In the absence of forces, a body moving through a centerless, directionless void continues to move with uniform speed in a straight line. Why should it do otherwise? But this is both the seed and the destruction of absolute space. For, one asks, a uniform path with respect to what? A body moving uniformly with respect to the sun will not be moving uniformly with respect to the earth. The assumption of an absolute space gives one a convenient platform from which to view all events; it is comforting to believe that it exists and has an independent reality.

Newton wrote:

> But because the parts of space cannot be seen, or distinguished from one another by our senses, therefore in their stead we use sensible measures of them. For from the positions and distances of things from any body considered as immovable, we define all places; and then with respect to such places, we estimate all motions, considering bodïes are transferred from some of those places into others. And so, instead of absolute places and motions, we use relative ones; and that without any inconvenience in common affairs.[5]

However, there is an ambiguity in this, recognized by Newton, that becomes the source of one of the most intriguing problems in the history of physics. Imagine that the first law of motion is valid from a certain platform. Viewed from that platform a body in the absence of forces continues to move uniformly without changing its speed or direction. It is easy to see that, given one platform where this is true, an infinite number of others exist—those which move uniformly with respect to the original platform. Suppose for convenience, but with no conviction, we call the first platform "the stationary center of the universe." This may be a fictitious concept, but it is so graphic that we do not wish to discard it prematurely. "The centre of the system of the world," said Newton, "is immovable. This is acknowledged by all, while some contend that the earth, others that the sun, is fixed in that centre."[6] However, as far as Newton's dynamics is concerned, as far as any observations one can make within the Newtonian system, whether the center of the universe is at rest, or whether it moves uniformly, makes no difference.

Imagine that this center of the universe is at rest in the center of absolute, uniform, and homogeneous space. From the point of view of this platform, the first law of motion is to be valid. Imagine, now, another platform (in the end we shall call all these platforms "frames of

---

* Although the concept of such a space—void—which contains matter seems congenial to us, our void is not as absolute as the one rejected in antiquity. Their void not only was empty but also could propagate no influences without the motion of actual particles (for how could an influence be propagated through nothing?). Therefore Posidonius' discovery that the motion of the moon was related to the tides on earth was interpreted as a refutation of the idea of a void.

reference"), which moves uniformly with respect to the center of the universe. It is a matter of elementary addition or subtraction to verify that a body moving uniformly with respect to the center of the universe will move uniformly with respect to the new platform with a different velocity. From the point of view of the new platform, all the laws of Newtonian dynamics would hold equally well. Or, what is more relevant, there would be no way to distinguish between the platform moving with respect to the center of the universe and the center of the universe itself. Thus the center of the universe must be legislated. It cannot be observed within the Newtonian system. If we can find one platform from which Newton's laws are valid, then Newton's laws are also valid from all other platforms moving uniformly with respect to the first.*

The immediate physical experience of the first law which we obtain in an enclosed space that is moving is also an immediate experience of the fact that we have no way of detecting our true or our absolute motion. How often have we been deceived on a train when we look up and see another train passing by outside the window, to discover that we are moving rather than the other train, or perhaps the other train is moving rather than us? Unfortunately, the railroad beds of our country have disintegrated too much in the last generation to make such a test really convincing. For if we are not to detect motion, there must be no rocking or vibration; the motion must be uniform. Perhaps it is easier to experience this in the interior of a ship, far from the engines, when on a quiet ocean, or perhaps in the interior of an airplane moving through smooth air. If we look out, we may see the clouds or the ocean going by. But we have no way of knowing whether we or the clouds are moving. The earth moves in its orbit with a speed of approximately 18 miles per second; the sun moves about the center of the galaxy; the entire galaxy moves. Yet, excluding small effects that we attribute to rotation, we are totally unaware of any of these motions; we might as well be turning on an axis through the stationary center of the universe—for all we know, we are.

FRAMES OF REFERENCE

At this point it is useful to introduce a very convenient technical device. Assume that a body occupies a certain position in space. How do we identify that position? We often refer to something else, saying "two miles from Times Square" or "half a mile from the Eiffel Tower, going toward the Seine." The absolute position of an object is of very little use to us. A friend broke his leg on a mountain in Val-d'Isère. At the time he did it, the earth was at a certain point of its orbit, the sun was at a point in its transit of the Milky Way, and so on. However, the doctor was interested in none of these things. His only concern was that the break was two inches above the ankle in the bone known as the tibia. One of the properties of our world is that he carried the break with him and its relative position with respect to the ankle and the knee was very well maintained. In this case a convenient frame of reference would be

* Collectively, all such platforms are called *inertial frames*.

that man's leg. The position of the break would be located for everyone's purpose as two inches above the ankle.

Manhattan is a good city for locating position because many of its streets are laid out in a grid. "Sixth Avenue and 42nd Street" defines a region that can be found by even the uninitiated visitor. It is clearly not a region in absolute space, but rather is a region with respect to the other streets and buildings of Manhattan and, because of the general rigidity of the earth, with respect to most other places on earth. Our space is three-dimensional, but, as we are mostly confined to the surface of the earth, "Sixth Avenue and 42nd Street" is sufficient. One presumes that the meeting will be at street level. However, it does not have to be. It could be on the fifth floor of a building on the corner of Sixth Avenue and 42nd Street.

This extremely simple idea* is formalized into the concept of a co-ordinate system. The three dimensions of space are laid out as the perpendicular axis, x, y, and z:

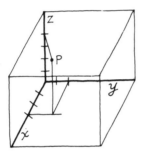

A point can be located by saying, for example, "Three along x, two along y, and four along z," or by a set of three numbers corresponding to x, y, and z [that is, $(3, 2, 4)$, meaning $x = 3$, $y = 2$, $z = 4$]. It is the same as saying "Sixth Avenue along x, 42nd Street along y, and the fifth floor along z."

Often one draws only two dimensions, because for many purposes what can be demonstrated in three dimensions can as easily be demonstrated in two. A typical two-dimensional coordinate system takes the form

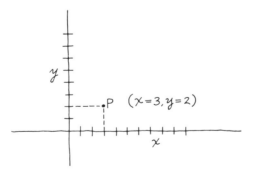

* This idea was first introduced by Descartes in his union of algebra and geometry (analytic geometry). To every point in three-dimensional space he could associate three numbers (x, y, z) and to every triplet of numbers a point in space. Then all the objects and theorems of geometry could be realized as algebraic objects and relations (see the Appendix).

These two perpendicular axes map out the points on a plane. The point in the figure above has the coordinates $x = 3$ and $y = 2$. If for some reason one said, "Meet me at $x = 3$ and $y = 2$," there would be no ambiguity, if one understood one was referring to this coordinate system.

Now, it is an elementary observation that the same point would have a different street and avenue number if referred from a different city, or from a different coordinate system, such as

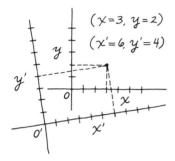

From the new grid, whose axes (or streets and avenues) are labeled x' and y', the coordinates would be $x' = 6$, $y' = 4$. The coordinates of the physical point P depend on which grid we have chosen to refer to.

Sometimes we would like to locate what we call an event, a point in space and time. This requires the three space dimensions and a clock:

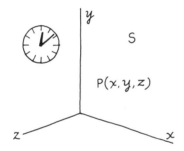

an event is located at the point in space P(x,y,z) and the time at which it occurs

We shall soon want to consider coordinate systems that move uniformly with respect to one another (designated for convenience "stationary" and "moving" frames of reference). Imagine that at the time $t = 0$ the x and y axes coincide and that the motion is along the x axis:

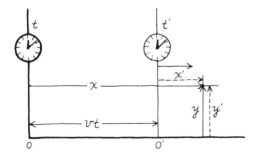

After the time t, the origin of the "moving frame," O', is the distance vt from O, the origin of the "stationary frame." Thus the event that occurs at the space-time point P is labeled (x, y, t) in the "stationary frame," and (x', y', t') in the "moving frame," where, according to the usual conventions,[*]

$$x' = x - vt$$
$$y' = y$$
$$t' = t$$

THE LUMINIFEROUS ETHER

Within the framework of the Newtonian system there is no way to locate an absolute center of the universe or to ensure that it is at rest; the concept of an absolute, immovable space is perhaps graphic and comforting, but it is intrinsically unobservable. The equations of electricity and magnetism and of the propagation of light as developed by Maxwell, however, do not maintain their form in going from one uniformly moving system to another and at least imply that a frame at rest with respect to the "light medium" (the ether) can be found and is to have a preferred status.

All the material waves, the waves that provide the prototypes from which we construct our abstract notion of a wave, are disturbances that propagate through a medium. The idea that light or electromagnetic waves could be abstract entities propagating through nothing does not seem to have been very congenial to the nineteenth-century mind. As Maxwell was to say in his *Encyclopaedia Britannica* article:

> The evidence for the existence of the luminiferous ether has accumulated as additional phenomena of light and other radiations have been discovered. And the properties of this medium, as deduced from the phenomena of light, have been found to be precisely those required to explain electromagnetic phenomena. Whatever difficulties we may have in forming a consistent idea of the constitution of the ether, there can be no doubt that the interplanetary and interstellar spaces are not empty, but are occupied by a material substance or body, which is certainly the largest, and probably the most uniform body of which we have any knowledge.[7]

Questions concerning the luminiferous ether and our relation to it grew of increasing concern, until one might say they reached an almost feverish pitch by the end of the nineteenth century. The reason was not entirely that one felt (as Maxwell felt) that there must be some medium through which electromagnetic and light waves propagated. For, as long as the ether could be done with or without, as long as it provided no more than a graphic visualization or a mechanical interpretation of the propagation of light, there would be no undue excitement. But what became increasingly exciting as the century neared its close was that the newly created electromagnetic theory of Maxwell, which already was proving

[*] This is the Galilean transformation of the space and time coordinates of a point between the moving (primed: x', y', t') and the stationary (unprimed: x, y, t) coordinate systems.

astonishingly successful, seemed to indicate that some effect of the ether should be observable.

The essence of the many experiments that were done in the latter half of the nineteenth century was this: According to Maxwell's theory of electromagnetic waves, light propagates with a speed of $c = 3 \times 10^{10}$ cm/sec. But with respect to what? In our everyday experience if we specify speed, we specify it with respect to a frame. An airplane distinguishes between its air speed and its ground speed. If it is flying in a tail wind of 100 miles/hr and its speed with respect to the air is 500 miles/hr, the airplane is moving at 600 miles/hr with respect to the earth. The same is true for ships moving in currents, rats running on treadmills, and so on. What, then, is light moving with the speed c with respect to?

The answer is really not given in Maxwell's theory; nor is it given in the theory of Young or Fresnel. Presumably, if light is a wave and if the wave moves through a medium, then light is moving with speed c with respect to the medium. And presumably, if electromagnetic waves are to be identified with light, then light or electromagnetic waves move with speed c with respect to the luminiferous ether.

Various interesting consequences follow. Suppose, for example, we are moving, ourselves, with respect to the ether. Does the fact that we are in motion with respect to the ether (with respect to which light moves with the speed c) affect our observations of optical phenomena? This question was asked in several different ways as the nineteenth century reached its close. And in every case no effect could be found—as though there was no motion of the earth with respect to the ether. As Maxwell was to say, with his usual prophetic vision:

> The whole question of the state of the luminiferous medium near the earth, and of its connection with gross matter, is very far as yet from being settled by experiment.[8]

23

THE MICHELSON-MORLEY EXPERIMENT

In 1881, Michelson conceived of a particularly simple and direct method for measuring the absolute motion of the earth with respect to the ether. According to everyday conventions, if the speed of light through the ether is c, then from the point of view of an observer in motion with respect to the ether, the speed of light should be different from c. Since the earth moves about the sun at a speed of approximately 18 miles/sec, it was reasonable to assume that at least at some part of the year the earth would be moving at a speed of the order of 18 miles/sec with respect to the ether. (One alternative would put the earth at rest and leave the rest of the universe flailing about it, a curious and ironic return to the Ptolemaic point of view.) Viewed at its simplest, and divested of the complications that arise due to technical difficulties, Michelson essentially measured the time it took a pulse of light to travel from one point to another a known distance away; he could then calculate its speed.

Nothing could be more elementary. But, as is easy to see, such a measurement is almost hopelessly difficult. As is often the case, it is not hard to see what one ought to do, but it is difficult to do it. The time it takes light to go from one point to another point on earth is much too short to measure such refinements as a possible speed of the earth with respect to the ether. What Michelson, and later Michelson in collaboration with Morley, did was to design with extraordinary ingenuity, using the latest devices available, an apparatus with which they could finally be confident that they could measure such a small thing as 18 miles/sec, as compared to the speed of light (186,000 miles/sec).

If the velocity of light with respect to the light medium is to be combined with the velocity of the observer with respect to that medium according to the usual ideas, then the time taken for a light pulse to complete a round trip (source to reflector back to source) will depend upon the speed of the apparatus with respect to the light medium and on the relative directions of the speed of the apparatus and the path of motion of the light pulse. The technical trick Michelson devised was to use the phenomenon of interference to detect the minute differences in transit time of light pulses due to the motion of the apparatus through the ether. To understand what he did, we calculate, according to his own conventions,* the time taken by a light pulse propagating in a medium from a source to a reflector and back. (We suffer a bit of algebraic complexity because the algebra itself is of interest in the theory of relativity to come.)

Parallel

We calculate first the time taken for a pulse† to move from a source to a reflector and back when the motion of the apparatus through the medium is parallel to the direction of motion of the pulse.

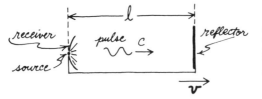

A source emits a single pulse which is reflected from the mirror a distance ℓ away back toward the source.

If the apparatus is at rest with respect to the light medium, a pulse produced at the source reaches the reflector in a time ℓ/c, and returns to the source in a time $2\ell/c$, the total time for the round trip being just the total length of the trip divided by the speed of the pulse.

* The Galilean transformation between a moving and a stationary coordinate system.
† We consider a pulse because it is graphic; it may be thought of as one wavelength of a periodic monochromatic wave, as shown here.

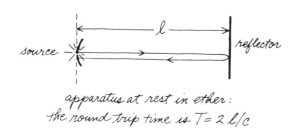

apparatus at rest in ether:
the round trip time is $T = 2\ell/c$

If the apparatus is moving through the light medium with a speed v, and we send out a pulse from the source, how long does it take this pulse to reach the reflector? It must travel the distance ℓ; but having traveled this distance, the reflector has moved away. So the pulse must travel farther to finally reach the reflector. If the reflector is moving faster than the pulse (v larger than c), the pulse will never reach it. If, however, the reflector moves more slowly than the pulse, the pulse will reach it, but will take longer to do so.

Call the time required for the pulse to reach the reflector t_{forward}. In the time t_{forward}, the reflector moves a distance vt_{forward}. Thus, in addition to the distance ℓ, the pulse must also travel the distance vt_{forward} to catch the reflector. The total distance is thus $\ell + vt_{\text{forward}}$. Since the pulse moves with speed c,

$$ct_{\text{forward}} = \ell + vt_{\text{forward}} \tag{23.1}$$

or

$$t_{\text{forward}} = \frac{\ell}{c - v} \tag{23.2}$$

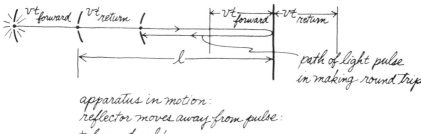

path of light pulse
in making round trip

apparatus in motion:
reflector moves away from pulse:
t forward $= \ell/c - v$
detector moves toward reflected pulse:
t return $= \ell/c + v$

Having been reflected, the pulse now returns. For the return trip, the source is moving into the pulse, shortening the distance the pulse must travel. If we call the time required for the return trip t_{return}, the receiver moves toward the pulse a distance vt_{return} in the time taken for the return trip, so that the return distance is just $\ell - vt_{\text{return}}$. Thus

$$ct_{\text{return}} = \ell - vt_{\text{return}} \tag{23.3}$$

or

$$t_{\text{return}} = \frac{\ell}{c + v} \tag{23.4}$$

If we now call the total time for the round trip when the motion of the pulse is parallel to the motion of the apparatus, $T_{||}$,

$$T_{||} = t_{\text{forward}} + t_{\text{return}} \qquad (23.5)$$

this becomes

$$T_{||} = \frac{\ell}{c - v} + \frac{\ell}{c + v} = \frac{2\ell c}{c^2 - v^2} \qquad (23.6)$$

Dividing numerator and denominator by c^2, this can be written in the conventional form:

$$T_{||} = \frac{2\ell}{c} \left(\frac{1}{1 - v^2/c^2} \right) \qquad (23.7)$$

If $v = 0$,

$$T_{||} = \frac{2\ell}{c} \qquad (23.8)$$

as it should when the apparatus is at rest. As v increases, $T_{||}$ becomes larger and larger. If v becomes as large as c, the pulse never reaches the reflector, and $T_{||}$ becomes infinite. With such an apparatus (if we could time the pulse in its round trip) we could determine our speed relative to the light medium.

In principle Michelson could have measured the total time for the round trip, $T_{||}$. However, a simple calculation is enough to show that this could not be easily measured directly. Suppose ℓ is 15 cm long. Then, when the apparatus is at rest in the ether, the trip time is

$$T_{||} = \frac{2\ell}{c} = \frac{30 \text{ cm}}{3 \times 10^{10} \text{ cm/sec}} = 10^{-9} \text{ sec} \qquad (23.9)$$

Imagine that the speed of the apparatus through the light medium (ether) is 3×10^6 cm/sec (about 18 miles/sec). Then the time for the round trip would be altered by only about 10^{-17} sec.

Measuring such intervals at the end of time periods 10^8 times as big is as difficult as trying to weigh an eyelash by weighing a man, pulling out the eyelash, weighing the man again, and subtracting to find the weight of the eyelash. The problem of the experimental physicist is to pull out the eyelash so that it can be weighed on a scale by itself—in this case to measure ΔT directly. Michelson's method was based on the fact that ΔT depends upon the angle the apparatus makes with the direction of motion. With the apparatus parallel to the direction of motion, as above, ΔT is a maximum. If it were perpendicular to the direction of motion, ΔT would be a minimum.

Perpendicular

We consider now the time of transit of a pulse moving perpendicular to the velocity of the apparatus through the ether. When the apparatus is stationary (see margin), the pulse must travel a distance 2ℓ, and if

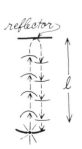

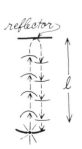*reflector*

the speed of the pulse in the medium is c, the time for the round trip is again $2\ell/c$.

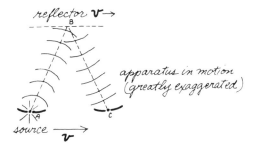

When the apparatus moves with a speed v as shown above, the pulse must travel the longer path $\overline{ABC}$; it will thus take longer for the pulse to complete the trip. Dividing the new distance $\overline{ABC}$ by the speed of the pulse in the medium c, we obtain the time $T_{\perp}$ for the round trip when the apparatus is perpendicular to its direction of motion:

$$T_{\perp} = \frac{2\ell}{c} \frac{1}{\sqrt{1 - v^2/c^2}} \tag{23.10}$$

If we now compare $T_{\perp}$ with $T_{\parallel}$, we find that the round trip takes less time when the apparatus is perpendicular to the direction of motion than when it is parallel:

$$T_{\parallel} = \frac{2\ell}{c} \frac{1}{1 - v^2/c^2} \tag{23.11}$$

while

$$T_{\perp} = \frac{2\ell}{c} \frac{1}{\sqrt{1 - v^2/c^2}} \tag{23.12}$$

Therefore,

$$T_{\perp} = T_{\parallel} \sqrt{1 - v^2/c^2} \tag{23.13}$$

The factor $\sqrt{1 - v^2/c^2}$, under normal circumstances, will make only the tiniest difference in the transit times of the two pulses. They will return about a thousandth of a trillionth of a second apart after a trip which itself takes only about a ten millionth of a second.

THE INTERFEROMETER

At first it would seem that we have only added to our difficulties. However, Michelson had already constructed a device, called an inter-ferometer, which essentially measured distances using the interference pattern produced by two light waves emitted in phase and recombined out of phase. By sending a pulse in two directions simultaneously, ex-tremely small time or distance measurements can be made by observing the interference patterns when the pulses recombine. Without going into all the details, we outline how Michelson was able to convert a very small time interval into an interference pattern considering a single pulse;

the principle is the same for the periodic waves of monochromatic light sources that are actually used.

The heart of the interferometer is a half-silvered mirror held at a 45° angle to the origin of the pulse. This divides the pulse so that half of it

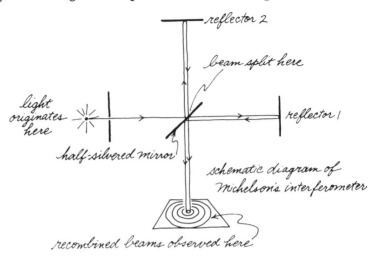

is reflected to R_2, half transmitted to R_1—the two pulses moving on perpendicular paths and beginning their trip in phase. When the pulses return (being reflected by R_1 and R_2), half of each reflected pulse will go through the mirror (one reflected, the other transmitted) and recombine at the screen. If the time of transit is identical, $T_{\parallel} = T_{\perp}$, then the two pulses will combine to give:

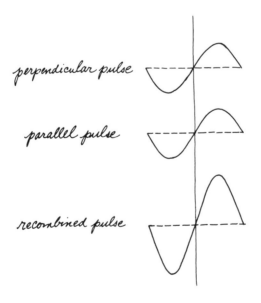

If, however, the times of transit are unequal, upon recombination the resulting pulse might have a different shape, as, for example,

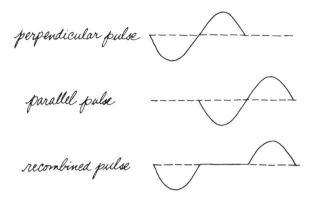

perpendicular pulse

parallel pulse

recombined pulse

and the viewer will see a change in the interference pattern. It was just this change in the interference pattern that Michelson proposed to use to measure the difference in the path time $T_\perp$ and $T_{||}$, which would enable him to calculate the earth's speed with respect to the ether.

THE RESULTS

When Michelson first performed the experiment in 1881, he used yellow light, and the shift to be expected was 0.04 of the distance between the fringes that made up the interference pattern. As he wrote later:

> In the first experiment one of the principal difficulties encountered was that of revolving the apparatus without producing distortion; and another was its extreme sensitiveness to vibration. This was so great that it was impossible to see the interference fringes except at brief intervals when working in the city, even at two o'clock in the morning. Finally, as before remarked, the quantity to be observed, namely, a displacement of something less than a twentieth of the distance between the interference fringes, may have been too small to be detected when masked by experimental errors.
> The first named difficulties were entirely overcome [in the second experiment] by mounting the apparatus on a massive stone floating on mercury;

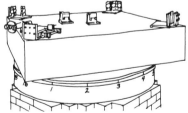

(After C. Kittel, et al., *Mechanics*, Berkeley Physics Course, I, McGraw-Hill, New York, 1962)

and the second by increasing, by repeated reflection, the path of the light to about ten times its former value. . . . Hence the displacement to be expected was 0.4 fringe [if the earth were traveling through an ether]. The actual displacement was certainly less than the twentieth part of this, and probably less than the fortieth part. But since the displacement is proportional to the square of the velocity, the relative velocity of the earth and the ether is probably less than one-sixth the earth's orbital velocity, and certainly less than one-fourth.

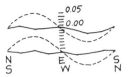

The vertical axis is the displacement of the fringes; the horizontal axis refers to the orientation of the interferometer relative to an east-west line.

The results of the observations are expressed graphically [in the figure]. The upper is the curve for the observations at noon, and the lower that for the evening observations. The dotted curves represent *one-eighth* of the theoretical displacements. It seems fair to conclude from the figure that if there is any displacement due to the relative motion of the earth and the luminiferous ether, this cannot be much greater than 0.01 of the distance between the fringes.[1]

The observations were continued over a long period of time—with the earth going in one direction in its orbit in May, and the opposite direction in November. It has since been repeated with variations over long periods of time and in various circumstances. And the result seems always to be the same—no speed of the earth with respect to the ether has ever been observed. Numerically the result as it stands today can be stated as follows: Given that light moves with the speed c with respect to an ether and given our analysis of the transit times, the earth's speed with respect to the ether never exceeds one thousandth of the earth's orbital speed around the sun.

24
THE PRINCIPLE
OF RELATIVITY

THE LORENTZ-FITZGERALD CONTRACTION

Secure in the position of junior patent examiner in the patent office at Berne, Albert Einstein, who as a student had caught the eye of no professor, who had been asked to leave the Gymnasium in Munich because of his adverse effect on the morale of fellow students,* and who

* He was advised by one of his instructors (the guidance counselor, no doubt) that he would never amount to anything and that his indifference was demoralizing both to his teachers and to the other students.

owed both his survival at the Swiss Federal Institute of Technology and his job at the patent office to his friend, Marcel Grossman, wrote in 1905:

> Examples of this sort, together with the unsuccessful attempts to discover any motion of the earth relative to the "light medium," suggest that the phenomena of electrodynamics, as well as of mechanics, possess no properties corresponding to the idea of absolute rest. They suggest rather that . . . the same laws of electrodynamics and optics will be valid for all frames of reference for which the equations of mechanics hold good. We will raise this conjecture (the purport of which will hereafter be called the "Principle of Relativity") to the status of a postulate, and also introduce another postulate, which is only apparently irreconcilable with the former, namely, that light is always propagated in empty space with a definite velocity* c which is independent of the state of motion of the emitting body.[1]

The two postulates, said Einstein, are "only apparently irreconcilable." Excluding, perhaps, his friend and colleague M. Besso, he was alone in the world at that moment in this knowledge.†

The failure of Michelson and Morley to observe the motion of the earth with respect to the ether was but one in a growing list of attempts made toward the end of the nineteenth century to determine the earth's velocity with respect to this light medium.‡ With each null result, theoretical physicists had to invent new and what became increasingly desperate explanations. But there was not much room to maneuver.

Before the turn of the century, Fitzgerald and Lorentz suggested that the dimensions of solid bodies might be altered by their motion through the ether, so that the length in the direction of motion shrunk by just the right amount to give a null result in the Michelson-Morley experiment. If, they proposed, the length parallel to the motion of the apparatus $\ell_{\parallel}$ shrinks, so that it is not "really" ℓ but rather $\ell_{\parallel} = \ell\sqrt{1 - v^2/c^2}$, then the trip time in the parallel direction will become precisely the same as the trip time in the perpendicular direction, and no fringe shifts should be observed as the apparatus is rotated.

As H. Minkowski was to say in 1908:

> According to Lorentz any moving body must have undergone a contraction in the direction of its motion, and in fact with a velocity** v, a contraction in the ratio $1:\sqrt{1 - v^2/c^2}$. This hypothesis

---

* The German word translated as velocity is an old usage that occurs frequently in Einstein's papers. Today we would use the word speed.

† It took a few years for the scientific world to appreciate what he had done. It is said that an unnamed professor of the University of Kracow one day brought this paper to the attention of his students with the words "A new Copernicus has been born."

‡ It seems incredible from our present point of view, but Einstein may not have known of the result of Michelson and Morley in 1905. He has been quoted as saying that he remembers learning of the experiment only after his 1905 paper. If this is the case, he based his belief in the two postulates on the lack of symmetry in electrodynamics and on the various other experimental results that had been obtained concerning the earth's speed and did not know of the one we now consider the most direct.

** Again, we would say speed today.

sounds extremely fantastical, for the contraction is not to be looked upon as a consequence of resistances in the ether, or anything of that kind, but simply as a gift from above — as an accompanying circumstance of the circumstance of motion.[2]

It was to provide a fundamental explanation of the shrinkage that was supposed to occur when a body was in motion that H. A. Lorentz attempted to develop a detailed theory of the behavior of rigid bodies when in motion.[3] The paper is startling because so many of the equations have the form that we have come to regard as correct. Almost everything is there except the striking change in point of view that we associate with this episode.

Lorentz considers the behavior of matter, for example, a rigid rod in motion, under the assumption that the forces that hold the rod together are electromagnetic or, if otherwise, behave as electromagnetic forces. Then, following Maxwell's electrodynamics, Lorentz shows that the forces between the charged particles making up the rigid rod will change if the rod is in motion with respect to an ether (between moving charges, for example, there exist magnetic forces that were not present when they were at rest). Lorentz then proposes that all forces behave as electromagnetic forces, that due to the change in these forces when the body is in motion the equilibrium position of the charges in the rod will change, and it turns out that when the rod is in motion, the rod will shrink.

In doing this he has to assume, for example, that electrons, which he takes to be spheres of radius R in their state of rest, have their dimensions changed by the effect of the translation in the same way as the rod itself. Further, he must assume that the forces between uncharged particles, as well as those between charged particles and electrons, are influenced by a translation in the same way as the electric forces. In other words, all forces behave the way electromagnetic forces do as far as behavior in translation is concerned. With these assumptions he is able to develop a physical picture of the shrinking of the rod. The rod shrinks because the forces between the particles that make up the rod change when it is in motion.

In the course of this paper he finds that the mass of the electron must change when it is in motion; in addition, he introduces a time, t', which he calls "the local time in the moving system." It is not the same as the time in the fixed system being related to it in a rather curious way, but using the "local time" his equations take a specially simple form.

The essence of Lorentz's investigation was thus that if material bodies are made up of charged particles and if the forces between charged particles behave as Maxwell's equations indicate, then, because of the change in forces between charged particles when they are in motion, one can conclude that a body would contract in the direction of its motion. Several puzzling assumptions were required. First, electrons must contract in the same way as the rigid body itself. Second, all other forces that exist (for example, gravitational forces) would have to behave in the same way as electromagnetic forces in the process of the translation. But if one were willing to assume these things, then one would have a physical picture of why a rod contracted when it was in motion.

Possibly our point of view might have developed in this way had it not been for Albert Einstein. Lorentz had found it convenient to introduce what he called the local time, in the moving system. It was convenient because using this local time, the equations of electrodynamics, the relations between the fields and the charges, retain the same form in the moving system as they had in the fixed system (something that is not true if one uses the same time in the moving system as in the fixed).

As Minkowski was to say,

> Lorentz called the t' combination of x and t the local time of the electron in uniform motion, and applied a physical construction of this concept, for the better understanding of the hypothesis of contraction. But the credit of first recognizing clearly that the time of the one electron is just as good as that of the other, that is to say, that t and t' are to be treated identically, belongs to A. Einstein.[*] Thus time, as a concept unequivocally determined by phenomena, was first deposed from its high seat.[4]

It was his attack on that most primitive of our dogmas, the dogma of time, that gives Albert Einstein's 1905 paper its special brilliance. And it is that which makes it so incomprehensible. For technically, what Einstein did is extraordinarily simple (much easier than the paper of Lorentz which preceded Einstein's, but which Einstein did not know), nothing more difficult than the statement that speed is equal to distance divided by time:

> . . . Any ray of light moves in the "stationary" system of co-ordinates with the determined velocity c, whether the ray be emitted by a stationary or by a moving body. Hence

$$\text{velocity} = \frac{\text{light path}}{\text{time interval}}.[5]$$

But if the speed of a pulse of light is the same viewed from one system or viewed from another which is moving with respect to the first, then our ideas about light path (distance) or time interval, or as it turned out both, will have to be revised.

THE CONSPIRACY

In the uniform space of Newton and Descartes, locating one's absolute position or motion was as difficult as it was irrelevant. If, as Newton did, one defined the sun as the immobile center of the universe, one could speak of one's distance from or motion with respect to the sun; but whether the sun were the center of the universe or somewhat off to one side, whether the sun was at rest or moving uniformly, was something that could never be determined within the framework of Newtonian theory. The electromagnetic theory and the ether with respect to which light moved added a new element: the speed of light was presumably 186,000 miles/sec with respect to the ether. If Maxwell's theory, Newton's mechanics, and ordinary notions of space and time were correct, then

---

[*] [Footnote in original:] A. Einstein, *Ann. Phys.*, **17**, 891 (1905); *Jahrbuch Radioaktivität und Elektronik*, **4**, 411 (1907).

it should be possible to determine the motion of an observer (in particular the motion of the earth) with respect to the light medium.

Attempt after attempt failed. And after every attempt, new and increasingly ingenious explanations had to be created. Perhaps the culmination of the effort came in the Michelson-Morley experiment, which posed the question of the earth's motion with respect to the luminiferous ether in a particularly direct and unambiguous fashion. (Various attempts at explanation, such as the hypothesis that the ether was dragged along by the earth near the surface of the earth, led immediately to other difficulties and had to be abandoned.*) Thus, however reluctantly, one seems forced to the conclusion that the earth does not move with respect to the ether. But this is too violent a return to the anthropomorphic universe whose debris lies at the foundation of what is called modern science. Although Aristotle could easily have accommodated the result, modern science could not. One could not seriously maintain at the end of the nineteenth century that the earth was the stationary center of the universe.

Like a detective, dogged at every step by clues that disappear, by fingerprints that vanish, by seemingly certain leads that evaporate, one begins to mutter darkly about a conspiracy. And "conspiracy" is the word the physicists began to use. There seemed to be a conspiracy on the part of nature herself to prevent us from learning the absolute motion of the earth through the luminiferous ether. In every case, when one tried what seemed to be a reasonable and obvious measurement, something overlooked prevented success. The Michelson-Morley experiment, which seems to be such a direct test of the speed of the earth through the ether, as a final irony if Lorentz' explanation is correct, is prevented from ever being successful by the almost miraculous shrinking of the arms of the apparatus in the direction of their motion. And it is thus with every experiment that was tried. In the words of Minkowski, something always happened "not to be looked upon as a consequence of resistances in the ether, or anything of that kind, but simply as a gift from above."[6] It was as though nature, in retribution for inconsistent theory and unwarranted dogma, had conspired to prevent earthbound physicists from determining their absolute motion, as in an earlier time, the gods intervened among men to alter the balance of fate.

Henri Poincaré pointed out that any such complete conspiracy must itself be regarded as a law of nature: It is not possible to discover an ether wind; it is not possible to discover the velocity of the earth with

---

* It was proposed that the velocity of light should be measured with respect to the source (if the source is moving toward us the speed of light would be greater). This was excluded by Miller, who redid the Michelson-Morley experiment using sunlight (not a lamp moving with the apparatus) as a source and again obtained a null result. Further, observations on the apparent motion of double star systems were shown by De Sitter to be inconsistent with the hypothesis that the velocity of light depended on the source.

The crucial observation to eliminate the possibility that the ether is dragged along by the earth near its surface was the observation of an extraterrestrial source of light—the stars. If the ether was dragged along at the surface of the earth, the starlight, being dragged with the ether, would move through a telescope at a different angle from the one in fact observed. It could then be proposed that the amount of ether drag varied with distance from the surface of the earth—a possibility eliminated when the Michelson-Morley experiment, repeated on top of a mountain and in a balloon, gave the same null result.

respect to the ether by any experiment. What had been explained away in a half a dozen particular cases by a variety of particular explanations were special cases of a more general rule: There is no way to determine one's absolute velocity with respect to the light medium. This must mean that the absolute velocity with respect to the light medium can never enter in any meaningful way in the equations of physics.

It is this that Einstein proposes* in his 1905 paper:

> . . . the same laws of electrodynamics and optics will be valid for all frames of reference for which the equations of mechanics hold good. We will raise this conjecture (the purport of which will hereafter be called the "Principle of Relativity") to the status of a postulate. . . .[7]

This is the famous principle of relativity. No physical process can depend on the absolute velocity of the system through space. For, if it did, one could, by observing the physical process, determine one's absolute velocity.

THE SPEED OF LIGHT

In spite of a certain complexity of algebra, the import of the Michelson-Morley experiment is extremely simple. Their result may be interpreted by asserting that the speed of a pulse of light is the same—about 186,000 miles/sec—whether one is moving or standing still while observing it.

Suppose we think of the experiment as an attempt to determine the speed from the point of view of the moving apparatus. Consider, for example, the trip time when the arm is parallel to the motion. We would say that on the forward trip, light's speed is $c - v$, and on the return trip, $c + v$. Then the trip times are

$$t_{\text{forward}} = \frac{\ell}{c - v} \qquad (24.1)$$

$$t_{\text{return}} = \frac{\ell}{c + v} \qquad (24.2)$$

giving, as before,

$$T_{||} = t_{\text{forward}} + t_{\text{return}} = \frac{2\ell}{c} \frac{1}{1 - v^2/c^2} \qquad (24.3)$$

However, the result of Michelson and Morley can be interpreted as saying

$$T_{||} = \frac{2\ell}{c} \qquad \text{(independent of } v) \qquad (24.4)$$

This would be the case if the speed of light from the point of view of the moving apparatus were always c.

---

* Probably without knowing of Poincaré's suggestion.

Consider one observer whom we will allow to relax in his favorite nightclub as he measures the speed of a light pulse passing by; he obtains 186,000 miles/sec. At the same time, another observer moving with respect to the first, even with a speed very close to that of light, regarding the same pulse of light, measures the same speed: 186,000 miles/sec. Both the stationary observer and the observer moving at a very high speed measure the same speed for the passing light pulse. (The martini is not the source of the paradox.)

light pulse

This is contrary to what we would normally expect. If, for example, the second observer were moving with a speed of 100,000 miles/sec in the same direction as the light pulse, we would expect that he would measure for the speed of light 186,000 miles/sec − 100,000 miles/sec = 86,000 miles/sec. If he were moving at the speed of light itself, we might expect that he would see the pulse as stationary since they both would be moving at the same speed.*

It is a matter of everyday experience that a car passing another on a highway can appear to move quite slowly in spite of the fact that each car is moving very rapidly with respect to the highway. Yet it is precisely this everyday perception that is being denied. A pulse of light passing a rapidly moving car, from the point of view of that car, is moving at 186,000 miles/sec, just as the same pulse of light, from the point of view of a person sitting by the roadside, is moving at 186,000 miles/sec. This is a strange idea, not something we would have arrived at in any obvious way if it had not been forced on us by what had been observed in the course of the nineteenth century. However, we cannot say that we have had any direct experience of speeds close to the speed of light and the revisions in our concepts of distance and time that we shall now be forced to make become significant only at such high speeds.

In measuring distances we have assumed that such things as rigid rods existed, that the rods could be marked to a standard length, and that this standard length would remain the same when the rod was moved about into new positions, when its orientation in space was changed, and, within reason, it would remain the same as time passed. In all this we make assumptions about the properties of the world. We know very well that rods do not have to retain their length; in time they may expand or contract. In a world where solid bodies did not exist, we might not have rods at all. Or, if our world were like the rubber sheet mentioned earlier,

---

* It is this, Einstein says, which first led him to a consideration of the entire question. For no such stationary pulse is a possible solution of Maxwell's equations.

twisting, expanding, and contracting, the notion of distance would become close to meaningless. Among other things, we have assumed that the length of the rod remains unaltered when the rod is in motion with respect to us. It is this last statement that we shall be forced to abandon. And with that we shall abandon the convention that an "interval of length" remains the same as measured from a stationary or a moving system—not because it is impossible to retain but because in a world where all rigid bodies (including those we would like to call rulers) shrink in the direction of their motion, the convention that "length" remains the same becomes confusing and inconvenient. We shall be left with the choice of saying that a length remains the same but the rulers to measure it shrink—or, what turns out to be a more convenient way to organize an experience, to say that length itself shrinks in a moving system.

The same will be true for our concept of time. And there is nothing that causes more trouble than this. We are wedded in a land where divorce does not exist to the idea of time flowing uniformly, ceaselessly, the same for everyone; thus the concept that in one system there could be one time, and in another system moving with respect to it there could be another, seems an absurd self-contradiction. However, if we insist on a definition of absolute time, defined with respect to some system, then an observer in a system moving with respect to the original system will be in a position of saying that all the physical processes in his system have unaccountably slowed down—heartbeats, chemical processes, all rhythmic motions including the motions of clocks themselves. Within the moving system, it is really only the relation of various rhythmic devices to others that are of significance. (If the heartbeat slows down, and the clock slows down also, by that clock the animal will live the same number of clock ticks.) We are thus tempted to redefine the time interval for the moving system.

TEMPS PERDU

Einstein's Definition of a Time Interval

We are a fairly adaptable species. Once, we are told, we swam in the oceans; now we trudge on land—unbearably hot and dry or too cold and damp. We suffer catastrophe, famine, violence, subway strikes, blackouts, shortages of water; but life goes on. We accept discomfort, irrationality, meaningless violence, and television commercials. But, time passing more slowly under some circumstances than others, events simultaneous for one person but occurring at different times for another—that is just too much.

Our language is so constructed that it is difficult to speak of time as anything other than that which flows uniformly and is the same for everyone. Among our most primitive concepts is that of the clock in outer space which ticks away the moments of true time. All other clocks are good only to the extent that they agree with this one. If we can construct no clocks at all that agree with it, we will admit that's a problem. But the fault is ours. Time flows on, but we cannot measure it properly.

"Absolute, true, and mathematical time," said Newton, "of itself, and from its own nature, flows equally without relation to anything external . . . just as relative time always more nearly approaches absolute time as we refine our measurements. . . ."

Einstein's incomparable insight was that none of us is in communion with this clock in outer space. To define a time interval, in fact, we choose an interlude between two events—a day bounded by two sunrises, a year by two astronomical happenings, or a second as the interval between one swing of a pendulum and another. It is from the observation of such rhythmic events that we have abstracted the concept of time, unconsciously adding the idea that everyone will see the same interval between the same two events and that two events which occur simultaneously for one observer occur simultaneously for all observers.

It is these additions that Einstein denies. Not of necessity but by choice. For he recognizes that in the comparison of time pieces that are separated from one another some assumption must be made.

> If at the point A of space there is a clock, an observer at A can determine the time values of events in the immediate proximity of A by finding the positions of the hands which are simultaneous with these events. If there is at the point B of space another clock in all respects resembling the one at A, it is possible for an observer at B to determine the time values of events in the immediate neighborhood of B. But it is not possible without further assumption to compare, in respect of time, an event at A with an event at B.* We have so far defined only an "A time" and a "B time." We have not defined a common "time" for A and B. . . .[8]

And at this point he introduces his own convention for the comparison: "We establish *by definition* that the time required by light to travel from A to B equals the time it requires to travel from B to A."[9] This simple, almost self-evident, definition—without exaggeration one of the most luminous insights ever achieved by the human mind—is the crux of his theory. Making the "time"† required for light to travel from A to B the same as that to travel from B to A is essentially taking the point of view of an observer at rest with respect to the light medium. (For if A and B are thought to be moving through the light medium, then A, for example, rushes into the light beam, whereas B rushes away from it; thus the time taken for light to go from A to B would be different from that taken to go from B to A, as, for example, in the analysis of the Michelson-Morley experiment.)

By convention, before Einstein, one would have agreed that for B and A at rest with respect to the "light medium" the "time interval" required for light to travel from A to B would be equal to that to travel from B to A for all observers (no matter what their state of motion with respect to B and A). However, for A and B moving with respect to the "light medium," the "time interval" required for light to travel from A to

---

* He agrees that one can determine simultaneous events at the same point in space. But to make this determination for points separated in space requires some communication—some signal—between them. What are the properties of existing signals?

† For "time" we might read "time interval."

B would have been different from that to travel from B to A—again for all observers.*

Einstein turns this around. He defines the "time interval" required for light to travel from A to B to be equal to that to travel from B to A from the point of view of the observer at rest with respect to A and B, *no matter what the state of uniform motion of A and B*. Thus all observers in uniformly moving systems may regard themselves at rest with respect to the "light medium." But now two observers (say one at rest and the other moving uniformly with respect to A and B) will no longer assign the same time intervals between events. For if by definition we assert that the time interval required for light to travel from B to A is the same as that from A to B for the observer at rest with respect to A and B (this is the definition of what we are to mean by a time interval), we have abandoned the concept of absolute time and allowed the local time in one frame of reference to be as good as the local time in another.

Simultaneity

Within any one reference frame everything is as usual; but the relations between two uniformly moving frames are changed. Whether, for example, two events happen at the same time depends on the frame from which they are viewed. Consider the situation shown here. A man

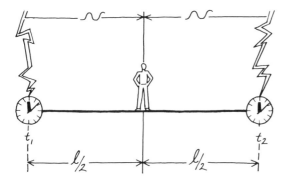

stands at the center, determined by Euclidian construction, between two clocks a distance ℓ apart. An event occurs at the left-hand clock that causes it to emit a light signal at its time t_1. An event occurs at the clock on the right at the time t_2 which causes it to emit a light signal. If the signals from both clocks reach the man stationed at the center simultaneously, he says that t_1 was equal to t_2. He says this because he has defined the time interval as the distance the light signal travels (that is, $\ell/2$) divided by the speed of light, which is the same in all systems. With this definition of time, if the signals arrive where he is simultaneously, he says the travel times were equal, so that t_1 is equal to t_2.

Another observer, with his own clocks moving with respect to the first, reads the situation differently. If, for example, he sees the man moving toward the right, he will see him moving into the light signal coming from the right, and away from the light signal coming from the left. The distances these signals travel, therefore, will be different. The

---

* Just the t_{forward} and t_{return} that occur in the analysis of the Michelson-Morley experiment.

speed (always the speed of light) will be the same. And since they arrive at the "center" simultaneously,* the time interval between the emission of light from the clock at the left and its receipt is different than the time interval between the emission of the light signal from the clock on the right and its receipt from the point of view of the moving observer. Therefore, he concludes that from his point of view t'_1 is not equal to t'_2.

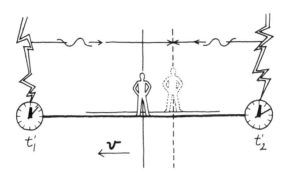

The distance moved by the light varies, depending on who watches it. Therefore the time interval, by definition, varies. And since the definition of simultaneous events separated in space depends on these time intervals, such events, which are simultaneous in one frame of reference, are no longer simultaneous in another.

Although the relations between one frame and another moving with respect to it are thus radically altered, in any given frame everything is perfectly normal. For example, if we are standing midway between A and B, we would usually assert that two events occurred simultaneously if we saw them at the same time (that is to say, if we see an event at A in coincidence with our seeing the event at B). However, if we did not agree with Einstein that the time interval for light to travel from A to B was the same as that from B to A, and if we believed we were moving with respect to the light medium, we would have to say the two events occurred at the same time at A and B if we saw them happen at different instants standing midway between A and B. Thus we can regard Einstein's definition of the time interval or of simultaneity as, given the nature of the world, the definition that is closest to our primitive notion of how two simultaneous events should appear to our eyes. If we accept this, it follows as a consequence that the two events separated in space which are simultaneous in one frame are not simultaneous in another.

> We see that we cannot attach any *absolute* signification to the concept of simultaneity, but that two events which viewed from a system of coordinates, are simultaneous, can no longer be looked upon as simultaneous events when envisaged from a system which is in motion relatively to that system.[10]

---

* Both observers can agree on the simultaneity of events that occur at the same point in space according to Einstein's conventions. If we attempt to deny this in any obvious way, there seem to be immediate problems. For example, if two signals reach the midpoint simultaneously, they might turn on a lamp. If not, the lamp might remain unlit. Presumably both observers should be able to agree whether or not the lamp is lit.

ON SPACE AND TIME INTERVALS AS REGARDED FROM SYSTEMS MOVING UNIFORMLY WITH RESPECT TO ONE ANOTHER

Speed is equal to distance divided by time. And the speed of light is measured to be the same by two observers moving with respect to one another. If we are to agree to both of these statements, we have to revise our notions of distance and time. It is difficult to do this unless we become almost painfully explicit; for our ideas of distance and time are rooted deep, and with them we have very successfully organized the world of our ordinary experience.

Consider a pulse of light sent out from the origin of what we call a *stationary coordinate system* (the coordinate system S). We do not mean by this that the system is at rest in any absolute sense, but rather that referred to this system "the laws of mechanics hold good" (it is an inertial frame) and we ourselves, as a matter of convention, wish to regard these coordinates as stationary.* In this coordinate system there are ordinary but expensive clocks located conveniently so we never have to walk very far to find one, synchronized according to Einstein's convention.

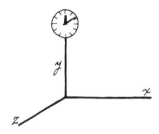

Imagine that the light pulse is emitted at the origin of the coordinate system, $(x = y = z = 0)$ at the time $t = 0$, and that it moves in the direction of the x axis. The distance it moves in a time t is ct, so that in this time it arrives at the point $P(x = ct, y = 0, z = 0)$ on the x axis.

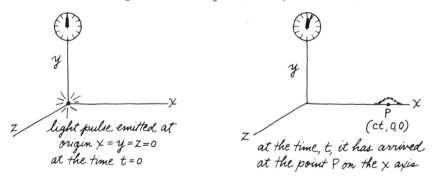

light pulse emitted at origin $x = y = z = 0$ at the time $t = 0$

at the time, t, it has arrived at the point P on the x axis

We now compare this result with a measurement made of the speed of the same pulse of light as viewed from a coordinate system S', moving

---

* Relative to the stationary coordinate system, the observer and his instruments are at rest. The moving coordinate system is in motion relative to the observer and his instruments.

uniformly with respect to the stationary coordinate system with the velocity v, along the x axis such that the origins of the two systems coincided when $t = 0$.

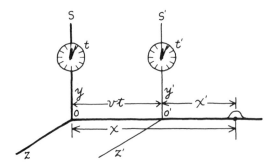

All measurements made from the moving coordinate system we will label with the coordinates x', y', and z'. The moving coordinate system also has its own beautifully synchronized and conveniently spaced clocks that run at their own rate, and we call that t'.

We would usually say[*] that y and z distances would be the same measured in the two systems and that the times t and t' would of course be the same if we had installed decent clocks, since all clocks measure absolute time. However, x' would not usually be thought to be the same as x, since the origin, O', of S' moves a distance vt in the time t. And a measurement of the physical point labeled P would give in the moving coordinate system an x' related to x by

$$x' = x - vt \tag{24.5}$$

Thus all together we would normally say that the relation between the coordinates of the *same* physical point P (that point where the light pulse is) as seen from the two coordinate systems is (the Galilean transformation)

$$\begin{aligned} x' &= x - vt & x &= x' + vt \\ y' &= y & \text{or the inverse} \quad y &= y' \\ z' &= z & z &= z' \\ t' &= t & t &= t' \end{aligned} \tag{24.6}$$

From the point of view of the moving coordinate system, the point P is not as far away from the origin as it is from the point of view of the fixed coordinate system, because the origin has moved toward P. As there is no motion in either the y or the z directions, $y' = y$ and $z' = z$. And, as always, $t' = t$. With all of this, the speed of the light pulse as seen from the moving system would be

$$\frac{x - vt}{t} = c - v \tag{24.7}$$

just what we expect, but contrary to what is observed.

What does this mean? The coordinate x' is the distance from O' to

[*] According to Galilean conventions.

P as measured in the moving frame—suppose we call it ℓ'. The coordinate x is the distance from O to P as measured in the stationary frame. This is the distance from O to O' (which is vt) plus the distance from O' to P as measured in the stationary frame—suppose we call that ℓ.

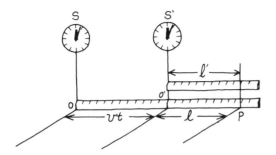

We wouldn't normally hesitate before asserting that $\ell = \ell'$, that the distance between two physical points (O' where someone might be sitting and P where the light pulse is) is the same whether measured from a stationary or a moving coordinate system. And there is no reason that we should; with that assumption we have been extremely successful in organizing an enormous variety of experience. We are not questioning its reasonableness; rather we wish to indicate that reasonable or not, an assumption has been made about a property of the world (in the same sense that one assumes that light rays travel in "straight lines") and the world might not turn out to be that way.

But if $\ell' = \ell$ and $t' = t$, then we obtain the Galilean transformation and (contrary to what is observed) the speed of light as measured in the moving system unequal to c. Under these circumstances, we have little choice other than to conclude that either ℓ' is not equal to ℓ, or t' is not equal to t, or both. That is to say, the primitive suppositions that the clocks move at the same rate as regarded from the moving and the fixed system, and that the meter stick retains its length when it is in motion, do not seem to be consistent with what is observed.

This is the point of view that Albert Einstein took in 1905. He chose to regard the conspiracy of nature that prevents one from measuring one's absolute velocity and the constancy of the velocity of light as postulates. "The following reflections are based on the principle of relativity and on the principle of the constancy of the velocity of light." The question he asked was essentially as follows. If, in the above, ℓ' is not equal to ℓ, and if time intervals in the moving system are not assumed to be the same as time intervals in the stationary system, what then is the relation to be? Given his convention for a time interval, that the speed of light is the same constant in a uniformly moving system, and that the laws of physics are independent of the uniform motion of the system to which they are referred, he could, as a consequence of these postulates, obtain the relations between ℓ' and ℓ and t' and t. These turn out to be

$$\ell' = \frac{\ell}{\sqrt{1 - v^2/c^2}} \tag{24.8}$$

$$t' = \left(t - \frac{vx}{c^2}\right)\frac{1}{\sqrt{1 - v^2/c^2}} \tag{24.9}$$

From the point of view of the stationary system (S) the length measured in the moving system as ℓ' is equal to

$$\ell = \ell'\sqrt{1 - v^2/c^2} \qquad (24.10)$$

which is smaller than ℓ'. Thus, for example, if in the moving system ℓ' measures 100 cm, and if v is one half the speed of light, then the same length measured from the point of view of the stationary system ℓ would be

$$\ell = 100\sqrt{1 - \tfrac{1}{4}} \simeq 87 \text{ cm} \qquad (24.11)$$

S' can take the point of view that he is at rest and that it is S that is moving (in the opposite direction). S' then concludes that it is the rods in the S frame that have shrunk. They come to these seemingly contradictory conclusions because their clocks do not agree. What is the "same time" in S' is no longer in the "same time" in S. Once they have given up agreement about their clocks they are forced to give up agreement about length.

Part of the confusion arises as follows. The rod is measured to have the length ℓ' in the S' frame. S measures it as it goes by. How? At the time $t = 0$ he observes that one end of the rod is at $x = 0$, the other at $x = \ell$. He then finds that

$$\ell = \ell'\sqrt{1 - v^2/c^2} \qquad (24.12)$$

A cry is heard from S': "You didn't mark both ends at the same time; you marked the far point too soon" ($t = 0$ at $x = 0$ and $x = \ell$ gives $t' = 0$ at $x = 0$ and $t' = -v\ell/c^2\sqrt{1 - v^2/c^2}$ at $x = \ell$). But by now S' is far away.

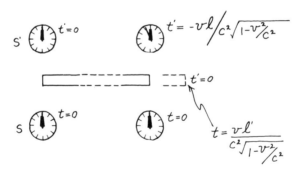

By analyzing in such detail the procedure one goes through to measure the length of a rod from the point of view of each system, one can finally be convinced* (often for no more than 30 seconds) that this new set of conventions is consistent. Why they are convenient we try to make clear in what follows.

A sphere at rest in the moving system S' appears† in the stationary

---

* One should try to think of rods so long that it palpably takes a while for the light signal to go from one end to the other.
† Appears is used loosely here. A photograph would (due to the time intervals separating the arrival of the light rays on the photographic plate from different parts of the body) not show a flattened object. But measurement of the object as it passed by would reveal the object to be flattened.

system as an ellipsoid, flattened in the direction of motion,

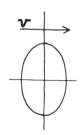

> . . . whereas the y and z dimensions of the sphere (and therefore of every rigid body of no matter what form) do not appear modified by the motion, the x dimension appears shortened in the ratio $1: \sqrt{(1 - v^2/c^2)}$, i.e. the greater the value of v, the greater the shortening. For $v = c$ all moving objects—viewed from the "stationary" system—shrivel up into plain figures. For velocities greater than that of light our deliberations become meaningless.[11]

The relation between t and t' is even stranger. Not only do the clocks in the moving system beat at a different rate, but they also do not seem to be synchronized properly. For the times they read seem to depend on their position; what is the same time in one system is a different time in the other.

TIME DILATION

That the interval between two events should appear different to different observers is so unconventional that we explore in detail how it comes about and its consequences. We choose as the first of the two events the emission of a light pulse from a source. The pulse is reflected from a mirror a distance ℓ away, and comes back to a receiver located at the source. The second event is defined as the arrival of the light pulse

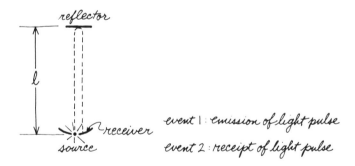

in the receiver. From the point of view of an observer who is at rest with respect to this apparatus, the so-called "time" (that is to say, the interval between the two events) is defined as the distance the pulse moves, 2ℓ, divided by the speed of light:

$$\text{time interval} = \frac{2\ell}{c} \qquad (24.13)$$

He could make such a device the heart of a clock. If, for example, the distance were 150 cm, he would say the time interval between emission and receipt of the pulse was

$$\frac{300}{3 \times 10^{10}} = 10^{-8} \text{ sec} \qquad (24.14)$$

And then, if he attached a counter to the receiver and arranged the device so that a new pulse was emitted at the same instant that the old one returned, by counting the intervals between emission and receipt of the pulse, he would have a clock or a record of time, just as an ordinary clock essentially counts the intervals between swings of a pendulum, or between oscillations of a balance wheel. The principle of all clocks is based on the hope that these swings of the pendulum, or the oscillations of the balance wheel, will occupy the same interval—a hope justified by the fact that various rhythmic patterns of this kind are consistent with one another.

Now imagine that this device is in uniform motion with a speed v in a direction perpendicular to the line between the source and the mirror and consider the interval between the two events: (1) emission of the pulse and (2) receipt of the pulse, as it appears to a "stationary" observer watching the device go by. From the point of view of the stationary ob-

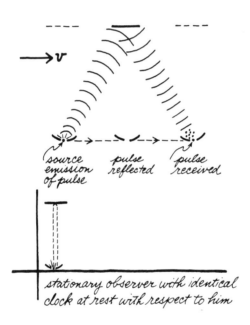

server the mirror has moved before the pulse reaches it, just as the source and receiver have moved before the pulse returns. Therefore the pulse follows the diagonal path rather than the previous path. Without doing any arithmetic, it is clear that from the point of view of the stationary observer the interval between the two events—pulse received and pulse emitted—will be longer, because the distance traveled is longer and the speed, which is the speed of light, we have agreed is the same from the point of view of all observers. How much longer can be answered with the same calculation done in analyzing the Michelson-Morley experiment. The time interval between the events 1 and 2 from the point of view of the stationary observer is

$$t = \frac{2\ell}{c}\frac{1}{\sqrt{1 - v^2/c^2}} \tag{24.15}$$

or

$$\begin{pmatrix} \text{time interval from the} \\ \text{point of view of the} \\ \text{stationary observer} \end{pmatrix} = \begin{pmatrix} \text{time interval from the} \\ \text{point of view of the} \\ \text{observer who moves} \\ \text{with the apparatus} \end{pmatrix} \frac{1}{\sqrt{1 - v^2/c^2}}$$

(24.16)

Thus from the point of view of the stationary observer, any clocks based on emission and receipt of the light pulses in the moving system would run slow because the fundamental interval on which they were based was too long. If, for example, v were one half the speed of light,

$$\sqrt{1 - v^2/c^2} = 0.87 \tag{24.17}$$

the interval recorded as

$$\frac{300 \text{ cm}}{3 \times 10^{10} \text{ cm/sec}} = 10^{-8} \text{ sec} \tag{24.18}$$

in the moving system would be recorded as

$$\frac{10^{-8} \text{ sec}}{\sqrt{1 - v^2/c^2}} = \frac{10^{-8}}{0.87} = 1.15 \times 10^{-8} \text{ sec} \tag{24.19}$$

in the stationary system.

If they wanted to argue, the stationary observer could say to the moving observer, "Your clock is slow. The interval between the events on which you've based your clocks is too long." However, the moving observer could equally well say, "My clock is right. It is your clocks that are slow." For, from his point of view having adopted Einstein's convention, he might as well be standing still, and the so-called stationary observer moving with the same speed in the opposite direction.

Before the argument gets too violent, we point out that what is being done here is a matter of definition. If either observer does not like the new convention, he is free to go back to the old one (we shall explore that possibility later.) However, we make no guarantee that the old convention will be either as convenient or as fruitful as the one we have just introduced. Essentially, following Einstein, we have defined the time interval to be distance divided by the speed of light. We have agreed, prodded by experimental results, that the speed of light is the same for all observers. And we have used Euclidian conventions for the combination of distances within a given frame of reference. Having accepted this, we have no further choice. We must agree that the interval appears different from the point of view of the two observers.

Now the exasperated objection: "That is a very amusing clock you've made, using expensive sources, receivers, and pulses of light. I followed you and agree with your conclusion. But suppose we unburden the taxpayer and use just a good old-fashioned clock, a pendulum, the watch on my wrist, or an hour glass. Can you show me that the interval between two swings of a pendulum (or the interval between the hour glass full and the hour glass empty) would appear differently to the two observers?"

We offer two answers. The first is somewhat pristine. It refers to the principle of relativity. ". . . The phenomena of electrodynamics, as well as of mechanics, possess no properties corresponding to the idea of absolute rest. They suggest rather that . . . the same laws of electro-dynamics and optics will be valid for all frames of reference for which the equations of mechanics hold good. We will raise this conjecture (the purport of which will hereafter be called the 'Principle of Relativity') to the status of a postulate. . . ."[12] Or, "the laws by which the states of physical systems undergo change are not affected, whether these changes of state be referred to the one or the other of two systems of coordinates and uniform translatory motion."[13] Or, to paraphrase Poincaré, there is a conspiracy on the part of nature that prevents us ever from learning our state of absolute motion with respect to the ether. If we accept these statements, which we summarize as the principle of relativity, we shall have to agree that all clocks—balance wheel, pendulum, or otherwise—must adjust themselves somehow to agree with the light clock we have just constructed.

Suppose it were otherwise. Suppose, for example, we had a handmade and incredibly expensive Swiss watch which measured absolute and true time. And suppose we compared it with our light clock, which is so evidently affected by its motion (Fig. 24.1). Then by observing the difference in rates of these two clocks, we would be able to deduce our absolute motion through the ether. For example, if the light clock ran at a rate only 0.87 that of the "true time" watch, we could deduce that

$$\sqrt{1 - v^2/c^2} = 0.87 \qquad (24.20)$$

or that

$$v = \tfrac{1}{2}c \qquad (24.21)$$

light clock

FIG. 24.1. Light clock compared with watch that measures "true time." If the two do not agree, the disagreement can be used to determine the absolute speed, contrary to the principle of relativity.

"That," says our inquisitor, "is precisely what should happen because, with some consideration, the light clock is so obviously defective." "Let us remind you," we say, crushingly, "that this is really just another attempt to measure our absolute motion. If we succeeded in doing it, there then would be no conspiracy on the part of nature to deprive us of the opportunity, contrary to the principle of relativity and to the many experimental results accumulated in the nineteenth century and contrary also, in a very direct way, to the results of Michelson and Morley, who can be thought to have compared two light clocks, one perpendicular, the other horizontal to the direction of motion. If these two clocks did not agree with each other, this would be revealed as the apparatus was rotated. But the Michelson-Morley experiment (on earth, on the sea, and in the air) gives always the same null result, which indicates that the two clocks beat at the same rate."

The inquisitor is silenced; but is he convinced? If he accepts the principle of relativity, the constancy of the speed of light, and Einstein's convention to regard a time interval as the distance a light pulse travels, divided by the speed of light, he is forced to the conclusion that the interval between two events separated in space will be different when viewed from two different frames of reference. He might agree further that in the case of the light clocks just constructed, it is clear why this happens. And that it follows from the principle of relativity that any other type of clock must share in this behavior.

But he looks unhappy. "How is it that this happens? Can it be analyzed or understood?" We descend from Olympus. It is difficult to analyze in detail because clocks are complicated machines, but a qualitative explanation can be given. We recall what Lorentz did before Einstein. Let us assume that any clock is constructed of rigid bodies, of atoms and so on, which hold themselves together via electrical forces. If this is so, then we already know from Maxwell's equations that electrical forces are motion dependent. Magnetic forces appear when charges move that were not there when the charges were at rest. Thus equilibrium positions can change. And therefore one can at least imagine that what is going on in the complicated watch due to its motion (from the point of view of the stationary observer past whom the watch is moving) is that, owing to the change in forces produced by the motion of the watch, there occur changes in equilibrium positions so that the various levers, pendula atoms, and so on, that make up the watch beat at a different rate, in fact, beat at a rate precisely to agree with the light clock.

This is a possible, but a less elegant point of view, because it denies the right of the observer on whose wrist the watch is fastened to consider his own local time as good as the time of any other system—the essence of the spiritual liberation provided by relativity theory. We can retain the notion of absolute time. But it becomes an encumbrance; the local time in any uniformly moving coordinate system is just as good as the local time in another uniformly moving coordinate system. There is no more reason to prefer the time as seen in one coordinate system to the time as seen in another than there is to state that it is the sun rather than Sirius that is at the center of the universe.

In our treatment of the wristwatch, there was a small codicil that we should now amplify. Lorentz assumed that what nonelectrical forces existed among the objects that made up the wristwatch would change their behavior when going from a stationary to a moving system in the same way as electrical forces. That is to say that electrical forces, as defined by Maxwell's equations, provided the pattern for the way force systems should behave in going from a fixed to a moving coordinate system. If this were not the case, then the various equilibrium positions of the watch would not be shifted properly, and one could not conclude that it would beat at the same rate as the time clock. Essentially, the electromagnetic forces of Maxwell are the model of how a force system must behave in order that the time rates be kept properly matched.

Now one might ask "Do Newtonian forces, such as gravitation, do Newton's equations themselves behave in the proper way?" The answer is "No." Therefore, either Newton's equations are correct as they stand, and the principle of relativity is incorrect, or vice versa. Both Einstein and Lorentz took the second point of view. And here, essentially, they initiated a program: that all the equations of physics must be written in such a form that the principle of relativity is obeyed. In those cases in which they do not, they must be modified.

This may or may not be the correct point of view. But one can state as a fact that it has been extraordinarily fruitful. We shall discuss in more detail later how Newton's equations are modified to make them consistent with the principle of relativity, although we might mention at the moment that when the equations are so modified, they are in remarkable agreement with experience. So much so that these relativistic equations have become part of the engineer's standard tools for the construction of devices in which particles move at speeds close to that of light.

THE GRANDFATHER CLOCK
IN OUTER SPACE

The theory of relativity is founded on the observed constancy of the speed of light, and our inability to detect our absolute motion with respect to the light medium. However, the primitive concept of absolute time, which has been discarded, is so deep and has been so successful for so long that it has become a part of our language—so constructed that it is difficult to even speak of time as being anything other than that which is the same for all people and all places. We say, "At that time . . . ," "In the time of Caesar . . . ," "At the same time, the astronaut . . . ," or "Meanwhile, back at the ranch. . . ." It is a concept so much a part of our thinking that it is difficult to change—to accept a definition of time that is different for different observers. And the mind always struggles, the question is always raised "Is it necessary? Is this what the experiments tell us?" The answer is "No, it is not necessary." And the fact that it is not necessary makes Einstein's achievement the more brilliant. For he has not shown us truth. Better, he has shown us a way to regard our experience so perfect that, having seen his canvas, we find it difficult to look at the world another way.

Suppose we attempted to construct a theory that would explain the same phenomena—the constancy of the speed of light and the inability to detect one's absolute motion through the light medium—without introducing the unconventional definition of time as done by Einstein. Suppose, for example, we took the point of view that there existed some frame of reference that really was preferred. Let us say a system of coordinates fixed in the sun. And suppose, by definition, we declared that this frame was at rest with respect to the luminiferous ether. We call it the *absolute rest frame*. And let us imagine a large number of grandfather clocks all very well made, synchronized with one another in this frame. We declare that these clocks read absolute time and, from now on, all clocks will be referred to them. Only those clocks which beat at the same rate are declared to be correct. For this frame we shall then have a conventional definition of time—the same for everyone. Simultaneous events will be defined as those which are simultaneous from the point of view of the grandfather clocks, which are by definition* said to be at rest with respect to the ether.

The Frame of Reference of the Grandfather Clocks

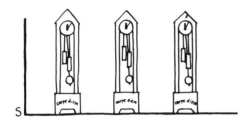

Now, from the point of view of the grandfather clocks, everything is fine. The clocks work well; they are synchronized; all rulers and watches behave as expected. The observer at rest in this frame, comfortably ensconced among the accurate clocks, aware that he is in the frame to which all measurements will be referred, feels a certain satisfaction.

But consider the situation as viewed by another observer at rest in a frame moving with respect to the grandfather clocks.

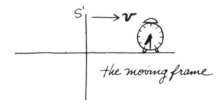

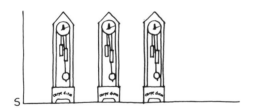

The observer in the moving (what we call the S') frame attempts to construct a clock, for example, a light clock, as described previously. He agrees that he is moving—the grandfather clocks are at rest—and therefore agrees that the path followed by the light beam between the source and the receiver of his light clock is

---

* We say "by definition" because, according to the principle of relativity, we can never actually know what frame is at rest with respect to the ether. Nor, for that matter, does it make any difference. But since it makes no difference, we can always define one frame to be the absolute rest frame.

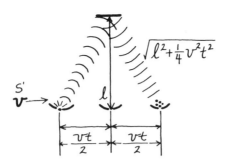

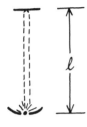

that is, the diagonal path, rather than the path which he would propose if he took the point of view that he was as much at rest as anyone else (Einstein's point of view).

By an argument identical to one before, he then calculates the interval between the emission and the absorption of the light pulse, the distance traveled divided by the speed,

$$t = \frac{2\ell}{c} \frac{1}{\sqrt{1 - v^2/c^2}} \qquad (24.22)$$

as compared with

$$\frac{2\ell}{c} \qquad (24.23)$$

which is the time interval between emission and absorption for the same device if he were to regard it as being at rest.

If he wishes to use $2\ell/c$ as his standard interval (let us say, for example, that $2\ell/c$ was 10^{-8} sec, which it would be if ℓ were 150 cm long), he would then say that the device as he was using it gave an interval larger than 10^{-8} sec by the factor*

$$\frac{1}{\sqrt{1 - v^2/c^2}}$$

Having done this, he would adjust all his clocks accordingly. Further, he would synchronize them so that they would read simultaneous times only if the two grandfather clocks at the same points in the rest frame read simultaneous times.

All of this he could do perfectly well. And by doing so he would have obtained a definition of time† in accord with some of our primitive and intuitive notions. If all observers in all frames did the same, time would be, in part, as we wish it to be—the same for everyone, absolute and unambiguous. Before we congratulate ourselves, however, we might explore some of the consequences of this convention for the observer in S′ and all the other frames moving with respect to the grandfather clocks.

---

* The speed v he could thereby measure is, of course, his speed relative to the frame of the grandfather clocks.
† Not Einstein's definition. For example, t_{AB} would not be equal to t_{BA} for the observer at rest in S′.

After making this adjustment for his clocks our friend in S' would find that all the physical processes in his environment had slowed down. If, for example, the ratio were

$$\frac{1}{\sqrt{1 - v^2/c^2}} = 2 \tag{24.24}$$

he would find twice as many seconds in a day—that is to say, the interval between sunrise and sunrise—twice as many seconds in a year—the interval between one spring equinox and the next—twice as many seconds in a lifetime—the interval between birth and death—and so on. Every process that occurs would be gratuitously expanded to take twice as many seconds. He might accept this as a nuisance, but perhaps not an intolerable nuisance. The relations between intervals between various events in his environment would remain the same. The year defined as the interval between one equinox and the next would go as many times—fourscore—into an average biblical lifetime. Only his fundamental unit of time, as defined by the interval between emission and receipt of the light beam—10^{-8} sec—would be cut in half. And this would involve no more than a change in his scale.

His definition of simultaneous events would be somewhat less convenient. If the S' frame were moving very rapidly with respect to the grandfather clocks (let us say close to the speed of light), then he would be forced to call events simultaneous when they did not look simultaneous at all. He would be forced to say that the turn of the signal light from red to green and the start of the taxicab through the intersection were not essentially simultaneous but were events separated by long periods of time. If, for example, the observer, signal, and taxi were moving with respect to the "true rest frame," as below,

because the observer agrees that he is moving into the light emitted by the taxi, and away from the light emitted by the signal, he will assert that the two events—the starting of the taxi and the change of the signal light—are simultaneous only if he sees the taxi begin to move before he sees the light change, a concept that would be difficult to explain to the local police.

Measurement of distance would be even more peculiar, since the observer in S' would agree that his measuring rods, his rulers in the direction of motion, are too short by the factor $\sqrt{1 - v^2/c^2}$.

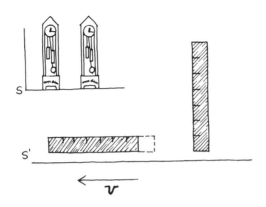

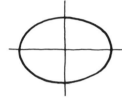

The meter stick S′ borrowed one time from the fixed frame, he agrees, is now too short when he lays it parallel to the direction of motion. However, it is just right as a meter stick when it is perpendicular to the direction of motion. Thus he is faced with the necessity of using a different rod to define a meter parallel to the direction of motion from the one he uses perpendicular to the direction of motion. If he wishes to draw a circle, he must draw an object that looks as shown; he agrees it looks funny. But with the peculiar definition of length that has been forced on him, it is this curve whose points are all equally distant from the center.

And as we catalogue the list of seeming absurdities into which he is forced, we might decide that the conventions he is using to describe his world are inappropriate. To preserve some of his primitive intuitions about time, he has been forced into a position where his meter stick changes length as he changes its direction, where things that appear to happen at the same time can no longer be thought to happen at the same time, and so on.

We might smile to ourselves and say "Here is a man who lives on a rubber sheet, continually stretching and twisting, attempting to describe his world as though he lived in a place where a rigid body maintains its shape." Or "Here is a man who lives on the surface of a sphere desperately trying to describe this surface as though it really were a plane." Our fundamental objection to his description is not that it is incorrect, because he can force it to be correct, he can force it to be an accurate description of his world. But rather that it is clumsy. It contains entities that serve no purpose. It conceals rather than reveals symmetries and, in the end, it is not an elegant visualization of the world. It has no aesthetic clarity.

The lonely splendor of Einstein's relativity is its austerity. He finds we can do as well without defining a frame that is at rest, without defining time so that it is the same for all observers. We do not need these crutches, other than emotionally, for "the phenomena possess no properties corresponding to the idea of absolute rest." Any inertial frame can be considered at rest; time or length, as defined in one frame, is just as good as time or length as defined in another, and such things as absolute rest and absolute simultaneity are intrinsically unobservable.

He thus discards the concepts of absolute rest and absolute time as things we neither can see, nor need—ghosts whose presumed existence may please or frighten us, but whose presence we never observe. Within a given frame of reference, matters are then relatively simple. Meter sticks do not change as we rotate them; circles look like circles; simultaneous events appear the way we believe they should; the lifetime of an individual will be fourscore years. In exchange for this the rules by which we go from our own system of reference to one that is moving with respect to us are altered. The alteration, although curious from our usual point of view, is not as curious as the alteration we would have to make otherwise.

How much Einstein in this respect is like a watchmaker. His concern is with the practical problem of building a machine that measures a time interval—the time that will be measured by a clock. Our atavistic memories object. A clock we build measures only imperfectly, "absolute, true, and mathematical time." "That may be so," the watchmaker responds. "The concept of absolute time may be pleasing, but we will never have anything but the clocks I build. And only the time that is measured on the clocks I build ever enters into the equations of physics, or the 'laws that govern' phenomena. If we insist on retaining the notion of true, absolute time, we may. But it becomes an encumbrance. It complicates our view of the world. We put it in one place, and it disappears before we reach another. In the end, we discard it as we discard a tool that has become useless and obsolescent."

"And in conclusion," Einstein writes, "I wish to say that in working at the problem here dealt with I have had the loyal assistance of my friend and colleague M. Besso, and that I am indebted to him for several valuable suggestions."[14]

Who was M. Besso? Was this friend and colleague another worker in that patent office in Berne, patiently listening to Herr Einstein explain, perhaps not for the first time, that the same laws of electrodynamics and optics are valid for all frames of reference for which the equations of mechanics hold good; that the principle of relativity is only apparently irreconcilable with the statement that light is always propagated in empty space with a definite velocity, c; that we see that we cannot attach any absolute signification to the concept of simultaneity, but that two events which viewed from a system of coordinates are simultaneous can no longer be looked upon as simultaneous events when envisaged from a system which is in motion relatively to that system; or that a rigid body which, measured in a state of rest has the form of a sphere, therefore has in a state of motion viewed from a stationary system the form of an ellipsoid? Did M. Besso listen? Did he smile? Who was M. Besso?*

* The author wishes to thank the reviewer who informed him that M. Besso was a colleague of Einstein's at the patent office, a very knowledgeable engineer, and that Besso became an in-law of Einstein when his wife's brother married Einstein's younger sister Maja.

25

THE UNION OF NEWTON'S LAWS WITH THE PRINCIPLE OF RELATIVITY

RELATIVISTIC MECHANICS

The principle of relativity and the constancy of the speed of light are made consistent by Einstein if one agrees to follow his prescription for the transformation of times and lengths between stationary and moving frames. The equations of electrodynamics—Maxwell's equations—then have the same form, as viewed from a stationary or moving system. For example, the speed of light (which is an electromagnetic wave, one of the solutions of Maxwell's equations) is the same in all uniformly moving systems. In addition, the principle of relativity requires that there be no way which permits one to decide if a system is at absolute rest: "the phenomena . . . possess no properties corresponding to the idea of absolute rest." This implies that all the equations of physics should be of such a form that one cannot utilize any of the phenomena they describe to discover the absolute motion of the system.

One is then led immediately to ask whether the equations of mechanics, Newton's equations, already more than 200 years old in 1905, themselves are of the proper form. The answer is "No." If lengths and times are transformed between one moving system and another according to Einstein, then Newton's equations, as Newton wrote them, are not consistent with the principle of relativity.

The logical situation is as follows. If lengths and times transform between one system and a system moving with respect to it according to the Galilean transformation, the way one would have believed prior to Einstein, then Newton's equations would not allow one to tell the difference between an observer at rest and another moving uniformly with respect to him. In all inertial frames Newton's laws would hold equally well. However, under such a transformation, Maxwell's equations would not maintain their form; in particular, the speed of light would be different for different observers.

The observation that the speed of light is the same for all observers in uniform motion lies at the root of the change of the transformations of lengths and time between fixed system and moving systems. With Einstein's transformations Maxwell's equations are the same in both systems. And the speed of light is a constant. However, now Newton's equations do not transform properly. (Without a large amount of analysis, it is not difficult to see that Newton's equations as they stand would produce problems. For example, if a uniform force acts on a body, according to Newton, the acceleration is a constant. A body subject to a constant accelerating increases its speed indefinitely. Thus from the point of view of Newton, a body can be accelerated until its speed is larger than that of light. However, an ordinary body moving at a speed greater than that of light involves the theory of relativity in probably insuperable problems.)

One then must either abandon the principle of relativity, or modify Newton's equations so that they conform with it. If we abandon the principle of relativity, we would say that by some mechanical experiment it should be possible to determine our absolute speed. However, the path taken by Lorentz, Einstein, and the physics of the twentieth century is to modify Newton's equations so that they are consistent with the principle of relativity. Which path is correct is not a matter that can be settled a priori. Nature could be either way; one has to look at the world to decide which way it is. There has been ample confirmation that it is Newton's equations that must be modified.

As we stated them originally, Newton's first two laws of motion were as follows:

1. In the absence of forces, a body moves with uniform speed in a straight line.

2. In the presence of forces, the body changes its state of motion in such a way that

$$\mathbf{F} = \frac{\Delta(m\mathbf{v})}{\Delta t} \qquad (25.1)$$

or

$$\mathbf{F} \, \Delta t = \Delta \mathbf{p} \qquad (25.2)$$

the impulse is equal to the change of momentum.

Before we modify these equations, we might as well answer the inevitable question. We have explored in some detail the structure that results when Newton's equations are developed for various force systems. In the case of planetary motion, for example, the consequences of Newton's equations, as applied to gravitational forces between bodies, are in remarkable agreement with the observed motion of the planets. We therefore have a structure that is in excellent correspondence with our observations of the world—which always inclines one to use colorful words, saying the structure is "correct," or "true," or "real." Now we assert that the equations and the structure that results are not consistent with the principle of relativity, so that they must be modified. In what sense can we do this?

It is clear that any system of equations that is to correspond with the same phenomenon must be very much like Newton's equations in that realm of experience for which Newton's equations are known to be valid, for example for the motion of the planets or for various mechanical systems on earth. We have no guarantee, however, that Newton's equations will continue to describe correctly phenomena that we have not observed. The equations we write down of necessity have a far greater generality than the phenomena from which they have been induced. And the question one really asks when one tries to modify Newton's equations to make them consistent with the principle of relativity is: Is it possible to modify Newton's equations in such a way that they are very much like Newton's equations, where we know they must be like Newton's equations

—that is, for planetary motion—but are different enough from Newton's equations in such a way as to be consistent with the principle of relativity?

It is always a delicate matter to modify a successful theory, because a theory such as Newton's implies a very elaborate structure of relations. Our experience indicates that a large portion of this structure is quite right, whereas further experience (with the principle of relativity) indicates that other parts of this structure are not. One then must go into the workings of the theory and modify things in such a way that that part of the structure which we know to be correct remains essentially the same (and by essentially we mean that we will allow modifications that are so small numerically that they do not fall under the observation of the senses) whereas that part of the structure which is modified greatly is usually a part with which we have had, as yet, no direct experience.

In the case of the relativistic modification of Newton's equations, for particles moving at small speeds the relativistic equations (we call them "relativistic equations") are Newton's equations with corrections that are too small to be easily observed. However, at speeds close to the speed of light, the modifications become important. Thus Newton's equations as he proposed them and their relativistic modification give an almost identical description of the phenomena in the range of motions of the order of the speeds of the planets. It is only at speeds close to the speed of light that the relativistic modifications become large enough to be easily observed. It is because Newton and his contemporaries dealt with the range of phenomena where the speeds were small compared to that of light that the equations as he proposed them were so successful. It is curious that often it has been possible to modify the fundamental postulates from which a theory was constructed in very radical ways, leaving great portions of the theory essentially unaltered. As though one were to take a cathedral, with its columns, arches, flying buttresses, gargoyles, and angels and transport it unblemished from the old world to the new.

The only modification necessary to make Newton's first and second laws consistent with the principle of relativity is to change the definition of the momentum, the quantity of motion. According to Newton, the quantity of motion is the mass times the velocity, $\mathbf{p} = m\mathbf{v}$. If we modify this as follows:

redefinition of the quantity of motion
to make Newton's equations consistent
with the principle of relativity:

$$\mathbf{p} = \frac{m_0}{\sqrt{1 - v^2/c^2}} \, \mathbf{v} \tag{25.3}$$

we have done the job. The quantity m_0 is called the *rest mass* of a particle, its mass when the particle is at rest, the mass that Newton would have assigned the particle. To obtain its *relativistic mass* we have divided the rest mass by the ubiquitous factor

$$\sqrt{1 - v^2/c^2}$$

which appeared first in our analysis of the Michelson-Morley experiment

—here, there, and everywhere in the theory of relativity. Rest, perturbed spirit. The relativistic form of Newton's equations then reads

1. A body in uniform motion continues in uniform motion if acted on by no forces.

2.
$$\mathbf{F}\,\Delta t = \Delta \mathbf{p} \tag{25.4}$$

where

$$\mathbf{p} = \frac{m_0}{\sqrt{1 - v^2/c^2}}\,\mathbf{v} \tag{25.5}$$

It is not difficult to see what some of the most important qualitative consequences of this are. We can regard the modification as changing the mass of the particle as its speed increases. As the speed approaches that of light, then v^2/c^2 approaches 1; $\sqrt{1 - v^2/c^2}$ becomes small and

$$\frac{m_0}{\sqrt{1 - v^2/c^2}}$$

(which is essentially the particle's inertia) grows large. Thus as the speed of the particle approaches that of light, its inertia increases indefinitely, becoming infinite when v is equal to c. As the inertia is a measure of the resistance of the particle to a change of motion, the greater the inertia, the greater force will be required to accelerate it. As the particle approaches the speed of light, it becomes more and more difficult to change its speed or to give it an acceleration. Thus we obtain the important result that is impossible, from the point of view of the relativistic equations, to accelerate a particle to a speed larger than that of light.

When the speed of the particle is small compared to that of light, the modification in Newton's equations becomes negligible. For example, at a typical planetary speed—something of the order of 18.6 miles/sec—v^2/c^2 is equal to

$$\frac{v^2}{c^2} = \left(\frac{18.6}{186{,}000}\right)^2 = 10^{-8} \tag{25.6}$$

This results in a modification of the mass of about one part in a hundred million. Thus we can say roughly that the planetary orbits, as predicted by Newton's equations unmodified, differ from the prediction of the relativistic equations by something of the order of one part in a hundred million.

It is in this sense that one says that the relativistic equations become the Newtonian equations in ranges of speed much smaller than the speed of light. As v approaches zero, $\sqrt{1 - v^2/c^2} \to 1$ and the relativistic equations become precisely Newton's equations. One of the ways of obtaining the relativistic equations is to ask: What is the system of equations that will become Newton's equations when the particle is at rest, and yet will satisfy the principle of relativity? One answer has been given above.*

---

* The answer is not unique. One usually chooses the simplest equations consistent with all of the known requirements.

We look again at Newton's equations in their relativistic form:

$$\mathbf{F}\,\Delta t = \Delta \mathbf{p} \tag{25.7}$$

$$\mathbf{p} = \frac{m_0}{\sqrt{1 - v^2/c^2}}\,\mathbf{v} \tag{25.8}$$

Other than the change in the definition of the momentum, they are precisely as in *Principia*—the force multiplied by the time interval in which it acts is equal to the change of momentum. For this reason many of the structural relations that we have already obtained for the nonrelativistic equations (in fact, all those relations that are a consequence of the second law and do not make explicit use of the fact that $\mathbf{p}$ is defined to be $m_0\mathbf{v}$) will also be valid in the relativistic case—an explicit example of how the structural relations within a theory may be preserved even if the interpretation of the symbols has been altered.

One of the most important results of Newtonian theory is that in the absence of forces the momentum remains constant. This is also true in the relativistic theory. In the absence of forces, $\Delta \mathbf{p}$, the change of momentum, is zero and therefore the relativistic momentum remains constant.* In a similar fashion, if Newton's third law is obeyed (action equals reaction) in the sense discussed above, then the relativistic momentum will be conserved in particle collisions. Thus we can again in the relativistic theory define a center of mass; we can prove the theorems that in the absence of external forces, the center of mass continues to move with a constant velocity, and so on.

In the development of Newtonian theory we saw that the change of the momentum, the impetus of medieval mechanics, was related to a force acting a given time. We introduced work and energy as quantities related to forces acting over a distance and were able to define kinetic energy in such a way that the work done on a particle was equal to its change of kinetic energy. We can again introduce the idea of a force acting over a distance, and again define work as previously. And if work is done on a relativistic particle, there is a something that changes. This something is related to the *vis viva* of medieval mechanics, and to the energy of Newtonian mechanics. And we call it again the kinetic energy. However, the form of the energy as a function of the speed is different than in the nonrelativistic case because the relation between the momentum and the speed has changed. In the nonrelativistic case we could write

$$\text{kinetic energy} = \tfrac{1}{2}m_0 v^2 = \frac{p^2}{2m_0} \tag{25.9}$$

since

$$p = m_0 v \quad \text{(magnitude)} \tag{25.10}$$

For a particle moving in the absence of forces the kinetic energy would be the total energy;

---

* This can be interpreted as meaning that among the relativistic equations it is possible to find a quantity—the relativistic momentum—which remains constant in the absence of forces.

$$E = \tfrac{1}{2}m_0 v^2 = \frac{p^2}{2m_0} \qquad (25.11)$$

The relativistic modification of this relation gives the following expression:

$$E = \sqrt{m_0{}^2 c^4 + p^2 c^2} \qquad (25.12)$$

which can also be written*

$$E = \frac{m_0 c^2}{\sqrt{1 - v^2/c^2}} \qquad (25.13)$$

That is to say, it is the quantity

$$\sqrt{m_0{}^2 c^4 + p^2 c^2}$$

which changes when a force acts on a relativistic particle through a given distance. For small values of the momentum the above expression becomes approximately

$$E \simeq m_0 c^2 + \frac{p^2}{2m_0} \qquad (25.14)$$

Thus for small velocities, the relativistic energy of a free particle becomes $p^2/2m_0$ (which is just the kinetic energy in the nonrelativistic case, as we expect) plus an additional term $m_0 c^2$, which is a constant. If the particle has zero momentum, the expression for the energy becomes

$$E = m_0 c^2 \qquad (25.15)$$

certainly the most famous formula in the lexicon of physics.†

In 1905, shortly after his original paper, Einstein published another called "Does the Inertia of a Body Depend upon Its Energy Content?" In this he analyzed the process by which a body would emit light. And he concluded as follows:

> If a body gives off the energy E‡ in the form of radiation, its mass diminishes by E/c^2. The fact that the energy withdrawn from the body becomes energy of radiation evidently makes no difference, so that we are led to the more general conclusion that the mass of

* We know that

$$p = \frac{m_0 v}{\sqrt{1 - v^2/c^2}} \qquad \text{(magnitude)}$$

Therefore,

$$E^2 = m_0{}^2 c^4 + \frac{m_0{}^2 v^2 c^2}{1 - v^2/c^2}$$

$$= m_0{}^2 c^4 \left(1 + \frac{v^2}{c^2 - v^2}\right)$$

$$= m_0{}^2 c^4 \left(c^2/(c^2 - v^2)\right)$$

and

$$E = m_0 c^2/\sqrt{1 - v^2/c^2}$$

† Since $m_0/\sqrt{1 - v^2/c^2}$ is sometimes denoted by m (the mass which changes with speed), from Eq. (25.13) it is also true that $E = mc^2$.
‡ Einstein used the symbol L for energy.

a body is a measure of its energy-content; if the energy changes by E, the mass changes in the same sense by $E/9 \times 10^{20}$, the energy being measured in ergs, and the mass in grams.[1]

He then concludes prophetically:

It is not impossible that with bodies whose energy-content is variable to a high degree (e.g. with radium salts) the theory may be successfully put to the test.[2]

In the region of small speeds what the theory of relativity has done is to add to the energy equation a constant energy associated with the rest mass of the particle, m_0c^2. We may regard this additional term as a kind of potential energy due to the rest mass of the particles. And, as in the case for any term in an energy equation, if this energy can be converted to kinetic energy, it can be converted to work. If not, then the maximum work the body can do is

$$E - m_0c^2$$

which can be called the *relativistic kinetic energy*. This, using Eq. (25.13), is equal to

$$\text{kinetic energy (relativistic)} = m_0c^2 \left(\frac{1}{\sqrt{1 - v^2/c^2}} - 1 \right) \quad (25.16)$$

If the mass of a body could be diminished somehow, we could convert the mass itself into energy and work. It is not necessarily possible to do this; it is not always possible to convert heat into work. However, there are some cases, as Einstein mentions—radium salts—and, as we know now, many other cases, in which matter, mass, is converted into energy. This possibility has removed the distinction between what seemed to be two separate concepts, mass and energy. One can no longer say that mass and energy are conserved separately. Therefore, one again generalizes the concept of energy and unites it with the concept of mass. Energy is conserved if energy includes the rest energy m_0c^2 of the system. Matter, however, is not freely convertible into energy. In some cases it will convert, and in other cases not. The rules that govern the conversions are the subject matter of the most currently active branch of physics.

26

THE TWIN PARADOX

Twins are born to be separated. They part at birth, lead widely different lives, and meet again in a surprised and joyful confrontation. It has been this way in drama for centuries. The twins of the twenty-first century will be no different. One remains on earth by the launching pad, while his brother leaves for a distant star on a rocket ship and travels at a very high speed. From the point of view of the earthbound brother, the clock moves quickly. Time flies. Ten years, twenty years, thirty years pass. His hair is gray, his eyesight weak, and he has, alas, developed a paunch; thus age comes to us all. But now he rejoices. His brother is returning. The rocket ship is sighted, and in a few days splashes down successfully. The brothers embrace in an emotional reunion. But what does the earthbound twin see? His brother is almost as he appeared thirty years before, his hair brown, his belly firm. He has hardly aged at all—five years, perhaps, but after a space of thirty, he looks almost younger than when he left. "And what have you done in the last thirty years?" he says to his brother. "Not very much; what did you do in the last five?" Is such a thing possible? We think yes.

The engineering may be difficult, perhaps too difficult, but if, if we could put the brother on a space ship that moves rapidly enough, then the clock on the space ship (and by clock we mean, of course, all physical processes, heartbeats, and so on) from the point of view of the brother on earth will beat more slowly than the clock on earth. Thus the brother on earth will age more rapidly. And when the brother returns, he will be younger; he will have lived less. Time intervals on the space ship will be related to the time intervals back on earth by

$$\text{(time interval on space ship)} = \sqrt{1 - v^2/c^2} \times \text{(time interval on earth)}$$
$$(26.1)$$

If the speed of the space ship is $0.99c$, then the square root becomes approximately

$$\text{for } v = \frac{99c}{100} \qquad \sqrt{1 - v^2/c^2} \simeq \frac{1}{7} \qquad (26.2)$$

and an interval of thirty years on earth would be equivalent to an interval of about four and a half years on a space ship; which means that the brother on earth would age thirty years, while the brother in the space ship would age only about four and a half.

We haven't yet subjected two brothers to this particular experience (although one can imagine a not too good comedy based on it[*]), so we haven't observed just this situation. But we have observed what we con-

[*] "Comedy of Cape Kennedy"?

sider essentially similar situations (those cases, for example, in which the interval between the birth and death of a particle is observed to be enhanced when it is in motion with respect to us), so perhaps we can believe that the brother will return having aged and lived less.

"Suppose," says the earthbound brother desperate to recoup for the years lost, "that I got into a space ship, too, and that we went off in opposite directions and then returned after a while. From the point of view of my space ship I am standing still and aging at the right rate. You are moving, and therefore you will age more slowly. However, from the point of view of your space ship, you might as well be standing still and I would be moving. Therefore, I will be aging more slowly. As I see it, we can both live thirty years and come back and have lived only five."

From the point of view of the earthbound observer, the brother in the space ship is in a moving frame of reference, and his clock moves more slowly. Therefore, when he has returned he has experienced fewer heartbeats, and he is younger. Why, however, couldn't one take the point of view of the brother in the space ship? He regards the earth as moving off and then returning toward him. If he takes this point of view, then the time is dilated on the earth and the brother on the earth experiences fewer heartbeats. And when they return, from his point of view, his earthbound brother should be the younger.

The assumption that produces this paradox is easy to make.° It is the assumption that either frame of reference is equally valid for viewing the experience. This is not what Einstein intended, nor is it in agreement with experience.

> The phenomena of electrodynamics as well as of mechanics possess no properties corresponding to the idea of absolute rest. They suggest rather that, as has already been shown to the first order of small quantities, the same laws of electrodynamics and optics will be valid for *all frames of reference for which the equations of mechanics hold good.*†[1]

All frames of reference are equivalent for which the equations of mechanics hold good. But the equations of mechanics do not hold good in *all* frames.

Consider the first law: A body continues at rest or in uniform motion unless acted upon by forces. This is not true in an arbitrary frame of reference. If a body is moving uniformly, that is, with constant speed and constant direction as we regard it, let us say, standing at a street corner, it will continue to move uniformly if we stand on another street corner, or if we turn around, or if we look at it standing on our hands. It will continue to appear to move uniformly if we put ourselves into uniform motion with respect to the street corner.

However, if we accelerate ourselves with respect to the street corner the object will no longer appear to move uniformly—it will appear to accelerate in the opposite direction—even though no forces are acting. Imagine an object standing still from one point of view. If we accelerate ourselves toward that object, then the object will appear to be ac-

° And is made quite frequently.
† Author's emphasis (L.N.C.).

celerating toward us. Now, if we agree that there are no forces acting on the object, then from the second point of view the first law of motion will not be obeyed by the object because it will be increasing its speed without a force acting on it. All frames of reference are thus not equivalent.

What are called "inertial frames" are all those frames for which Newton's laws of motion hold. And one way of stating the content of Newton's dynamics is that there exists one frame of reference in which Newton's laws are valid. And from this frame all others can be obtained by changing motion uniformly.

The problem as far as the twins are concerned is that if the earth and the rocket ship move uniformly with respect to one another (that is, imagine the rocket ship has now attained its speed), either of these frames is equally legitimate, is an inertial frame. If one is an inertial frame, so is the other. But if the rocket is to return to the earth, then the rocket must slow down and stop and come back to earth or the earth must slow down and come back to the rocket. And as the rocket slows down it is no longer an inertial frame—a frame in which the laws of mechanics hold good. And therefore we can no longer view phenomena from the point of view of the rocket without making changes in the laws of mechanics and electrodynamics.

The changes are not too difficult to understand. Additional forces appear as one decelerates. These additional forces change the vibration time of pendula, and so on. Therefore, if it is the rocket that stops and turns around and if we wish to calculate time intervals and do it the easy way, we calculate them from the point of view of the earth, which is an inertial frame. If we wish to do it from the point of view of the rocket, then we must make various changes in the laws of mechanics. The confusion comes from attributing too high a degree of symmetry to the motion, to asserting that any frame of reference is equivalent and that one can calculate in the same way from *any* frame.

27

THE GENERAL THEORY
OF RELATIVITY
(Einstein's Theory of Gravitation)

The general theory of relativity can be thought to arise out of an attempt to generalize the special theory in such a way that, in some sense, all frames of reference (the observer on earth and the one in the space ship) can be used equally (but not equally conveniently) to regard experience. Or, from what appears to be a quite incredible coincidence. In Newton's system the force (gravity) exerted by one body on another—say the earth with mass M_e—on a small lead ball of mass m is

$$F = \frac{GM_e\, m}{R^2} \qquad \text{(magnitude)} \qquad (27.1)$$

proportional to the mass m of the lead ball. From the second law, the acceleration of the lead ball

$$a = \frac{F}{m} \qquad \text{(magnitude)} \qquad (27.2)$$

is inversely proportional to its mass. Thus, subjected to the earth's (or any) gravitational forces, all balls—iron, ivory, or lead—fall with the same acceleration:

$$a = \frac{GM_e}{R^2} \qquad (27.3)$$

the result attributed to Galileo but known to Simon Stevin and John Philoponus. This equality of acceleration has been confirmed to great accuracy by Eötvös and Dicke.

Why two seemingly disparate things—the gravitational force and the acceleration—should be so intimately connected puzzled Newton (see footnote, page 59) and has puzzled physicists ever since. Other forces, electromagnetic, frictional, nuclear, and so on, do not display this property.[*] There might be something very special about the gravitational force. This point of view was developed by Einstein in the years between 1905 and 1915. Rather than attributing the numerical identity to an

---

[*] That is, two charged bodies in the same electric field will in general experience different accelerations.

accident, Einstein proposed that it was the result of an identity of the two seemingly different concepts—the gravitational force and the acceleration itself. If they were one, then the numerical identity could be understood as the result of a maladroit theory that had said the same thing twice.

THE PRINCIPLE OF EQUIVALENCE

But in what sense is the gravitational force to be made equivalent to acceleration?

Imagine, said Einstein, that we are in an elevator (the elevator has become famous) so that we cannot see outside. If the elevator is at rest (or moving uniformly) in a constant gravitational field, then all objects released inside it will fall to the floor with the acceleration of gravity.

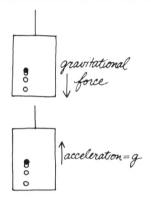

Imagine a similar elevator near no massive objects, in a zero gravitational field. But imagine that the second elevator accelerates upward with respect to an inertial frame with the acceleration g. All objects released in the second elevator will, so to speak, be left behind, so that from the point of view of an observer inside they would appear to accelerate uniformly toward the floor with acceleration of magnitude g. The same "real" motion can be regarded equally as the result of (1) the acceleration of the elevator upward or (2) a gravitational force downward. This observation is formalized as the principle of equivalence: a uniform gravitational field is equivalent to a uniform acceleration.

If this were all we would be finished. One would say the laws of physics could be equally well written in all uniformly accelerating frames (with the addition of the gravitational force), or could be written with no gravitational forces by choosing the right frame. One could take the point of view that a particle seen accelerating down with acceleration g

was doing so because it was acted upon by a gravitational force or because our elevator (frame of reference) was accelerating upward with acceleration g:

A ray of light, for example, if it moves in a straight line in an inertial frame, would appear from the point of view of an observer in the uniformly accelerating elevator shown in the margin to follow the curved path also shown. An observer inside a fixed elevator in a gravitational field would see the same curved path.

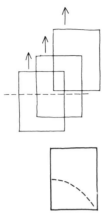

We might even grow tired of speaking of gravitational forces or of insisting that we were accelerating and generalize the principle of inertia to replace "A particle moves with uniform motion in a straight line unless acted upon by a force" by "The natural motion of all bodies is to accelerate toward the floor with acceleration g unless acted upon by a force."

But the problem is made more complicated by the fact that "real" gravitational fields are not necessarily uniform and that we can have non-uniform accelerations. Consider the gravitational fields due to the earth at points a and b below:

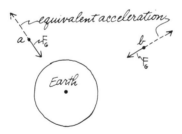

Both the magnitude and direction of the field change. We could extend the principle of equivalence to say that locally (near point a or point b) the earth's gravitational field is equivalent to a uniform acceleration, but that the uniform acceleration varies from place to place. Therefore, to regard the gravitational forces produced by bodies (particles, planets, galaxies, or the entire universe) as equivalent to a local acceleration, the magnitude and direction of the acceleration must be a function of position in space. The generalization of the principle of inertia to the non-uniform gravitational field of the earth, for example, might now be stated: The natural motion of all heavy bodies is to accelerate toward the center of the earth, to which no doubt *Il maestro di color che sanno,*[*] by now also having classified the resistance of air as a violence, would

[*] Aristotle among his pre-Christian friends as he appears in Canto IV of the *Divine Comedy.*

long ago have agreed.

In contrast to Newton's theory, the deviations from straight-line paths due to gravitation are not attributed to forces but to a property of the space. Light rays, for example near large massive objects, do not travel in Euclidian "straight lines":

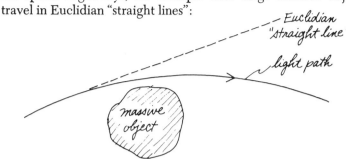

And since all bodies, light included, travel curved paths (in the absence of forces), one can call the space itself curved. For we then have the choice of saying that the space in the vicinity of the object is Euclidian but nothing travels in a straight line or of attributing a curvature to the space itself. Einstein chose the latter.

Other forces (electromagnetic and nuclear forces—to be introduced later) seem at first not to fit into this system at all well. Since they do not affect bodies in proportion to their mass (new entities—electric change or nuclear charge—must be introduced), they appear most naturally as somewhat extraneous forces in a system in which the gravitational force has been replaced by a curved space. Many attempts have been made, by Einstein and others, to unify the Maxwell field with Einstein's general theory (unified field theories).*

It has recently become fashionable to assert that the general theory is almost without experimental verification. What is meant is that there are few tests to distinguish Einstein's theory from either Newton's theory or other possible relativistic generalizations of Newton's theory. For in the nonrelativistic limit (low speeds, ordinary strength gravitational force) it gives all the results of Newton's theory (Kepler's laws, and so on). This is true to such a high degree of accuracy that for the fifty years since it has been proposed, only a handful of experimental results (two or three) exist that can distinguish the predictions of general relativity from Newton's gravitation theory. However, if one believes that Newton's theory should satisfy the principle of relativity and that in some sense the principle of equivalence should be included, then Einstein's form is perhaps the simplest and the most elegant. In any case Einstein's theory is as well verified as that of Newton (better) and that is very well verified indeed.

Traditionally there are three† so-called tests of general relativity,

* Recently a completely geometric theory of classical electromagnetism and gravitation was given by C. W. Misner and J. A. Wheeler, *Ann. Phys.*, **2**, 529 (1957). In this formulation the shape of the space determines the electromagnetic field as well as the gravitational forces.

† The preliminary results of a fourth test of general relativity are now available. According to the general theory, there should be a time delay for a ray of light passing near a star. This can be measured by the radar tracking of Mercury and Venus as the signal passes near the sun and is consistent with the prediction of Einstein's theory [Shapiro, Pettengill, Ash, Stone, Smith, Ingalls and Brockelman, *Phys. Rev. Letters*, **20**, 1265 (1968)].

three instances in which the general theory predicts a result slightly different from that of Newton but which is amenable to observation (or almost amenable to observation).

Bending of Light

When a ray of light passes through a gravitational field, it should be bent (according to the principle of equivalence alone). In particular, when starlight passes close to the limb of the sun it should appear to be displaced:

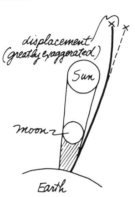

The displacement can be seen during an eclipse but is otherwise not visible. An international expedition during the total eclipse of the sun in 1919 photographed the field of the stars as the sun was in eclipse and studied this photograph compared with others of the same portion of the heavens when the sun was not there. They observed a displacement (probably regarding it as somewhat of a miracle that any displacement was observed) that was interpreted to be a verification of Einstein's prediction (1.75 seconds of arc) for a ray of light passing close to the sun.

The measurement is very difficult to make and to interpret (two astronomers looking at the same photograph we are told do not necessarily interpret it in the same way). Recently the question has been reopened and new attempts are being made to make the measurement. Everyone would agree that there is some bending of the light, but whether the amount is numerically in agreement with that to be expected from the general theory is not certain.

The Red Shift

The rate at which a clock beats depends on the local gravitational field. (This is the explanation of the twin paradox from the point of view of the general theory. The twin on the rocket stopping and returning can be regarded as subject to gravitational fields not experienced on earth, and it is these that distinguish between the two.)

The most elementary clock beat is that of the vibration of an atom or of light particles (photons, to be introduced more formally in Chapter 31). The shift in the frequency of vibration of a photon produces a change in color of the light toward the red and is thus known as the (gravitational) red shift. The photon has both an energy and a frequency

$$E = h\nu \tag{27.4}$$

where $h = 6.7 \times 10^{-27}$ erg-sec* and

$$\lambda\nu = c \tag{27.5}$$

If energy is related to mass by

$$E = mc^2 \tag{27.6}$$

and if all masses are acted on in the same way by gravitational forces (Fig. 27.1), then near the surface of a large star of large mass M_s and radius R_s the gravitational potential energy of such a photon would be

$$-G\,\frac{E}{c^2}\,\frac{M_s}{R} \tag{27.7}$$

The kinetic energy of the photon decreases as it climbs (so to speak) from the low gravitational potential of the star to the higher one of the earth. This gives a corresponding change in frequency,

$$\frac{\Delta\nu}{\nu} = \frac{\text{change in gravitational potential}}{c^2} \tag{27.8}$$

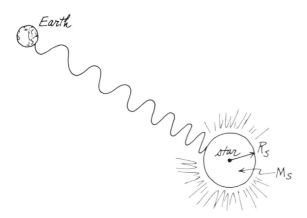

FIG. 27.1. If the photon travels from the star to earth, its change in gravitational energy is $-G\,\dfrac{E}{c^2}\left(\dfrac{M_s}{R_s} - \dfrac{M_e}{R_e}\right)$.

Although the effect is observed astronomically, its magnitude is rather uncertain. Recently it was measured near the surface of the earth using a new technique of remarkable sensitivity.[1] (The expected shift for a photon of wavelength $\lambda = 3000$ Å that falls 10^4 cm is

$$\frac{\Delta\nu}{\nu} = \frac{gL}{c^2} \simeq 10^{-15}) \tag{27.9}$$

The measurement and prediction are compared below:

$$\frac{(\Delta\nu)_{\text{experimental}}}{(\Delta\nu)_{\text{calculated}}} = 1.05 \pm 0.10 \tag{27.10}$$

* Planck's constant—also to be introduced in Chapter 31.

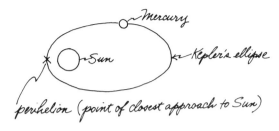

perihelion (point of closest approach to Sun)

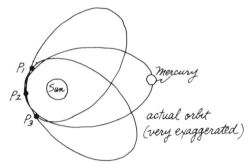

actual orbit
(very exaggerated)

Kepler's joy was founded in part on elliptical orbits. But the planets, perturbed by their neighbors to some extent, do not travel precisely in ellipses. For Mercury in particular, this manifests itself most visibly in what is called the advance of the perihelion—the point of the orbit closest to the sun. According to Kepler the planet should return to the same perihelion each year. What is observed, however, is a very small advance of this closest point (with respect to the fixed stars about 540 seconds of arc per century).

The effect of all other seen, known planets, as was first realized by Leverrier, is to produce an advance of the perihelion of about 500 seconds of arc per century. Something like 43 seconds of arc per century is the difference between the prediction of Newton's theory and what is observed. No obvious explanation seemed to do (a new planet—named in anticipation Vulcan—does not seem to be there). Here, surely, were it necessary, the hand of God could adjust, once a century, the orbit of the errant planet. But it is not necessary; Einstein's theory, which gives Newton's results for the other planets, gives for the perihelion of Mercury just this extra 43 seconds of arc per century.

General relativity thus is in agreement with all present observations from the motion of the tides to the advance of the perihelion of Mercury. It can be thought of as a modification of Newton's gravitational theory to make it consistent with the principle of relativity which incorporates the principle of equivalence, thus unifying the concepts of acceleration and gravitation. In it the concept of a force is not necessary, because the motion of all bodies is determined by the curvature of the space, which itself is determined by the presence of the bodies.

It is the Parthenon of physics—the work of a single architect who "once leaned down over the side of his hammock strung between Aries and the Circlet of the Western Fish . . ."[2] and gave us that structure, dominating, not by its size or utility, but by the perfection of its proportions, the Acropolis of human invention. If in the future, according to the rules by which we have agreed to play, it becomes necessary to raze this temple and replace it with a middle-income housing project whose efficient use of space is demonstrably greater, no doubt there will be available eager hands to do the job. *Sic transit gloria mundi.*

STRUCTURE OF THE ATOM

28

SILVER THREADS

Relativity, from one viewpoint the beginning of twentieth-century physics, is from another the capstone of classical physics, the final and most elegant variation of that world view initiated by Galileo and Newton. In a sense, after Einstein, classical physics, just as after Mozart classical music, could go no further. But within the theory of relativity there is a point of view—a reanalysis of fundamental ideas and an explicit display of the wisdom of retaining only concepts amenable to measurement (discarding those, such as absolute time, which were not), an emphasis on the equivalence of one observer and another which should be reflected in the laws of physics—developed with extraordinary variations as the twentieth century progresses.

That remarkable inclusiveness of classical physics, from the motion of the planets to the behavior of gases, could justify, at least psychologically, the attitude that the view of the world it presented was essentially complete. Surely things existed that were not understood. But would they be understood within the classical picture? Had the framework been outlined completely, and was what remained only the filling in of the details? Or, were there important new elements still to be created? The great successes of classical theory—mechanics, energy, electromagnetic theory, and the electromagnetic nature of light—make understandable the opinion that all that remained was the filling in of the sixth decimal place.[*] There was evidence which, in retrospect, enables us to see that all was not well. In retrospect, one is able to see much; but for those in the middle of events, it was like one of those puzzles where by moving squares one is required to arrange the numbers in an order, from 1 to 15; one finds that the first thirteen fall easily into place, but the last one or two cannot be made to fit properly. Yet, one would rather endlessly and hopelessly manipulate the last two rather than go back to the beginning and admit that it is another order, right from the beginning, which will fit all the numbers into their place. When one has successfully constructed a theory that encompasses so much of the world, it is hard to admit that it will have to be razed so that its foundations can be rebuilt completely. Rather, one struggles for a very long time to adapt, to modify, to cut, and to alter so as to include even the most unfriendly and alien facts into the compass of a point of view which one knows, which one has been taught, and whose success is demonstrable and great.

[*] This famous remark about the "sixth decimal place" was made, according to Millikan in his autobiography, by Michelson at the dedication of the Ryerson Physical Laboratory at the University of Chicago in June, 1894. Michelson believed he was quoting Kelvin, and later told Millikan that he regretted ever having said such a thing.

In the tapestry that was classical physics, if one could have looked properly, one would have discerned those threads of a new color that did not fit the many themes and variations of the Newtonian woof and warp. But whether these alien threads would disappear, an error of the weaver's hand, or whether they would continue and develop into new and rich themes of their own was something the spectator, or even the weaver, could not know too well at the time.

DISCRETE SPECTRAL LINES

Newton, "to try therewith the celebrated *Phænomena of Colours*,"[1] darkened his chamber, made a small hole in his window shutters, placed his prism at the entrance and produced a "pleasing divertisement to view the vivid and intense colors produced thereby."[2] This was in 1666. In 1802 W. H. Wollaston noted, among the vivid and intense colors produced thereby, some dark lines. In 1814, Fraunhofer, by combining a prism with a small-viewing telescope and observing a distant narrow slit through this combination, created what we call a spectrometer (Plate 28.1). Viewing the sun's spectrum through this instrument, he saw

> . . . an almost countless number of strong or weak vertical lines which are darker than the rest of the colored image; some appeared to be almost perfectly black.

Announcing the result in 1817 he added:

> I have convinced myself by numerous experiments and by various methods that these lines and bands are due to the nature of sunlight and do not arise from diffraction, optical illusion, etc.

The dark lines fall among the various colors. And since one can assign to each color a wavelength, a specific wavelength could be assigned to each of the dark lines. Fraunhofer, emphasizing his preference for Latin rather than Greek, labeled them A, B, C, D . . . (Plate 28.2). In particular, two lines in the yellow part of the spectrum which were close to each other and had been labeled D by Fraunhofer led Gustav Robert

PLATE *28.1*. A spectrometer. (Courtesy of J. A. Lubrano and D. Scales)

Kirchhoff and Bunsen (so well known for his burner) in 1859 to an explanation of the dark lines in a spectrum. Kirchhoff writes:

> While engaged in a research carried out by Bunsen and myself in common on the spectra of colored flames, by which it became possible to recognize the qualitative composition of complicated mixtures from the appearance of their spectra in the flame of the blow pipe, I made some observations which give an unexpected explanation of the origin of the Fraunhofer Lines and allow us to draw conclusions from them about the composition of the sun's atmosphere and perhaps also of that of the brighter fixed stars.[3]

He continues:

> Fraunhofer noticed that in the spectrum of a candle flame two bright lines occur which coincide with the two dark lines D of the solar spectrum. We obtain the same bright lines in greater intensity from a flame in which common salt is introduced. I arranged a solar spectrum and allowed the sun's rays, before they fell on the slit, to pass through a flame heavily charged with salt. When the sunlight was sufficiently weakened there appeared, in place of the two dark D lines, two bright lines. . . .[4]

and further:

> We may assume that the bright lines corresponding with the D lines in the spectrum of a flame always arise from the presence of sodium; the dark D lines in the solar spectrum permit us to conclude that sodium is present in the sun's atmosphere.[5]

Shortly afterward, Kirchhoff announced what can be called the two fundamental laws of spectroscopy: (1) To each chemical species there corresponds a characteristic spectrum; and (2) every element is capable of absorbing the radiation which it is able to emit. If one introduces the element sodium into the flame of a Bunsen burner, the flame glows a brilliant yellow. (The fact that salt is present in so many materials is the explanation of the yellowness of most flames.) This yellow, analyzed, is dominated by the two bright D lines. However, while the heated sodium emits the two bright lines, a cool sodium vapor is capable of absorbing these same two colors. Thus, if a bright source of light that contains all colors is passed through a cool sodium vapor, the two D lines appear as dark lines against the bright background of the colors of the spectrum.

Kirchhoff then proposed that the reason for the dark lines in the sun's spectrum was the presence of relatively cold vapors of sodium and other elements in the outer atmosphere surrounding the sun. The continuous light coming from the interior of the sun, passing through these relatively cool vapors, had absorbed the lines characteristic of the various elements. Thus, by a study of the dark lines in the spectrum of the sun or in the spectrum of a star, one could deduce what elements were present in its outer atmosphere—a result that might have surprised the French positivist Auguste Comte, who, in 1825, offered the chemical composition of the stars as a perfect example of knowledge permanently inaccessible to man.

The essence of the discovery of Fraunhofer, Bunsen, and Kirchhoff can be summarized as follows: If one heats a pure element to a sufficiently

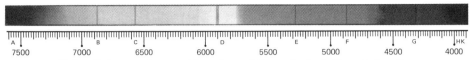

PLATE 28.2. Solar spectrum with a few of the principal Fraunhofer lines. (Courtesy of Sargent-Welch Scientific Company)

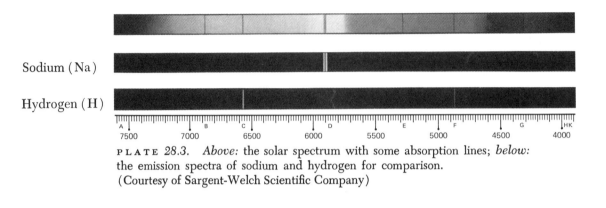

Sodium (Na)

Hydrogen (H)

PLATE 28.3. *Above:* the solar spectrum with some absorption lines; *below:* the emission spectra of sodium and hydrogen for comparison. (Courtesy of Sargent-Welch Scientific Company)

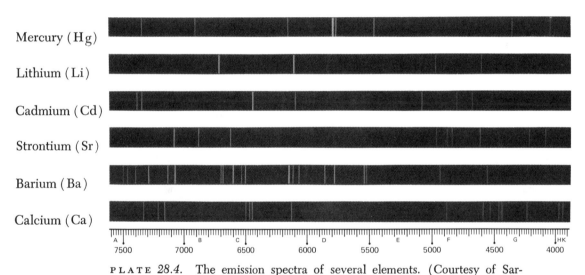

Mercury (Hg)

Lithium (Li)

Cadmium (Cd)

Strontium (Sr)

Barium (Ba)

Calcium (Ca)

PLATE 28.4. The emission spectra of several elements. (Courtesy of Sargent-Welch Scientific Company)

PLATE 28.5. Part of the Fraunhofer spectrum of the star Sirius. (Courtesy of J. A. Lubrano and D. Scales)

For color versions of the above plates, see Insert 2.

high temperature, which can be done, for example, by placing that element in a relatively colorless flame of a Bunsen burner, a color will usually appear, a color characteristic of the element: sodium-yellow, strontium-red, and so on. If this light is analyzed in a spectroscope into the various wavelengths of which it is composed, one finds for each element a characteristic series of lines, always the same for the same element. No two elements are known to have the same set of lines.

An immediate and obvious use of this observation permits one to analyze complicated chemical species into their constituent elements. One has only to place a substance in a flame and compare the observed spectrum with the known spectra of the various elements to pick out the elements present in the unknown. And, from the intensities of the various lines, one can even determine how much of each element is there. Thus, as Kirchhoff and Bunsen say,

> . . . it became possible to recognize the qualitative composition of complicated mixtures from the appearance of their spectra. . . .[6]

—a striking solution to the age-old problem of determining the chemical composition of a given substance.

But from the physicist's point of view, the fact that each separate element had associated with it its own distinct spectrum of sharp lines— that each atom had its own signature—was very striking and mysterious. It was soon realized that some of the lines were not visible but lay in the ultraviolet or in the infrared and had to be analyzed by special techniques since they could not be seen in an ordinary spectrometer. In particular, hydrogen, the lightest element, had an especially simple spectrum, and this spectrum was subjected to a searching analysis.

In 1885, Johann Jakob Balmer published a formula that fitted the four most prominent lines of the hydrogen spectrum; a comparison of the expression with other lines just then being observed indicated that it fitted almost the entire spectrum very accurately. All subsequent analyses of the spectra of the various elements had their origin in this formula. He arrived at it with no theoretical preconceptions, in an attempt to find a simple algebraic expression that would yield the observed lines.

> The wave lengths of the first four hydrogen lines are obtained by multiplying the fundamental number $b = 3645.6$ in succession by the coefficients 9/5; 4/3; 25/21 and 9/8. At first sight these four coefficients do not form regular series; but if we multiply the numbers in the second and the fourth by 4 a consistent regularity appears and the coefficients have for numerators the numbers $3^2, 4^2, 5^2, 6^2$ and for denominators a number that is less by 4.[*][7]

He did not, however, add that the elements sing—a measure of the progress of science since the time of Pythagoras.

Balmer's formula can be written

$$\lambda = b \, \frac{n^2}{n^2 - 4} \qquad (28.1)$$

where, if λ is the wavelength of the spectral line in angstroms,

[*] That is, $9/5 = 9/(9 - 4)$, $16/12 = 16/(16 - 4)$, $25/21 = 25/(25 - 4)$, and $36/32 = 36/(36 - 4)$.

$$1 \text{ angstrom (written } 1 \text{ Å}) = 10^{-8} \text{ cm}$$

then b is the numerical constant to which Balmer refers,

$$b = 3645.6$$

and n is an integer which takes the values

$$n = 3, 4, 5, 6, \ldots$$

where each value of n gives a different value of λ and corresponds to a line of the observed spectrum of hydrogen.

Was this numerology? Astrology? Black magic? It was, perhaps, a combination of all three. But that was possibly not as important as the incontrovertible fact that λ, as given by the formula

$$\lambda = b \, \frac{n^2}{n^2 - 4}$$

did, to a remarkable degree of accuracy, correspond to the observed lines. And it provided the initial organization of the raw data, which, just as Kepler's had done for Newton, would provide the algebraic relations for which the new theories must strive.

Example. Verify that a red line exists in the hydrogen spectrum for $n = 3$.

$$\lambda = 3645.6 \, \frac{3^2}{3^2 - 4} \text{ Å} = 3645.6 \, \frac{9}{5} \text{ Å}$$

$$= 6562.1 \text{ Å} = 6.5621 \times 10^{-8} \text{ cm}$$

which is in the red portion of the spectrum.

X RAYS

Few discoveries have had so immediate and enormous a psychological impact as Röntgen's discovery of

> . . . an active agent [which] passes through a black card-board envelope, which is opaque to the visible and the ultra-violet rays of the sun or of the electric arc. . . .[8]

He wrote further:

> We soon discover that all bodies are transparent to this agent, though in very different degrees. . . . Paper is very transparent; behind a bound book of about one thousand pages I saw the fluorescent screen light up brightly, the printer's ink offering scarcely a noticeable hindrance. In the same way the fluorescence appeared behind a double pack of cards; a single card held between the apparatus and the screen being almost unnoticeable to the eye. A single sheet of tin-foil is also scarcely perceptible; it is only after several layers have been placed over one another that their shadow is distinctly seen on the screen.[9]

He called this penetrating agent, produced when the discharge of a fairly large induction coil was made to pass through a vacuum tube, *X rays* (Fig. 28.1) and investigated its properties in a classic series of experiments.°

Within three weeks these X rays were being used to photograph broken bones. Few discoveries have captured the popular imagination more quickly. The news was reported in that most extravagant fashion with which we are familiar today and swept the world. The discovery of this penetrating agent, which made objects visible through opaque screens, soon made the X ray a household word with which to conjure, replacing magnetism, which had been equally fashionable a century before. Spinsters trembled, lest their modesty be invaded by a voyeur equipped with the most up-to-date scientific apparatus. Ever responsive to the market, the entrepreneur, the moving force of the burgeoning American economy, offered for the consumer X-ray-proof vests and clothes. Proust would write:

> And Françoise answered, laughing, "Madame knows everything. Madame is worse than the X-rays." (She pronounced X with an affectation of difficulty and with a smile in deprecation of her, an unlettered woman, daring to employ a scientific term.)

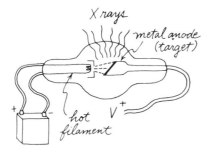

F I G. *28.1.* X-ray tube. The current heats the filament to a temperature at which it emits electrons copiously. When the potential difference is large, the electrons are attracted to and hit the target with high speeds, producing X rays. (After G. Holton, *Introduction to the Concepts and Theories of Physical Science*, Addison-Wesley, Reading, Mass., 1952)

RADIOACTIVITY

Coming hard on the discovery of X rays, in 1896 Henri Becquerel described the discovery of radioactivity from uranium. He writes:

> Some months ago I showed that uranium salts emit radiations whose existence has not hitherto been recognized, and that these radiations possess remarkable properties, some of which are similar to the properties studied by M. Röntgen.[10]

The discovery of this active agent in uranium led to that most famous search by the Curies. They conclude:

° Another approach was provided by F. Smith at Oxford. When he found that his photographic plates became fogged if they were kept near the Crooke's tube, he had his assistant move them to a cabinet farther away.

The various reasons which we have presented lead us to believe that the new radioactive substance contains a new element, to which we propose to give the name *radium*.[11]

An analysis of the penetrating power of the radioactive emissions from radium led to the discovery that there seemed to be three components; in the great classical tradition they were denoted alpha, beta, and gamma rays.° As we shall see later, the alpha rays, the least penetrating, were identified as being positively charged, relatively heavy, and related intimately to helium. The beta rays (light, negatively charged) were later to be identified as "electrons." And the gamma rays (with no observable mass or charge), closest in their properties to the X rays of Röntgen, were later identified as a type of an X ray. The amount of energy carried off from the radioactive salts by these radiations was enormous compared to the amounts of energy usually involved in chemical processes—a fact recognized by Einstein when he wrote, in 1905, in his article connecting matter and energy:

> It is not impossible that with bodies whose energy-content is variable to a high degree (e.g. with radium salts) the theory may be successfully put to the test.[12]

There were other strands. The kinetic theory and statistical mechanics, which were successful in explaining so many of the average equilibrium properties of matter, had led to several difficulties of a fairly profound nature. About one of these† Maxwell was to say in a lecture late in his life: "I have now put before you what I consider to be the greatest difficulty yet encountered by molecular theory." As usual, Maxwell was right. The problem was never explained within the context of classical kinetic theory. But it was only one question of that type which classical theory could not encompass. In general, the detailed properties of materials—their magnetic and optical properties; why some are conductors, why others are not; what is the nature and internal constitution of matter —these questions never would be resolved within the domain of classical physics. These threads that were beginning to appear—the problems that troubled Maxwell, a paradox first observed by J. Willard Gibbs, radioactivity, X rays, the discrete spectral lines—single threads at first, did not disappear. The future would be dominated by themes which, reading backward, one could see to begin with these lonely strands. But in the end it was on the trail of that old and elusive concept, the atom, that the world created by Newton and Descartes, the world of Democritus, Epicurus, Lucretius, and Gassendi came to the end of its usefulness.

° Rutherford (1899) identified alpha and beta rays by differences in penetration. Villard discovered gamma rays (1900). Rutherford, who worked with a magnetic field, showed that the rays consisted of positive, negative, and neutral parts. The terms alpha and beta were introduced in his paper.
† The heat capacities of nonmonoatomic gases.

29

DISCOVERY OF THE ELECTRON

THOMSON'S EXPERIMENT

The hypothesis of the atom, that irreducible entity, patterns of which in the void form the objective world as we see it, is as old as our civilization:

> Nature resolves everything into its component atoms.[1]

For Newton, hard, massy, and indivisible; the atom of kinetic theory, whose average kinetic energy is what we read as temperature; the atom of the chemist, whose uniform combinations reveal its presence in chemical reactions; the hydrogen atom of Prout, combinations of which make up all the elements. Often in disrepute, often in the background, for at least 25 centuries the concept of the atom had existed.

But what was the atom? What meaning did it have to ask the question? By the turn of the century the elaborate development of classical theory and the introduction of new techniques made possible an increasingly insistent and detailed return to this question: What is the nature of the atom? That theme and its variations develop into a major movement of twentieth-century physics.

In the later part of the nineteenth century, many studies were made of the discharge of electricity through rarefied gases. These discharges (produced with an induction coil or electrostatic machine—either capable of creating large potential differences) were passed between a negative terminal, called a *cathode,* and a positive terminal, called an *anode,* both of which were sealed into a glass tube from which most of the air had

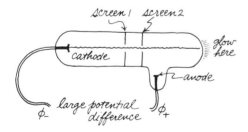

been exhausted. When the air is sufficiently exhausted from the tube, a dark region around the cathode, known as the *Crooke's dark space,* grows longer and longer until finally it extends toward the other side of the tube, which is then observed to glow—the color depending upon the kind of glass of which the tube is made. If various screens are introduced, such as screen 1 and screen 2 here, the glow is confined to a spot at the end of the tube, as though something coming from the cathode is going through the holes in the screen, reaching the glass on the other side to make it glow. These somethings were christened *cathode rays.*

About the nature of these rays, there was intense speculation at the end of the nineteenth century. They were thought by some, like light, to be due to a process in the ether; others thought that they were produced by particles with an electric charge. In 1895 Jean Perrin managed to catch these rays in an insulated container and, by measuring the charge of the container, to demonstrate that they carry a negative charge. Soon afterward, J. J. Thomson performed the classic experiment in which he first identified these cathode rays as what were to be called electrons. He wrote:

> The experiments discussed in this paper were undertaken in the hope of gaining some information as to the nature of the Cathode Rays. The most diverse opinions are held as to these rays; according to the almost unanimous opinion of German physicists they are due to some process in the ether to which—inasmuch as in a uniform magnetic field their course is circular and not rectilinear—no phenomenon hitherto observed is analogous: another view of these rays is that, so far from being wholly etherial, they are in fact wholly material, and that they mark the paths of particles of matter charged with negative electricity.[2]

He then went on to describe an experiment by which he measured the ratio of the mass of the particles to their charge, assuming that they were charged particles.

By placing an electric field across the plates labeled d and e in Fig. 29.1, or a magnetic field perpendicular to the direction of motion of the rays, Thomson was able to deflect the spot of light on the end of the tube; this deflection was larger the stronger the electric or magnetic field. After convincing himself that what was happening was independent of the gas in the tube, he wrote:

> As the cathode rays carry a charge of negative electricity, are deflected by an electrostatic force as if they were negatively

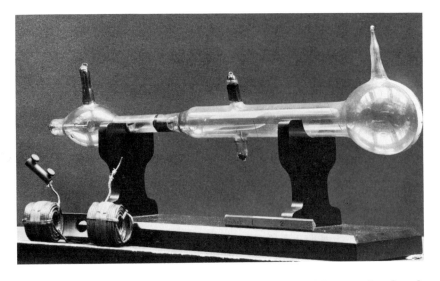

PLATE 29.1. Thomson's apparatus. (Lent to Science Museum, London, by the late Sir J. J. Thomson, Trinity College, Cambridge)

electrified, and are acted on by a magnetic force in just the way in which this force would act on a negatively electrified body moving along the path of these rays, I can see no escape from the conclusion that they are charges of negative electricity carried by particles of matter. The question next arises, What are these particles? are they atoms, or molecules, or matter in a still finer state of subdivision? To throw some light on this point, I have made a series of measurements of the ratio of the mass of these particles to the charge carried by it.[3]

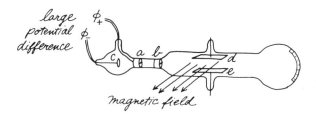

FIG. 29.1. Diagram of Thomson's apparatus. (Redrawn by permission of the publishers from W. F. Magie, A Source Book in Physics, Harvard University Press, Cambridge; copyright 1935, 1963 by the President and Fellows of Harvard College)

The analysis proceeded as follows. The force on the charged particles (let us assume their charge is q) due to an electric field, E, between the plates d and e is

$$F = qE \qquad \text{(magnitude)} \qquad (29.1)$$

At the same time, the force on the charged particles due to the magnetic field **B** perpendicular to their direction of motion is

$$F = \frac{q}{c} vB \qquad \text{(magnitude)} \qquad (29.2)$$

If, for example, the charge on the particles is negative, then the force due to an electric field from e to d will be directed downward. At the same

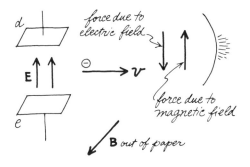

time, the magnetic force, due to the charged particles moving through the magnetic field, with the arrangement shown above, will be directed upward for negative particles. Therefore, by adjusting the electric and magnetic fields until the spot at the edge of the tube remained undeflected, Thomson could determine that the forces due to the magnetic and electric fields were equal, and obtain the relation

$$qE = \frac{q}{c} vB \qquad (29.3)$$

or

$$v = \frac{cE}{B} \qquad (29.4)$$

This enabled him to determine the speed of his postulated particles. Then, turning off the electric field and adjusting the magnetic field, he was able to vary the deflection of the particles at the end of the tube. Since the time during which the particles were exposed to the magnetic field was now known (because he knew their speed), he knew how long the particles were accelerated, due to the magnetic field acting upon them. With this, from their deflection he could determine the ratio of their charge to their mass.

He obtained in the end a ratio of the mass to the charge of his postulated particles of[*]

$$\frac{m}{e} \simeq 1.3 \times 10^{-7} \text{ g/coulomb} \qquad (29.5)$$

He concludes:

> From these determinations we see that the value of m/e is independent of the nature of the gas, and that its value 10^{-7} is very small compared with the value 10^{-4}, which is the smallest value of this quantity previously known, and which is the value for the hydrogen ion in electrolysis.
> Thus for the carriers of the electricity in the cathode rays m/e is very small compared with its value in electrolysis. The smallness of m/e may be due to the smallness of m or the largeness of e, or to a combination of these two.[4]

This carrier of electricity, the active constituent of the cathode rays, was eventually christened the *electron*, the first twentieth-century elementary particle.

Thomson later wrote:

> My first attempt to deflect a beam of cathode rays was to pass it between two parallel metal plates fastened inside the discharge-tube, and to produce an electric field between the plates. This failed to produce any lasting deflection. . . . The absence of deflection on this view is due to the presence of gas—to the pressure being too high—thus the thing to do was to get a much higher vacuum. This was more easily said than done. The technique of producing high vacua in those days was in an elementary stage.[5]

Not the first time that the greatest difficulty in doing a crucial experiment lay as much in developing necessary techniques as in its conception.

---

[*] In Gaussian units the modern value is

$$\frac{m}{e} \simeq 1.90 \times 10^{-18} \text{ g/esu.}$$

It now became a very important question to learn either the charge or the mass of these particles separately. The charge of gaseous ions had been measured in Thomson's laboratory to be about 6.5×10^{-10} esu. Assuming that the charge on these ions was the same as that on his cathode particles, it followed that the mass of the cathode particles was extremely small:

$$m \simeq 10^{-27} \text{ g} \tag{29.6}$$

The cathode particles he at that time called "corpuscles," or primordial atoms; the word electron was used to denote the amount of charge carried by a corpuscle. However, the usage finally arose of calling the particle itself an electron. It was later (in 1909) that Millikan, by measuring the charges on oil droplets, determined that the elemental charge, the charge that was assumed to be the same as that of the electron, was about 4.77×10^{-10} esu. The modern values of the charge and mass of the electron are

$$e = 4.803 \times 10^{-10} \text{ esu} \tag{29.7}$$

$$m = 0.9107 \times 10^{-27} \text{ g} \tag{29.8}$$

ELECTRICAL CONSTITUTION OF MATTER

Somewhat before Thomson's work on cathode rays, it occurred to Pieter Zeeman that

> . . . the period of the light emitted by a flame might be altered when the flame was acted upon by magnetic force . . .[6]

He performed the experiment and wrote

> . . . It has turned out that such an action really occurs. I introduced into an oxyhydrogen flame, placed between the poles of an . . . electromagnet, a filament of asbestos soaked in common salt. The light of the flame was examined with a . . . grating. Whenever the circuit was closed both D lines were seen to widen.[*]
>
> Since one might attribute the widening to the known effects of the magnetic field upon the flame, which would cause an alteration in the density and temperature of the sodium vapour, I had to resort to a method of experimentation which is much more free from objection. . . . It thus appears very probable that the period of sodium light is altered in the magnetic field.[7]

"A true explanation of this phenomenon," he then writes,

> . . . appears to me to be afforded by the theory of electric phenomena propounded by Prof. Lorentz.
>
> In this theory, it is considered that, in all bodies, there occur small molecular elements charged with electricity, that all electrical processes are to be referred to the equilibrium or motion of these "ions"[†] and that the undulations of light are vibrations of the ions. It seems to me that in the magnetic field the forces directly acting on the ions suffice for the explanation of the phenomena.

[*] With sufficient resolution the lines are seen to split; he was able to detect only a broadening.

[†] He is speaking of what will be called electrons.

Prof. Lorentz, to whom I communicated my idea, was good enough to show me how the motion of the ions might be calculated, and further suggested that if my application of the theory be correct there would follow these further consequences. . . .[8]

The calculation by Lorentz, based on the assumption that it was the altered orbit of a charged corpuscle due to the magnetic field which was producing the broadening, yielded the ratio of charge to mass of this charged particle precisely the same as that to be found by Thomson, and led to predictions concerning the Zeeman splitting, which Zeeman verified soon after in a second series of experiments.

Such results, plus the obvious fact that electricity could be obtained from matter by rubbing, that charged particles, like cathode rays, were emitted from matter under the proper circumstances, led to a belief, almost universally accepted at the turn of the century, in the electrical constitution of matter. If matter was somehow composed of electrical material, what was the arrangement of this material within matter? To answer such questions there began an incessant probing of matter by X rays, electrons, alpha particles, and so on, and an accompanying theoretical activity to attempt to construct from the materials then known an atom that manifested the observed properties. As late as 1897, William Thomson (Lord Kelvin) would still consider seriously the possibility that "electricity is a continuous homogeneous liquid." The work of J. J. Thomson made this unlikely, and the resulting electrons became immediately a building block for the construction of the atom.

From a variety of experiments, including those involving the passage of electrons through matter, Thomson concluded that, roughly, the number of electrons per atom should be of the order of the weight of the atom, the atomic weight of the chemist. (It was later shown by Barkla that the number of electrons per atom, for the lighter atoms at least—excluding hydrogen—was more nearly one half the atomic weight.) Now, the normal atom must be electrically neutral, since, if matter is a collection of atoms and if matter is electrically neutral (we have already discussed the forces that would result if matter were not neutral), then the individual atoms that make up matter must themselves be neutral. If this is the case, then the quantity of positive electricity per atom must be the same as the quantity of negative electricity.

Thomson had concluded that the mass of the electron was probably of the order of one two-thousandth of the mass of the hydrogen atom. Assuming that hydrogen, the simplest atom, contains a single electron and a single positive charge (making the entire atom electrically neutral), the positive charge should then have associated with it a mass 2000 times as great as the mass of the electron. Thus, if one assumed that the masses were definitely associated with the charges, one would conclude that practically the entire mass of hydrogen (and perhaps other atoms as well) was associated with its positive charge.

THOMSON'S ATOM

In 1902 Kelvin proposed an atomic model in which the positive charge was to be distributed throughout some small region, perhaps a sphere, with the electrons embedded in this distributed charge much as raisins in a cake. Since vibrating electrons would emit light, according to Maxwell's theory, presumably the electrons and the positive material were at rest when the atom was undisturbed. (There was a problem here because it was known that a distribution of charges of this kind could not remain at rest in a stable configuration under electrical forces alone; however, it was possible that in the interior of the atom there were other forces as well.)

J. J. Thomson developed this idea and investigated in particular those arrangements of corpuscles (electrons) that would lead to stable configurations (given a fixed distribution of positive charge).* He conjectured that perhaps particularly stable configurations of electrical matter resulted in the chemically inactive elements (such as the noble gases), whereas other, less stable, configurations produced the more active elements. He thus hoped to account for the periodic table.

When such an atom was disturbed (in the heat of a flame, for example) it would presumably be the electrons that vibrated, since they were light, while the heavy positive material remained at rest. And these vibrations possibly would produce the spectral lines that had been observed, the different arrangements of electrons in each atom producing the characteristic signature of that atom—its own spectral lines. From the observed wavelengths of the radiation emitted from atoms, Thomson could estimate the spread of the positive electricity. In order that wavelengths in the visible spectrum be emitted, he estimated that the sphere of positive electricity should be about 10^{-8} cm in radius, which was in remarkable agreement with estimates of the size of the atom already available from the kinetic theory. Although he was not able to fit the details of the observed spectra, and although there were serious problems implied in its construction, Thomson's results were a strong indication that it might be possible to go further in this direction.

This enterprise on which Thomson and his colleagues were embarking was still very much within the classical system. It was generally assumed that Maxwell's equations described the behavior of electricity and magnetism and that Newton's equations, when necessary modified according to the theory of relativity (we shall always understand by Newton's equations their relativistic modifications), governed the motion of matter. The important forces between subatomic particles were presumed to be electrical; gravitational forces are too small to be of significance compared to electrical forces. Thus the concepts were all those of classical physics and this beginning attempt to construct an atom was very much like previous attempts to construct a gas or a rigid body. One used known or newly discovered materials and combined them according to known rules in an attempt to obtain an object that had the needed properties.

"*Hypotheses non fingo*," Newton had said. "About the nature of gravity I make no assumptions." About the nature of the atom, however, as should be clear from the path we are traveling, many assumptions were made. The ratio of the charge to the mass of the particle associated with cathode rays seemed secure. Yet, that this charge was the same as that found on ions was assumed; that the massive part of the atom was associ-

---

* The positive charge would presumably be held together by nonelectrical forces.

ated with its positive charge was assumed; that an atom with a dispersed positive charge, electrons spread throughout like raisins, could exist in equilibrium was assumed; and so on. Some of these assumptions, it turned out later, were justified; others were not. But in an enterprise such as that of the construction of an atom, one must be daring. In retrospect it might have saved time not to have made assumptions about the nature of the fluid through which electric and magnetic fields penetrate. For the atom, however, once one believed it was there, it was necessary to make assumptions, and these assumptions had consequences. When the consequences were in disagreement with experience, they were abandoned. And when the most likely assumptions were clearly contradictory to classical ideas, it was classical physics that was abandoned.

30

RUTHERFORD'S NUCLEAR ATOM

The situation in 1910, as pieced together from these various experiments and conjectures, seemed to be somewhat as follows: atoms, electrically neutral, consisted of electrons (of very small mass) and positive material with the major part of the mass. When undisturbed, the electrons were presumably at rest so as not to radiate energy; when excited, the electrons vibrated and thus gave off light. One possible distribution of electrical material had been extensively investigated by Thomson, but no direct evidence for the actual arrangement of electrical material was yet available.

Such evidence could be obtained using a method that had been developed during the first decade of the new century. This consisted in firing various particles through thin pieces of matter and studying their deflection to attempt to probe the structure of the object that did the deflecting.

> Since the α and β particles traverse the atom, it should be possible from a close study of the nature of the deflexion to form some idea of the constitution of the atom to produce the effects observed. In fact, the scattering of high-speed charged particles by the atoms of matter is one of the most promising methods of attack of this problem.[1]

If, for example, one had a thin foil of gold, one might fire particles through this foil and observe their deflection. This type of process (called *scattering*), in which one observes the scatter of the incident particles from the targets, to study the nature either of the incident particles or

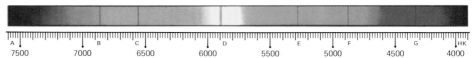

PLATE 28.2. Solar spectrum with a few of the principal Fraunhofer lines. (Courtesy of Sargent-Welch Scientific Company)

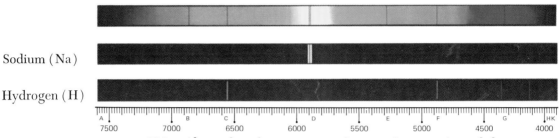

Sodium (Na)

Hydrogen (H)

PLATE 28.3. *Above:* the solar spectrum with some absorption lines; *below:* the emission spectra of sodium and hydrogen for comparison. (Courtesy of Sargent-Welch Scientific Company)

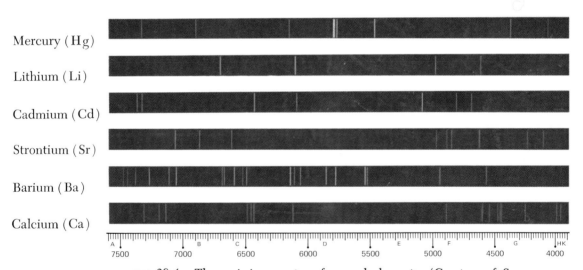

Mercury (Hg)

Lithium (Li)

Cadmium (Cd)

Strontium (Sr)

Barium (Ba)

Calcium (Ca)

PLATE 28.4. The emission spectra of several elements. (Courtesy of Sargent-Welch Scientific Company)

PLATE 28.5. Part of the Fraunhofer spectrum of the star Sirius. (Courtesy of J. A. Lubrano and D. Scales)

the target, has become one of the primary tools in the study of atoms, nuclei, and fundamental particles.

The arrangement of all scattering experiments (without concerning ourselves at the moment with the formidable technical variations) is

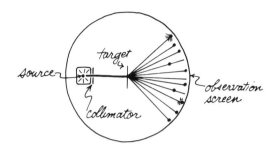

In principle nothing could be simpler. One begins with a source of bombarding particles. A collimator eliminates strays, so that one has finally a narrow pencil of bombarding particles possibly moving at some known speed. One places in their path a target and observes the number of particles scattered to the various angles by the use of some observation screen.

As is often so, these things are easier said than done. It is difficult to obtain rapidly moving particles (radioactive materials provided the first source) and the construction of an observation screen sensitive to the bombarding particles is itself not trivial:

> The development of the scintillation method of counting single alpha particles affords unusual advantages of investigation, and the researches of H. Geiger by this method have already added much to our knowledge of the scattering of alpha rays by matter.[2]

Before 1910, theoretical and experimental investigations had been done for the scattering of incident radiation and various particles from matter—X rays, electrons or beta rays, and alpha particles. The heavy alpha particles provided the best opportunity for probing the detailed nature of the atomic structure. In order, as a practical matter, to study as much as possible single encounters between the bombarding particles and the target atoms, one wanted to bombard a target as thin as possible; otherwise the scattering from one atom after another would eventually mask the effect of single collisions. A gold foil has the happy property that it can be hammered to an extremely fine thinness which, with a knowledge of the weight of the gold atom and the weight of the gold foil per cubic centimeter, enables one to estimate that foils with a thickness of only 400 atoms are feasible.

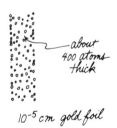

about 400 atoms thick

10^{-5} cm gold foil

$M_\alpha = 6.62 \times 10^{-24}$ g

$V_\alpha = 1.6 \times 10^9$ cm/sec

collimated alpha particles emitted from a radioactive polonium source

The alpha particle was known at that time to have a mass of about 6.62×10^{-24} g, just the mass of the helium atom. It was further known to have a positive charge twice the magnitude of the charge on the electron. The alpha particles emitted from a radioactive polonium source were known to move with a speed of 1.6×10^9 cm/sec. It could be assumed (and this assumption was made) that the alpha particles were helium from which the electrons had somehow been stripped in the process of emission. This view was corroborated when Rutherford and Royds detected helium when alpha particles were collected in a container. Geiger had directed such alpha particles at a thin gold foil about 4×10^{-5} cm thick and had observed their deflection as recorded on a zinc sulfide screen. The zinc sulfide scintillated (emitted a dot of light) when an alpha particle impinged upon it. Thus, by watching various sections of the screen with a microscope, he could count the number of scintillations per minute. And this gave the relative number of particles scattered through that particular angle.

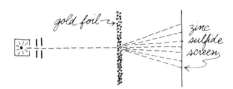

In the earliest experiments using gold as a target and alpha rays as bombarding particles, the first thing noted was that, in spite of the fact that the gold foil was approximately 400 atoms thick, almost all of the particles went right through with barely any deflection, as though the gold foil and the atoms that compose it were almost completely transparent to the bombarding particles.

Rutherford relates:

> I had observed the scattering of α-particles, and Dr. Geiger in my laboratory had examined it in detail. He found, in thin pieces of metal, that the scattering was usually small, of the order of one degree. One day Geiger came to me and said, "Don't you think that young Marsden, whom I am training in radioactive methods, ought to begin a small research?" Now I had thought that, too, so I said, "Why not let him see if any α-particles can be scattered through a large angle?" I may tell you in confidence that I did not believe that they would be, since we knew that the α-particle was a very fast, massive particle, with a great deal of [kinetic] energy, and you could show that if the scattering was due to the accumulated effect of a number of small scatterings, the chance of an α-particle's being scattered backward was very small. Then I remember two or three days later Geiger coming to me in great excitement and saying, "We have been able to get some of the α-particles coming backward . . ." It was quite the most incredible event that has ever happened to me in my life. It was almost as incredible as if you fired a 15-inch shell at a piece of tissue paper and it came back and hit you.[3]

Because of the large mass of the alpha particle (about 8,000 times the mass of an electron), it could be presumed that a collision of an alpha particle with an electron would have a negligible effect on its trajectory, the heavy alpha moving through a cloud of light electrons as a 15-in. shell through a swarm of mosquitos. However, the mass of the entire gold atom was known to be about 50 times the mass of the alpha particle—a mass associated with the positive charge in the gold atom. If the force between the positive charge and the alpha particle were sufficiently great, a collision between the two could deflect the alpha particle from its course, as a 15-in. shell striking a cannonball. Thus one might guess that inside the thin tissue of paper were cannonballs.

P L A T E *30.1.* Alpha particles cross the view field from left to right. One has hit an oxygen atom and has glanced off to the upper right corner, while the oxygen atom recoils toward the lower right corner. (G. Holton, *Introduction to Concepts and Theories of Physical Science*, Addison-Wesley, Reading, Mass., 1952)

On consideration, I realized that this scattering backward must be the result of a single collision, and when I made calculations I saw that it was impossible to get anything of that order of magnitude unless you took a system in which the greater part of the mass of the atom was concentrated in a minute nucleus. It was then that I had the idea of an atom with a minute massive center carrying a charge.[4]

The problem with Thomson's distribution of positive charge is that it is not sufficiently concentrated to produce a force great enough to deflect the alpha particle by more than a small part of a degree. If the 79 positive charges of gold or the Z charges of any nucleus are distributed uniformly over a sphere that has a radius of 10^{-8} cm, outside that sphere, the force on the alpha particle is

$$F = \frac{2Ze^2}{r^2} \qquad \text{(magnitude)} \qquad (30.1)$$

Inside the positive sphere, the force and the potential are altered. It is a straightforward calculation to find the precise variation of the force and the potential as a function of distance inside the sphere. The results obtained are shown in Fig. 30.1.

The essence of Rutherford's argument is contained in the observation that for a uniform charge distribution, both the potential and the force reach a maximum value; one can calculate what the maximum possible deflection of an alpha particle will be in a single encounter with such a gold atom and one finds that this maximum deflection depends critically on the size of the positive charge distribution. A deflection through angles larger than 90°, such as had been observed by Geiger and Marsden, could not possibly be produced by positive charge distributions spread out over a sphere as large as 10^{-8} cm; rather a sphere of charge of radius smaller than 10^{-11} cm would be required.

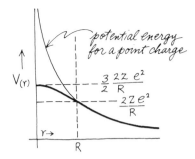

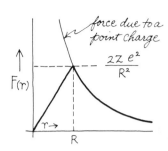

F I G. *30.1.* Qualitatively, we might look at the variation in force as follows: Outside the sphere, the alpha particle views the charge distribution as a point and experiences the Coulomb repulsion due to a point charge. At the center of the sphere, the force must be zero, because the alpha particle would be equally repelled by all parts of the sphere. What is significant is the fact that, in contrast to what would result if all the charges were located at a point (in which case the potential energy and the force would increase without limit as one approached the center), if the charge is smeared over a sphere of radius R, the potential does not get too much larger than its value at the edge of the sphere, and the repulsive force between the alpha particle and the nucleus reaches its maximum at the edge of the uniform sphere of charge distribution, and at this edge is just

$$F_{\text{max}} = \frac{2Ze^2}{R^2} \qquad \text{(magnitude)} \qquad (30.1)$$

A method of making this estimate follows. Imagine an alpha particle making a head-on collision with a gold atom and rebounding backward:

In these circumstances the alpha particle will stop completely for an instant before it returns. At this turning point its original kinetic energy will be converted completely into potential energy. For this to happen, the potential energy must be large enough in magnitude so that it can equal the original kinetic energy (the potential barrier must be high enough so the alpha particle does not penetrate). This means that the positive charge distribution must be small. The maximum potential energy of the alpha at the center of a uniform charge distribution is given by

$$\frac{3}{2} \frac{2Ze^2}{R} \qquad (30.2)$$

In order that the alpha particle come to rest, all its initial kinetic energy when it left the polonium source must be converted and must be equal therefore to this maximum potential energy, or

$$\tfrac{1}{2}mv^2 = \frac{3}{2} \frac{2Ze^2}{R} \qquad (30.3)$$

This gives a value for the radius of the charge distribution of

$$R = \frac{3}{2}\frac{2Ze^2}{\frac{1}{2}mv^2} \qquad (30.4)$$

$$= \frac{(3)(2)(79)(4.8 \times 10^{-10}\text{esu})^2}{(2)(\frac{1}{2})(6.62 \times 10^{-24}\text{ g})(1.6 \times 10^9 \text{ cm/sec})^2}$$

$$\simeq 6.5 \times 10^{-12} \text{ cm}$$

Rutherford calculated that from the large Thomson positive charge distribution particles should never be deflected more than 0.03 degrees in a single collision; in undergoing multiple collisions they should have about an equal chance of being deflected one way as another. Therefore, large deflections as a result of many single deflections in the same direction were very improbable. (It had been calculated on the basis of the Thomson model that a total deflection greater than 90° in traversing the gold foil would have only one chance in 10^{3500} of occurring.)

To increase the deflection, the only thing that could easily be done was to decrease the size of the charge distribution; as R decreases, the maximum force will go up. Finally, it will be large enough to produce, for head-on collisions, the large deflections that are observed. To obtain the observed deflection R has to be reduced from 10^{-8} cm to about 6×10^{-12} cm. Thus Rutherford could conclude:

> The theory of Sir J. J. Thomson . . . does not admit of a very large deflexion of an a particle in traversing a single atom, unless it be supposed that the diameter of the sphere of positive electricity is minute compared with the diameter of the sphere of influence of the atom.[5]

Instead of a distribution of positive charge, with electrons embedded like raisins in the positive cake, as in the theory of J. J. Thomson, Rutherford considered an atom with a central and highly concentrated positive charge (soon to be known as the nucleus) with the electrons spread out over a very large region beyond it.

> We shall first examine theoretically the single encounters with an atom of simple structure, which is able to produce large deflexions of an a particle, and then compare the deductions from the theory with the experimental data available.[6]

He then calculated the orbit of an alpha particle subjected to the Coulomb force due to a heavy massive charge concentrated essentially at a point. This was no more than the classical planetary problem already solved by Newton; in this case (the energy being larger than zero) the orbit was a hyperbola.

Rutherford's calculation went as follows. A large number of alpha particles bombard the many gold nuclei in a foil. Some come relatively close to a point positive nuclear charge and others relatively far away. The trajectory of any individual alpha particle depends on how closely it approaches a gold nucleus; the further away, the less it is deflected. Several characteristic trajectories are shown here.

He then averaged the trajectories over all the paths of incoming par-

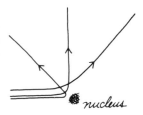

nucleus

ticles and compared his results with Geiger and Marsden. They agreed fairly well.

> The angular distribution of the α particles scattered from a thin metal sheet affords one of the simplest methods of testing the general correctness of this theory of single scattering. This has been done recently for α rays by Dr. Geiger, who found that the distribution for particles deflected between 30° and 150° from a thin gold-foil was in substantial agreement with the theory.[7]

He found further that he could predict the dependence of the number of scatterings on the thickness of the gold foil, on the magnitude of the central charge, and on the energy of the incoming alpha particle. To the extent that experiments at that time could check these predictions, they seemed to be corroborated.

31
ORIGINS OF THE QUANTUM THEORY

THE DILEMMA POSED BY RUTHERFORD'S ATOM

About Rutherford's atom, Niels Bohr was to say:

> I remember, as if it were yesterday, the enthusiasm with which the new prospects for the whole of physical and chemical science, opened by the discovery of the atomic nucleus, were discussed in the spring of 1912 among the pupils of Rutherford. Above all, we realised that the localisation of the positive electrification of the atom within a region of practically infinitesimal extension allowed a great simplification in the *classification of the properties of matter*. In fact, it permitted a far-reaching distinction between such atomic properties as are wholly determined by the total charge and mass of the nucleus and those which depend directly on its internal constitution.[1]

He added:

> Rutherford's model of the atom puts before us a task reminiscent of the old dream of philosophers: to reduce the interpretation of the laws of nature to the consideration of pure numbers.[2]

In Rutherford's atom the massive positive charge was confined to a minute region at its center, while the electrons circulated about it in a far-removed cloud, accounting for the atom's chemical properties and protecting the nucleus from ordinary encounters. Since atoms are neutral, one expected that there would be as many electrons as positive charges. A natural classification of the atoms would thus be according to the number of positive charges (and therefore the number of electrons). The weight of the positive matter in the interior might be secondary. So grew the concept of the atomic number, Z : 1 for hydrogen, 2 for helium, 3 for lithium, and so on.

Rutherford felt, in 1911, that ". . . the stability of the atom proposed need not be considered at this stage . . . ,"[3] but in 1912 and 1913 these considerations became paramount. The positive charge, according to Rutherford, was believed to be confined within a sphere of radius of the order of 10^{-12} cm. The electrons were believed to be distributed over a region of the order of 10^{-8} cm from the center. (By way of comparison, this would place the electrons, if everything were magnified evenly, farther than the earth from the sun.) If one assumes only electrical forces (gravitational forces are much too small, and one is not inclined at this moment to assume anything else since it was the assumption of electrical forces that allowed one to arrive at this picture of the atom), the question of how the atom is held together now arises. The first answer that suggests itself is so natural, so straightforward, and conceptually so economical that one hesitates to make it. The electron cannot stand still without support; it would fall into the nucleus, just as a still earth would fall into the sun. But the electron can revolve about the nucleus—a charged solar system, an absurd analogue of the planetary solar system. Could nature be so economical? Could the world be so made that an unimaginative repetition on an atomic scale of the planetary solar system would be the basis of the atomic constitution of matter?

The forces are electrical rather than gravitational, which makes them much stronger, but their form is identical. It is perfectly feasible to construct a little charged solar system in which the electron revolves about the proton. Can we then propose this as the dynamics for the Rutherford atom? If it were possible, there would not have ensued after Rutherford's proposal that crisis which resulted in the complete upheaval of classical physics and its replacement by quantum mechanics. For there is one flaw in our construction. It is unavoidable, irremediable, and fatal; it comes directly from Maxwell's theory.

The special glory of Maxwell's theory was the prediction that an accelerating charged particle, for example, an electron moving in a circle, would radiate electromagnetic waves. Maxwell had used these waves to unify electromagnetism and light. Hertz had produced them; Marconi had sent them across the Atlantic Ocean. According to Maxwell's theory, an electron orbiting around a positive charge to make a charged solar system should radiate light of a frequency equal to its frequency of revolution. In radiating this light, the electron would lose energy. As it lost energy it should spiral closer and closer to the positive charge, radiating more and more light until finally it reached the positive center. This distinction between the charged and the planetary solar system lay at the heart of Maxwell's theory as much as the second law lies at the heart of Newton's mechanics. If an electron moving in a circular orbit did not radiate electromagnetic waves, what was producing the

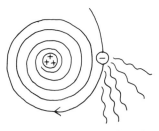

An electron in an orbit about a positive charge radiates energy spiraling into the positive charge as its energy decreases.

electromagnetic waves coming from antennas in which just such electrons were rushing back and forth?

Thus, one could not easily construct a charged solar system consistent with Maxwell's electrodynamics. Neither was it easy to imagine static electrons pinned in space with a heavy positive charge attracting them without imagining that they would fall toward it. One might then be tempted to introduce some other force which maintained the electron in some static position far from the positive charge. But there was no evidence for such a force, and on grounds of economy it seemed preferable not to have to introduce it.

The requirement of Maxwell's theory that an electron accelerating about a positive central charge should radiate energy undermined the possibility for the stability of the Rutherford atom from the point of view of classical physics. The time it takes for an electron to spiral into the nucleus from a typical orbit is very short—of the order of billionths of a second—uncomfortably inconsistent with our sense of the stability of the atomic matter of which we are made. Further the radiation that would be emitted in such a collapse would be continuous, the frequency increasing as the radius decreases. The spectrum would be a continuous band of color contrary to the observed discrete lines, characteristic of the individual atoms that had been made the basis of the chemical analysis of various materials in the course of the nineteenth century. It would be hard to say why any two atoms would be alike, for there would be no reason for two hydrogen atoms, even if they each consisted of a single electron in orbit about a single concentrated positive charge, to have their electrons in exactly the same orbits. Yet it is one of the elementary facts of spectral observation that hydrogen gas—any hydrogen gas—when excited always emits light of the same discrete frequencies.

This collection of facts, interpreted via existing theory, seemed to lead to a dead end at every turn. One might say, in retrospect, that Rutherford's atom was the ultimate application of purely classical principles in the atomic domain. As though he had opened the seventh seal, there was a silence . . . and the angels were given seven trumpets. . . .

In 1913 Niels Bohr proposed his famous theory of the hydrogen atom. One cannot say that he resolved the problems raised by Rutherford. In a sense he crystallized the dilemma in an even more dramatic form. Focusing his attention entirely on the construction of a nuclear atom, Bohr took what principles of classical physics he needed and added several nonclassical hypotheses almost without precedent; the mélange was not consistent. But they formed a remarkably successful theory of the hydrogen atom. It would be years before it could be said that one had a consistent theory again.

The essence of the problem as Bohr saw it was as follows: From the considerations of Rutherford and those before him, it seemed clear that a nuclear atom was required, a heavy positive charge lying at the center with electrons revolving about it—in the case of hydrogen a single positive charge and a single electron. If we ignore electromagnetic radiation, it is possible to construct such an atom classically. The orbits do not possess any individuality and the electrons can have any period and frequency, just as for the planets.

Considering the circular orbits that an electron could describe about a positive charge (it is possible to treat elliptical orbits, but the main features of the Bohr theory can be illustrated using the circular orbits which he himself employed), Bohr proposed that:

1. Of all the possible classical circular orbits only certain ones are allowed.

2. When an electron is in one of these allowed orbits, contrary to Maxwell's theory, it does not radiate energy.

3. It is when the electron makes a transition from one allowed orbit to another that it radiates energy.

There was a certain presumption in asserting what was contrary to Maxwell's electrodynamics and Newton's mechanics, but Bohr was young. The correctness of his vision was its closeness to what was in fact being observed. And in the end the problem of theoretical physics was to make consistent his assertions. Bohr's postulates were radical; yet they already had some precedent. The restriction of electrons to particular orbits is a theme that had its origin in the work of Max Planck at the close of the nineteenth century.

MAX PLANCK'S QUANTUM OF ACTION

The quantum of action was first introduced as the result of an analysis of what we may now consider a rather obscure phenomenon. At the turn of the century there was great interest in the distribution of radiation from what is known as a "black body." (A heated enclosure containing electromagnetic waves, in particular light, in thermal equilibrium: a little oven which has been so designed that any light emitted from its inside surface bounces back and forth many times before it is permitted to leave the small hole in the oven.) What was of particular interest was that for such an oven one should be able to deduce from the laws of statistical mechanics alone, independent of the material out of which the oven was made, what the distribution and intensity of the radiation should be. We know as a matter of common experience that an object heated turns from a dull to a bright red, then white, and if heated to a high enough temperature, it turns bluish; the coil that heats a pot is bright red, the tungsten filament that lights a bulb is yellow or white. Experimental studies of black-body radiation revealed no special surprises. The hole, as the oven was heated, glowed first dull, then bright red, then white, and so on.

The theory, however, was almost intrinsically paradoxical. According to the classical analysis, electromagnetic radiation can occur in such

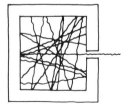

"black body"

a cavity in any of the possible standing waves. For a one-dimensional cavity these standing waves would appear as in Fig. 31.1. The longest standing wave that can fit onto this line has the wavelength

$$\lambda_{\max} = 2\ell \tag{31.1}$$

In general, the possible wavelengths are

$$\lambda = \frac{2\ell}{n} \qquad n = 1, 2, 3, \ldots \tag{31.2}$$

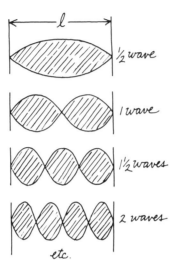

F I G. *31.1.* The four longest standing waves in a one-dimensional cavity.

And there is no limit on how short they can get. From the point of view of Maxwell's electromagnetic theory, each of these possible standing waves represents a degree of freedom of the electromagnetic field (just as for a many-particle system in one dimension each additional particle adds another degree of freedom). From the point of view of statistical mechanics, since the system was in thermal equilibrium (that was the essential point of constructing the black body), what energy it possessed should be equally divided among all excitable degrees of freedom. Since there were an infinite number of such degrees of freedom (all the possible standing waves) and only a finite amount of energy to be distributed, each degree of freedom should have zero energy. Thus no light would come out at any color or, if it came out at all, it should always be in the extreme violet or blue (because most of the standing waves occur at the shorter wavelengths)—something that is clearly contrary to the fact that the coil that heats the pot on our stove glows a dull red. This result, first obtained in 1900 by Lord Rayleigh and James Jeans, became known as the *ultraviolet catastrophe.*

A great deal of thought was devoted to the problem with no real success until Max Planck introduced the rather startling and completely *ad hoc* hypothesis that the light was emitted in bundles in the black-body enclosure and that the amount of energy in each bundle was related to the frequency of the light by

$$E = h\nu \tag{31.3}$$

In this way, h, *Planck's constant*, the quantum of action, was introduced for the first time into the body of physics. By matching the resulting theory with the observations it was determined to be of the order of

$$h \simeq 6.6 \times 10^{-27} \text{ erg-sec} \qquad (31.4)$$

It is completely unprecedented; in classical theory the energy of a wave is related to the amplitude of that wave: large ocean waves have a large energy. The frequency, an independent quantity, depends on the number of vibrations per second of the disturbance producing the wave. There is no necessary relation in classical physics between the energy and the frequency. One can have waves of low energy and high frequency or of high energy and low frequency. However, if one agrees with Planck that each such standing wave emitted into the enclosure carries the minimum energy $h\nu$, which can be written

$$h\nu = \frac{hc}{\lambda} \qquad (31.5)$$

rather than, as for the classical wave having as small an energy as one wishes, then, as λ grows very short (high frequencies: blue and violet colors), the minimum energy to excite a standing wave grows very large —so large that the wave cannot be excited and is no longer available as a degree of freedom. Thus, instead of having an infinite number of degrees of freedom,

$$\lambda = \frac{2\ell}{1}, \ell, \frac{2\ell}{3}, \ldots \qquad (31.6)$$

the number of degrees of freedom would be finite, those between $\lambda = 2\ell$ and λ such that $h\nu = hc/\lambda$ is larger than the average energy available to excite a degree of freedom.

The energy in the container would now be divided among a finite number of standing waves, thus avoiding the catastrophic disappearance of all the energy into the ultraviolet. The average energy would increase with temperature, shifting the emitted light to the violet as the temperature increased, in agreement with observation.

Planck's assumption was *ad hoc*; it was unprecedented; it was askew to everything that had occurred before in classical theory; but it had one virtue: the resulting theoretical distribution of radiation was almost identical with what was seen coming out of the hole in the oven (Fig. 31.2).

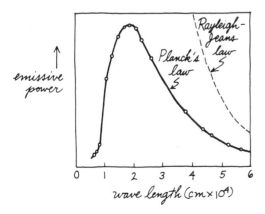

FIG. *31.2.* Comparison of Planck's law and the observed black-body radiation. The white circles are the observations. (After F. K. Richtmyer and E. H. Kennard, *Introduction to Modern Physics*, 4th ed., McGraw-Hill, New York, 1947)

Was the situation clarified by Planck's hypothesis? With it, it was possible to construct a theoretical curve in agreement with observation. But an understanding of its underlying basis was elusive. And so one might say Planck's relation, $E = h\nu$, remained in the wings—an intrusion into the main sweep of classical physics that Planck, among others, hoped would somehow disappear.

Planck says he

> . . . tried immediately to weld the elementary quantum of action h somehow into the framework of the classical theory. But in the face of all such attempts, this constant showed itself to be obdurate.[4]

Then,

> The failure of every attempt to bridge this obstacle soon made it evident that the elementary quantum of action plays a fundamental part in atomic physics. . . .[5]

And finally,

> My futile attempts to fit the elementary quantum of action somehow into the classical theory continued for a number of years, and they cost me a great deal of effort. Many of my colleagues saw in this something bordering on a tragedy. But I feel differently about it. For the thorough enlightenment I thus received was all the more valuable. I now knew for a fact that the elementary quantum of action played a far more significant part in physics than I had originally been inclined to suspect. . . .[6]

In 1905, the year that he published "On the Electrodynamics of Moving Bodies," Einstein also published a paper showing that if one treated the light trapped in the interior of a black-body oven as a gas of particles of energy $E = h\nu$, he could obtain Planck's result. In addition, he could provide an explanation of a phenomenon known as the *photoelectric effect*. Here again there was a problem somehow associated with a peculiarity in the absorption or the emission of electromagnetic radiation by matter.

While investigating the nature of electromagnetic waves in 1887, Heinrich Hertz found that his electrodes more easily discharged when exposed to ultraviolet light. In a demonstration of the same phenomenon, light is shined on the cathode of an evacuated tube, as shown in Fig. 31.3, and electrons are ejected. They move toward the positive electrode, the anode, and thus a small current flows. It is the relation of this current to the intensity and the color of the incoming light that is peculiar from the point of view of classical theory. The energy and the number of emitted electrons could be determined by a measurement of current and voltage as in Fig. 31.3. One would expect, for example from Maxwell's theory, that a stronger source of light would result in more energetic electrons being emitted. But this is contrary to what is observed.

The total current observed, a measure of the total number of electrons emitted, depends on the intensity of the light. But the maximum energy of individual electrons for a given cathode depends only on the

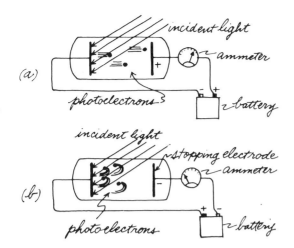

FIG. 31.3. (a) A method of observing the photoelectric effect. The photoelectrons ejected from the metal plate being irradiated are attracted to the positive collection electrode at the other end of the tube, and the current that results is measured with the ammeter. (b) A method of detecting the maximum energy of the photoelectrons. As the stopping electrode is made more negative, the slower electrons are repelled before they can reach it. Finally a voltage will be reached at which no photoelectrons whatever are received at the stopping electrode, as indicated by the current dropping to zero; this voltage corresponds to the maximum photoelectron energy. (After A. Beiser, *The Science of Physics*, Addison-Wesley, Reading, Mass., 1964)

color of the incoming light, that is, on its frequency. Thus, even an extremely weak source of light can occasionally result in a high-energy electron being emitted if the light is of high enough frequency. On the other hand, light of a lower frequency never results in emitted electrons—no matter how intense it is. This hardly makes sense from the point of view of Maxwell's theory, because from that point of view the light, an electric field vector, an oscillating electromagnetic wave, transfers its energy due to the force exerted by the electric field on the charged electron. If the source of light is very weak, then one has a weak electromagnetic electric vector spread over all of space. For such a weak electric field to produce a high-energy electron should take a long time. But the production of photoelectrons is observed to be practically instantaneous with the arrival of light on the cathode—no matter how weak the light.

Picking up the theme introduced by Planck, Einstein proposed in 1905 that light was not only emitted in units of energy, $E = h\nu$, but it was also absorbed in such bundles—bundles that came to be known as photons. Again, the energy of absorption was equal to the mysterious h, Planck's constant, multiplied by the frequency.

Nothing could have been more contrary to the classical idea of how a wave transfers energy. Consider, for example, a cork on the surface of a pond. If the height of the wave is small, the cork bobs weakly—as weakly as the wave is small. Einstein was proposing that this was not the method by which light—presumed to be a wave—transferred its energy. No matter how weak the intensity of the light, as long as it arrived at all, it arrived in a bundle whose energy content was $E = h\nu$, and if the energy was transferred at all, it was transferred in these bundles. Thus the energy transferred to an individual electron had nothing to do with the intensity of the incoming light, being related only to the frequency, as observed. If the frequency was too low, sufficient energy could not be transferred to an electron by one bundle to eject it from the surface of the cathode—no matter how intense the light. (The possibility remained that an electron might receive more than one bundle, a process which, when all the details are finally fitted into place, occurs—but not very frequently.) Further, these bundles arrived with the arrival of the light (since they were the light); thus there would be no expected delay in the emission of the photoelectrons no matter how weak the intensity.

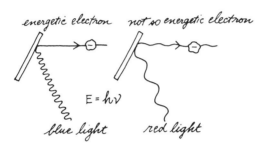

Einstein's proposal had the virtue of explaining very nicely the observations concerning the photoelectric effect. But again this relation could only stand somewhat askew to the body of classical physics. On the one hand, there was the entire classical tradition and on the other, as a somewhat recalcitrant player, stood this relation of both Planck and Einstein—that electromagnetic energy is absorbed and emitted in bundles, $E = h\nu$.

Example. What is the lowest frequency of light that will cause the emission of photoelectrons from a surface that needs 1.9 ev to eject an electron?

$$1.9 \text{ ev} = (1.6 \times 10^{-12} \text{ erg/ev}) \times (1.9 \text{ ev})$$
$$= 3.0 \times 10^{-12} \text{ erg} \tag{31.7}$$

Therefore,

$$\nu = \frac{E}{h} = \frac{3 \times 10^{-12} \text{ erg}}{6.6 \times 10^{-27} \text{ erg/sec}} = 4.6 \times 10^{14} \text{ cycles/sec} \tag{31.8}$$

This is light in the red part of the spectrum. All higher frequencies will also produce photoemissions of electrons.

NIELS BOHR'S HYDROGEN ATOM

What a difference 20 years had made! In 1890 the canvas of physics seemed sketched completely by the old masters. Every line of classical development seemed so successful that in their continued application to the various phenomena of nature it seemed only a matter of endurance and patience to put in the final touches of the brush. But by 1911 the situation could not have been more confusing. Coming right out of the great classical theories, electrodynamics and mechanics, were those curious and undigestible phenomena: the discrete emission and absorption of electromagnetic radiation, the stability of the Rutherford atom, the energy-frequency relation for light, and Planck's quantum of action. It seemed a time ready for great change, a time charged with adventure, with new possibilities, with the excitement of a potential upheaval, now about to appear and to command the scene.

In 1913 Bohr proposed his atom. It was the atom of Rutherford—a tiny, centrally concentrated positive charge, the electron moving about it in a circular orbit according to Newton's second law, attracted toward the positive nucleus according to Coulomb's law of force. Its acceleration (like that of any body moving uniformly in a circle) could be written

$$\mathbf{a} = \frac{v^2}{R} \qquad \text{magnitude}$$
(31.9)

directed toward center of the circle

The force between electron and nucleus was attractive, directed toward the nucleus, and had the magnitude (for hydrogen, one positive charge)

$$F = \frac{e^2}{R^2} \qquad \text{(magnitude)}$$
(31.10)

Therefore, from the second law

$$\frac{e^2}{R^2} = \frac{mv^2}{R}$$
(31.11)

or

$$mv^2 = \frac{e^2}{R}$$
(31.12)

Twice the kinetic energy is equal in magnitude to the potential energy. The total energy of the electron in its orbit (kinetic plus potential energy) is:

$$E = T + V = \frac{1}{2} mv^2 - \frac{e^2}{R}$$
(31.13)

which, using Eq. (31.12), becomes

$$E = -\frac{1}{2} \frac{e^2}{R}$$
(31.14)

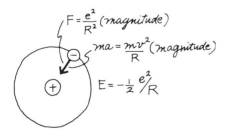

The total energy is just half the potential energy in magnitude. This is as true of a planet in the solar system as it is for an electron rotating about a positive nucleus. We are assuming that the electron moves in a circular orbit, that it obeys Newton's laws of motion, and that it is attracted toward the positive center, via Coulomb's law. Any circular orbit is possible and at the moment we have not asked any questions about radiation from such orbits. There is so far nothing unclassical in our assumptions. And so far we have not avoided any of the difficulties.

Now Bohr introduced his famous hypothesis—only certain of all the possible orbits were to be allowed, he declared, those orbits for which a quantum condition related to the condition of Einstein and Planck was satisfied. The quantum condition, as Bohr phrased it, restricted the angular momentum of the various possible orbits. For a circular orbit, the magnitude of the angular momentum of the electron (linear momentum multiplied by lever arm) is

$$L = mvR \qquad \text{(magnitude)} \qquad (31.15)$$

The linear momentum of a particle of mass m was defined in Chapter 4 as

$$\mathbf{p} = m\mathbf{v}$$

a vector whose magnitude is mv. We define the angular momentum of this particle about some point O for the simple case that the particle moves in a plane circular orbit about the point O. (We will need nothing more complicated.)

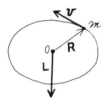

It is a vector directed as shown, perpendicular to the plane of motion, whose magnitude is

$$L = mvR$$

Only those orbits, Bohr proposed, in which the angular momentum satisfied the condition

$$L = mvR = \frac{nh}{2\pi} \qquad n = 1, 2, 3, \ldots \qquad (31.16)$$

only those orbits for which the angular momentum was an integral multiple (n is an arbitrary integer) of Planck's constant, divided by 2π, were to be allowed.* As we shall see, the quantum condition restricts the electron to certain possible circular orbits. However, in each of these orbits it should still, according to Maxwell's theory, radiate energy. This problem Bohr solved (if one can use that word) by fiat. He asserted that an electron, contrary to Maxwell's theory, contrary to Hertz's experiments, contrary to what everyone had believed until that time, does not radiate

* The quantity $h/2\pi$ occurs even more frequently than h itself, and so has been honored with a symbol of its own:

$$\frac{h}{2\pi} = \hbar$$

electromagnetic energy when it is in a stationary orbit—an orbit allowed by the quantum condition. When, then, does it radiate energy? This was the origin of Bohr's third postulate. Only when the electron makes a transition from one allowed orbit to another, Bohr asserted, does it radiate electromagnetic waves. And the precise amount radiated was to be given by the Einstein-Planck relation

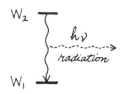

$$W_2 - W_1 = h\nu \qquad (31.17)$$

Thus, the energy of the light radiated was equal to the difference in energy of the two levels between which the electron leaped.

Each of these assumptions—the quantization condition, the lack of radiation while in one of the quantized orbits, and the radiation in the leaps between the orbits—was contrary to what was then known of classical theory. However, it was necessary to postulate the stability of the atom in some way. The radiation in jumps seemed to be consistent with what had already been revealed by Einstein and Planck. And the quantization condition was not too different from the original condition of Planck. Let us follow this mixture of classical and nonclassical postulates to see the kind of atom that Bohr obtained.

The quantization condition (31.16) gives a relation between the speed and radius of a possible orbit. Newton's second law, plus the Coulomb force, also gives a relation between the speed and the radius of an orbit (31.12). It is possible to combine these—the classical and the new quantum conditions—to give certain, and only certain, allowed radii:

$$R_n = \frac{h^2}{4\pi^2 me^2}\, n^2 \qquad n = 1, 2, \ldots \qquad (31.18)$$

The smallest possible radius occurs when $n = 1$, giving a value called the *Bohr radius*:

$$R_{\min}\ (\text{called } a_0) = \frac{h^2}{4\pi^2 me^2} \qquad (31.19)$$

Its value is calculated by substituting the values of Planck's constant (already known), the mass, and the charge of the electron. These yield

$$R = a_0 \simeq 5.3 \times 10^{-9}\ \text{cm} \qquad (31.20)$$

When the electron is in this orbit the atom has its lowest total energy (largest negative value); it is in what is called its ground state and can make no further transitions. According to Bohr's postulate it does not radiate. Thus it is stable. Further, all hydrogen atoms will eventually make transitions into this lowest state and therefore will be identical.

The energy levels of the various Bohr orbits can be computed as follows: From purely classical considerations, we have shown that the energy for an arbitrary circular orbit is equal to

$$W_n = -\frac{1}{2}\frac{e^2}{R_n} \qquad (31.21)$$

We therefore find, for the radii permitted according to Bohr's conditions, the allowed energy levels

$$W_n = -\frac{2\pi^2 m e^4}{h^2} \frac{1}{n^2} \qquad n = 1, 2, \ldots \qquad (31.22)$$

The energy of the atom for the electron in the lowest Bohr orbit (the ground state) is:

$$W_{\text{lowest}} = -\frac{2\pi^2 m e^4}{h^2} \simeq -2.2 \times 10^{-11} \text{ erg}$$
$$\simeq -13.6 \text{ ev} \qquad (31.23)$$

Finally, the electromagnetic energy radiated in transitions between one level and the next is given by Bohr's third condition:

frequency radiated in transitions: $\nu = \dfrac{W_2 - W_1}{h} = \dfrac{2\pi^2 m e^4}{h^3}\left(\dfrac{1}{n_1^2} - \dfrac{1}{n_2^2}\right)$
$$(31.24)$$

When he inserted the values then available for the various constants, c, h, and the charge of the electron, he found for the coefficient[*]

$$\frac{2\pi^2 m e^4}{h^3} = 3.1 \times 10^{15} \text{ sec}^{-1} \qquad (31.25)$$

As calculated from the frequencies of the lines of the Balmer series, its value was

$$3.290 \times 10^{15} \text{ sec}^{-1}$$

Bohr considered that:

> The agreement between the theoretical and observed values is inside the uncertainty due to experimental errors in the constants entering in the expression for the theoretical value.[7]

Since both n_1 and n_2 are integers $(1, 2, 3, \ldots)$, this gives immediately the result that in transitions between atomic levels light is radiated in various discrete frequencies. If we now let $n_1 = 2$, and let n_2 take any integral value from 3 up, we find a formula precisely of the Balmer type, written as a function of the frequency rather than the wavelength. To make the comparison we convert frequencies to wavelengths, using

$$\lambda\nu = c \qquad (31.26)$$

Eq. (31.24) becomes

$$\lambda = \frac{2ch^3}{\pi^2 m e^4}\left(\frac{n^2}{n^2 - 4}\right) \qquad (31.27)$$

which is to be compared with the empirical Balmer formula

$$\lambda = b\,\frac{n^2}{n^2 - 4} \qquad (31.28)$$

Using modern values $2ch^3/\pi^2 m e^4 = 3.6448 \times 10^{-5}$ cm, compared with Balmer's $b = 3.6456 \times 10^{-5}$ cm.

[*] The modern values are a bit different.

Other lines were also predicted by the Bohr model. The Balmer formula gave the series in which the electrons made transitions from the various levels to the level $n = 2$. One might also expect to observe series of lines in which electrons made transitions to the level $n = 1$, to the level $n = 3$, and so on. Such series were identified in the course of spectroscopic investigations and are known now as the Lyman, Balmer, Paschen, and Brackett series, among others (Fig. 31.4).

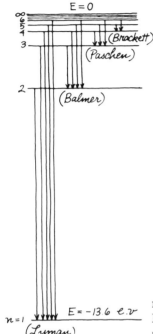

FIG. 31.4. Transitions between levels of Bohr's hydrogen atom producing the various series.

The connection with Maxwell's theory Bohr attempted to construct as follows. Electrons, he had assumed, emitted radiation only in transitions between orbits. As one went to higher and higher orbits, if he assumed further that the major electron transitions were between n values differing by 1, the frequency emitted would become closer and closer to the classical frequency of an electron moving in that particular orbit. Thus for the larger orbits (let us say those approaching more closely macroscopic conditions), the radiation emitted according to Bohr's theory would approach more and more closely the radiation as expected from Maxwell's theory. It would arise, however, from transitions rather than from accelerated motion alone.

Frequency of radiated light in a transition between Bohr orbits:

$$\nu = \frac{W_1 - W_2}{h} = \frac{2\pi^2}{h^3} me^4 \left(\frac{1}{n_2^2} - \frac{1}{n_1^2} \right) \qquad (31.29)$$

If $n_2 = n_1 - 1$,

$$\left(\frac{1}{n_2^2} - \frac{1}{n_1^2} \right) = \frac{n_1^2 - n_2^2}{n_1^2 n_2^2} = \frac{n_1^2 - (n_1^2 - 2n_1 + 1)}{n_1^2(n_1 - 1)^2} = \frac{2n_1 - 1}{n_1^2(n_1 - 1)^2} \simeq \frac{2n_1}{n_1^4}$$

$$(31.30)$$

The last is approximately true as n_1 becomes much larger than 1: for very large orbits. Thus the frequency radiated in such a transition is approximately

$$\nu \simeq \frac{4\pi^2}{h^3} \, me^4 \, \frac{1}{n_1{}^3} \tag{31.31}$$

But what is the frequency of an electron in an orbit? It is the number of revolutions about the nucleus per second, or

$$\nu = \frac{\text{electron speed}}{\text{circumference of orbit}} = \frac{v}{2\pi R} \tag{31.32}$$

and this is determined by the various classical and quantum conditions, as:

$$mvR_n = \frac{nh}{2\pi} \tag{31.33}$$

Therefore,

$$\left(\frac{v}{2\pi R_n}\right) 2\pi m R_n{}^2 = \frac{nh}{2\pi} \tag{31.34}$$

so that

$$\nu = \frac{v}{2\pi R_n} = \frac{hn}{4\pi^2 m} \frac{1}{R_n{}^2} \tag{31.35}$$

but

$$R_n = \frac{h^2}{4\pi^2 m e^2} \, n^2 \tag{31.36}$$

Therefore,

$$\nu = \frac{hn}{4\pi^2 m} \frac{(4\pi^2)^2 \, m^2 e^4}{h^4 n^4}$$

$$= \frac{4\pi^2 m e^4}{h^3 n^3} \tag{31.37}$$

This is the frequency of the electromagnetic radiation emitted by an electron orbiting the nucleus in a "classical" atom—the same as that emitted by the electron making transitions from levels n to $n - 1$ (for n large) in Bohr's atom, Eq. (31.31).

The Bohr theory was remarkably successful. He produced a stable, Rutherford-type atom, incorporated the quantum idea of Einstein and Planck, and obtained a series of discrete spectral lines characteristic of the hydrogen atom, in agreement with the Balmer series and other series that had recently been or were yet to be discovered. (A series of lines observed in the spectrum of helium—the Pickering series—was identified as being due to Bohr-type transitions in ionized helium. Since ionized helium consists of a nucleus with two positive charges plus one electron in orbit, its analysis is identical to that of hydrogen, replacing e by $2e$ for the nuclear charge.) With some further assumptions he could give at least a crude qualitative account of some features of the periodic table (the chemical properties) of the elements. The achievement was remarkable; but the theory was not complete. In a sense it raised more questions than it answered. The electrons were permitted only in their allowed orbits. In these orbits, contrary to Maxwell's theory, they did not radiate; but radiation was emitted where they made transitions, from one orbit to another. Where were they during the transition? Did the electron exist at all in between orbits? Did the electron exist in a usual sense when it was in its allowed orbit? Could one ask such questions?

In the period following 1913, elaborate efforts were made to understand what Bohr had done. Although his theory had been so successful for hydrogen, it could not be applied with any quantitative success to the other elements. Bohr worked for many years attempting to construct a theory of atomic helium. And although he was able to make a limited connection with classical theory, this combination of classical, neoclassical, and contrary-to-classical principles Bohr had put together, which successfully for the first time produced a hydrogen atom, remained almost as much an enigma as before Bohr had begun to work.

THE
QUANTUM
THEORY

32

THE ELECTRON AS A WAVE

Bohr published his results in 1913—results at once a sensation and a mystery for the world of physics. But England, Germany, and France, the separate sources of the new physics, were soon otherwise occupied. Einstein was at work creating his theory of gravitation (one of whose consequences would be tested in 1919 by an international expedition that measured the bending of light from a star passing by the sun in the course of an eclipse). In spite of the extraordinary success of Bohr's theory in accounting for the spectrum and other aspects of the hydrogen atom, attempts to extend it to helium and to other atoms were not very fruitful. And although the evidence continued to accumulate for the particle behavior of light when it interacted with matter, the apparent inconsistency of Bohr's postulates—the mystery posed by Bohr's atom—remained unsolved.

During the 1920s there began several lines of attack that would eventually lead to that reconstruction called quantum physics. Although they seemed completely alien at first, by 1930 it had been shown that they were all equivalent—different statements of the same theme. We follow one.

In 1923 Louis de Broglie, then a graduate student, introduced the conjecture that particle-like objects (say electrons) should display wave properties. "It would seem," he wrote,

> . . . that the basic idea of the quantum theory is the impossibility of imagining an isolated quantity of energy without associating with it a certain frequency.

Wavelike objects show particle properties (that is, light is emitted and absorbed very much as a particle). This was demonstrated by Planck and Einstein and was utilized by Bohr in his atom. Why not, then, those objects that we normally conceive of as particles? (Let us say electrons.) Why should they not display wave properties? Indeed, why not? What circular orbits were for Plato, harmonies among whole numbers for Pythagoras, regular solids for Kepler, or the solar system centered about the giver of light for Copernicus, this symmetry between wave and particle was for de Broglie.

What would the wave properties be? De Broglie proposed the following. The photon had been observed to be emitted and absorbed in discrete bundles in such a way that the energy was related to the frequency by

$$E = h\nu \qquad (32.1)$$

At the same time, the relation between the energy and momentum of a relativistic light quantum (a particle with zero rest mass) is[*]

$$E = pc \qquad (32.2)$$

Together these give

$$h\nu = pc \qquad (32.3)$$

But

$$\lambda\nu = c \qquad (32.4)$$

With this he could obtain a relation between the wavelength and momentum,

$$\lambda = \frac{h}{p} \qquad \text{(photons)} \qquad (32.5)$$

for a supposedly wavelike object, the photon, which seemed to be emitted and absorbed in definite bundles.

De Broglie now proposed that all objects—those thought to be wavelike and those thought to be particle-like—should have associated with them a wavelength related to their momentum in just this way. An electron, for example, should be accompanied by an associated wave whose wavelength would be

$$\lambda = \frac{h}{p} \qquad \text{(all particles)} \qquad (32.6)$$

What this associated wave was, de Broglie was not prepared to say at the time. But if one assumed that an electron did, in some sense, have a wavelength associated with it, there were consequences.

Consider the Bohr quantization condition for the stable electron orbits. Suppose the stable orbits were those in which an integral number of wavelengths could be fitted into the circumference of the orbit—those orbits for which one could set up a standing wave. Standing waves—those on a spring or in an atom—form a pattern that remains constant and retains its features in time. Only certain special wavelengths are possible in given situations:

Let us suppose, de Broglie said, that only those hydrogen atomic orbits are possible for which a standing-wave pattern can be set up. In order that such a wave pattern exist, an integral number of wavelengths must fit into the circumference. Thus

$$n\lambda = 2\pi R. \qquad n = 1, 2, 3, \ldots \qquad (32.7)$$

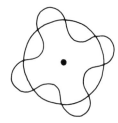

But the associated wavelength of the electron is related to its momentum by

[*] This is the general, relativistic energy-momentum relation in that case in which the rest mass is zero:

$$E = \sqrt{m_0^2 c^4 + p^2 c^2}$$

If m_0 (the rest mass) is equal to zero,

$$E = \sqrt{p^2 c^2} = pc$$

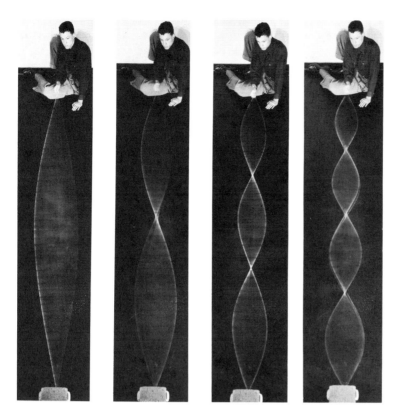

PLATE *32.1.* Only certain definite frequencies or wavelengths will produce fixed patterns. (Physical Science Study Committee, *Physics*, D. C. Heath, Boston, 1967)

$$\lambda = \frac{h}{p}$$

With this Eq. (32.7) becomes

$$\frac{nh}{p} = 2\pi R \tag{32.8}$$

or

$$pR = L = \frac{nh}{2\pi} \tag{32.9}$$

Bohr's quantization condition. Thus, if one could associate a wavelength with an electron, then the Bohr quantization condition essentially was the condition that an integral number of standing waves should fit into a given orbit in order that the orbit be stable. And the quantization condition now was not a special property of the atom but was a consequence of a property associated with the electron (and finally all particles) itself.

THE DAVISSON-GERMER EXPERIMENT

Astonishment was the reaction that greeted de Broglie when he presented his thesis to the faculty of sciences of the University of Paris in

1924. For what was the meaning of the wave associated with the electron? In what sense was the electron that Thomson had identified as an object with charge $-e$ and mass m a wave? An answer was not long in coming. In the United States C. J. Davisson and L. H. Germer, scattering low-energy electrons off the surface of a metal crystal, observed that the scattered electrons came off in rather peculiar peaks. In 1926, it was suggested to Davisson, when he took some of his preliminary data to a conference at Oxford, that these peaks might be interpreted as diffraction of the associated electron wave.

The essence of the Davisson-Germer experiment[1] is that the atoms of the nickel crystal they used form a regular array that behaves like a diffraction grating. The maxima in the scattering of the electrons from the crystal they interpreted as diffraction maxima when the condition

$$n\lambda = d \sin \theta \qquad (32.10)$$

was satisfied. But this is just the condition for diffraction maxima for a wave.

In a diffraction grating, light passes through the many slits a distance d apart. Thus each slit is a source of the transmitted wave.

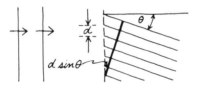

At the angles such that $n\lambda = d \sin \theta$, the waves originating at each opening reinforce each other on the distant screen and one obtains a maximum.

The atoms of a good crystal behave in an analogous way: Each atom is a source of the reflected wave. If the atoms are spaced a distance d apart, then again at angles such that $n\lambda = d \sin \theta$, the waves originating at each atom reinforce each other and on the distant screen one obtains a maximum:

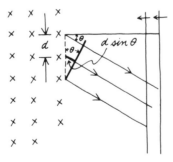

Since d, the distance between the atoms, could be calculated and $\sin \theta$ was directly measurable, one could, assuming the electron behaved like a wave, calculate the wavelength from the diffraction maxima. And this agreed within 1 percent with the wavelength proposed by de Broglie.

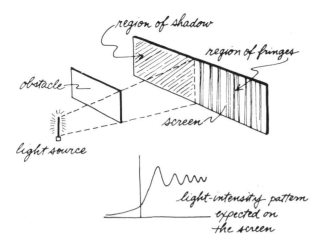

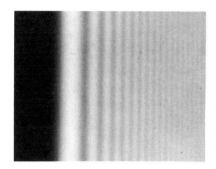
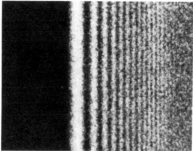

PLATE 32.2. Electron diffraction compared with the corresponding fringes made by visible light in the shadow of an edge. *Top:* Diagram of the setup used with visible light. An ordinary carefully made straightedge will serve; the fringes are closely spaced and must be studied a meter or so behind the obstacle. The small graph *(center)* shows the expected pattern of light intensity on the screen. *Lower left:* The pattern painted by light. (J. Valasek, *Introduction to Theoretical and Experimental Optics,* Wiley, New York, 1949) *Lower right:* The pattern painted by electrons. For the electrons, the straightedge is a small cubical crystal of MgO, much smaller than 10^{-4} cm, and the fringes have been photographed with the aid of an electron microscope. (H. Raether, "Elektroninterferenzen," *Handbuch der Physik,* XXXII, Springer, Berlin, 1957)

Since that time there has been the opportunity to do a large number of experiments to attempt to verify whether or not one can associate a wavelength with any particle. The results are unambiguous. Electrons, protons, all particles can be made to display precisely the wave properties—interference, diffraction, and so on—displayed by light. In Plates 32.2 and 32.3 we compare diffraction and interference patterns made by light and by electrons. Even if electrons arrive one at a time, the interference pattern builds up—the electron seems to interfere with itself. And for electrons, as for light, if one path is blocked, the interference pattern disappears.

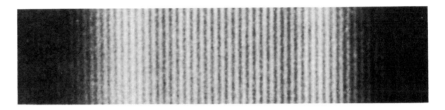

PLATE *32.3.* Comparison of the interference patterns from an arrangement analogous to a double slit produced by *(above)* light and *(below)* electrons. (Above: J. Valasek, *Introduction to Theoretical and Experimental Optics,* Wiley, New York, 1949. Below: photograph by Professor G. Mollenstedt, University of Tubingen, Germany)

33

SCHRÖDINGER'S EQUATION: THE LAW OF MOTION FOR QUANTUM SYSTEMS

Shortly after de Broglie introduced the idea of the associated wave for an electron, Erwin Schrödinger proposed an answer to the question of what happens to the associated wave if a force acts on it.° Contained in what is now known as Schrödinger's equation, it is the heart of quantum physics. In 1926 Schrödinger published a series of papers containing his now famous equation and its applications to many of the basic problems of quantum theory.

° It is related that, giving a colloquium on the subject, he encountered a persistent questioner: "But what will happen to the matter wave if a force acts on it?" The questioner is said to have been Peter Debye, and Schrödinger was sufficiently troubled to find an answer.

Schrödinger's equation gives the de Broglie wave associated with an electron, any other particle, or finally any quantum system. Given the mass of the particle, and given the forces to which it is subjected, let us say gravitational, or electromagnetic, then Schrödinger's equation gives the possible waves associated with this particle; the waves (functions of position and time) give a number associated with any position in space at an arbitrary time. And they are designated by the hardest working symbol in twentieth-century physics: the wave function

$$\psi(x, y, z; t) \tag{33.1}$$

The essence of the Schrödinger equation is that, given a particle, and given the force system that acts, it yields the wave function solutions for all possible energies. The wave function satisfies the most fundamental property of waves—the property of superposition. If under a given set of circumstances two solutions for the Schrödinger equation exist, the sum of the two solutions will also be a solution for the same circumstances. Just as in classical theory, this means that a trough and a crest can be added to cancel one another. Thus, one can have interference, that most characteristic wave phenomenon. And this now is associated with what were thought of previously as particles: electrons or protons and finally even with entire systems.

The classical dynamics of a body can be summarized as follows. Given a Newtonian particle of mass m which at the time t_0 is at the position $\mathbf{r}_0$ with the velocity $\mathbf{v}_0$, and given the forces that act on this particle, from Newton's second law its position and velocity at all future times, the orbit it will trace, can be calculated. In the dynamics of the corresponding "quantum particle," given its wave function at the time t_0 (the wave function thus containing all the information possible from the quantum point of view) and given the forces that act on the "particle," from the Schrödinger equation (the second law of quantum physics) the form of the wave function at all future times can be calculated.

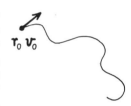

FREE SOLUTIONS

For a free particle (one on which no forces act) the Schrödinger equation yields again the results proposed by de Broglie. Solutions exist for any energy larger than zero; the energy is related to the momentum in the nonrelativistic theory by

$$E = \frac{p^2}{2m} = \tfrac{1}{2}mv^2 \tag{33.2}$$

And associated with a wave of momentum $\mathbf{p}$ is the de Broglie wavelength

$$\lambda = \frac{h}{p} \tag{33.3}$$

In the absence of external forces the momentum remains unchanged; the wave maintains its form and thus may be said to possess an inertial property.

At boundaries, the interface between two media, or in passage around barriers, the waves display very much the same properties as the waves we have previously studied—displacements of a water surface, a

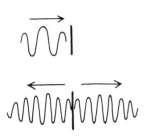

spring, or the waves identified with light. A Schrödinger wave, for example, is reflected from an impenetrable barrier, or partially reflected and partially transmitted at a barrier that is penetrable. At a hard, mirror-like boundary the wave is reflected in such a way that the angle of incidence equals the angle of reflection. At the boundary between two penetrable media there is an incident, a reflected, and a refracted wave, which satisfy rules of reflection and refraction such as those discussed in Chapter 15.

By looking at photographs of the behavior of light or water waves, we see displayed some of the qualitative properties of the Schrödinger wave. A Schrödinger wave of wavelength λ passing through a barrier with the opening d will be diffracted (as will a light or a water wave),

the angle to the first minimum being given by $d \sin \theta = \lambda$. The same wave passing through a barrier with a double-slit opening gives the diffraction patterns shown before.

BOUND SOLUTIONS

For unbounded systems, just as classically, there are solutions of the Schrödinger equation for all energies; it is possible for a particle to move at arbitrary speed. For bound systems, corresponding to a particle confined to some definite region of space,* one finds a characteristic feature of quantum mechanics quite different from classical mechanics. As a general rule, for such bound systems, not all energy levels are possible. The reason is related closely to that proposed initially by de Broglie: an integral number of waves must fit into a closed orbit in order that the orbit be stable.

This has been illustrated previously for the Bohr orbits. We would like to display this property in more detail in a situation (very characteristic in quantum physics) in which a particle in one dimension is trapped in a container whose walls confine it always to the interior. The particle of mass m is constrained to move on the line between the two walls separated by a distance ℓ.

From the Newtonian point of view the situation is very simple. The

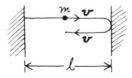

---

* For the Newtonian system of a planet subject to the gravitational force of the sun, closed bound elliptic or circular orbits result when the energy is smaller than zero, open unbound (hyperbolic) orbits when the energy is larger than zero.

particle can move at an arbitrary speed; it strikes one wall, bounces back, strikes the other, bounces back, and so on. Its energy is related to the speed or to the momentum by

$$E = \frac{p^2}{2m} = \frac{1}{2}mv^2 \qquad (33.2)$$

(One can, for example, calculate the momentum transferred at the walls to the particle—as was done in the analysis of the model for a gas in the kinetic theory.) There is nothing to restrict the energy, which goes continuously from zero, when the speed is zero, to any arbitrarily large amount, with the provision that the speed not be so great that the ball or the wall is destroyed.

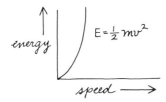

In the Newtonian view, the energy is a quadratic function of the speed and is continuous, going from zero to some arbitrarily large value as the speed of the particle increases.

From the point of view of quantum physics, the situation is quite different. The possible solutions of the Schrödinger equation are just those in which a standing de Broglie wave can be fitted between the constraining walls on the two sides of the container. The condition that the walls constrain the particle is translated, from the point of view of the Schrödinger equation, into the condition that the amplitude of the waves is zero at the walls. (This corresponds to elastic scattering in the Newtonian case.) Thus, the first four possible standing-wave solutions of the Schrödinger equation for this situation are shown in Fig. 33.1.

It is an essential feature of the solutions of the Schrödinger equation that only special wavelengths are allowed in such a situation—wavelengths intimately related to the size of the container (the length of the line) just as in the case of the Bohr orbits the possible wavelengths were related intimately to the size of the orbits. This limits the possible solutions of the Schrödinger equation and the possible energies. To form standing waves, the wavelength must satisfy the condition that between the walls of the container there are an integral number of half-wavelengths. The longest wavelength is then

$$\lambda_{\text{longest}} = \frac{2\ell}{1} = 2\ell \qquad (33.4)$$

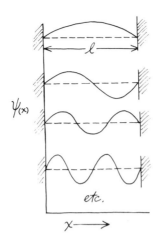

$\psi_{(x)}$

$x \longrightarrow$

FIG. 33.1. The first four standing-wave solutions of the Schrödinger equation for a particle contained along the line of length ℓ. The first wave has a single peak; the second two, and so on. Waves are possible with an arbitrarily large number of peaks. The condition that the walls are impenetrable requires that the wave function go to zero at $x = 0$ and $x = \ell$.

The next is

$$\lambda_{\text{next longest}} = \frac{2\ell}{2} = \ell \qquad (33.5)$$

And, in general,

$$\lambda_n = \frac{2\ell}{n} \qquad n = 1, 2, 3, \ldots \qquad (33.6)$$

This precise condition on the wavelength is not as important as the fact that only certain wavelengths are allowed.

Since the wavelength is related to the magnitude of the momentum by

$$\lambda = \frac{h}{p}$$

the restriction on the wavelength implies a restriction on the possible magnitudes of the momenta of the particle:

$$\text{possible momenta:} \quad p = \frac{h}{\lambda} = \frac{h}{2\ell} n \qquad n = 1, 2, \ldots \qquad (33.7)$$

And since the momentum is related to the energy by

$$E = \frac{p^2}{2m}$$

one has a restriction on the possible energies that this quantum system can have:

possible energies: $E = \dfrac{p^2}{2m} = \dfrac{1}{2m}\left(\dfrac{h^2}{4\ell^2}\right) n^2 \qquad n = 1, 2, \ldots$

$$(33.8)$$

Again, the precise form of the relation is not as significant as the fact that only certain energies are allowed. If we now draw a graph, as previously, of the relation between energy and momentum, for the quantum system we would find again that the energy is a quadratic function of the momentum (Fig. 33.2). But not all possible values of energy are allowed. The allowed energies are points that fall on the curve but do not fill in a continuous curve.

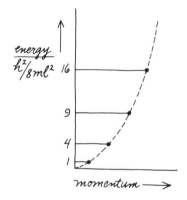

FIG. 33.2. Allowed kinetic energy values for a quantum particle confined to a one-dimensional container. One often indicates their values by drawing lines across the energy axis—called the energy levels of the system. They form a discrete rather than a continuous set, as in the Newtonian case. The lowest level is not zero—a peculiarly quantum phenomenon of great significance.

Thus, from the point of view of the Schrödinger equation for a particle that is in a bound state (confined to a finite region of space), not all energies, not all momenta, and not all wavelengths are allowed. Only a discrete set, a small portion of the energies that would have been allowed in Newtonian mechanics, occurs. One gets analogous results, although differing in quantitative detail, for particles contained in three dimensions, for particles contained by walls that are not rigid, and for particles contained by a potential such as that which produces the hydrogen atom. This, then, from Schrödinger's point of view, is the origin of discrete levels of the Bohr atom, and the discrete levels characteristic of all bound quantum mechanical systems.

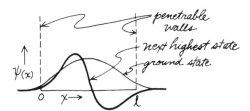

F I G. 33.3. If the walls are not impenetrable, the wave function does not go to zero at $x = 0$ and $x = \ell$. However, the various solutions are standing waves with one, two, three, . . . nodes. The numerical value of the energy levels is not exactly the same as those for the rigid walls, of course, but the qualitative properties, the discreteness of the energy levels, and the relation between the number of nodes and the number of the level are maintained. Further, those parts of the wave function that penetrate the walls die out (go to zero) rapidly as the distance of penetration increases.

Example 1. A particle of mass ½ g confined to a line 1 cm long has the allowed energies

$$E = \frac{h^2}{8m} \frac{n^2}{\ell^2} \qquad n = 1, 2, \ldots \tag{33.9}$$

$$\simeq 10^{-53} n^2 \text{ erg}$$

The speed of the particle in the lowest state is about 6×10^{-27} cm/sec, stationary by ordinary standards. In contrast, an electron, mass 9.1×10^{-28} g, confined on a line 10^{-8} cm long, has the allowed energies

$$E \simeq 5 \times 10^{-11} n^2 \text{ erg}$$

The speed of the electron in its lowest state is about 10^8 cm/sec.

The wavelength of a particle of energy E and mass m is

$$\lambda = \frac{h}{p} = \frac{h}{\sqrt{2mE}} \tag{33.10}$$

If the particle is an electron with an energy of 1 ev,

$$\lambda \simeq \frac{6.6 \times 10^{-27}}{\sqrt{2 \times 9 \times 10^{-28} \times 1.6 \times 10^{-12}}} = 1.2 \times 10^{-7} \text{ cm}$$

A mass of 1 g moving 1 cm/sec has a wavelength

$$\lambda = \frac{h}{mv} = 6.6 \times 10^{-27} \text{ cm} \tag{33.11}$$

Example 2. A 1-g ball bearing is confined to a 10-cm-long line. What would be its change in speed as it went from the lowest to the next quantum state?

$$p = mv = \frac{h}{2\ell} n \tag{33.12}$$

$$v = \frac{h}{2\ell m} n$$

when n increases from 1 to 2:

$$\text{change in } v = \frac{h}{2\ell m}(2-1) = \frac{h}{2\ell m}$$

$$\simeq \frac{6.6 \times 10^{-27} \text{ erg-sec}}{(2)(10 \text{ cm})(1 \text{ g})} = 3.3 \times 10^{-28} \frac{\text{cm}}{\text{sec}}$$

a small-enough change in speed to escape detection. When subjected to a force, no "jumps" of speed will be seen, and the ball seems to gain speed in a smooth Newtonian fashion.

If, instead of a ball bearing, we have an electron on an atomic "line" of length 10^{-8} cm, then

$$\text{change in } v \simeq \frac{6.6 \times 10^{-27} \text{ erg-sec}}{(2)(10^{-8} \text{ cm})(9.1 \times 10^{-28} \text{ g})}$$

$$= 3.6 \times 10^8 \text{ cm/sec}$$

a large enough change in speed to have noticeable effects.

34

WHAT IS THE ASSOCIATED WAVE?

If there is some confusion now as to the meaning of these waves associated with matter, it is not surprising. In the early 1920s this confusion was universal.

The problem had arisen previously for light. After Young and Fresnel, light became a wave. The fundamental wavelike property was interference: A crest could combine with a trough to produce a tranquil place, light could cancel other light; it was difficult to envision a particle as having this property. But with Planck's quantum, Einstein's photon, and Bohr's atom, by 1920 it was clear, in spite of its confirmed wavelike properties in interference and diffraction, that when it came to transfers of energy or momentum, light behaved in many respects like a particle. With de Broglie and Schrödinger, perhaps the situation had become more symmetrical, but the confusion was almost complete. For now, not only was light to be treated sometimes as a wave and sometimes as a particle, but matter itself—the ultimate, the final repository of atomic, corpuscular properties—the atoms of Democritus, Gassendi, and Newton —now had associated with them in some mysterious way a wave.

But, like a sculptor who knows the angle at which to hold his chisel, a painter who knows how the colors will mix on the palette, like any craftsman who knows his art with his fingertips but cannot verbalize it, the physicist of the 1920s had become accustomed to treating light or matter as a wave in diffraction or interference and as a particle in emission, absorption, or transfer of energy. Accustomed, resigned, used to, experienced with—perhaps these were the words. It was a procedure involving art and craft. It was hard to be logical, for the postulates were not there. And because of that long (and in retrospect misleading) debate concerning waves or particles, the situation seemed even less satisfactory than it was. It was as though light or matter displayed sometimes a particle-like, other times a wavelike, behavior which were mutually contradictory. As some texts stated in exasperation, light (or matter) is sometimes a particle and sometimes a wave.

DENSITY OF MATTER INTERPRETATION

Perhaps the most natural interpretation of the wave associated by Schrödinger and de Broglie with the electron would be to say (as Schrödinger attempted to do) that in some sense this wave represents the density of matter. With this interpretation the electron, rather than having its entire mass and charge concentrated at a point, would have that same mass and charge smeared over some region of space, the amount of

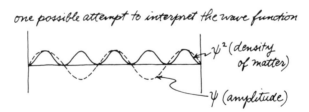

mass or charge in a given region being proportional to the magnitude of the square of the wave function.* (It is ψ^2 rather than ψ, the wave function itself, that is so interpreted because ψ^2 is always positive whereas ψ is not. The amount of matter in some region of space can be zero, but one cannot find an easy interpretation for a negative amount of matter.)

From this follows the requirement that the total area under the curve ψ^2 as a function of x be equal to the mass of the electron—that is, if one adds up all the electron matter distributed over a region one should arrive finally at that quantity of matter associated with the electron.

* For technical reasons, the wave function must be a complex number. And this complex number times its conjugate, written $\psi^*\psi$, is a number always larger than zero. For real wave functions $\psi^*\psi = \psi^2$. We shall avoid the problem by dealing explicitly only with situations in which the wave function is real.

For the ground state of a particle contained between two rigid walls:

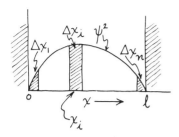

If ψ^2 is interpreted as a density of matter, then the amount of matter in the interval Δx_i about the point x_i as above, would be (the shaded area)

$$\psi^2(x_i)\Delta x_i$$

The amount of matter in the intervals at either end is small because ψ^2 is small there. The total quantity of matter must be equal to the mass of the electron. Thus,

$$\text{area under the curve} \simeq \psi^2(x_1)\,\Delta x_1 + \cdots + \psi^2(x_n)\Delta x_n = m$$

For the wave function above, we see that the density of the electron matter is small at the edges of the container and large in the center. Thus, presumably, if one could measure (*if one could measure*) one would find much of the electron in the center and little at the edges. This interpretation is not entirely incorrect. However, it has at least one consequence, so embarrassing it has to be abandoned completely.

Imagine, as shown, that an electron with its associated wave approaches a barrier—perhaps a small force field that might be due to a certain number of negative charges. Imagine that the negative charges of the barrier are fixed to some very heavy material, so they are essentially immovable. The electron approaching from the left will be repelled by the negative charges; classically its path is easily and completely determined. If it is moving fast enough (if its energy is large enough), it penetrates the barrier and moves onward to the right; if its energy is not large enough, it is reflected and bounces back to the left.

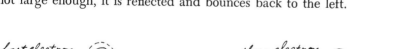

The solution of Schrödinger's equation for this kind of a situation, however, gives the following typical wavelike effect. Part of the associated wave penetrates the barrier and continues moving toward the right. Another part is reflected and goes back toward the left (Figs. 34.1 and 34.2). (This is not too different from what we have seen previously in the case of a one-dimensional wave, at a boundary between a lighter and a heavier spring.)

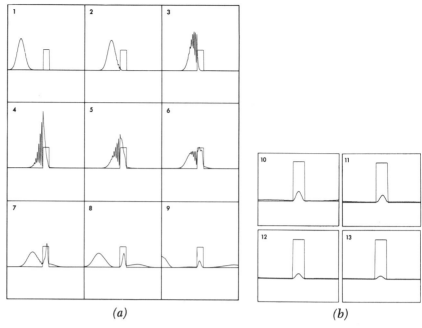

(a)   (b)

F I G. *34.1.* Here we show the behavior of an actual solution of Schrödinger's equation as it strikes a barrier: *(a)* Part of the wave penetrates the barrier and continues to move to the right. Part of it is reflected. The average energy of the wave is equal to the barrier height. *(b)* The decay of that part of the wave function left in the barrier is shown.

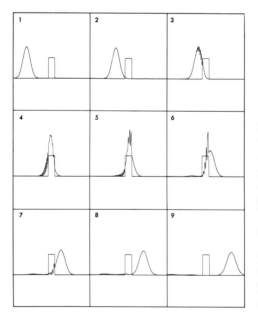

F I G. *34.2.* This wave penetrates almost completely. Its average energy is twice the barrier height. (Fig. 40.1 and this one courtesy of *American Journal of Physics*; from A. Goldberg, H. M. Schey, and J. L. Schwartz, "Computer-Generated Motion Pictures of One-Dimensional Quantum-Mechanical Transmission and Reflection Phenomena," *American Journal of Physics,* **35**, March, 1967)

The result is amusing. It demonstrates once more that the wave associated with the electron behaves very much like any other wave. A one-dimensional wave on a spring at a boundary is partially reflected and partially transmitted. This presents no problem for a spring. It presents no problem technically, as far as the solution of the Schrödinger equation is concerned. But conceptually it is baffling. For if the square of the function is interpreted as the density of matter of the electron, then what has happened in this process is that the electron has been divided—part of

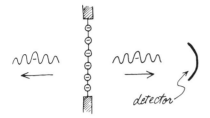

its matter on the right, and part on the left. If we now capture the part of the electron that has been transmitted with some device, presumably we should capture only that part of the electron's charge and mass that has been transmitted. Thus we should measure a charge and mass some fraction of the total electron charge and mass. But this is precisely contrary to all known experience with electrons.* No one has ever observed an electron divided in half, in a third, or in a quarter; it always comes in a bundle of charge $-e$ and mass m. It is this that makes the interpretation of the square of the wave function as the density of the distribution of mass and charge of the electron impossible. For it implies, contrary to experience, that the charge and the mass of the electron can be divided.

PROBABILITY INTERPRETATION

This led Max Born to propose an interpretation of the wave function whose consequences are possibly as revolutionary as those of any other idea of the twentieth century. He wrote later:

> Again an idea of Einstein's gave me the lead. He had tried to make the duality of particles—light quanta or photons—and waves comprehensible by interpreting the square of the optical wave amplitudes as probability density for the occurrence of photons. This concept could at once be carried over to the [Schrödinger] ψ-function: $|\psi|^2$ ought to represent the probability density for electrons (or other particles). It was easy to assert this, but how could it be proved?[1]

The probability interpretation, as developed by Born, now is the standard interpretation of the associated wave. The square of the wave function, ψ^2, represents a possibility density.† Thus the particle is likely to be found at the place where the wave function is large and not at the place where the wave function is small. Again, ψ^2 is chosen rather than ψ, because ψ itself can be negative and an interpretation for a negative probability is hard to find. With this interpretation, one arrives at

---

* If one actually attempted to do what has been described above, in the detector on the right we would sometimes observe an electron and sometimes not. But when it did reach the detector it would always be with the mass m and charge $-e$.

† Note again: The square is correct only if the wave function is a real number—true in the cases with which we deal.

the requirement that the total area under the curve ψ^2 as a function of x must be equal to 1, because that represents the probability that the electron be some place—which is taken to be certain.

Consider again the ground state of the particle contained between rigid walls. If ψ^2 is interpreted as a density of probability, the probability

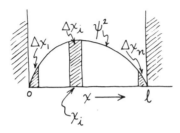

that the particle be in the interval Δx_i is

$$\psi^2(x_i) \, \Delta x_i$$

The probability that it be in the intervals at either end is small, because ψ^2 is small there. The total probability must be equal to 1 (since we agree that the electron is certainly between the walls). Thus

$$\text{area under the curve} \simeq \psi^2(x_1) \, \Delta x_1 + \cdots + \psi^2(x_n) \, \Delta x_n = 1$$

This leads to the general condition, consistent with the probability interpretation, that the total area under the curve is equal to 1.

The interpretation of the barrier experiment would now be as follows. Since we agree that the electron begins on the left, before its collision with the barrier all its associated wave would be on the left. After the collision with the barrier, part of the wave function comes back and part goes through. The square of the part that goes through—the part to the right—when added up represents the probability that the electron has penetrated while the part that comes back represents the probability that the electron has been reflected.[*] To the question, Will the electron strike the detector on the right? the answer we give is that the probability that it strike the detector on the right is the weight of the square of that part of the wave function that has gone through. However, if it does strike the detector on the right, it strikes with its full charge and mass.

A single electron might either strike or not. But if a large number of electrons are fired at the barrier, if the experiment is repeated over and over, or if in the experiment many electrons are fired at once, the pattern

---

[*] The probability of an event that is certain is, by definition, 1. For example, the probability of a coin landing either a head or a tail is 1. The probability of tossing a head is $\frac{1}{2}$, meaning that about $\frac{1}{2}$ the time a head appears. On a roulette table there are 37 numbers 0, 1, . . . , 36. The probability of landing on any one is $\frac{1}{37}$. Thus, about 1 in 37 times a given number turns up. However, anyone who has played roulette knows there are fluctuations.

that initially would seem to be random would begin to acquire a uniformity. Some fraction of the electrons would go through and strike the detector, and the others would be reflected and go back. But each one that went through would carry its complete charge and mass.

The probability interpretation of the wave function saves one from the embarrassment of an electron split into pieces. However, one now cannot say where the electron really is, whether a given electron will strike a detector or not. One can only say, for example, that the probability is one in five that the electron strike the detector; and four out of five that it is reflected back. We object: But this is no more than roulette. Is this the best that can be done? The answer, according to quantum theory and the probability interpretation is: Yes, that is the best.

Like a mountain range that divides a continent, feeding water to one side or to the other, the probability concept is the divide that separates quantum theory from all of physics that preceded it. For classical theory, as envisioned by Descartes or Newton, concerned the motion of palpable particles: corpuscles, hard, massy, and indivisible, which could be clearly located in space. The essence of Newton's system is: Given the position and velocity of a particle at some instant of time, and given the forces to which it is subjected, its path for all future times can be computed and determined. The traditional and classical example is that of the planets in their motion about the sun. Given the force of gravitation, and Newton's laws of motion, the orbits of the planets can be computed for all times in the future and all times in the past. One can say there will be an eclipse of the sun 25 centuries hence, which has the meaning that the earth, the moon, and the sun will occupy a specific position 25 centuries from the time of the calculation—or 25 centuries before. So much so that one can date historical events using backward calculations of the motions of the planets and their satellites.

But it is just this possibility that quantum mechanics relinquishes; for assuming that the sun, the moon, and the earth behaved as quantum mechanical particles (they do in our present view, but the magnitude of the quantum effects for such objects is too small to fall under the observation of our senses), one could only say that the probability that the sun, the moon, and the earth would be in a position so that an eclipse would be observed, let us say 5 or 25 centuries previously, would be one third or one half or one quarter. The whole notion then of a certain dating of an eclipse or an historical event would evaporate. It is this that causes the most anguish. Although there is no reason that we can expect that we shall be able to predict what will happen in the world with certainty, or even with probability, the idea that there are laws governing motion and these laws operate with certainty has been thoroughly ingrained, and successfully employed, at least since the time of Newton, and was implicit in much that went before. And what has replaced it has not been to everyone's taste. Einstein objected that "God does not play dice with the universe." Schrödinger himself never accepted the statistical interpretation, reprimanding Max Born at one time for continuing to state that the probability interpretation was generally accepted when he knew that besides Schrödinger, Einstein, Planck, de Broglie, and others were not satisfied with it. ("You, Maxel, you know that I am very fond of you, and nothing can change that, but I must once and for all give you a basic scolding. . . .")

What makes the situation even more curious is that there is nothing undetermined about the wave function itself. Given the wave function for a system at some time, and the forces to which that system is subjected, the precise form of the wave function for all future time is determined by the Schrödinger equation. It is that the wave function—which from the quantum point of view contains all the information—does not contain as much as we have been trained to expect. The only consistent available interpretation seems to leave us in the midst of a lottery rather than a well-organized retirement plan. But we have learned that the mathematical structure (the relation between postulate and theorem, and between one theorem and another) is often the most important element of a theory. The interpretation of the meaning of a theory has changed often, without changing the structural relations or its accord with reality. Our interpretation of Newtonian mechanics, for example, has changed radically between the nineteenth and the twentieth centuries. Yet a structural relation, such as that between the gravitation force and the elliptical orbits of the planets, remains essentially unaltered.

POSSIBLE ALTERNATIVES

It may happen in the future that an alternative to the probability interpretation will be developed. It has been proposed, for example, that there is some internal structure of the electron which as yet we do not know and, just as in statistical mechanics, in quantum mechanics one is assuming that variables which describe this structure are essentially random. To make this more precise, let's return to the electron, which sometimes penetrates the barrier and other times not. Let us agree that there is internal to the electron something we as yet have not discovered, which, with a failure of imagination, we represent as an arrow below:

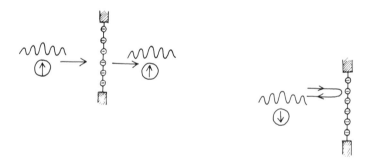

If the arrow points upward, let us say the electron penetrates the barrier. If the arrow points downward, let us say the electron is reflected. Now the argument might go that we as yet have no knowledge of this arrow. (We are humble enough to be willing to believe that things still exist in the world of which we have no knowledge.) The wave function is a cover for this lack of knowledge, replacing a precise knowledge of the position of the arrow with a statistical statement, just as statistical mechanics replaces a precise knowledge of the positions and momenta of 10^{23} particles with a statistical statement about certain average behavior. Thus quantum mechanics gives us a statistical statement only because it is incomplete; it is only when the internal arrows are found that the theoretical structure will be complete again.

We should emphasize that such a theory has never been successfully created. There are problems in the assumption of an arrow internal to an electron, involving the localization of the effects of the arrow, and so on. However, no one can really say that such a theory, or some other theory, could not be constructed which would allow us to abandon the probability interpretation. What we can say at the moment is that no such theory has been constructed, the probability interpretation has been extremely fruitful and extremely successful, and at the moment we are not forced by the phenomena to produce a theory that explains the probability interpretation.

It is always difficult to say whether the construction of an underlying explanation is a fruitful course or not. Maxwell's attempts to produce a machine—wheels, gears, and so on—to interpret the electromagnetic field do not seem to have been fruitful. The reason, presumably, is that the mechanical explanation is less fundamental than the field concept itself and what has happened is that the field concept has become primary and the mechanical explanations have become secondary, or have disappeared entirely.

The kinetic theory of gases, on the other hand, is an excellent example in which a mechanical interpretation is extremely fruitful. There, rather than saying that the irreducible concepts were temperature, volume, and the pressure of the gas, if one assumed that the gas was composed of mechanical entities—hard, massy spheres—and one applied Newton's laws of motion to the spheres, one could construct such concepts as temperature, pressure, and entropy and deduce relations among them in correspondence with experience. With that independent evidence for the existence of the atom, the whole structure achieved a unity that it would never have had if one had always thought of heat, temperature, and energetics as some separate subject. And since what we are essentially attempting to do is to construct a picture of the world using as few concepts as possible, the kinetic theory of gases must be considered a very fruitful direction to have taken.

One cannot say at the moment whether it is fruitful to attempt to interpret the wave function in terms of something deeper, or whether this is useless and that the wave function itself is to be considered the primitive entity, more fundamental than any interpretation we might make of it. We cannot even say if it is useful to worry about the question in this form. If in the future we are required by the necessity of subnuclear phenomena to modify the structure of quantum mechanics, the reinterpretation may come by itself.

One statement of the classical principle of causality is perhaps "similar causes produce similar effects." If all the causes are known and similar, the effects are always the same. From the quantum point of view this no longer seems to be so. Consider the electron headed toward a barrier (as treated before):

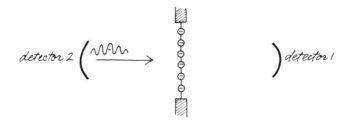

Classically the electron would either penetrate the barrier and arrive at detector 1 or be reflected and arrive at detector 2. Given its initial energy it can be calculated to do one or the other.

From the quantum point of view as we have seen, the most that can be calculated is the probability that the electron arrive at detector 1 or 2. This means that given all the causes (the wave function of the electron at $t = 0$ and the forces at the barrier), the effects are not always the same. One is tempted to say that all the causes are not known (that is, the wave function does *not* contain all the information; there are internal variables), but as yet there is no evidence for such unknown causes. If we agree for the moment that all knowable causes are known, then similar causes from the quantum view yield different effects:

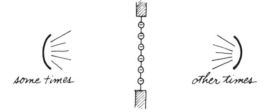

The microscopic event (the arrival of the electron at the detector) can be magnified by various means to produce macroscopic effects. For example: arrival at 1 activates a device that poisons a cat imprisoned in a cage; arrival at 2 deactivates this device. When we look at the cage we find either a live or a dead cat—a macroscopic event. (The cat is variously reported to belong to Klein or Schrödinger.)

Although macroscopic, this is not what we would call a normal experience. If it were, we would probably not be so agreeable to the convention that "similar causes produce similar effects"—an explicit example that this is a convention introduced to order our experience of the world (as first pointed out by David Hume) and not a property of the phenomena themselves.

35

ON THE CONSISTENCY OF
THE QUANTUM VIEW

WHERE IS THE QUANTUM?

If one wants paradoxes, one can now have them. The interpretation of
the associated wave as a probability amplitude leads to results that seem
extraordinary and curious, results that confound the intellect and amaze
indeed the very faculties of eyes and ears. But what is amazement other
than the disappointment of our expectations? And our expectations, what
are they? The classical organization of motion so successful in ordering
the movement of the tides and the planets misleads us entirely when
extrapolated to that domain of phenomena—involving distances of the
order of 10^{-8} cm, masses of the order of 10^{-27} g— in which the wavelike
properties of particles become apparent, in which the probability interpretation gives results different from those of the world of which we have
direct experience.

Possibly most surprising and most talked about is the appearance of
what seems to be a limitation on the kind of knowledge one can have
concerning some things that classically would have been considered completely knowable.

Implicit in the classical formulation of motion is the presupposition
that those elemental corpuscles out of which the world is to be constructed, those idealizations of a billiard ball—hard, massy, and indivisible, although one might not be willing to say whether they were
smooth or rough, colored yellow or blue—occupy some definite place in
the world: At any given time one can say where they are, how fast they
are moving, and what path they have followed. The fundamental problem
of classical physics is just—given the force system—to determine the path
a particle traces in time.

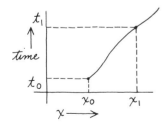

At the time t_0, the classical particle is located
at some point x_0. At the later time t_1, it is located at the point x_1. In between it has traced
out some definite path moving at some definite
speed.

The classical particle of mass m moving with speed v, subjected to no forces and contained between two walls, continues to move with the same speed in a straight line according to the law of inertia, until it strikes the wall. It is then reflected (elastically) and returns at the same speed in the opposite direction; we can follow its path in detail.

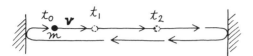

From the quantum point of view, the same particle, mass m, if it is to have some definite speed, has associated with it a periodic wave of wavelength

$$\lambda = \frac{h}{mv} \qquad (35.1)$$

which has crests a distance λ apart, repeated from one end of the container to the other.

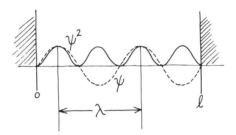

If one asks where the quantum particle is, the answer is that the probability of its appearance is given by the square of the wave function. For the wave function above, the probability would have maxima and minima periodically from one wall to the other. One therefore can hardly speak of the particle being at the point x_0 at t_0, at the point x_1 at t_1, and so on. The nature of the wave function for a particle of given momentum is such that the particle is equally likely to be at many places at any time. The particle has a momentum mv but cannot really be said to move from one place to another. This leads to a situation that is almost incomprehensible from the point of view of the classical corpuscle. For if there is anything that is clear about a billiard ball other than its hardness and its massiness, it is that it is at some given place on the table at every given time. This is something we can watch with our own eyes, amazed as we may be by the trajectory it may take under the cue of an expert.

To make matters possibly more confusing, this does not mean that a quantum particle cannot be localized. We may if we wish localize a quantum by associating with it a wave function that is localized in space, for example,

Or, even more extreme,

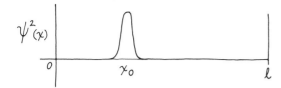

At the time t_0, this wave function, squared, is almost completely localized at the point x_0. Thus the probability for finding the particle at this point is close to 1 (certainty), whereas the probability for finding the particle any other place is zero.

The momentum of this quantum localized at x_0 we find from the de Broglie relation

$$p = \frac{h}{\lambda} \qquad (35.2)$$

But what is the wavelength of such a wave function? It has no single wavelength. The only wave function that has a single wavelength is a periodic wave function that repeats itself from one end of the container to the other. A wave function of the type shown above must be constructed out of a large number of periodic wave functions of different wavelength.* To analyze the momentum of such a quantum, one essentially decomposes the localized wave into the different periodic waves that have gone into making it up.†

The answer to the question, What momenta, what wavelengths are involved if we wish to completely localize an electron or a quantum at x_0? is that such a wave function is an equally weighted sum of waves of all wavelengths. The wave function that localizes a particle at a point is constructed by summing the periodic waves of all wavelengths from the very longest to the very shortest. This means, from the point of view of the quantum theory, that the momentum, if one can speak of it at all, of a particle localized at a point is equally likely to be zero or infinitely large.

The major surprise that quantum physics introduces in this regard is that the quantum, which has the particle's property of discreteness, lacks the particle's property of having at the same time a simultaneous position and a velocity. It is sometimes said that the problem is that we

* As mentioned in the discussion of waves, Chapter 15, an arbitrary wave form can be constructed of the sum of periodic waves of different wavelengths. This is a result that is completely independent of quantum theory; it has only to do with the nature of waves.

† One uses here the fact that the solutions of the Schrödinger equation can be constructed out of the sums of other solutions. Each periodic wave is a solution. And thus we may construct a solution, a localized wave, by making a sum of different periodic waves. However, since each periodic wave of different wavelength represents a different momentum, such a localized wave does not correspond to a single momentum but rather to a large number of momenta.

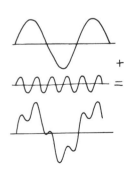

cannot measure simultaneously the position and velocity of a quantum particle. And, as we shall see, we cannot. But the fact that we cannot make this measurement is no reason for denying the existence of a simultaneous velocity and position or, what is perhaps more precise, for not constructing a theory in which the quantum particle has a simultaneous velocity and a position. The situation is really deeper. As we have seen from our constructions, our interpretation of the wave function does not allow us to attribute to it, in the classical sense, simultaneous position and momentum. If we attempt to construct a wave function that has a definite momentum, the particle associated with it does not have a specific position. If we attempt to construct one that has a specific position, it does not have a definite momentum. Thus the object that, from the quantum view, contains all the information cannot simultaneously be assigned these two classical properties.

HEISENBERG'S UNCERTAINTY PRINCIPLE

Although strange, this is not very complicated. In quantum theory, position is associated with the square of the amplitude, momentum with the wavelength of the wave function, and it is no more than a fact about waves that they cannot be completely localized and have a single wavelength. Part of the curiosity arises because one begins with the classical image of a particle and then is surprised that the quantum particle differs from its classical predecessor in a rather unusual way.

If one insists on thinking of the quantum particle in classical terms (in particular, if one attempts to give it both a position and a momentum), then the maximum possible precision of position and momentum is limited by a remarkably simple relation, the uncertainty or indeterminacy principle, proposed first by W. Heisenberg:

$$(\triangle p)(\triangle x) \simeq h \qquad (35.3)$$

where $\triangle p$ and $\triangle x$ are the imprecision or uncertainty in the momentum and position of the particle. The product of uncertainty in momentum and uncertainty in position is of the same order as Planck's constant. Contrasted with the classical case, one cannot both localize a quantum particle ($\triangle x = 0$) and assign to it a precise momentum ($\triangle p = 0$); therefore, it follows no path or orbit in the sense of a classical particle. The uncertainty is not meant in a psychological sense. It is intended to describe the nature of an object which does not possess both of the properties—position and momentum together, an object very loosely comparable to an atmospheric storm: spread over large distances, the winds are gentle zephyrs; confined in a small region it is a hurricane or a tornado.

The uncertainty principle puts in a remarkably simple form what has been difficult to say using the Schrödinger wave. If one has a wave function of a given wavelength or a given momentum, the position is completely undetermined, because the probability is equal that the particle may be any place. On the other hand, if the particle is completely localized, then one must sum over all periodic waves to produce such a localization and the momentum or wavelength is completely unspecified. The precise numerical relation between the lack of definition of position and momentum (which can be derived from the wave theory itself and is not specifically related to quantum mechanics being a part of the nature of all waves—sound waves, water waves, and the waves we have seen on a spring) is given in very simple form by the Heisenberg uncertainty principle.[*]

Imagine the particle we previously considered restricted to motion along a line and contained between two walls separated by a length ℓ. The uncertainty in its position is no larger than the distance between the two walls, since we know that it is between them. Therefore, $\triangle x$ is equal to or smaller than ℓ:

$$\triangle x \leq \ell \tag{35.4}$$

The particle might of course be even more localized. But given that it is constrained between the two walls, it can be no *less* localized than between 0 and ℓ. Thus, our "uncertainty" or lack of definition of its x coordinate can be no larger than ℓ. Given this, our "uncertainty" in its momentum is at least as large or larger than h/ℓ:

$$\triangle p \gtrsim \frac{h}{\ell} \tag{35.5}$$

The momentum is related to the speed by

$$p = mv \tag{35.6}$$

so we have an uncertainty in the speed equal to

$$\triangle v = \frac{\triangle p}{m} \gtrsim \frac{h}{m\ell} \tag{35.7}$$

If the particle is an electron and the distance between the walls is 10^{-8} cm, $\triangle v$ exceeds

$$\triangle v \gtrsim 7 \times 10^8 \text{ cm/sec} \tag{35.8}$$

Therefore, if one localizes an object with the mass of an electron to a distance of the order of 10^{-8} cm, then the extent to which one can speak of its speed can be no better defined than within about 7×10^8 cm/sec.

We can use this result to estimate the lowest energy the system can have, which, because the momentum is of necessity undetermined is in general not zero—a striking difference from classical physics. There the energy of the above particle is just its kinetic energy, and the lowest energy the system can have is that when the particle is at rest: $E = 0$. For the quantum system, as just calculated, the uncertainty in momentum for the confined particle is

$$\triangle p \simeq \frac{h}{\triangle x} \simeq \frac{h}{\ell} \tag{35.9}$$

[*] If we associate with the quantum a wave function localized in some region and we then ask, What is the momentum to be associated with this quantum? then, since in the Schrödinger theory the magnitude of the momentum is to be found from the wavelength by

$$p = \frac{h}{\lambda}$$

and since such a localized wave contains not one wavelength but rather a sum of waves of different wavelength, then associated with this quantum is not one momentum but rather a spread of momenta corresponding to the various wavelengths of periodic waves that are required in the sum to produce the localized wave. The spread of momenta required (which can be obtained from the wave theory) is given by the uncertainty relation.

The momentum of this particle cannot be completely defined, varying over an interval of size h/ℓ; the smallest momentum we can assign to the particle occurs when this interval is centered around zero:

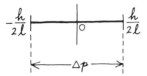

so that the momentum varies in magnitude between zero and $h/2\ell$. Thus it would be consistent with the uncertainty principle to assign the particle a momentum of magnitude as small as

$$p = \frac{h}{2\ell} \qquad (35.10)$$

Any smaller and we might be in trouble. The energy corresponding to such a momentum is

$$E = \frac{p^2}{2m} = \frac{1}{2m}\frac{h^2}{4\ell^2} \qquad (35.11)$$

which can be compared with the lowest energy we actually calculated from the Schrödinger equation fitting the smallest possible standing wave into the container:

$$E_0 = \frac{1}{2m}\frac{h^2}{4\ell^2} \qquad (35.12)$$

The point is not the exact numerical agreement but is that we can make a rough estimate of the minimum energy using the uncertainty principle. We also obtain an understanding from this point of view of why the lowest kinetic energy for a quantum mechanical system (opposed to a classical system) is never zero. The corresponding classical particle confined as above would have its lowest kinetic energy, zero, when it was at rest. The quantum particle, however, can never be at rest if it is confined. It has an intrinsic uncertainty in its momentum or velocity which shows up as an energy—closely related to the energy it can have in a precise calculation using the Schrödinger equation.

This very general situation has consequences especially in the analysis of the quantum mechanical equivalent of kinetic theory or statistical mechanics. It has become a part of popular lore that temperature, as is the case in kinetic theory, is related to the intrinsic motion of the atoms that make up a system. For quantum systems at high temperatures, something very similar to this is the case. However, at very low temperatures, quantum systems cannot come to complete rest. The lowest temperature corresponds to the lowest possible state of a system—classically, a state in which the particles are at rest, but from the quantum point of view, as above, a state for which the energy is given by Eq. (35.12) and which does not represent a particle at rest.

From all that is said above there appears a too large preoccupation with electrons between two walls. Our concern with electrons is perhaps understandable. But the walls? If we examine what has been done, we see that the force system, container or otherwise, that has constrained the electron to a small region of space is not terribly relevant. Two walls, a central force, or various barriers give about the same results. The means by which the electron is constrained are not as significant as the fact

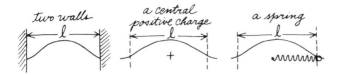

that it *is* constrained—its wave function is localized. If this is so, it is composed of a sum over periodic waves and the momentum is undefined as

$$\triangle p \, \triangle x \simeq h \tag{35.13}$$

What about ordinary large particles? Are they quanta or are they particles in the Newtonian sense? Does one have Newtonian mechanics for objects on an ordinary scale and quantum mechanics for objects on a small scale? We can consider all particles, all objects, even the earth, as quanta. However, the quantum phenomena, the wavelike properties, the uncertainties in position and velocity, become too small "to fall under the observation of the senses" when particles reach anything like the masses to which we are normally accustomed in macroscopic phenomena.

Consider, for example, the particle we have treated above. Suppose it is a ball bearing, with a mass of a thousandth of a gram (a very small ball bearing). If we constrain it in a distance that we might see, say, a thousandth of a centimeter, then we would have

$$\triangle x = 10^{-3} \text{ cm} \tag{35.14}$$

so that

$$\triangle p \simeq \frac{h}{\triangle x} \simeq 6 \times 10^{-24} \frac{\text{g-cm}}{\text{sec}} \tag{35.15}$$

or

$$\triangle v = \frac{\triangle p}{m} = \frac{6 \times 10^{-24}}{10^{-3}} = 6 \times 10^{-21} \frac{\text{cm}}{\text{sec}} \tag{35.16}$$

Thus there is a lack of definition of the speed of the order of 6×10^{-21} cm/sec.

This one illustration contains the heart of the answer to the question as to whether one has many theories or just one. The quantum theory we are describing becomes the classical theory so closely that one can hardly observe any deviations when the masses and the distances are of ordinary size. It is only with particles of the order of the mass of the electron and distances of the order of atomic distances that quantum phenomena, the wavelike character of matter, begin to manifest themselves. Even for an object 10^{-3} g in mass confined to 10^{-3} cm, the uncertainty in speed is much too small to be noticed in an ordinary observation.

The Heisenberg uncertainty relations connect not only position and momentum but also other variables of a system that classically would be thought to be independent. One of the most interesting and useful for our purposes is a connection between the uncertainty in energy and in time. This is usually written

$$(\triangle E)\,(\triangle t) \simeq h \tag{35.17}$$

If we can keep a system in a certain state for a very long time, then we can know its energy very precisely; however, if it remains in that state for only a very short time, then there can be a sizable lack of definition of its energy; the precise relation is given above.*

This relation is typically applied to the decay of one quantum state into another. For example, suppose a certain particle lives for 10^{-8} sec; that is, between the time it is produced and the time it decays to something else, a time of 10^{-8} sec elapses. Then the greatest precision with which we can know the energy of this particular particle is given by

$$\triangle E = \frac{h}{10^{-8}} \simeq 6 \times 10^{-19}\,\mathrm{erg} \simeq 4 \times 10^{-7}\,\mathrm{ev} \tag{35.18}$$

which, in this case, is extremely precise. As we shall see later, certain so-called *elementary particles* live for only something of the order of 10^{-21} sec—the time between their creation and their annihilation. Therefore, the time interval in which the particle is in a given state is very short, and the uncertainty in the energy is

$$\triangle E \simeq \frac{6 \times 10^{-27}}{10^{-21}} \simeq 6 \times 10^{-6}\,\mathrm{erg} \simeq 10^{6}\,\mathrm{ev} = 1\,\mathrm{Mev} \tag{35.19}$$

This is a rather large uncertainty in the energy, 10^{6} ev (a million electron volts being abbreviated 1 Mev), and reveals itself, as we shall see later, in the fact that such so-called elementary particles, sometimes called resonances, rather than being sharply defined with a precise energy, have an energy that is spread out over a rather large range.

This relation also gives what is called a natural width to the levels of a quantum system. If an atom, for example, decays from one level to

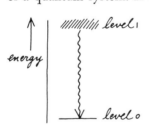

another, then level 1, from which it decays, is not precisely defined in energy because it lives only a certain finite time. For example, suppose as for a typical atomic system the time of decay of a level is 10^{-9} sec. In that case, the spread in the energy would be given by

---

* If the wave function maintains its shape—does not change in time—one can associate a definite energy with it. But if it does not change in time, the system remains the same; in this sense Δt would be infinite, and $\Delta E = 0$. If, however, the wave function changes in time, it is a mixture of several energies according to uncertainty relation.

$$\triangle E = \frac{h}{\triangle t} \simeq \frac{6.6 \times 10^{-27} \text{ erg-sec}}{10^{-9} \text{ sec}} \simeq 6.6 \times 10^{-18} \text{ erg} \simeq 4 \times 10^{-6} \text{ ev}$$

$$(35.20)$$

This is the natural width of the energy levels of an atomic system.

HEISENBERG'S THOUGHT EXPERIMENTS

The point of view being developed was not entirely pleasing. During the 1920s questions were posed, which were made as embarrassing as possible. The argument that perhaps summarizes all of them went as follows: In the new quantum theory a particle, such as an electron, has associated with it a wave. But the wave is such that one cannot define a position and a momentum simultaneously. Such a theory may be consistent, but it is incomplete. For example, let us imagine an electron described by a periodic wave of wavelength λ, so that it has a specific momentum:

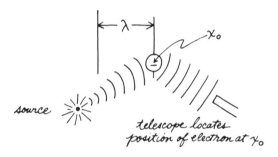

Now, suppose we investigate this wave carefully, perhaps by shining light on it in the gentlest manner possible. Then after a while we will find a light reflected from that point at which the electron "actually" is. Thus we locate the "actual" position of the "real" electron, which, because it has associated with it the wave of definite wavelength λ, has a definite momentum. For the "real" electron there is no uncertainty. The uncertainty occurs because the quantum theory associates with the electron only a wave. Thus, the classical argument goes, the quantum theory is incomplete, for it does not give us a complete description of the "actual" electron. It perhaps successfully describes some of the phenomena, but there are certain phenomena, for example the simultaneous position and momentum of the electron, that it cannot describe.

The argument begins essentially from the Cartesian or Newtonian notion of a particle that is discrete and follows a definite orbit, which we can imagine ourselves watching. Since we can watch the orbit and see it traced out, the quantum theory, which denies the existence of such an orbit, which speaks only about a wavelength, or, if it speaks about a position, does not speak about a definite momentum, gives at best an incomplete description of the phenomenon. It is as though, if the motion of the planets were in the quantum domain, one had a theory that related the average radius of the planetary orbit to its period but did not give the detailed position of the planet in time in its course around the sun. Therefore, the theory, while giving some relations (for example, it might give Kepler's third law), would not give the elliptical orbits and, therefore, would have nothing to say about this aspect of the data of Tycho Brahe and other astronomers. One could say about such a theory that it was partially successful, that it contained within itself some of the relations but that it did not contain a most important thing, the actual planetary orbits.

It was to answer such arguments that Heisenberg devised his famous Gedanken (thought) experiments—experiments by which he attempted to show that the process described above, by which the "real" electron is located, is not a possible process in our world. These Gedanken experiments (marvelously treacherous) fall in that class of constructs in which one attempts to prove that something is not possible. We are always left with that feeling of frustration experienced after the demonstrations of the impossibility of perpetual motion machines. Every time it is shown that a particular device has not succeeded, we respond by hoping that perhaps the next one will.

The brunt of Heisenberg's argument is that it is not possible, with the materials in fact available, to measure the position and momentum simultaneously; it is the nature of the world that the experiment conceived by the classicists, and just described, is not actually something that can be done. If it could, quantum theory would in fact be inadequate. Let us then reanalyze the classical experiment above to see where, according to Heisenberg, it fails.

Its basic presupposition is that the electron can be located by a means so gentle that it is essentially undisturbed and that it remains with the momentum that it had initially. With the classical light wave as conceived by Maxwell, Young, or Fresnel, this could be done. For if one could shine light of very short wavelength, that is, in the high blue or even in the ultraviolet (if one had special detectors), and of very, very low intensity (the low intensity meaning that the waves had a very low amplitude), then, from the classical point of view, because the wavelength was short, the position of the electron could be located very precisely. And because the amplitude was small, very little disturbance would be imparted to the electron—as though the electron, a cork on the surface of the water, was disturbed very little by a tiny ripple in the water. However, if the ripple had a short wavelength, one could determine fairly precisely where the electron was.

But this, says Heisenberg, is precisely what Planck and Einstein have taught us cannot be done. For the energy transfer of the light wave (photon) is not related to its amplitude but to its frequency. Thus, as we attempt to make its wavelength short so that the "electron" can be located, the energy it transfers increases. If a photon collides with the electron and bounces back, it may enable us to locate the position of the electron. But in so doing it transfers momentum to the electron; the more precisely we try to locate the electron, the shorter we wish to make the wavelength of the light, but the shorter we make the wavelength, the larger the momentum of the photon becomes and the more momentum we can transfer to the electron; and the more precisely we attempt to locate the rebounding photon, the less we are able to calculate how much momentum it has transferred to the electron. Thus, says Heisenberg, rather than the classical view in which one locates the electron so gently that one disturbs nothing, in fact, if one shines light on the electron, and if one locates it very precisely, one changes its momentum by an indeterminate amount, so that it is no longer represented by a periodic wave of wavelength λ.

[From the point of view of the associated wave, what is happening is as follows. If the electron wave function is initially a periodic wave, it is not localized. Now, if we attempt to "find" it (with photons, for example), what results is nothing much of the time—since the photons pass through without interacting. When a photon interacts with the "electron" violently enough so that the electron can be localized, the interaction has changed the electron wave function so that it looks perhaps as shown. It is now very localized but has no definite wavelength

(or momentum). Nothing is specially mysterious about this. Before the interaction, the electron wave function had a definite wavelength and was not localized. After, it is localized and does not have a definite wavelength. It is completely normal for the electron wave function to change in time with interactions. The essence of Heisenberg's argument is to show that the electron cannot be located by using another object (say the photon) without this accompanying change in the electron wave function. And it cannot, because the photon itself is a quantum object. To "locate" the electron it must interact with it (otherwise it just passes through), and when it interacts it changes the electron's wave function.]

Of course, if one particular device does not work one tries another. Heisenberg and others constructed a whole series of devices; in each case they failed. About all these thought experiments, there is the same elusive quality. Somehow, if one just tried a little harder, one could do it; people have tried hard, but no one has succeeded. If there were a classical wave in the world which had the property that one could locate with infinite gentleness a particle without disturbing it whatsoever, then in fact the quantum theory would not be complete. However, we know of no such classical wave. In particular, the most natural wave to use, light, has already displayed its particle properties. It was the interaction of radiation with matter that prompted the beginning of the quantum theory. Thus, considering those things which exist in the world, it is not possible, even in thought, to make a simultaneous location of position and velocity, for light particles, electrons, all particles have the fundamental quantum properties. In an attempt to locate the position of one quantum with another, the first quantum suffers uncontrollable changes in some of its variables. If one locates its position very precisely, one must do it with a particle of a very short wavelength which changes the momentum in a very large manner. In working out the details of each of these thought experiments, Heisenberg showed that the best one could do within the given structure of the quantum theory was to locate a position and a momentum with the maximum accuracy according to his uncertainty principle.

Thus, says Heisenberg, the thing envisioned by the classicists, as far as we know, is not in fact possible. One cannot observe an actual orbit of an electron moving around a proton. One cannot follow it as one can follow a billiard ball. And if an orbit or position and velocity cannot simultaneously be observed, a physical theory does not have to contain the possibility of defining them.

ON THE ROLE OF UNOBSERVABLE QUANTITIES IN A PHYSICAL THEORY

The time of quantum physics was also the time of positivism, and since in the positivist view great stress is laid upon using only those constructs which are directly observable and avoiding those which are not, much has been said about quantum physics growing out of the attempt to discard what is unobservable. If an electron orbit, for example, is not observable or if position and velocity are not simultaneously observable, they should not, it is said, appear in a theory. However, in all physical theories, including quantum theory, there is only a very small part which is directly observable. To call a physical theory a modern cathedral, it is as though only the façade were comparable to the face of reality, whereas the entire interior structure, the columns, the arches, and so on, were somewhat removed and not directly observable. In Newtonian theory, the first and the second laws of motion can be thought of as conventions. It is only constructions and theorems far removed from the original postulates that are actually put into correspondence with experience. That is, under the influence of an inverse-square law, a planet moves in an elliptic orbit, and so on. In the kinetic theory of gases, those hard, massy objects that we call the molecules are so far removed from experience that it does not matter that they do not look like atoms at all; the kinetic theory gives one an adequate description of such concepts as temperature, pressure, and the volume of a gas. In the quantum theory such things as potential energies (even the wave function) are not directly observable; it is the square of the wave function, the probability, that in a sense is observable.

What one requires of a theory is not that it contain no elements that are not directly observable, but rather that it contain something that can be put into correspondence with what is directly observable. When the correspondence breaks down, when it becomes clear that the theory is no longer in accord with reality, as was the case in an attempt to construct the atom, the easy and the normal thing to do is to alter it as little as possible. It is only when pressed very hard that one will finally alter the internal structure and perhaps the entire foundation to once again achieve a façade in agreement with reality.

What Heisenberg said, in effect, was that a theory *need not* contain elements if they are not observable. It *may* contain them but it *need not*. Such elements, for example, as the position and the velocity of a particle which would be required in a classical theory with Cartesian particles *need* not be retained if the world is a quantum world. The Cartesian electron that classical man senses hidden in the shadow of its associated Schrödinger wave, he senses only in his imagination—trained from childhood like a truffle hound to sniff it out. But if he separates himself from the images in his mind and looks at the world, he will find, at least as far as we know now, that there is no Cartesian electron obscured in the mist of its associated wave. The associated wave is all there is.

But the fact that a Cartesian electron is not observable means only that there is no *necessity* that it be present in a physical theory. The classical physicist cannot be denied the privilege of constructing a theory in which a Newtonian corpuscle appears in such a way that it is unobservable by all presently known techniques. But this cannot be made a requirement, for a physical theory is required to describe only the world as we observe it. What Einstein said in 1905 was that we can dispense with absolute time; quantum physics in the 1920s said that we can dispense with simultaneous position and velocity. In both cases, these things are unobservable. And, since they are unobservable, we are permitted to construct a physical theory in which they do not occur. Whether they are there or not, even though they are unobservable, we cannot say. But if something is unobservable, and we still insist that it is there, the insistence may be due to a strong habit of mind. Such concepts, ghosts we can never observe, amiable, friendly, perhaps comforting, in the end, like Marley's chains, encumber us, so we cannot move freely until we have cast them off.

36

THE TRANSITION FROM THE
QUANTUM TO
THE CLASSICAL VIEW

Having accepted the quantum theory, as a conservative with reluctance or as a radical with enthusiasm and abandon, both conservative and radical can legitimately ask what has happened to Newtonian mechanics and Maxwell's electrodynamics. What has happened to the structure of nineteenth-century physics, so comprehensive and complete that it seemed that only the decimal places were left to be filled in? In constructing a bridge, will we now solve Schrödinger's equation? Does one recompute the orbits of the planets using the uncertainty principle? Must we rediscover Neptune and Pluto using wave mechanics? And, if one wishes to put on a complaining face, Why did we learn Newtonian mechanics and Maxwell's electrodynamics at all, since scientists change their opinions so rapidly, since in one breath they tell us a theory is firmly established and in the next they tell us, usually with some amusement, of a large new realm of phenomena in which the "so firmly established and so closely in agreement with experience" theory now fails so completely that even its fundamental concepts must be abandoned?

We were convinced by interference phenomena that light was a wave, but then we are shown that it behaves like a particle. We were convinced that the Newtonian theory was essentially correct, and then we are shown that it is so inadequate in the quantum domain that one even had to abandon the concept of a corpuscle that had an orbit. What then is it that we are supposed to believe when presented with what is called a scientific theory?

The relations and the structure of any theory—the theory of the external world, the theory a detective constructs, or Newton's mechanics —are abstracted from a small portion of experience and essentially amount to the introduction of a number of conventions that have a certain internal structure and that enable us to relate one phenomenon to another. Even so simple a thing as repeating the same experience, let us say dropping a stone from one day to the next, takes us from one experience to another, and we have no guarantee that the repeated experience will be the same.

However, the underlying belief of the scientific enterprise is that it will be possible to construct a theory that not only is in agreement with the experience we have had but will also be in agreement with similar experience under similar circumstances (and we presumably can agree on what are similar circumstances) and which has a validity beyond the precise experience from which it has been extracted. The successful scientific theories of the past, such as Newtonian mechanics and classical electrodynamics, have shown an (what can only be called astonishing) ability to predict and to relate phenomena—even those unknown before the theories were proposed. It would be somewhat surprising if the rules proposed on the basis of experience with everyday objects were valid in the atomic domain. It is already somewhat of a surprise that the Newtonian rules are valid equally for a planet and for a small projectile on the surface of the earth. Since in the domain of atoms this is no longer the case, to understand atoms one has introduced the new theoretical structure of quantum physics built around the quantum, the object that sometimes behaves like a particle and sometimes like a wave, the object whose position and momentum cannot be simultaneously defined. What then is the relation of the quantum theory to the classical theory we have already learned?

The answer is straightforward—in a sense the only one possible. The quantum theory is a generalization of the classical theory; it is a larger, wider, and broader theory. It is a theory that applies both in the atomic domain and the domain of the planets. If we calculated, using the Schrödinger equation, the orbit of a planet—the earth, for example— about the sun, one would obtain Newton's result. But no one ever does this and the reason is obvious. The structure of Newtonian mechanics already, it is clear, correctly describes the orbit of the earth about the sun. Therefore, if the quantum theory is to be in correspondence with experience, it must have essentially the same structure in this domain. And what is in fact the case is that the quantum theory essentially becomes Newtonian theory and/or classical electrodynamics when one enters the classical domain. This is meant in the following sense. If one begins with the postulates of the quantum theory, one can deduce as a theorem that as masses become large, as relevant wavelengths become small compared to the distances of the problem, the answers given by the quantum theory become precisely the answers given by Newtonian theory or by classical electrodynamics. When one says precisely, one means that the difference numerically becomes vanishingly small—numbers of the kind we have already seen: for example, uncertainties in speed of the order of 10^{-21} cm/sec.*

There are several ways to tell where the transition is made and where quantum effects become important. One is to consider the wavelength of the object involved and compare it to the distances that are typical in the problem, using the fact that wavelike properties do not become apparent until the wavelength is of the same order as a typical distance in the problem. Let us compare, for example, an electron in its orbit about the nucleus, and the earth in its orbit about the sun. The wavelength of an electron in its lowest Bohr orbit about a proton is precisely equal to

---

* Essentially what happens is that as a particle mass grows larger and larger, it becomes possible to localize the wave that is associated with the particle more and more without producing a great uncertainty in the momentum. Then, under the influence of a force system, the Schrödinger equation, which is the quantum domain, tells us how the wave will develop in space and time, tells us finally, as the particle mass grows large enough, that the way a force will affect this particle is just according to Newton's second law.

the distance the electron travels in a single circuit about the proton. The ratio of the wavelength to the typical distance is 1, so we might expect quantum effects to be very large. They are. The de Broglie wavelength of the earth compared to the circumference of the earth's orbit is

$$\frac{\lambda_{\text{earth}}}{2\pi R} \simeq 3 \times 10^{-75} \qquad (36.1)$$

The number is small. So small that any wave or quantum effects would be obscured by the collision of the earth with a dust particle. So small that the probability that the earth is at any place, other than its classically calculated position, is absurdly close to zero.

One can calculate the wave function of Mars, using Schrödinger's equation, to find that the very localized wave that accompanies Mars follows an elliptical Newtonian orbit with remarkable precision. Given an uncertainty in Mars's speed, let us say of the order of 10^{-20} cm/sec, then the uncertainty in Mars's position would be about 10^{-33} cm—not detectable by the observations of Tycho Brahe, as accurate as they were.

Thus one would say that the existing physical theory is the quantum theory. Its domain begins in the nucleus and extends at least as far as the solar system. Whether it extends to infinitesimally small distances, infinitesimally small times and masses, we do not know. Whether the generalization of Newtonian mechanics, general relativity, applies to cosmic distances, to the universe as a whole, we do not know. The usual, the natural, the sensible procedure is to ask oneself what the current theory would predict in a new domain and to attempt to understand the new domain using the current theory. Sometimes we are successful, as in the kinetic theory of gases, and sometimes we are surprised.

Somewhere between the region of phenomena such as the planets in their orbits and electrons in their atomic orbits there is what we might call a transition between the quantum and the classical domain. Although the quantum theory applies everywhere, one would not be likely to use its more intricate concepts and technical machinery to deal with a situation such as that of the planets in their orbits; for one can prove, as a general theorem, that more accuracy than one would ever require can be obtained using the Newtonian equations. Thus engineers who design bridges, aerodynamicists who design planes, are not called upon to employ the concepts of quantum physics in their calculations. In the same way, radio engineers who design antennas and power engineers who design transformers can confine themselves to using classical electrodynamics. For it is a general theorem of the quantum theory that all the relations of classical theory hold good when one enters the classical domain.

However, the point at which one enters the classical domain can often be subtle. One can see fairly clearly where the transition is in going from an electron about a proton to a planet about the sun. However, in more complex phenomena, there are sometimes quantum effects that persist to a surprisingly large scale. The anomalies in a specific heat, pointed out by Maxwell, to his mind one of the most serious defects of the classical theory, turn out to have been the intrusion of a quantum effect into the seemingly classical domain. The paradox, observed by Gibbs, in the heart of statistical physics, again turns out to be a similar intrusion. In the same way, as we shall have some opportunity to see further, many of the properties of materials cannot really be understood using classical concepts. In a rough way, though, one can say that somewhere between the hydrogen atom and the planetary orbits systems begin to behave in a typically classical fashion; the boundary is not completely well marked; there are subtle penetrations of quantum effects into what might be thought to be completely classical domain. Thus, although the quantum theory becomes the classical theory, the region of transition, somewhat amoebalike, has offered surprises to those who explore its edge.

THE
QUANTUM
WORLD

On What Is Called Common Sense

There have been bitter complaints from some of our contemporaries that physics in the twentieth century has become too abstract, has lost touch with those things ordinary people can understand, has lost contact with common sense and substituted instead constructs so abstract that the ordinary mind can never attain them. The complaint seems particularly ironic when we remind ourselves that very much the same thing has been said to painters and composers. Each of us who must view the world as we see it is accused of doing violence to the view held by the previous generation, which has by now been absorbed into the everyday mentality as "common sense." A painter, just as a physicist, imposes his own vision of the world on what are essentially the raw data, available to everyone at least in principle. The completed canvas is the painter's view of the world; it is the mark of important paintings or physics that a generation later we find ourselves looking at the world in the same way. The ridicule that greeted the first impressionist exhibitions has turned now to ridicule of their imitators' calendar stereotypes.

Common sense, as Einstein said, is "that layer of prejudices we acquire before we are 16." It is a cliché that the common sense of the new generation is formed from concepts laboriously constructed by their elders, that what is avant-garde for one generation is common sense and prosaic for the next. It seems dubious that the Newtonian conception of the world would have been common sense to the Greeks in the time of Aristotle or, for that matter, to the scholastic scholars. It was not even so for many of his contemporaries. And those so enamored of common sense (at present Newton's world) are often just those who complain that the mechanical Newtonian view destroyed the magical Medieval world.

For the present generation of physicists, it is quantum physics that is common sense. There they have that fingertip awareness of what is and is not right. Quantum physics is their craft and they have a feeling for its ins and outs the way a painter has a feeling for what is possible on a canvas or a sculptor for what is possible with marble and a chisel. Any physicist can write down Maxwell's equations or use Newton's laws to derive the elliptical orbits of a planet about a sun. But when it comes to the subtler relations within classical physics, matters that might have been called common-sense knowledge for the nineteenth century, the twentieth-century physicist has all but forgotten.

For the physicist trained since the 1930s, the calculation of the transition probability from one atomic level to another is a straightforward and unemotional matter. He will sit down, pencil in hand, and, given his level of concentration, within an hour or two will come up with the result. If he were asked, however, to calculate the disturbance of a planetary orbit due to a small planet nearby, although he would know in principle what he must do, he would not, as a rule, sit down and complete the calculation in a straightforward way. He would think, refer to some book, think some more, and then probably get the wrong answer.

The world of the quantum is the world in which physicists have lived since the 1920s. So much so that when, as the expression goes, the space age began and it suddenly became of great interest to calculate orbits and the most effective way to change orbits of spaceships, the directions in which to point thrusters, and so on, one had almost to train a new generation to, so to speak, remind them of relations that must have been commonplace to those who once read Laplace's book. For physicists, in the sense that they are physicists, operate on that boundary between what is known and what is not. And, since the 1920s, that boundary has been so much within the quantum domain that it has become for them the arena of ordinary experience.

The domain of the quantum has its own surprises. Viewed with a classical eye, it often seems paradoxical. Yet the symbolic surprises have so far always been confirmed by what actually occurs in nature. There seems little doubt that the world of the atom is essentially that of the quantum, and it is this world we now go on to explore in more detail.

37

THE HYDROGEN ATOM

The hydrogen atom is for quantum physics what the problem of the planets was for classical physics. It is that problem whose solution is possible and the results of which can be compared in a most immediate way with experience—that problem where the rules can be clearly applied, where there is no friction and where the consequences are in such close agreement with experience as to convince us that there must be a sense in which they are essentially correct. Through an analysis of the level structure of hydrogen, some of the deepest modern discoveries have been made. And by a comparison of the predictions of quantum theory with the observations on hydrogen, some of the most accurate agreements between experiment and theory have been obtained.

As we analyze this atom we see unfolded before us the kind of information obtained easily from the quantum theory and that is of interest in the atomic domain. It is necessary for consistency and completeness that the quantum theory be able to answer for any of the experiences that one can have with the hydrogen atom. It is in the same sense that the concept of a chair, a summary of our experience of a chair, which answers for its hardness, its color, its continuing presence in a room if we surprise it by suddenly turning around, is the bringing together of all the experience we have of a chair.

GROUND-STATE ENERGY

The dominating feature of any bound quantum system is that only certain energies are allowed; the energy levels, as opposed to the classical case, are discrete. In the classical theory the lowest energy of the electron-proton system would be that for which the electron were precisely superimposed on the proton and standing still. It would thus have zero kinetic energy and minus infinity potential energy. In the quantum theory this is not so, and we now have several ways of understanding why not. Using the uncertainty principle, for example, if we attempted to localize the electron on the proton, by that localization we would produce a very large uncertainty in its speed that would be reflected as a very large kinetic energy.

As an illustration of the way such quantum ideas can be put to work, we use very simple concepts to derive again the ground-state energy of the hydrogen atom. The total energy of an electron in a circular orbit about a proton is

$$E = T + V = \frac{p^2}{2m} - \frac{e^2}{r} \qquad (37.1)$$

Using the de Broglie relations,

$$p = \frac{h}{\lambda} \qquad (37.2)$$

$$n\lambda = 2\pi r \qquad (37.3)$$

and assuming that the electron is in its lowest orbit $(n = 1)$ gives the result

$$\lambda = 2\pi r \qquad (37.4)$$

We obtain the lowest energy of the system as a function only of the distance of the electron from the proton:

$$E = \frac{h^2}{8\pi^2 m r^2} - \frac{e^2}{r} \qquad (37.5)$$

What is amusing is that contrasted to the classical expression, where the momentum and the distance are completely independent so that the lowest energy occurs as mentioned previously for $p = 0$ and $r = 0$, in the quantum theory it is just this that is not possible. There is a connection between p and r: If one tries to reduce r to zero, in Eq. (37.5) it is true the potential-energy term becomes negative and large in absolute value but the kinetic-energy term increases at the same time, and faster. Therefore, the energy becomes very large as r goes to zero; as r goes to infinity it becomes negative and very close to zero. The point at which the energy is minimum occurs at the distance called a_0. This value of r for which the energy is a minimum can be calculated. It is that place at which a ball rolling in such a trough would come to rest. Or it can be estimated from a good graph. In any case the answer is given below:[*]

$$E(r) \text{ is a minimum when } r = a_0 = \frac{\hbar^2}{me^2} \qquad a_0 \simeq 5.3 \times 10^{-9} \text{ cm}$$
$$(37.6)$$

This distance, when the energy is a minimum, defines what is known as the Bohr radius of the hydrogen atom: 5.3×10^{-9} cm. Further, the energy for this value of the radius

$$E \ (r = a_0), \text{ called } E_0, = -13.6 \text{ ev} \qquad (37.7)$$

is precisely the value of the energy of the hydrogen atom in its ground state. If one wants to ionize the atom, to liberate the electron from the proton, it will take just 13.6 ev of energy to do this.

It may seem somewhat surprising that this calculation is sufficient to give one the ground state of the hydrogen atom; things, as one might

---

[*] The combination $h/2\pi$ occurs so frequently it is honored with a symbol of its own: $\hbar$.

guess, do not always work out this simply. But the technique we have employed is the basis of a very general and powerful method for the solutions of problems in the quantum theory. It relies on the fact that the ground-state quantum level occurs at that value where the energy of the system is a minimum consistent with the constraints.

ANGULAR MOMENTUM QUANTIZATION

The level structure of a particle on a line confined between two walls was produced by fitting standing waves of shorter and shorter wavelengths between the two walls. For the hydrogen atom the level structure comes about in a somewhat analogous fashion; there is greater complexity, however, because (1) it is a system in three dimensions, and (2) there is not a simple container but rather an attractive force that varies with the distance. The two extra dimensions introduce two new quantum numbers. (One can think of standing waves in each of the three directions.) However, due to the spherical symmetry of the Coulomb force it is more convenient to introduce quantum numbers associated with the angular momentum.

In classical physics the angular momentum of an electron about a proton is a vector of arbitrary magnitude and direction. This is defined in one simple situation in Chapter 31. In quantum theory this vector is restricted as follows. To the magnitude and the direction of the angular momentum correspond what are essentially the angular momentum de Broglie waves and what are called the angular-momentum quantum numbers. As in the Bohr theory and in a manner quite analogous to that of the particle in the cubical container, only certain de Broglie standing waves (angular momenta—magnitude and directions) are possible solutions of the Schrödinger equation. These are characterized by the quantum numbers

$$l = 0, 1, 2, \ldots \quad \text{.(related to the magnitude of the angular momentum)}$$

$$m_l = l, l - 1, l - 2, \ldots, 0, -1, -2, \ldots, -l \quad \text{(related to the component of the angular momentum in a spatial direction—called the \textit{magnetic quantum number})}$$

The first is related to the angular momentum of the system by

$$l(l + 1)\hbar^2 = (\text{angular momentum})^2 \qquad (37.8)$$

l	$(\text{angular momentum})^2 = l(l+1)\hbar^2$
0	0
1	$2\hbar^2$
2	$6\hbar^2$
:	
:	

The second m_l is restricted to all integer values between $+l$ and $-l$, giving $2l + 1$ values of m_l for a given l.

l	the $2l + 1$ possible values of m_l
0	0
1	−1 0 +1
2	−2 −1 0 +1 +2
3	−3 −2 −1 0 +1 +2 +3
⋮	

The quantum number m_l is interpreted as the component of the angular momentum on some spatial direction such as that of an applied magnetic field.

$$\left(\begin{array}{l}\text{component of angular momentum}\\ \text{in chosen spatial direction}\end{array}\right) = \frac{m_l h}{2\pi} = m_l \hbar \qquad (37.9)$$

Thus, contrasted with the classical angular momentum vector, from the quantum point of view both the magnitude and direction of the angular momentum (just as for the de Broglie standing waves in a container) take on discrete values characterized by whole numbers (Fig. 37.1). Would Pythagoras be pleased?

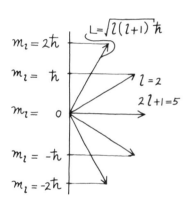

FIG. 37.1. The magnitude of the quantum angular momentum vector is characterized by the whole number l (here $l = 2$) and its component with respect to the vertical direction by the whole number m_l (above $m_l = 2, 1, 0, -1, -2$). One can interpret m_l as the component of the angular momentum in the vertical direction, because it never exceeds l (the component cannot be larger than the vector itself). Contrasted with the classical angular momentum vector, the directions in space are restricted. (This historically was referred to as space quantization.)

THE LEVEL STRUCTURE OF HYDROGEN

The wave-function solutions and thus the energy levels of the Schrödinger equation for an electron subject to the Coulomb force of the proton (hydrogen) are specified by three quantum numbers

$$n \quad l \quad m_l$$

l and m_l are the angular-momentum quantum numbers, and n, the principal quantum number, picks out the de Broglie standing wave characterizing the radial dependence of the wave function of the electron subjected to the Coulomb force. The three quantum numbers (characteristic of a particle in three dimensions) are not independent, as were those for the particle in the cubical container (n_x, n_y, and n_z could take any integral values independent of one another), but are limited in the following way.

The principal quantum number, n, can take the values

$$n = 1, 2, 3, 4, \ldots$$

The quantum number that characterizes the total angular momentum can take all integral values smaller than n,

$$l = n - 1, n - 2, \ldots, 0$$

while m, the magnetic quantum number, can take the values as described above,

$$m_l = l, l - 1, \ldots, -l$$

The energy levels depend only on the principal quantum number and are the same as those obtained by Bohr:[*]

$$E = \frac{-2\pi^2 me^4}{h^2} \frac{1}{n^2} \qquad n = 1, 2, 3, \ldots \qquad (37.10)$$

Thus the energy-level diagram of the hydrogen atom from the point of view of the Schrödinger equation is shown in Fig. 37.2.

F I G. 37.2. The energy levels are classified according to their energy and their angular momentum. The actual labels used—$S\,(l = 0)$, $P\,(l = 1)$, $D\,(l = 2)$, and so on—come from the historical spectroscopic classification. The current notation and usage are a mélange of symbols that have grown up over almost half a century. (After R. T. Weidner and R. L. Sells, *Elementary Modern Physics*, rev. ed., Allyn and Bacon, Boston, 1968)

[*] This is so in spite of the fact that the angular momentum assignments are not the same. In the Bohr atom the ground-state circular orbit had angular momentum $L = \hbar$. The ground state of the Schrödinger atom (quantum numbers $n = 1$, $l = 0$, $m_l = 0$) has angular momentum zero.

The S states—states of zero angular momentum—are more or less uniformly distributed over a sphere. The comparable classical motion—directed toward or away from the center—would be somewhat as shown. $P, D, \ldots$ states have angular momentum $1, 2, \ldots$ For a given value of n the highest angular momentum permitted, $l = n - 1$, corresponds to a classical circular orbit.

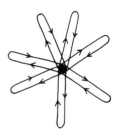

For the other atoms, helium through uranium and beyond, the details of the level structures are somewhat different, as we shall discuss; in general, one can draw a pattern of levels similar to that above but with different energy spacings.

Now, when we speak of an unexcited sample of such atoms in a cooled container or drifting in space, it is a sample in which each atom finds itself in its ground state. Thus collections of atoms are as identical as two physical systems can be. This provides some understanding for the fact that as far as one can see, one hydrogen or helium atom is identical to any other. A situation difficult to understand from the classical point of view where, considering for example the solar system, one would not expect one planetary orbit to be identical with another or, for that matter, two solar systems to be identical even though the laws of motion are the same and the forces between the sun and the planets are the same. In a classical system all orbits are possible. There are no discrete states, and therefore solar systems, as far as we know unlike atoms, should vary widely among themselves.

IN A MAGNETIC FIELD

In 1896 Zeeman had found it "very probable that the period of sodium light is altered in the magnetic field," that in the presence of a magnetic field "both D lines were seen to widen."[1] Upon closer analysis the broadened lines of the spectra of such elements as hydrogen or sodium were seen to be split into finer lines called *multiplets*. For example, the line identified as due to the $2P \longrightarrow 1S$ transition of hydrogen ($n = 2, l = 1$ to $n = 1, l = 0$) is seen to broaden in a magnetic field and to split into three lines.

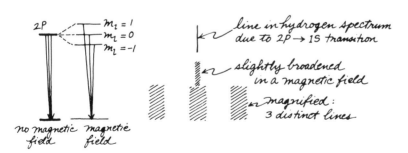

Classically, this effect could, as shown by Lorentz, be produced by the alteration of the electron orbits due to the magnetic field. From the point of view of the quantum theory it comes about due to the splitting of those energy states of the hydrogen atom of the same angular momentum into states having slightly different energies. This is an interesting and very typical illustration of how the degeneracy related to a symmetry property is removed when the symmetry is removed. (For the particle enclosed in a cubical container some of the degeneracy would be removed, if one of the sides—say ℓ_z—were changed in length, but the lower degeneracy typical of a square would remain.)

The degeneracy of all the energy levels corresponding to the $2l + 1$ directions of the angular momentum or the plane of motion of the electron is a direct consequence of the fact that the energy of the atom is not dependent on the direction of the angular momentum (space is the same in all directions).

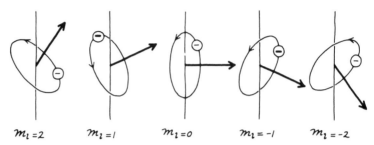

$m_l = 2$ $m_l = 1$ $m_l = 0$ $m_l = -1$ $m_l = -2$

F I G. 37.3. In empty space the plane of rotation of the orbital electron has no effect on the energy of the atom. Thus the levels characterized by the $2l + 1$ magnetic quantum numbers corresponding to the $2l + 1$ directions of the angular momentum vector have the same energy.

In the presence of a magnetic field the situation is fundamentally altered; there is now a special direction: the direction of the magnetic field; and the direction of the angular momentum vector with respect to the magnetic field does affect the energy of the atom, because the electron circulating in its orbit can be regarded as a current, and the energy of interaction of this current with the external magnetic field depends upon the relative directions of the two.

The splitting for a given value of ℓ is:

$$\Delta E = \left(\frac{eh}{4\pi mc} \right) m_l B \qquad (37.11)$$

The amount of splitting is quite small. Thus, for example, in a magnetic field of 10,000 gauss (quite a strong magnetic field) the broadening of the $2P$ levels of hydrogen would be about 10^{-4} ev (Fig. 37.4).

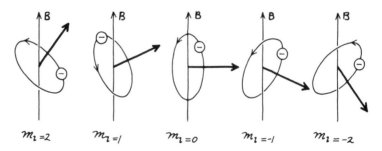

$m_l = 2$ $m_l = 1$ $m_l = 0$ $m_l = -1$ $m_l = -2$

F I G. 37.4. In an external magnetic field the direction of the angular momentum vector does affect the energy. In the illustration above $(l = 2)$ the magnetic quantum number can take its usual five values $(-2, -1, 0, 1, 2)$. The energy of the system now depends on this quantum number —thus the name. The splitting of the levels is

$$m_l = \quad 2$$

$$m_l = \quad 1 \qquad \qquad \left(\frac{eh}{4\pi mc}\, B \right)$$

$$m_l = \quad 0$$

$$m_l = -1$$

$$m_l = -2$$

In a magnetic field, therefore, the triply degenerate P level splits into three levels. The five-times degenerate D level splits into five levels, and so on. The energy difference between the split levels just corresponds to the different energies due to the various angles of the current loop with the external field (see Fig.(37.4).

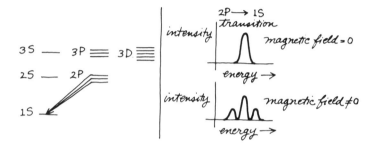

FIG. 37.5. Splitting of the various m_l levels in a magnetic field; the transitions—between, for example, the three $2P$ levels and the $1S$ level—now have slightly different energies. Rather than a single line, one now sees a triplet with slightly separated energies. Observation of spectra of atoms in magnetic fields is one method to determine how many states are associated with a single level.

ELECTRON SPIN

More detailed study of the structure of individual spectral lines fairly early revealed that they were split in a complicated manner that came to be called the *anomalous Zeeman effect*. In particular, there occasionally occurred a doubling of lines which seemed to indicate a doubling of energy levels beyond all those corresponding to the $2l + 1$ degeneracies of l. The problem was particularly puzzling because this seemed to indicate (as had been suggested quite early) the existence of a new quantum number that could take only two values. But the origin of such a quantum number was quite elusive. All those readily available seemed to be accounted for. Considering the electron in the cubical container, for example, having specified n_x, n_y, and n_z, or for the hydrogen atom having specified n, l, and m_l, the state is completely determined. The observed doubling of lines seemed to indicate that in addition to the quantum numbers n_x, n_y, and n_z (which are associated with the momenta in the three spatial directions) or n, l, and m_l, one had to add another, which could take only two values—where did it come from?

It is said that Wolfgang Pauli, who was very much interested in the problem, had thought about it for so long, with such intensity, with such preoccupation, and apparently with such sadness in his face that he was stopped in the street in Copenhagen, where such things happen, by an old woman overwhelmed by his expression of sorrow. When she asked him what was the matter, Pauli could only shake his head, shrug his shoulders, and mumble, "Madam, I cannot understand the anomalous Zeeman effect."

The doubling was finally explained by Goudsmit and Uhlenbeck in what is one of the most important proposals made in the period. They suggested that the electron, in addition to having a charge and a mass, possessed another internal property—this they called a *spin*. Each electron, they proposed, possesses a spin or an internal angular momentum —just as, for example, the earth has an angular momentum due to its rotation about its axis, in addition to the angular momentum due to its revolution about the sun. The spin angular momentum of the earth, however, can easily be attributed to the rotation of a spherical distribution of matter. For the electron, thought to be a point, the origin of this spin was more difficult to visualize. As Pauli objected, the introduction of such an idea seemed to be an atavistic association of a classical concept with what was clearly a quantum object.

If one, however, agrees to attribute a spin* to the electron that corresponds to an internal angular momentum number of

$$l_{\text{spin}} \text{ (denoted by } s) = \tfrac{1}{2} \qquad (37.12)$$

(for the internal–spin–angular momentum the associated quantum number can be half-integral; for the space angular momenta it is restricted to integral values), the number of projections of this spin in a given spatial direction is two:

$$m_s = -\tfrac{1}{2} \text{ or } +\tfrac{1}{2} \qquad (37.13)$$

and this results in the required doubling of the levels. Further, if one associates a magnetic moment with the spin (one might think of the magnet produced by current due to the supposedly rotating electron),

$$
\begin{array}{l}
\text{magnetic moment} \\
\text{of electron "due} \\
\text{to current produced} = \dfrac{e\hbar}{2mc} \qquad (37.14) \\
\text{by its spinning} \\
\text{charge"}
\end{array}
$$

* The square of the spin angular momentum is $s(s+1)\hbar^2$.

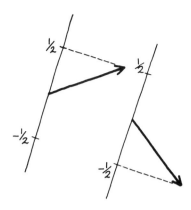

F I G. 37.6. The spin angular momentum has two projections in a given spatial direction corresponding to the angular momentum component

$$m_s = -\frac{1}{2}\frac{h}{2\pi} \text{ or } m_s = \frac{1}{2}\frac{h}{2\pi} \tag{37.15}$$

Since in the absence of an external magnetic field the energy of the system is independent of the direction of the spin, we have a doubling of levels.

then, in a magnetic field the two directions of spin would result in a doublet of levels separated by the energy*

$$\frac{e\hbar}{mc}B$$

Although it was difficult to account for the origin of electron spin and although it was some time before this internal property was made completely consistent with what else was known, it was soon clear that the proposal very well accounted for the observed structure of the spectral lines. And, as we shall soon see, the doubling of levels, due to electron spin, plays a critical role in constructing the periodic table of elements.

---

* A spinning ball of charge produces a magnetic field not unlike that shown for a current loop in Chapter 18. In an external magnetic field, as for a compass needle, the spinning electron has forces exerted on it that tend to line it up with the field.

38

MANY-PARTICLE QUANTUM SYSTEMS

THE PAULI EXCLUSION PRINCIPLE

One of the honored and self-evident precepts of common sense is that two things cannot occupy the same place at the same time, a statement whose denial seems in itself self-contradictory. Yet, for quanta this statement, if it has any meaning, is not always true. To the extent that one can define place and time for photons, two, three, or any large number of photons can occupy the same place at the same time—or, as is said, can be described by identical wave functions. It is this possibility that allows one to build up the classical electric and magnetic fields from photons.

It is possible to superimpose identical wave functions for as many photons as one wishes. Thus the probability that the first photon be at point x_0 is identical to the probability that the second photon be at x_0 or that the Nth photon be at the point x_0. All quanta which possess such a property are called *bosons*,* and all bosons have integral spin quantum numbers: 0, 1, 2, 3, and so on. The photon itself has spin 1.

There exists another class of quanta, called *fermions*,† which have half-integral spin quantum numbers: $\frac{1}{2}$, $\frac{3}{2}$, and so on, for which it is not possible to superimpose the wave functions in the manner described above. This impossibility is the quantum equivalent of the classical statement that two things cannot occupy the same place at the same time. First introduced into quantum physics by Wolfgang Pauli in 1925, it is known as the *Pauli exclusion principle*, perhaps a compensation for his unhappiness over his inability to explain the anomalous Zeeman effect and his even greater unhappiness with the explanation.

From the point of view of the Schrödinger theory, which begins by speaking of one-particle systems, the exclusion principle is an additional postulate that restricts the choice of possible wave functions for many-electron systems. From the point of view of relativistic quantum theory, the exclusion principle is a consequence of the postulates of relativity and the quantum theory for spin $\frac{1}{2}$ particles. It is one of the triumphs or mysteries of modern quantum theory that all fermions—particles that obey the exclusion principle—are required to have half-integral spin while all those that do not, bosons, must have integral spin.

The content of the exclusion principle can be stated by saying that if one has two electrons, it is not possible that the wave function describing one electron is identical to the wave function describing the other (that the quantum numbers of the two electrons cannot be the

---

* After the Indian physicist Bose.
† After Enrico Fermi, who first discussed their properties.

same). Thus it is not possible to have two electrons, each with identical momentum and with identical spin orientation. (It is this that makes the concept of spin so extremely important; for it *is* possible to have two electrons with identical momentum if their spins are opposite; this gives a critical doubling of levels.) It is further not possible to have two electrons localized in exactly the same positions with exactly the same spin (Fig. 38.1).

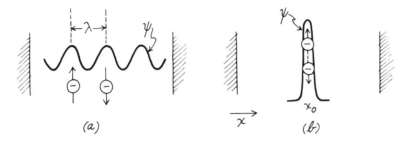

(a) x (b)

F I G. *38.1.* It is possible to have two electrons described by wave functions with identical momentum (a), or by wave functions concentrated at a particular position x_0 (b), only if their spins are opposed.

This restriction on the wave functions of many-electron systems, combined with the basic level structure intrinsic in the quantum theory, is responsible for the structure of the periodic table, the nature of chemical bonding, and most of the properties of matter.

APPLICATION TO NONINTERACTING ELECTRONS IN A CONTAINER—PRIMITIVE MODEL OF A METAL

As an explicit illustration we consider a system of N quantum particles in a one-dimensional container of length ℓ. Extended to a three-dimensional container—a cube—this is a simple model of electrons in a metal and can be made to produce, in at least a qualitative way, many of the properties of normal metals—response to electric or magnetic fields, specific heat capacity, and so on.

As we have discussed before, the wave functions for the quantum particle in such a container are characterized by the standing de Broglie waves

$$\lambda = \frac{2\ell}{1}, \frac{2\ell}{2}, \frac{2\ell}{3}, \dots, \frac{2\ell}{n}, \dots \qquad (38.1)$$

and the energies are

$$E = \left(\frac{h^2}{8m\ell^2}\right) n^2 \qquad (38.2)$$

For a single particle the lowest state of the system is that corresponding to $n = 1$.

$$\lambda = 2\ell \qquad E = \frac{h^2}{8m\ell^2} \qquad (38.3)$$

For several particles in such a container, if the particles do not interact with each other—no forces between them—the total wave function can be constructed by assigning to each particle one of the above de Broglie standing waves.

Now imagine two quantum particles in the container. If the two particles are bosons, then the lowest state is that in which both particles have the de Broglie wavelength 2ℓ:

$$\lambda_2 = 2\ell$$
$$\lambda_1 = 2\ell$$

$$E = \frac{h^2}{8m\ell^2} + \frac{h^2}{8m\ell^2} = 2\frac{h^2}{8m\ell^2} \tag{38.4}$$

If they are fermions, then the exclusion principle requires that they *not* have the same quantum numbers. Thus, if they are to have the same de Broglie wavelength, they must have opposite spin:

$$\lambda_1 = 2\ell, \uparrow$$
$$\lambda_2 = 2\ell, \downarrow$$

$$E = \frac{h^2}{8m\ell^2} + \frac{h^2}{8m\ell^2} = 2\frac{h^2}{8m\ell^2} \tag{38.5}$$

Adding a third, fourth, or Nth, boson only increases the number of particles in the state with de Broglie wavelength $\lambda = 2\ell$ (for the lowest state of the system)

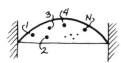

$$\lambda_1 = 2\ell$$
$$\lambda_2 = 2\ell$$

$$E = N\frac{h^2}{8m\ell^2} \tag{38.6}$$

$$\vdots$$
$$\lambda_N = 2\ell$$

(This is the basis on which a classical wave can be built from bosons. As more and more bosons are piled into the same wave function, the probability of finding a large number in a given quantum state becomes appreciable, which gives the system the property of a classical continuous field, an electromagnetic field, for example.)

For a third, fourth, or Nth fermion, the situation is quite different. A third fermion cannot be assigned the de Broglie wave $\lambda = 2\ell$ because whether its spin were ↑ or ↓ it would duplicate in all its quantum numbers one of the first two—and this is what the exclusion principle forbids. Thus all that remains is to choose the next highest de Broglie wave. For three fermions the assignment for the ground state would be

$$\lambda_1 = 2\ell, \uparrow$$
$$\lambda_2 = 2\ell, \downarrow \qquad E = \frac{h^2}{8m\ell^2}(1+1+4) = 6\frac{h^2}{8m\ell^2} \tag{38.7}$$
$$\lambda_3 = \ell, \uparrow$$

For N fermions* the lowest state would have the quantum numbers:

* N has arbitrarily been chosen to be even.

$$\lambda_1 = 2\ell, \uparrow \qquad \lambda_3 = \ell, \uparrow \cdots \lambda_{N-1} = \frac{4\ell}{N}, \uparrow$$

$$\lambda_2 = 2\ell, \downarrow \qquad \lambda_4 = \ell, \downarrow \cdots \lambda_N \ = \frac{4\ell}{N}, \downarrow \qquad (38.8)$$

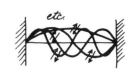

and its energy would be

$$E = \frac{h^2}{8m\ell^2}\left(1 + 1 + 4 + 4 + \cdots + \left(\frac{N}{2}\right)^2 + \left(\frac{N}{2}\right)^2\right) \quad (38.9)$$

The effect of the exclusion principle on the wave function for a system of many fermions is thus to prevent all the fermions from occupying the lowest single-particle state (Fig. 38.2). The lowest allowed energy of

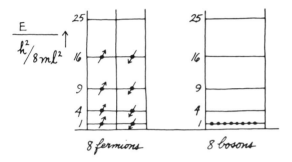

F I G. 38.2. A comparison of the single-particle levels filled for eight fermions and eight bosons. For the bosons $E = 8$ $(h^2/8m\ell^2)$ (ground state), while for the fermions $E = 60$ $(h^2/8m\ell^2)$ (ground state).

the system is for this reason much larger than that for the same number of bosons. For electrons in a metal the effect is to make the energy of the ground state very large and to leave only those states near the fermi surface available for further interactions. For interacting systems of particles the wave functions are modified due to the interparticle forces. However, in many qualitative respects the situation is unchanged, as we shall see next in a discussion of the periodic table.

THE PERIODIC TABLE

It is in its effect on the construction of atomic systems that the exclusion principle plays one of its most dramatic roles. In the absence of the exclusion principle, as one built up the periodic table, the ground-state wave function for the first few atoms would look as shown. The shaded area indicates that region of space where the wave function squared, and thus the probability of finding the electron, is large. Thus, in the absence of the exclusion principle, as one increased the number of positive charges, each of the electrons would try to find its way to the lowest, the 1S state. For an atom, with let us say N positive charges on the nucleus, each of the N electrons would be the equivalent of the 1S state for this atom.

hydrogen

helium

lithium etc.

Such a series of atoms would have completely different chemical properties from those actually observed—properties that would hardly be expected to vary much from one atom to the next. It is observed, however, that helium with two positive charges and two electrons is a noble gas—so stable it is almost impossible to compound with another element.[*] Lithium, on the other hand, with three positive charges, is an extremely active alkaline metal. To go from the noble gases to the alkaline metals, from the least active of elements to the most, requires only the addition of a single positive charge to the nucleus and a single electron to the outer structure. One goes from fluorine (so active that its acid—hydrofluoric acid—dissolves even glass) to inert neon, the second noble gas, by adding a single positive charge to the nucleus and a single electron to the outer structure. Adding one more, one has sodium, so violently active that it bursts into flame on contact with water.

None of this, nor the other observed regularities within the periodic table, would be very comprehensible, if electrons fitted themselves about the central nucleus by essentially entering the lowest state, as above.

In an attempt to account for the periodic table, Bohr, before the exclusion principle and before the Schrödinger equation, proposed that only two electrons could be in the lowest state, eight in the second state, and so on. The conjecture is not entirely in agreement with what we believe now, but is close enough so that it resulted in the construction of a series of atoms that at least roughly had the qualitative features of the periodic table (Fig. 38.3).

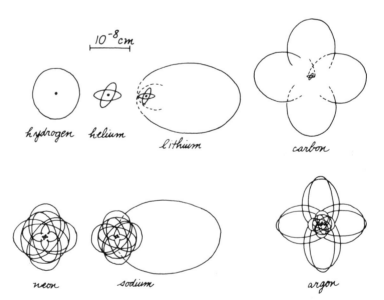

F I G. 38.3. The filling in of the periodic table according to Bohr. The diagrams give a crude idea of the relative distances of the outer electron from the nucleus. The angular momentum assignments, however, are not the same as those that result from the Schrödinger equation. (After J. Orear, *Fundamental Physics*, Wiley, New York, 1961)

---

[*] Recently, after at least a century of effort, chemists have succeeded in producing noble gas compounds.

From the point of view of the Schrödinger equation and the exclusion principle, one can understand the filling in of the periodic table in the following way. The first atom, hydrogen, consists of a single proton with an electron in its 1S state. The second atom, helium, with its two positive charges, can have its two electrons both in the 1S state, because they can be of opposite spin. When one goes to lithium, however, with three positive charges, the 1S quantum numbers have been filled, and it is necessary to put the next electron farther away from the nucleus in the 2S state. Because the binding energy of the 2S state is less than that of the 1S state, this last electron is less strongly bound to the nucleus than the first two. Helium is among the most nonreactive of the noble gases; to remove an electron from helium requires about 24.6 ev. To remove the outer electron from lithium requires only about 5.4 ev. Thus, in chemical reactions— which essentially involve the transfer of electrons from one element to another—helium is much less reactive than lithium.

hydrogen

helium

This very straightforward analysis would be exact if the electrons didn't interact with one another. They would just fill in the states with the various quantum numbers characteristic of the hydrogen atom with corresponding energies, just as for the one-dimensional container. The fact that they do interact with each other modifies the force acting on each of them in the following way. The third electron added to the lithium atom, for example, when far away from that atom, sees essentially one positive charge (the triply charged nucleus shielded by the two 1S electrons) but when very close (inside the 1S electrons) it feels the much

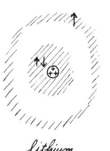

lithium

1S orbital electrons

force here
due to 3 charges

force here
due to 1 charge

stronger force due to the triply charged nucleus. It thus feels about the same force of attraction as one proton for one electron (in hydrogen) when far away, and a stronger force when very close to the nucleus.

r_0

$r \rightarrow$

1S orbital
electrons

$F(r)$ force between ionized
lithium atom and an electron

$\dfrac{e^2}{r^2}$ (force between one proton
and electron—hydrogen)

Although the electron-lithium ion force has a dependence on r different from Coulomb's law, it is approximately spherically symmetric so that its degenerate level structure will be the same as that of hydrogen, even though the placement of the energy levels may be shifted a bit. Thus one can use the hydrogen quantum numbers (n, l, m_l, and m_s) which are now to be thought of as characteristic of the spherically symmetric force system due to the ionized lithium atom, if we recall that the exact placement of the energy levels is not necessarily that of hydrogen.

Adding another positive charge to the nucleus and another electron to the 2S state results in the next element, beryllium. The ionization potential (the energy needed to remove a single electron) is 9.3 ev. This energy has increased over that for lithium due to the increase in nuclear charge, which draws all the electrons in closer. Now both the 1S and 2S states are filled. The next levels are the 3×2 degenerate levels of the 2P state.* Six electrons can go into these altogether, three electrons with spin up and three electrons with spin down. (The electron spin here plays a critical role in doubling the levels.) The 2P state is bound somewhat more weakly than a third electron in the 2S state would have been if permitted, so that the ionization potential of boron, the five-electron atom, is only 8.3 ev, somewhat less than the 9.3-ev potential for beryllium.

Beyond boron the ionization potential resumes its upward march, as the increasing nuclear charge continues to pull all the electrons closer: 6 electrons, carbon, 11.3 ev; 7 electrons, nitrogen, 14.5 ev; falters at 8 electrons, oxygen, 13.6 ev; and continues at 9 electrons, fluorine, 17.4 ev; and 10 electrons, neon, 21.6 ev—the end of this series and a noble gas. The next electron cannot go into a 1S, 2S, or 2P state. All these levels are filled. It must therefore take the principal quantum number $n = 3$; it is farther from the nucleus and thus less strongly bound than the $n = 2$ levels. It goes into the 3S level. The ionization potential of sodium, which has 11 electrons, is 5.1 ev.

One proceeds through the rest of the table in more or less the same manner. The similarities among elements in the periodic table can be attributed to similarities in the wave function of the outer electron. The *alkali metals*, for example, all violently active, mark the beginning of a new shell (principal quantum number), the outer electron in an S state far enough away from the nucleus-inner electron core so that its binding energy is comparatively small. Lithium, sodium, potassium, rubidium, cesium, and francium all have this structure and are all highly reactive—francium the most highly reactive element in the periodic table. The halogens (the counterparts of fluorine) are the elements fluorine, chlorine, bromine, iodine, and astatine, which need one electron to fill their P shell. And the *noble gases*, helium, neon, argon, krypton, and xenon (they rarely associate with either themselves or the other elements) are all characterized by tightly drawn-in filled shells and no close levels for the addition of another electron.

In this way at least the qualitative properties of the periodic table can be understood. It is clear from what we have done that there are a certain number of conjectures involved. The exact solution of the Schrödinger equation for even the two-electron nucleus system, helium, has

---

* The three levels $m_l = 1, 0, -1$, each of which can have the two possible spin directions $m_s = \frac{1}{2}$ or $-\frac{1}{2}$. Therefore, $3 \times 2 = 6$ levels.

never been obtained. For many-electron systems exact solutions of the Schrödinger equation, using the largest computers, with any methods now known, are out of the question. Thus, as in classical systems or in any logical system where the consequences of the rules are too complex to be followed from the postulates, we introduce solutions consistent with everything that is known. It seems reasonable to believe from the general consistency of the results that quantum physics as it is now understood contains within it the periodic table of elements.

39

THE ATOMIC NUCLEUS

WHAT IS IT MADE OF?

From the scattering of alpha particles Rutherford had concluded that the atom contained its positive charge in a small, massive center of radius, smaller than 10^{-12} cm. He could not know immediately whether this nucleus was a single positively charged particle or an aggregate of many positively charged particles—in 1911 the total charge of the nucleus could only be estimated. Rutherford did not need to comment on the structure of the nucleus in his 1911 paper. However, he could not resist the opportunity to venture an explanation of alpha-particle radiation.

> It may be remarked that the approximate value found for the central charge of the atom of gold ($100\,e$) is about that to be expected if the atom of gold consisted of 49 atoms of helium, each carrying a charge of $2e$. This may be only a coincidence, but it is certainly suggestive in view of the expulsion of helium atoms [a particles] carrying two unit charges from radioactive matter.[1]

The idea of one element being constituted from others was not new. In 1815 William Prout had suggested that all elements were compounded of hydrogen atoms. They were to be the πρώτη ὕλη, the "first matter" out of which all matter was made. The conformity of the weights of almost all the elements to multiples of the weight of the hydrogen atom reinforced this idea. In 1886 Sir William Crookes had proposed the primary substance *protyle* out of which all atoms were formed. The problem in all such attempts is to find the primary substance that economically and fruitfully unifies the experience we have.

We now believe that the nucleus is not made up either of alpha particles in the sense that Rutherford proposed or of hydrogen nuclei as was proposed by Prout. It is thought to contain at least two different types of massive particles, the proton, which is the positively charged nucleus of the hydrogen atom, about 1800 times as massive as the electron, and the neutron, a neutral particle of about the same mass as the proton, both called nucleons. The helium nucleus consists of two protons and two neutrons, and this is what is emitted in alpha radiation.

In 1902 Rutherford and Frederick Soddy discovered that when radioactive elements decay, their chemical properties change. Soddy spent the next decade studying this phenomenon and discovered that the products of radioactive transformations were often of different atomic weight but had the same chemical properties. For example, in 1910 he tried chemically separating radium from mesothorium, the decay product of thorium. In his 1921 Nobel Prize lecture he stated:

> From this date [1910] I was convinced that this nonseparability of the radioelements was a totally new phenomenon . . . that the relationship was not one of close similarity but of complete chemical identity.[2]

At the time Soddy did not know what to make of his evidence. Rutherford had guessed that ". . . the central charge is proportional to the atomic weight, . . ."[3] an approximate, although not exact, relation. Since the central charge determines the chemical properties (by the number of electrons it attracts), Rutherford's guess was contradicted by Soddy's work: Soddy had discovered elements of different weight that were chemically identical; by Rutherford's hypothesis they should have differed.

What is now considered correct was first proposed by a somewhat amateur physicist, the Dutchman Van der Broek, who pointed out that ". . . according to Rutherford, the ratio of the scattering of a particles per atom divided by the square of the charge must be constant."[4] Yet, the ratios, computed by Geiger and Marsden by putting the charge proportional to the atomic weight, showed a considerable variation. To remedy this, Van der Broek proposed instead that the charge was proportional to the number of the element in the periodic table, called the atomic number. Rutherford was not amused, complaining: ". . . a lot of guesses for fun without sufficient foundation."

But Soddy showed how fruitful, with or without foundation, this suggestion was and in so doing coined the word *isotopes*, meaning elements of different weight that have the same position in the periodic table and therefore the same chemical properties. Soddy comments in Van der Broek's support:

> The successive expulsion of one a and two β particles in three radioactive changes in any order brings the intra-atomic charge of the element back to its original place in the table, though its atomic mass is reduced by four units.[5]

On this basis, Soddy concludes: ". . . the central charge of the atom on Rutherford's theory cannot be a pure positive charge. . . ."[6] Soddy was not ready to introduce the neutron. Instead he postulated a nucleus of alpha particles (and maybe hydrogen) and electrons. The electrons, naturally enough, were emitted as beta radiation and would account for the fact that the nuclear charge on this basis was usually less than half the atomic weight, as it would have been if the nucleus contained only alpha particles.

So in 1913 this was how the nucleus stood: enough alpha particles to make it sufficiently massive, and enough electrons to make the nuclear charge low enough to fit into the periodic table. The system worked reasonably well; but there were peculiarities—such as the knocking out of hydrogen nuclei from the nucleus. In 1919 Rutherford reported the first artificial transmutation of an element. He had bombarded nitrogen with high-energy alpha particles and, to his surprise, a new radiation that he soon identified as hydrogen nuclei was emitted, and the nitrogen was turned into oxygen. He concluded that

> . . . the nitrogen atom [nucleus] was disintegrated under the intense forces developed in a close collision with a swift a particle, and that the hydrogen atom which is liberated formed a constituent part of the nitrogen nucleus.[7]

Thus there arose in the 1920s the belief that there were hydrogen nuclei (called *protons* by Rutherford in 1920) in the nuclei of atoms. To account for the mass, there were A protons, where A is the atomic weight of the atoms, and to account for the charge, there were $(A - Z)$ electrons in the nucleus also, where Z is the atomic number of the atom. This theory or similar theories in which some alpha particles were also included in the nucleus were fairly generally accepted, although suggestions were made from time to time that the proton-electron pairs might actually be single massive neutral particles, what we now call the neutrons. This was suggested by Rutherford in 1920, but since no experimental evidence for it existed (and the emission of beta particles seemed to contradict it), the neutron was not yet taken seriously.

In 1925 Goudsmit and Uhlenbeck proposed that the electron has an intrinsic spin of quantum number $\frac{1}{2}$. Soon afterward the proton was also shown to have a spin of $\frac{1}{2}$. If one now attempted to put both electrons and protons into a nucleus, there were problems. For example, the nitrogen nucleus, atomic weight 14 and atomic number 7, would presumably have 14 protons and 7 electrons. Its net spin could not be zero, because there is no way that the minimum spin of the 21 spin-$\frac{1}{2}$ particles in the nucleus could be less than $\frac{1}{2}$. (Two spins of $\frac{1}{2}$ can add to give 1 or zero; three to give $\frac{1}{2}$ or $\frac{3}{2}$; four to give 0, 1, or 2; five to give $\frac{1}{2}$, $\frac{3}{2}$, cr $\frac{5}{2}$; and so on. In general, an odd number of $\frac{1}{2}$ spins can add to give an odd spin —never zero.) Yet measurement showed the spin of the nitrogen nucleus to be zero. This and the seeming difficulty of containing light-weight electrons in so small a space as the nucleus, where by the uncertainty principle they would have very large kinetic energy could be resolved with Chadwick's new particle, the neutron.

Chadwick's introduction of the neutron (a neutral massive particle —about the same mass as the proton and also with a spin quantum number of $\frac{1}{2}$) resolved the difficulties mentioned above. Nitrogen (atomic weight 14, atomic number 7) would now be composed of 7 protons and 7 neutrons which should yield a spin of zero. Electrons with embarrassingly large kinetic energy need not be imagined contained in the nucleus, and by the addition or subtraction of neutrons one could easily understand why an atomic weight could change while the atomic number (number of charges) remained constant. So with the neutron the nucleus took the form it is thought to have today (Fig. 39.1).

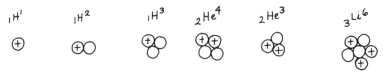

$_1H^1$ $_1H^2$ $_1H^3$ $_2He^4$ $_2He^3$ $_3Li^6$

F I G. *39.1.* In addition to the chemical symbol, a subscript before the letter indicates the charge on the nucleus and a superscript after represents the atomic weight of the nucleus. Thus the deuteron is represented by $_1H^2$, two isotopes of helium by $_2He^4$ (helium 4) and $_2He^3$ (helium 3). $\oplus$ ≡ proton; $\bigcirc$ ≡ neutron.

HOW DOES IT STAY TOGETHER?

In this way, before the atom had been completed, the nucleus was actively in construction. If the nucleus was a single charge Z with mass A, perhaps no additional questions need have been asked. But this would require hundreds of different and unrelated entities. On grounds of economy, as well as on the basis of observations (one nucleus could be transmuted into another, protons, neutrons, alpha particles, electrons, and gamma rays were ejected from nuclei), it seemed inevitable that the nucleus be made of more primitive objects held together in some way and capable by rearrangement of turning into other nuclei.

Without too much calculation it is evident that something new must be added, if one is to regard the nucleus as made of protons and neutrons held together by forces between them. The two classical forces—gravitation and electromagnetism—cannot account in any conventional fashion for the stability of nuclear matter. The force of repulsion due to the positive charge on each proton is

$$F_{coulomb} = \frac{e^2}{R^2} \quad (\text{magnitude}) \qquad (39.1)$$

the gravitational force of attraction is

$$F_{grav} = \frac{GM_pM_p}{R^2} \quad (\text{magnitude}) \qquad (39.2)$$

and the ratio of the two is

$$\frac{F_{grav}}{F_{coulomb}} = \frac{GM_p{}^2}{e^2} \simeq 10^{-36} \qquad (39.3)$$

The smallness of this number (among nuclear particles it evidently depends upon the normal elemental charge and mass) is the origin of the classification of gravitational forces as insignificant in comparison with electrical forces.

Having begun this way, one is led to the idea of some new force exerted between the particles of nuclear matter (protons, neutrons) which is attractive and stronger than the repulsive electromagnetic forces; so strong that in spite of the electromagnetic repulsion of one positive charge for another, stable nuclei can exist—stable enough so that positive evidence for the transmutation of one nucleus into another did not become explicit until the end of the nineteenth century.

The energies of atomic transitions are measured in electron volts. The decay products of nuclear disintegrations, the alpha, beta, and gamma rays, typically have energies on the order of millions of electron volts. The energy differences in nuclear processes are so large by ordinary standards that the resulting mass differences can be measured using

$$\text{change in mass} = \frac{\text{energy released}}{c^2} \qquad (39.4)$$

The comparison of the masses and kinetic energies of initial and final nuclei in nuclear processes gives probably the most accurate and embracing confirmation of this relation.

The measured masses of the protons and neutrons can be compared to the masses of the nuclei that contain them. The difference in mass (mass defect) can be compared with the energy actually released when protons and neutrons are combined or the energy required to separate them. For illustration, consider the deuteron-nucleus of that isotope of hydrogen which contains one proton and one neutron; we follow a procedure like that of balancing a bank statement.

mass of proton	1.67243×10^{-24} g
mass of neutron	1.67474×10^{-24} g
(1) total	3.34717×10^{-24} g
(2) observed mass of deuteron	3.34321×10^{-24} g
mass defect [difference: (1) − (2)]	0.00396×10^{-24} g

$$\begin{aligned} \text{binding energy} \quad &= \quad (\text{mass defect}) \times c^2 = 3.564 \times 10^{-6} \text{ erg} \\ (E = mc^2) \quad & \qquad\qquad\qquad = 2.225 \text{ Mev} \end{aligned}$$

This energy, 2.225 Mev, is the energy that should be emitted when the neutron and proton combine to form a deuteron or is the energy required to separate the deuteron into its components. This can be tested, for example, by measuring what energy photon is required to produce a disintegration (photodisintegration) of the deuteron (Fig. 39.2).

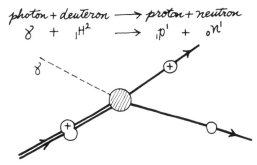

$$photon + deuteron \longrightarrow proton + neutron$$
$$\gamma \quad + \quad {}_1H^2 \quad \longrightarrow \quad {}_1p' \quad + \quad {}_0n'$$

FIG. 39.2. If the deuteron is initially at rest, the minimum energy photon that can cause this disintegration is just equal to the binding energy of the deuteron;° mass of photon + mass of deuteron = mass of proton + mass of neutron. This minimum is measured to be 2.226 ± 0.003 Mev.

In general, the binding energy of the nucleus (the ground-state energy of the nuclear system, if the energy of the nucleons separated from each other is taken as zero) can be measured either as the energy required to break the nucleus into its components (protons and neutrons) and to separate them, or as the difference in mass between the bound

° The small center-of-mass connection is ignored.

nuclear system and that of the protons and neutrons out of which it is made. These two are related by the mass-energy relationship (as with the deuteron above).

NUCLEAR FORCES AND NUCLEAR MODELS

In spite of the consistency of the evidence indicating the existence of nuclear forces, their presence is not obvious on a macroscopic scale. One is thus presented with an initially paradoxical situation in which forces of the order of 100 times stronger than the electrical forces may be presumed to exist, forces that are attractive between all nucleons and yet are not seen in any ordinary experience. (Gravitational forces between two ordinary-sized bodies are just barely visible with elaborate preparations; electrical forces are easily visible with simple preparations—small imbalances produce enormous forces; but such macroscopic manifestations of nuclear forces have never been seen.) This, in addition to evidence from the scattering of one nucleon with another, led to the idea of nuclear forces of immense strength but very short range—so strong they dominate all others at close range, but so short in the range of their effect that little trace of their existence is visible beyond distances of the order of 10^{-12} cm.

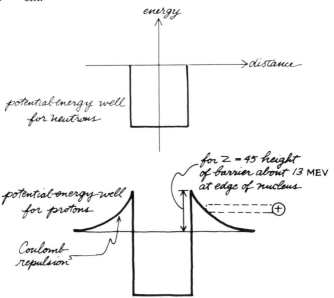

FIG. 39.3. For protons the Coulomb potential energy is superimposed on the square nuclear-potential well, giving a shape as shown. This results in a barrier against proton penetration of the nucleus.

Compared to the atomic problem, that of the nucleus is extraordinarily difficult. But the dominating feature of nuclear forces—the fact that they have a very short range and that they are very strong—can be summarized in a very simple manner by picturing the nucleons in a nucleus as confined in a small spherical container with high (but not impenetrable) walls.

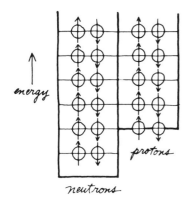

FIG. 39.4. The neutron potential-energy well is deeper than that of the protons, but since the exclusion principle allows only two neutrons in each level, eventually it becomes energetically favorable to add protons.

Nuclear models based on just such a simple idea—neutrons and protons each with internal angular momentum (spin) quantum number $\frac{1}{2}$, each with its own magnetic moment (each a little magnet), each obeying the exclusion principle (they are both fermions) confined to a spherical container of approximate radius 10^{-12} cm (growing larger slowly in proportion to the number of nucleons)—thus a quantum system with quantum numbers not unlike those of the hydrogen atom and not completely different from the periodic table of elements, but with a force whose radial dependence is different, can account at least roughly for many of the observed properties of nuclei.

NUCLEAR PROCESSES AND STABILITY

When a nucleus is formed with too many neutrons or too many protons, or when for a given number of nucleons, it is in an excited state rather than in its ground state, it behaves very much like an atom and often makes a transition to another state of lower energy emitting either light or particles. What transitions it can make depends upon what kind of particles besides photons it can emit; we are thus led to ask what processes are possible among nuclei.

The first observations of nuclear processes on earth, radioactivity, revealed the alpha, beta, and gamma rays. From our present point of view, gamma emission (the emission of photons) in nuclear processes is due to the transition of the nucleus from one nuclear level to another with the emission of a photon due to electromagnetic interactions in complete analogy with atomic transitions. The larger energies involved in nuclear transitions result in the more energetic photons—called gamma rays.

The gamma rays emitted from excited nuclei can be classified into series and, as for atomic spectra, can be used to classify the various nuclear-energy levels. One can identify discrete levels, allowed and forbidden transitions, and so on. The analogy with atomic spectra provides evidence for the discrete nuclear-energy levels and support for the treatment of the nucleus as a quantum system.

Beta decay—the process in which electrons or positrons are ejected from the nucleus with an accompanying change in the nuclear charge—is now presumed to be the result of the fundamental process

$$\text{neutron} \longrightarrow \text{proton} + \text{electron} + \text{antineutrino (electron type)}^*$$

$$n \longrightarrow p + e^- + \bar{\nu}_e$$

(The place of the antineutrino, the massless, chargeless antiparticle of the massless, chargeless neutrino, will be discussed later.) This process has a strength intermediate between electromagnetic and gravitational interactions (10^{-11} weaker than electromagnetic interactions), and because of it the neutron in empty space lives for about 15 minutes (an enormous length of time on the nuclear scale of events). A nucleus with Z protons and $A - Z$ neutrons may by beta decay go to a nucleus of $Z + 1$ protons and $A - Z - 1$ neutrons:

$$(Z, A) \longrightarrow (Z + 1, A) + e^- + \bar{\nu}_e$$

The neutrino was originally proposed by Pauli to maintain the conservation of energy in such decays. Since that time it has been more or less directly observed.

The process of alpha decay is both of importance for understanding of nuclear transitions and an excellent illustration of the application of quantum mechanical ideas to nuclear phenomena. In 1928 Gamow in Germany and Condon and Gurney in the United States independently showed how the new quantum theory could provide a natural mechanism for alpha emission and could yield at least order-of-magnitude estimates of how often an alpha emission would take place.

The measured energies of alpha particles emitted from nuclei in the process of alpha decay are too small from the classical point of view. A doubly charged particle ejected from thorium, for example, should have an energy of about 26 Mev, disregarding any additional energy with which it is thrown out of the nucleus.

Yet alpha particles are never observed to have such energies. Their energies are typically on the order of 5 Mev. It is as if the alpha particles originated at a point quite distant from the edge of the nucleus, where the repulsive potential is lower, so that the nucleus ejects the unwanted alphas less energetically.

"If we consider the problem from the wave mechanical point of view," Gamow wrote,

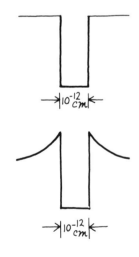

> the above difficulties disappear by themselves. In wave mechanics a particle always has a finite probability, different from zero, of going from one region to another region of the same energy, even though the two regions are separated by an arbitrarily large but finite potential barrier.[9]

Gamow proposed that an alpha particle might be confined inside the nucleus (being a part of the nucleus) by a potential that in the absence of the Coulomb force would look as shown (short range and very strong). Superimposing the Coulomb force on this, we get a picture that looks not unlike a volcano with a deep crater. Somewhere in that crater at some energy, the alpha particle moves. For instance, it might have

---

* Two distinct neutrinos have been identified: one associated with electrons (above), the other with muons.

enough energy so that it could exist outside the nucleus but not enough energy to get there classically.

However, the solution of the Schrödinger equation for a potential of the volcano form, above, for a particle with energy larger than zero but smaller than the maximum height of the potential, yields a wave function (as we have seen previously) that goes continuously from one classically allowed region to the other—thus allowing the quantum particle to "tunnel through" the classically forbidden region.

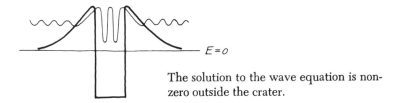

$E = 0$

The solution to the wave equation is non-zero outside the crater.

The general stability of nuclear matter—why some nuclei are stable while others are radioactive—is understood as follows. If a nucleus by alpha, beta, or gamma transitions can get to another state of lower energy, it generally does so. If not, it is stable. Consider, for instance, the nuclei of carbon and boron, which have mass number 12. Boron, element 5, has 5 protons and 7 neutrons. Carbon, element 6, has 6 protons and 6 neutrons. The mass of the boron nucleus is, however, slightly greater than the mass of the carbon nucleus. The difference is 2.476×10^{-26} g, so that the binding energy of the carbon nucleus is about 14 Mev greater than that of boron. It is, then, energetically favorable for a boron nucleus to decay into carbon. What is needed is a mechanism. In the case of boron and carbon, the boron can emit a beta particle and convert its seventh neutron into a sixth proton. (It is equally possible for the process to go the other way, but for this a beta particle must be available.) Carbon-12 is stable because it has no way to get to another combination of 12 nucleons that is energetically more favorable than that of the carbon nucleus.

Sometimes greater stability can be obtained by giving off alpha particles. For instance, of all *single* nuclei, $_4Be^8$ is more stable than either $_3Li^8$ or $_5B^8$. Yet $_4Be^8$ is not stable, because it is not the most energetically favorable combination of 8 nucleons. The most favorable combination is that of two alpha particles (helium nuclei):

$_3Li^8$	13.3174×10^{-24} g	$_4Be^8$	13.2880×10^{-24} g
$_5B^8$	13.3183×10^{-24} g	$2(_2He^4)$	13.2878×10^{-24} g

Thus beryllium-8 eventually breaks down into two alpha particles.

If we put the stable nuclei on a graph of number of neutrons against number of protons, we find that as the nucleus becomes heavier, the percentage of neutrons rises from 50 percent in helium, to about 59 percent in barium, to about 61 percent in uranium (see Fig. 39.5). This means that if a heavy nucleus decays by emission of an alpha particle which has only 50 percent neutrons, it will have to lose some neutrons to get to a stable configuration (see Table 39.1). It can usually do this by converting neutrons into protons by beta decay.

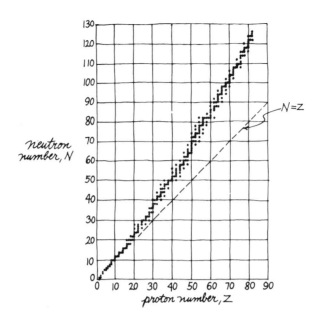

FIG. 39.5. Neutron number versus proton number for the stable nuclides. (After R. T. Weidner and R. L. Sells, *Elementary Modern Physics*, rev. ed., Allyn and Bacon, Boston, 1968)

TABLE 39.1

Decay Scheme: Uranium to Lead
(times shown are half-lives)

$_{92}U^{238}$	$\longrightarrow$	$_{90}Th^{234} + {}_2He^4$	in 4.51×10^9 years
$_{90}Th^{234}$	$\longrightarrow$	$_{91}Pa^{234} + {}_{-1}e^0$	in 24.1 days
$_{91}Pa^{234}$	$\longrightarrow$	$_{92}U^{234} + {}_{-1}e^0$	in 1.14 minutes
$_{92}U^{234}$	$\longrightarrow$	$_{90}Th^{230} + {}_2He^4$	in 2.50×10^5 years
$_{90}Th^{230}$	$\longrightarrow$	$_{88}Ra^{226} + {}_2He^4$	in 80,000 years
$_{88}Ra^{226}$	$\longrightarrow$	$_{86}Em^{222} + {}_2He^4$	in 1620 years
$_{86}Em^{222}$	$\longrightarrow$	$_{84}Po^{218} + {}_2He^4$	in 3.825 days
$_{84}Po^{218}$	$\longrightarrow$	$_{82}Pb^{214} + {}_2He^4$	in 3.05 minutes
$_{82}Pb^{214}$	$\longrightarrow$	$_{83}Bi^{214} + {}_{-1}e^0$	in 26.8 minutes
$_{83}Bi^{214}$	$\longrightarrow$	$_{84}Po^{214} + {}_{-1}e^0$	in 19.7 minutes
$_{84}Po^{214}$	$\longrightarrow$	$_{82}Pb^{210} + {}_2He^4$	in 1.64×10^{-6} seconds
$_{82}Pb^{210}$	$\longrightarrow$	$_{83}Bi^{210} + {}_{-1}e^0$	in 22 years
$_{83}Bi^{210}$	$\longrightarrow$	$_{84}Po^{210} + {}_{-1}e^0$	in 5 days
$_{84}Po^{210}$	$\longrightarrow$	$_{82}Pb^{206} + {}_2He^4$	in 138.3 days
$_{82}Pb^{206}$		stable	

NOTE: In this table, $_{-1}e^0$ means an electron (charge $-e$, nuclear mass number 0).

But there are other nuclear processes. In December, 1938, Hahn and Strassman observed with surprise that one of the products of bombarding uranium (atomic number 92) with neutrons was a radioactive isotope of barium (atomic number 56). Lise Meitner, who along with so many others had fled the thousand-year Reich, proposed a month later the explanation that the uranium nucleus had been split apart. Likening the nucleus of uranium to a liquid drop with every little surface tension, she explained:

> It seems therefore possible that the uranium nucleus has only small stability of form and may, after neutron capture, divide itself into two nuclei of roughly equal size. . . . These two nuclei will repel each other and should gain a total kinetic energy of [about] 200 Mev., as calculated from nuclear radius and charge.[9]

At Columbia University in New York, Enrico Fermi, who had used his 1938 Nobel Prize trip to Stockholm to escape from Italy, realized that if the fission produced neutrons among its reaction products, the reaction could be self-sustaining, and began his own experiments on the splitting of uranium in what is now known as nuclear fission. Einstein, alerted by Leo Szilard, a co-worker of Fermi, wrote to President Roosevelt and obtained a few thousand dollars for Fermi's work, which soon became known as the Manhattan Project; its object was to produce a fission bomb before the Nazis did.

The process that Hahn and Strassman had observed was probably the following reaction:

$$_0n^1 + {}_{92}U^{235} \longrightarrow {}_{92}U^{236} \longrightarrow {}_{56}Ba^{146} + {}_{36}Kr^{90}$$

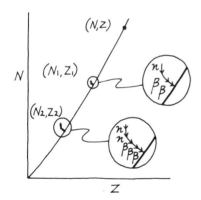

F I G. 39.6. Transformations in the nuclear-fission reaction as they appear on a neutron-proton diagram. (After R. T. Weidner and R. L. Sells, *Elementary Modern Physics*, rev. ed., Allyn and Bacon, Boston, 1968)

Both $_{56}Ba^{146}$ and $_{36}Kr^{90}$ are extremely unstable. The problem is that they have too many neutrons—an excess so great that some of these neutrons are released almost instantaneously ($\sim 10^{-14}$ sec):

$$_{56}Ba^{146} \rightarrow {}_{56}Ba^{145} + {}_0n^1$$
$$_{56}Ba^{145} \rightarrow {}_{56}Ba^{144} + {}_0n^1$$

and
$$_{36}Kr^{90} \rightarrow {}_{36}Kr^{89} + {}_0n^1$$

These nuclei still have too many neutrons, but they can now reach stability by beta decay, in which neutrons turn into protons (see Figs. 39.6 and 39.7).

As Lise Meitner had guessed, the average amount of energy released in the fission reaction is about 200 Mev. Most of it is in the kinetic

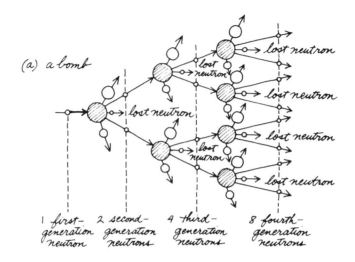

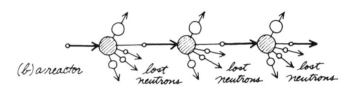

F I G. 39.7. Fission chain reactions. (a) An uncontrolled chain reaction—the principle of a nuclear bomb. (b) A controlled chain reaction—the principle of a nuclear reactor. (After K. R. Atkins, *Physics*, Wiley, New York, 1965)

energy of the products, and therefore is seen as heat. Furthermore—as finally determined by Fermi—the reaction produced enough neutrons so that it could be self-sustaining. On the average each uranium disintegration produced 2.5 neutrons. These neutrons could then be used to produce more disintegrations. In times of the order of 10^{-14} sec a fantastically large number of uranium nuclei can be split, and each splitting produces about 200 Mev of energy. (In contrast, a typical chemical reaction will produce less than 10 ev of energy per atom.) Thus, if the reaction is pushed to the limit, the result is a bomb. On the other hand, if the reaction is controlled by letting only one of the resultant neutrons strike a new uranium nucleus, then the reaction will not expand uncontrollably and can be used in an atomic reactor to produce heat energy (Fig. 39.7).

An even greater potential source of energy is that in which light nuclei are fused together into heavier nuclei with the production of energy. The idea of nuclear fusion developed separately from that of nuclear fission, but it took the advent of nuclear fission to make fusion possible on earth. Nuclear fusion was first suggested in 1929 by Atkinson and Houtermans as the source of energy in the sun and the stars. A decade later, Hans Bethe elaborated on this idea and proposed a sequence of nuclear reactions that could release large amounts of energy and is similar to what we believe occurs in the sun today. Ten years later a fusion reaction was produced in the explosion of the first hydrogen bomb.

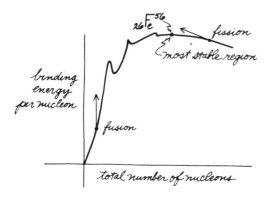

FIG. 39.8. Both the fission and fusion lead to the more stable nuclei nearer the center of the curve.

If four protons (hydrogen nuclei) could combine to form a helium nucleus [charge being conserved by the emission of positively charged beta particles (positrons)], then great amounts of energy should be released.

$$
\begin{array}{lll}
\text{mass of } {}_1\text{H}^1 & 1.6724 \times 10^{-24} \text{ g} & \\
(1) \quad \text{mass of 4 } ({}_1\text{H}^1) & & 6.6896 \times 10^{-24} \text{ g} \\
\text{mass of } {}_2\text{He}^4 & 6.6439 \times 10^{-24} \text{ g} & \\
\text{mass of 2 positrons} & 0.0018 \times 10^{-24} \text{ g} & \\
(2) \quad \text{sum} & & 6.6457 \times 10^{-24} \text{ g} \\
\quad \text{difference} \quad (1) - (2) & & 0.0439 \times 10^{-24} \text{ g} \\
\text{which is equivalent to about 25 Mev} & &
\end{array}
$$

The reaction is more efficient than the fission reaction since about 1 percent of the initial mass is converted to energy in fusion, as against 0.09 percent in fission. (A more likely process is deuteron + deuteron → helium, or tritium + hydrogen → helium.) Such reactions are energetically favorable, yet they are so difficult to produce that they occur almost exclusively in stellar interiors. It is natural to ask why.

The answer is that the Coulomb repulsion of nuclei keeps them so far apart at normal temperatures that such reactions rarely occur. For the reactions to occur, the nuclei must approach close enough to be within the range of the nuclear forces, and for this to occur, the energy of the nuclei must be on the order of 10^5 ev. The problem, then, is to produce and contain such energetic nuclei.

From this one can see why nuclear fusion has become entwined with nuclear fission. There are at present no known means on earth to produce temperatures of millions of degrees for long enough periods other than by the heat given off in a nuclear explosion. [The problem is similar to that of maintaining an ordinary chemical fire. One wants the energy produced by the chemical fusion of some of the atoms to heat other atoms so they will fuse also (that is, one wants to stay above the kindling temperature so the reaction will sustain itself). This requires a container, so the energy released can be directed in part to the other atoms, and it requires a match to start the process.] For nuclear fusion a match could possibly be some source of high-energy protons (some of these will be described later), but a container to hold material at temperatures of millions of degrees together for a long enough time (research is being done today on methods of holding plasmas* by means of magnetic fields) has not yet been invented.

On the sun and in the stars such processes proceed normally, as the interior temperatures are high enough to maintain one of the various fusion cycles, which release enough energy to keep the stars very hot. (The material of the stars holds itself together by its gravitational attraction.) It is in this way that we finally can make sense of the enormous energy production of the stars.

Doubt thou the stars are fire; . . .

40

DIRAC'S RELATIVISTIC ELECTRON

The quantum theory discussed above does not satisfy the principle of relativity. The first successful attempt to create a relativistic wave equation was made by Paul Adrien Maurice Dirac in 1928. Dirac's equation, though it was successful partially because it could be interpreted in a simple manner, led finally to the complexity and richness that are intrinsic in the relativistic quantum theory. To unite these two twentieth-century principles, Dirac returned to the problem of a free electron, an electron moving through empty space in the absence of forces. In the Newtonian theory, such a particle is an entity with the charge $-e$, mass m, and, after the 1920s, an intrinsic spin and magnetic moment. In the absence of external forces, such a Newtonian particle would move along an inertial path (with uniform velocity). To specify the object completely, one needs at some time (call it t_0) its velocity, its position, and the direction of its spin.

Its energy is given by:

* Completely ionized gases; each atom is charged (ionized), but the entire system is neutral.

$$E = \frac{p^2}{2m}$$

The direction of the spin and the magnetic moment are arbitrary and independent of its position in space and time. Thus for every complete specification of the path of the electron in space and time, there are an infinite number of internal states of the electron (internal meaning not connected with its space and time properties) associated with the arbitrary direction of the spin.

The Schrödinger electron, without spin, has its state completely specified by its wave function at a given time (say t_0). In the one-dimensional container analyzed previously, this could be done by choosing the single quantum number that determined the de Broglie wave

$$\lambda = \frac{2\ell}{n} \qquad n = 1, 2, \ldots \tag{40.1}$$

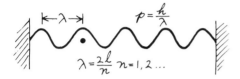

Its energy is given by:

$$E = \frac{p^2}{2m} = \frac{1}{2m} \left(\frac{h}{\lambda}\right)^2 \tag{40.2}$$

In the presence of spin, for a given de Broglie wave, the Schrödinger electron does not acquire an infinite number of internal states as the Newtonian electron, but only two—the result of the quantum condition on angular momenta.

The Schrödinger electron with spin is characterized by the associated

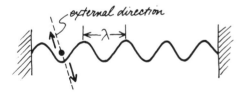

wave of wavelength λ and by the direction of the spin. It has two internal states that can be interpreted as the two possible spin directions. In the absence of a magnetic field these are degenerate. (Their energy is the same.) The effect of the introduction of spin is therefore to double the number of possible wave-function solutions to Schrödinger's equation. In the absence of spin, one might have a certain wave function of space and time that satisfied the Schrödinger equation, as below:

$$\psi(x, y, z; t)$$

In the presence of spin, there would be two such solutions: two functions with precisely the same space-time properties but with opposite spins:

$$\psi_\uparrow (x, y, z; t) \quad \text{and} \quad \psi_\downarrow (x, y, z; t)$$

(It was, of course, the doubling of certain spectral lines, which suggested that there were twice as many levels in the hydrogen atom as those obtained for a spinless electron, which led initially to the introduction of the concept of spin.) The most important effect of the spin is this doubling of the levels (with the exclusion principle), allowing two electrons to occupy the same spatial state if their spins are opposed. In the presence of a magnetic field, the two spin states have different energies, depending on whether the magnet with which they are associated lines up with or opposed to the magnetic field.

Dirac attempted to write down a wave equation for the electron analogous to Schrödinger's equation, but one that was consistent with the principle of relativity. He found, if he wanted to obtain as solutions de Broglie waves with a probability interpretation and with a relativistic energy-momentum relation[*]

$$E^2 = c^2 p^2 + m^2 c^4 \tag{40.3}$$

rather than the nonrelativistic energy-momentum relation

$$E = \frac{p^2}{2m} \tag{40.4}$$

as occurs for the solutions of Schrödinger's equation, that he was obligated to give the electron four internal states. As opposed to a doubling of the states, as had occurred when spin was added to the Schrödinger equation, he found a quadrupling of all states:

$$\psi_1(x, y, z; t) \qquad \psi_2(x, y, z; t) \qquad \psi_3(x, y, z; t) \qquad \psi_4(x, y, z; t)$$

Dirac was able to interpret two of these internal states as those that would occur if the electron has a spin—the first remarkable success of the Dirac theory. The spin of the electron was a consequence of the attempt to write down a wave equation for the electron, consistent with the

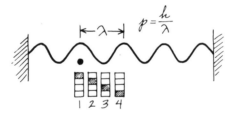

For the Dirac electron, characterized by the de Broglie wave with wavelength λ, $p = h/\lambda$:

$$E^2 = c^2 p^2 + m^2 c^4 \tag{40.5}$$

There are four internal states (above as 1, 2, 3, or 4). One can think of the electron as being characterized by its de Broglie wave and by another quantum number that can take on four values.

principle of relativity, contrasted with the nonrelativistic theory of Schrödinger, in which the spin was an additional hypothesis proposed to explain initially the doubling of certain spectral lines. In the Dirac theory the spin was a consequence of the original postulates: the quantum theory

[*]In this section m is the observed rest mass of the electron ($\simeq 10^{-27}$ g).

and relativity. In essence, if one has a relativistic electron it must have spin; the spin in Dirac's theory was ½. In addition, the magnetic moment associated with that spin was just that observed. Thus, for the first time, an internal property of a particle arose as a consequence of the attempt to write down a consistent equation for its space-time properties.

external direction

Two solutions of the Dirac equation could be interpreted as corresponding to spin-up or spin-down for the electron. What do the other two solutions correspond to?

The other two solutions of the Dirac equation arose, curiously enough, in a very simple way. Because the energy is related to the momentum in a relativistic theory by

$$E^2 = c^2 p^2 + m^2 c^4 \qquad (40.6)$$

there are two solutions for the energy as a function of momentum. (If one takes the square root of both sides, one has two solutions due to the elementary fact that a quadratic equation has two roots.) In this case the solutions are easy enough to find:

$$E = \begin{cases} \sqrt{c^2 p^2 + m^2 c^4} \\ -\sqrt{c^2 p^2 + m^2 c^4} \end{cases} \qquad (40.7)$$

The upper is a solution of positive energy, the lower of negative energy. If the momentum of the particle is zero, we obtain

$$E = \begin{cases} mc^2 \\ -mc^2 \end{cases} \qquad (40.8)$$

In the Dirac theory there are two such solutions for each direction of the spin. Thus the total number of solutions for a given spatial state was $2 \times 2 = 4$, the four internal states Dirac had found.

In relativistic classical mechanics, one might point out, these negative energy solutions would also occur. However, they cause no special problem there because they can be ignored. One can say that particles simply never appear in negative energy states in our world; in essence, in the classical theory, it is possible consistently to exclude the negative energy solutions without doing any violence to the theory. In the quantum theory, however, as Dirac realized very soon, it was not possible to do this. The negative energy solutions would have physical effects; there was no consistent way to exclude them. If one wanted to enjoy the relationship between the relativistic equation and the spin and magnetic moment of the electron, one must deal with the negative energy solutions. However, if an electron could actually exist in such a negative energy state, it would behave in a most extraordinary way. Rather than slowing down as it collided with other particles and finally settling down to rest, it would speed up, going faster and faster with no limit, until its speed was the speed of light. . . . A summary investigation of what the properties of these solutions would be made clear that they had never been observed, nor did it seem likely that they ever would be. This led Dirac to the following rather remarkable suggestion.

DIRAC'S VACUUM

For the Schrödinger electron with spin, the energy, as a function of momentum, is given by

$$E = \frac{p^2}{2m} \tag{40.9}$$

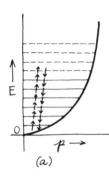

(a)

and for an electron confined in a container, the energy levels are closely but not continuously spaced. Diagram (a) shows the energy-level spectrum of a Schrödinger electron with spin. Two spin states are permitted in each level. The lowest energy level is as close to 0 as the uncertainty principle will allow it to get (depending upon the size of the container), and each level can be occupied by two electrons, one of spin-up and the other of spin-down. According to the exclusion principle, no more than two electrons are allowed for each de Broglie wave. In the Schrödinger picture, if a given electron is initially in some state of momentum **p** with energy E and collides with other objects (say electrons at rest) to which it can give its energy, it gradually makes transitions to lower energy levels until finally it is at the lowest level. See diagram (b). The energy-level structure of the Dirac electron, in contrast, appears as in (c). Again two spin states are permitted in each level.

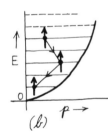

(b)

The spectrum for energies larger than zero begins with mc^2 (the rest energy of the electron) plus whatever energy the electron must have as a consequence of the uncertainty principle and the fact that it is confined in a container. Above this, we have discrete but closely spaced levels satisfying the relativistic energy-momentum relation; each of these levels can be occupied by two electrons with opposite spin. The radically new feature is the set of levels with energy less than zero. These begin at $-mc^2$ and descend downward to $-\infty$, a mirror image of the energy levels larger than zero, which begin at mc^2 and go upward to ∞. The difficulty and the opportunity in the quantum theory lies in the fact that one cannot isolate the minus states from the plus states; electrons can make transitions from the positive energy states to the negative energy states, as indicated in (d), and an electron that began in a positive energy state, if it could collide with other objects and lose energy, would eventually make a transition into a negative energy state and then to another, continuing to move more and more rapidly as its energy went to $-\infty$.

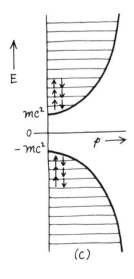

(c)

To prevent this catastrophic electron descent, Dirac proposed that what we usually think of as the vacuum is not a void and empty, but rather, that we regard the vacuum as that state in which every one of the negative energy levels is filled with two electrons. This, of course, places a large (infinite) number of electrons in what we usually call the vacuum. Thus, when an electron finds itself in what is labeled empty space, the situation is not, as we might think, that of diagram (e), in which the electron would find it possible to make a transition to the rather uncomfortable negative energy states, but rather as in (f), where all the negative energy states are already occupied by electrons. The additional electron cannot, by the exclusion principle, make a transition to one of the occupied negative energy levels and therefore remains where we would like it—among the positive energy states. If there were no exclusion principle for electrons, it would not have been possible to do this. (This is the basis of the relationship in relativistic quantum theory between spin and statistics—negative energy states and the exclusion principle.)

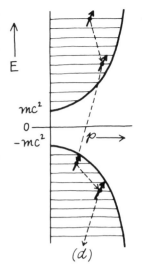
(d)

At first, this suggestion raises more problems than it seems to answer. The vacuum, in a somewhat startling leap, has been transformed from the void of Democritus, Gassendi, and Newton to what is possibly more like the filled plenum of Aristotle or Descartes. It no longer contains nothing but is a highly complicated state containing, in a certain sense, half of everything—that state in which all the negative energy levels are occupied by electrons. (The conception of a vacuum has been refined somewhat in the interim and the apparent asymmetry between the filled and the empty states removed, but the dynamics of our present vacuum and its great complexity are intimately related to the vacuum originally proposed by Dirac.)

There are several obvious problems. In the first place, each of the electrons has a charge $-e$ so that an infinite number of them would possess an infinite charge. Thus, the vacuum, as visualized by Dirac, with all the negative energy states filled, would possess an infinite charge. Dirac does not disagree. However, he says, we read this seemingly infinite charge as the normal state of affairs. It is only deviations that we perceive. In the same way the vacuum would have an infinite mass and infinite negative energy. Each of these, says Dirac, is the normal state of affairs, and we see only deviations. What is worse is that the charged electrons filling the negative energy states, because they interact via Coulomb and other electrical forces, form an extremely complex interacting system. Thus the solutions we have put down, which correspond to free electrons, that is, electrons in the absence of forces, would hardly seem to apply. Again, Dirac does not know the answer, but he says, let us assume that the level structure which results is something like the level structure one would obtain for free particles.

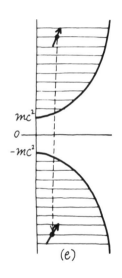
(e)

If one is willing to make all these assumptions, one obtains the following:

1. It is possible to obtain wave functions for an electron corresponding to an arbitrary momentum, satisfying the relativistic energy momentum relation, where the electron automatically has a spin quantum number $\frac{1}{2}$ and the magnetic moment $eh/4\pi mc$.

2. The dynamics of this electron are very much what one would expect for a relativistic modification of the Schrödinger equation. Its energy is always larger than zero, and it can make transitions to only positive energy states, since the negative energy states are all filled; in this respect it is quite normal.

3. The vacuum is redefined to be that state in which all the negative energy levels are filled and all the positive energy levels are empty.

But consider what happens if one of the electrons that occupies a

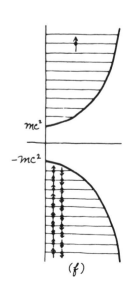
(f)

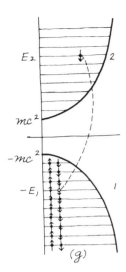

negative energy state is excited to a positive energy. This process is just as possible as the reverse, in which an electron in a positive state jumps down to a negative state. If such a process occurs, one has an electron now in a positive energy state and a *hole* among the negative energy state. How are we to interpret this?

[In diagram (g), the Dirac electron that normally occupies the negative energy state $(-E_1)$ has made a transition to the positive energy state E_2. The energy after the transition is

$$E_2 - (-E_1) = E_1 + E_2]$$ (40.10)

Compared to what, by convention, we call nothing, there appears an electron in a positive energy state and the absence of an electron in the negative energy state.

ANTIMATTER

The hole in the otherwise filled sea of negative energy states is the origin of the concept of antimatter; for if the normal situation, the vacuum, is that state in which all the negative energy levels are filled, then how does that state appear—shown in diagram (g)—in which level 1, a negative energy level, is empty, and level 2, a positive energy level, is filled with the electron that had occupied 1? We expect to see an electron with energy E_2 (spin↑), for that level is occupied. But how does the hole appear? What we accepted as normal was that situation in which the hole, with all its negative energy associates, was filled with electrons. When state 1 is empty we see, so to speak, the absence of a negative energy electron, as Odysseus, according to Kafka, might have heard the absence of the song (the silence) of the sirens.

This absence of the negative energy, negatively charged electron, says Dirac, we interpret as a positive energy and positive charge. For if we are used to seeing five negative charges in a certain region, so used to seeing them that when we see them we see nothing, then, if suddenly only four charges are there, the absence of a negative charge appears as a positive charge. Not only that, but the energy of this particle will be positive, for if the normal situation is to have the negative energy state filled, a hole will mean less negative energy or more positive energy. If the normal situation is that in which helium-filled balloons tugged upward on our arm, we might, having become habituated, reinterpret this tug upward as a zero tug. Then, if one of the balloons disappeared, we would feel less of a tug upward, which we could interpret as a tug downward. If the energy of the vacuum is taken as zero (this is a standard and universal convention), then the energy of the state in which level 2, a positive level, is occupied, and level 1, a negative level, is empty, is given by

$$E_2 - (-E_1) = +\sqrt{c^2 p_2^2 + m^2 c^4} - (-\sqrt{c^2 p_1^2 + m^2 c^4})$$

$$= \sqrt{c^2 p_2^2 + m^2 c^4} + \sqrt{c^2 p_1^2 + m^2 c^4}$$ (40.11)

Thus, compared to the vacuum, the hole appears as a positive particle with a positive energy. If there ever were a case in which such a hole state had been opened up, then a positive energy electron could fall into the hole state (Fig. 40.1). This would give the appearance of a positive particle and a negative particle annihilating each other. Thus the designation *antimatter,* or for this particular particle, *antielectron.*

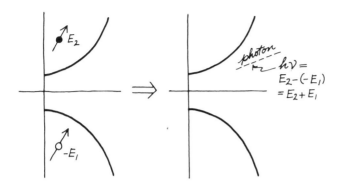

F I G. *40.1.* If a hole is available, the positive energy electron can make a transition and fall into it, leaving the positive energy states empty and the negative energy states filled—the vacuum. Thus the negative and positive charge would appear to annihilate one another and disappear. In the process a photon energy $= E_2 - (-E_1) = E_2 + E_1$ is produced.

Does one ever see such objects? "We are therefore led to the assumption," Dirac writes,

> . . . that *the holes in the distribution of negative-energy electrons are the protons.* When an electron of positive energy drops into a hole and fills it up, we have an electron and proton disappearing together with the emission of radiation.[1]

His conjecture was that the positive particle was the proton. The proton has a spin of $\frac{1}{2}$; it has the charge of the electron, except that it is positive. The fact that its mass was almost 2000 times greater than that of the electron made one a little uncomfortable, but, Dirac pointed out, perhaps the difference in mass was due to the tremendous electrical interaction between all the electrons occupying the negative energy states. One did not know, but if this could be accepted, not only would Dirac's theory have as a consequence the existence of an electron with two spins, but it would also predict the existence of the proton, a truly remarkable unification.

However, Oppenheimer quickly pointed out that if the proton were the antiparticle of the electron, if the hole in the negative energy sea could be interpreted as a proton, then an ordinary electron should be able to fall into this hole. That is, it should be able to annihilate with the proton, leaving behind nothing but light. He then calculated the rate at which this annihilation would occur (the transition probability: electron + hole → vacuum) and found that the lifetime of ordinary matter (atoms made of electrons in orbits about protons) would be about 10^{-10} sec, "a numerical discrepancy," with the observed stability of the universe. "Thus," says Oppenheimer,

. . . we should hardly expect any states of negative energy to remain empty. If we return to the assumption of two independent elementary particles, of opposite charge and dissimilar mass, we can . . . retain the hypothesis that the reason why no transitions to states of negative energy occur, either for electrons or protons, is that all such states are filled.[2]

One can sense the momentary confusion and hesitation. But almost immediately Carl D. Anderson, working in California, reported:

On August 2, 1932, during the course of photographing cosmic-ray tracks produced in a vertical Wilson chamber (magnetic field of 15,000 gauss) designed in the summer of 1930 by Professor R. A. Millikan and the writer, the tracks shown in [Plate 40.1] were obtained, which seemed to be interpretable only on the basis of the existence in this case of a particle carrying a positive charge but having a mass of the same order of magnitude as that normally possessed by a free negative electron.[3]

The particle Anderson had discovered, later called the *positron,* had the same mass and spin as the electron and a charge and magnetic moment identical in magnitude but opposite in sign. When it

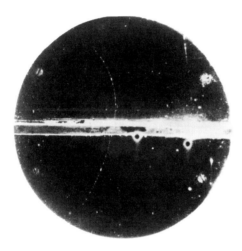

PLATE *40.1.* Anderson's original photograph of a positron. The particle enters from the bottom (high speed, small curvature), is slowed down in the metal plate, and leaves it with a lower speed (larger curvature). From the direction of curvature one can deduce that its charge is positive. (Courtesy of Professor C. D. Anderson, California Institute of Technology)

met an electron it annihilated, leaving photons in its wake. Further, it could be created together with electrons if there was sufficient energy available in a collision. All these things were in fact implied by Dirac's theory, if the absence of a negative energy electron were interpreted as a positron; thus was born the first antiparticle.

Currently we find it convenient to regard the situation in a more symmetrical light. The vacuum is considered the state of lowest energy (zero), in which neither particles nor antiparticles are present; the charge is zero. An electron, negatively charged, and its antiparticle, the positron, positively charged, each can be thought to have positive energy, and each satisfies the relativistic energy-momentum relationship. They are completely symmetrical.

When sufficient energy is concentrated in a small volume, electrons and positrons can be created in pairs. (They must be created together for every electron that is raised out of the negative energy sea leaves behind a hole.) In the presence of a strong magnetic field, the positive charge circles one way and the negative charge circles the other. This characteristic pattern is seen over and over again in photographs of high-energy nuclear collisions in a magnetic field.

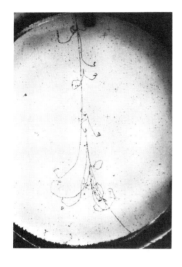

PLATE *40.2*. Electron-gamma shower formation, An electron with momentum of 1 Bev/c (and therefore energy of ∼1 Bev) entered a methyl iodide propane bubble chamber, initiating a shower of gammas and pair electrons. The entire development can be seen because of the high density and high Z of the medium. (Cambridge Bubble Chamber Group—Brown University, Harvard University, Massachusetts Institute of Technology, Padua University)

A positron is very much a stranger in the presence of ordinary matter, and in its progress through ordinary matter it is only a short time before it meets up with an electron and annihilates, leaving only photons.

It is one of the consequences of the relativistic quantum theory that all particles with spin quantum number $\frac{1}{2}$ (all particles with half-integral spin quantum number $\frac{1}{2}$, $\frac{3}{2}$, . . .) should have antiparticles that can be distinguished from the particles and can annihilate with them. Among these are the proton and the neutron (spin quantum number $\frac{1}{2}$); their antiparticles have recently been observed.[*] Their observation was either a triumph or an anticlimax, depending on one's point of view. Theoreticians have believed implicitly for years that the antiparticle of the proton and the neutron must exist. It was, so to speak, a requirement of their entry into the guild. However, we have learned that it is just those results which we expect the most which we should attempt to observe directly. The failure to have observed an antiparticle of the proton or the neutron or for any of the other spin $\frac{1}{2}$ particles that exist would necessitate a catastrophic reevaluation of the foundation of theoretical physics.

---

[*] More details will be given later on the machines and the methods by which this was done.

FIRST
MATTER

41

WHAT IS AN
ELEMENTARY PARTICLE?

The voyage in search of that enduring, unchanging, primordial material, the first matter ($\pi\rho\dot{\omega}\tau\eta$ $\ddot{\upsilon}\lambda\eta$), from combinations of which all things are made, like the pursuit of the golden fleece, begins in the heroic past. But, more difficult than Jason's quest, it continues still. The earliest hero of whom we hear is Thales of Miletus, perhaps for that reason called the Father of Philosophy. He spoke of unity and multiplicity; we are told he believed that everything must come from a single substance—water. A contemporary, Anaximenes, also of Miletus, agreed that there was a single substance, but his choice was air; Heraclitus preferred fire. Parmenides of Elea rejected water, air, and fire, and rejected even change. He postulated that being (which he conceived as a concrete substance) was the foundation of the universe and proposed that everything which *is* forms the being and that being cannot change. This conception of unchanging and eternal being produced a crisis, because if there was no change, how could the physicist conceive of movement—an issue that divided the next generation. (It is possible that Zeno's paradoxes were meant to demonstrate the illusory nature of motion.)

Democritus, about 80 years later, in attempting to reconcile Parmenides' ideas with the seemingly obvious fact that change does take place, divided the being into a number of beings, or atoms, each of which was not further divisible and was not capable of change. He maintained that nothing, just like being, did exist, and that the atoms were separated by nothing. (Being was that which is full of atoms, and nothing was the void.) This void provided space in which the atoms could move. The atoms of Democritus were in perpetual motion; they could not fuse with one another nor could they be created or destroyed, but they could change their positions. This he suggested is sufficient to explain all change that we perceive. Thus he proposed (and possibly he was the first) that all our experience is a result of rearrangements of atoms.

It was from the arrangements and combinations of a limited number of such atoms that the world of Epicurus, Lucretius, and Gassendi was constructed; for Newton those fundamental atoms were hard, massy, and indivisible:

It seems probable to me that God in the beginning formed

Matter in solid, massy, hard, impenetrable Particles . . . and that these primitive Particles, being Solids, are incomparably harder than any porous Bodies compounded of them; even so very hard, as never to wear or break in pieces; no ordinary Power being able to divide what God himself made one in the first Creation . . . should they wear away, or break in pieces, the Nature of Things depending on them would be changed. Water and Earth, composed of old worn Particles and Fragments of Particles, would not be of the same Nature and Texture now, with Water and Earth composed of entire Particles in the Beginning. . . .[1]

Somehow in a manner never precisely defined or successfully worked out, the combinations of these atoms would produce matter as we see it. The chemists of the nineteenth century changed the concept of an atom and attached it to those elements, those constituents of matter, now counted at something close to 104, which are fundamental enough as far as the classification of matter goes, but which are today universally believed to be composite objects constructed from other objects more elementary.

The introduction of the electron and the nuclear atom at the beginning of the twentieth century resulted in a list of elementary objects as follows: the electron, the photon, the proton (hydrogen nu-

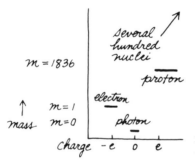

FIG. 41.1. Candidates for elementary particles before 1920.

cleus); what one was to do with the several hundred other nuclei was not yet clear (Fig. 41.1).

Chadwick's discovery of the neutron in 1932 made possible, at least in principle, the construction of all the atomic nuclei from protons and neutrons: the hydrogen nucleus a single proton, the deuteron a proton and a neutron; helium two protons and two neutrons, and so on, through uranium 92 protons and 146 neutrons. The atoms were constructed by the addition of electrons to the nuclei in the manner prescribed by Bohr or by the quantum theory. This itself allowed a remarkable unification: the 92 elements of the chemists, the artificial elements, and all the isotopes need no longer be considered separately as elementary entities, but could be visualized as being constructed from those relatively satisfactory building blocks, the proton, the neutron, and the electron.

The neutron, it is true, is not a completely stable object; left to itself in empty space it decays in about 15 minutes into a proton, an electron, and a new particle, the neutrino (mass = 0, charge = 0, spin = $\frac{1}{2}$), introduced so that conservation of momentum and energy could be retained. Thus by 1933 the list of elementary particles might have appeared as shown in Fig. 41.2. Now there were six: the neutron by itself is not completely stable, but from these could be constructed at least qualitatively a picture of all the known world.

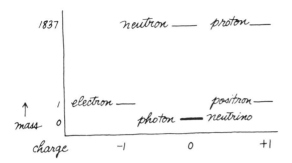

FIG. 41.2. Elementary particles (1933).

In what sense are they particles? One does not expect them to display the properties of a Newtonian or a Cartesian corpuscle, because in the atomic domain one cannot trace out orbits as one would for a planet. They are particles, particularly in the sense that they carry certain properties that are neither divisible nor separable. If one considers the electron, for example, it carries a mass m which is about 10^{-27} g, a charge of about 10^{-10} esu, a spin of $\frac{1}{2}$, and so on, and these are completely inseparable. The charge cannot be separated from the mass; the spin cannot be removed and neither the mass nor the charge divided in two. Further, this object interacts with the rest of the world through electrical, gravitational, and other forces in a well-defined way. Thus we might say that what we consider elementary objects are certain combinations of properties—charge, mass, spin, and others to be treated below, which are associated with a single entity, and which are transferred together.

One would like to be able to say that all such elementary objects are eternal in the sense beloved by the Greeks. (How can an object be fundamental if it does not live forever?) It would be nice to be able to retain the notion that these "Particles" were "never to wear or break in pieces," but, as will become evident, it is extremely inconvenient to attempt to organize things this way. For example, the helium atom, consisting we would say of two protons, two neutrons, and two electrons, is stable; by itself the helium atom will persist forever. Yet one hardly wishes to call the helium atom an elementary object.

In contrast, the neutron, which one would very much like to call an elementary object, although even this will probably not be possible, lives for only about 15 minutes. As mentioned, left by itself it decays into a proton, an electron, and a neutrino. The word "decays" introduces an entirely new concept, for, as has been revealed to us from the 1920s on, it is not only atoms or nuclei that can be transmuted from one to the other* but also the objects sometimes called elementary. Some, like the neutron, make this transmutation without any assistance if left by themselves.

---

* A process somewhat more difficult than that envisioned by the ancient alchemists but all the same possible; it was not, after all, absurd to try to turn lead to gold; it just was something that turned out not to be possible using chemical fire.

If all known matter could be constructed from these particles according to some consistent set of rules, one might accept the decay of the neutron and consider the situation relatively satisfactory. But since the 1930s and with increasing frequency in the 1950s and 1960s, there have been created and discovered, at a somewhat frightening rate, more and more of these so-called elementary entities. So that at present, depending on the classification, one might speak of 30, 50, or even 100 of them. With each is associated a fairly definite mass, charge, spin, and so on. Most of them live very short times—of the order of millionths or trillionths of a second or less.* Yet stability alone cannot be used as a criterion for classifying an elementary particle. They are produced in nuclear collisions of sufficient energy; they decay into other particles that again decay. Perhaps in some sense one possibly can say that the matter we know is constructed of these objects, but precisely how is not known. To consider such a large number of objects all elementary seems at the minimum unaesthetic; one again wishes to say that there exists a sense in which all these objects are manifestations of some more fundamental property or themselves are constructed out of some more elementary object. Whether such a thing can be done in a straightforward way is not known; whether the concept of elementary objects or elementary

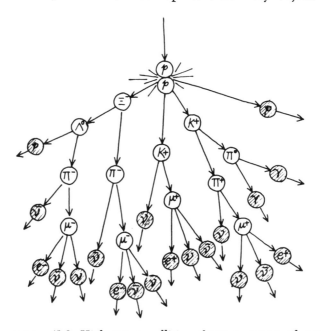

F I G. *41.3.* High-energy collision of two protons produces a shower of particles. One of the original protons darts away unchanged (upper right). The other gives rise to a negative Ξ particle and two positive *K* mesons. These particles, being unstable, then decay into other particles, some of which in turn decay. At the end only stable particles remain, some 20 of them in the event depicted here. (After S. B. Treiman, "The Weak Interactions." Copyright © by *Scientific American,* Inc., **200**, March, 1959, p. 77. All rights reserved.)

* Although short by human standards, these are long times on a nuclear scale.

particles will turn out to be fruitful is itself not known. It is the understanding of this question toward which one is groping right now, that is one of the major preoccupations of current physics.

One would, for example, like to be able to say that there is a sense in which a neutron or a proton were elementary. However, if two protons collide at high energies, what results often is a relative riot of new particles of all kinds, as though the protons themselves were elaborate conglomerations of a whole variety of other particles (Fig. 41.3). The meaning of these symbols, the method of production and observation of these particles, and the current attempts to classify them are what will concern us next.

42

HOW ARE THEY SEEN?

The motion of the planets against the background of the fixed stars, the motion of projectiles or of colliding particles, can be observed with our own eyes. Electrons in atomic orbits are not seen; it is the light emitted that is observed and is interpreted as resulting from the transition of an atom from one level to another. Almost all atomic theory has been constructed to bring order into the emission of light from excited atoms. Subnuclear events are even further removed from one's immediate experience. We do not see nuclear particles directly; they emit little light in an organized way. If one wants to study how they scatter when they collide, how long they live, and into what objects they decay, this must be done with observations based on the traces or the tracks they leave in various instruments. The concepts and hypotheses introduced in particle physics are designed to order the responses of these instruments.

It is a fact that energetic charged particles moving through the proper media leave trails. In many cases, although one can guess, we do not know precisely why. However, the hypothesis that such trails are left by charged particles and that the thickness of such trails depends on their charge, speed, energy, and so on, has brought order into such a large variety of observations that we can be as assured, relatively speaking, that the tracks we observe are the traces of charged particles as we can be that the images we see with our eyes are the traces of external objects.

In September, 1894, C. T. R. Wilson spent a few weeks in the observatory of Ben Nevis, the highest of the Scottish hills, studying what he called "the wonderful optical phenomena shown when the sun shone on the clouds surrounding the hill top." His interest was excited especially by "the colored rings surrounding the sun or surrounding the shadow cast by the hill top or observer on mist or cloud," and made him "wish to imitate them in the laboratory."[1]

> At the beginning of 1895 I made some experiments for this purpose—making clouds by expansion of moist air after the manner of Coulier and Aitken. Almost immediately I came across something which promised to be of more interest than the optical phenomena which I had intended to study.[2]

What had distracted Wilson's attention led to the discovery of the cloud chamber, a standard apparatus for the observation of nuclear particles almost till the present time. He had observed that the passage of X rays and ultraviolet light through his cloud chamber would produce a fog that took minutes to dissipate and soon determined that this was due to ion formation in the air by action of the rays. In the spring of 1911, 15 years after the initial discovery, the idea came to Wilson that he could follow the path of a charged particle moving through his chamber, because small cloud droplets would form on the ions formed along the particle's path. His first test with X rays revealed the path of electrons ejected from the atoms of the air by the rays.

The cloud chamber that Wilson had constructed, in its simplest form is a chamber containing vapor with a glass plate on one end and a piston on the other. In its nonsensitive position, the vapor is kept in a saturated state (for water vapor in air we would say the relative hu-

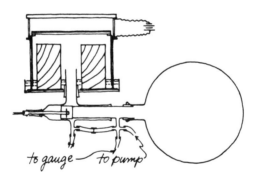

FIG. *42.1*. Schematic diagram of Wilson's cloud chamber. (After C. T. R. Wilson, *Proceedings of the Royal Society*, A87, 1912, p. 278; courtesy of the Royal Society)

midity was 100 percent; the air cannot hold any more vapor). To sensitize the chamber, the piston is suddenly pulled out, increasing the volume, decreasing the temperature, and leaving the vapor in what is called a supersaturated state. (The air in the chamber is holding more vapor than it properly should.) Clouds (condensed water) should form, but they do not immediately unless there is something for the water droplets to condense on: dust particles, ionized atoms, and so on. [A supersaturated vapor will usually not condense unless it has these impurities to condense onto.] The cloud chamber (Plate 42.1), however, is kept dust-free and is cleared of ions by the use of an electric field. The only impurities that do occur in the sensitized chamber are the ions produced by charged particles passing through the chamber. An energetic charged particle passing through the supersaturated vapor knocks electrons out of the atoms in its path and leaves a track of ions in its wake (Plate 42.2). The condensation of cloud droplets on these ions takes place all along this path, and thus the track of the ionizing particle can be seen

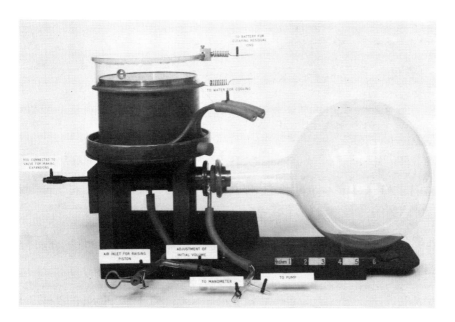

PLATE 42.1. Wilson's original cloud chamber. (The Science Museum, London; courtesy of the Cavendish Laboratory, Cambridge)

and photographed. Under proper illumination the droplets appear as bright spots on a dark background. The "trails" appear somewhat broadened, as a result of diffusion of the ions and droplets through the gas.

The chamber itself may be as large as 1 ft³, and many gas-vapor combinations, such as air-water vapor or argon-alcohol, may be used. The ionizing particles may be of many types, the only requirement being that they must be charged. Neutral particles and photons, not being charged, do not ionize a line of atoms and do not leave tracks. The tracks are usually photographed (often from two angles so that a three-dimensional reconstruction can be made), and the results can then be studied at leisure.

The length of the track and the density of the droplets in it considered together give a measure of the energy of the particle. The charge

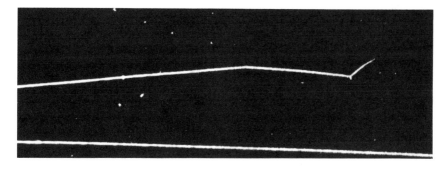

PLATE 42.2 This photograph was taken by C. T. R. Wilson in 1912 with his original apparatus. It shows the final portion of an alpha-ray track in air, enlarged 5.4 times. The track shows two characteristic kinks due to the alpha-particle having collided with nuclei of oxygen or nitrogen atoms in the air. (Courtesy of the Science Museum, London)

and momentum of a particle can be determined by placing the chamber in a magnetic field and measuring the curvature of the track. Lead plates and other similar materials are often placed in the chamber to provide materials with which the particles can interact (to slow them down, for example).

As a tool the cloud chamber has been of enormous value, yet it is inconvenient because the density of the gas in it is so low that there is relatively little chance of a high-energy particle undergoing a nuclear interaction in the chamber. The bubble chamber (Plate 42.3), invented by D. A. Glaser in 1952 (legend has it that Glaser thought of the idea for the bubble chamber while watching bubbles grow on some rough points on the side of a glass of beer), might be said to work like a cloud chamber turned inside out. Vaporization, like condensation, apparently must have a place to start. Without impurities on which vapor bubbles can form, a liquid becomes *superheated,* but remains for a time unvaporized. When a charged particle passes through the liquid, it leaves a track of ions on which, like the rough points on a beer bottle, bubbles can form. The high density of the liquid in a bubble chamber as compared to the gas in a cloud chamber makes interactions of high-energy particles in the chamber much more likely and thus increases the number of events that can be observed.

The liquid is heated under pressure to a temperature above its normal boiling point. If the pressure is suddenly released, the liquid remains in a superheated state. As an example, liquid hydrogen (which, although dangerous, is often used, because its simple structure makes interpretation of results easier since the target particle is a single proton essentially at rest) is maintained at 27°K at 5 atm. It normally boils at atmospheric pressure at 20°K. When the pressure is suddenly released the hydrogen becomes superheated, tracks are formed on the ions left by charged particles, and the results are photographed.[*] The bubble

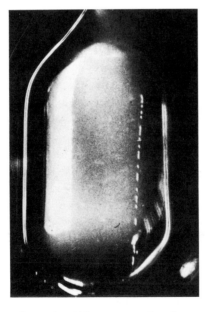

PLATE *42.3* An early bubble chamber. (Courtesy of Professor Donald A. Glaser, University of California, Berkeley)

[*] The seed bubbles are now thought to grow due to energy transferred by the high-energy particle to atoms along its path.

chamber is very well suited for performing experiments using high-energy particle accelerators, because the recycling time is less than 1 sec, and the sensitivity of the chamber can be timed to coincide with each pulse of the accelerator.

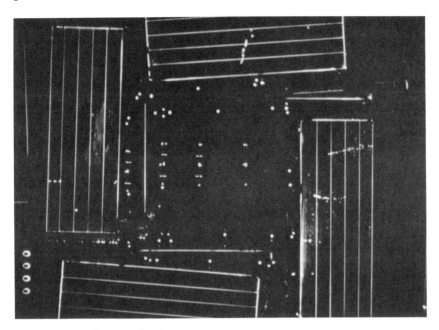

PLATE 42.4. Photograph of an event in a spark chamber. (Courtesy of Professor Robert Lanou, Brown University)

Recently another device, the *spark chamber,* has come into use. This utilizes the fact that air containing ions is more conductive than ion-free air. Thus, if an air gap is subjected to an intense electric field which is just below the level that causes the air to break down and conduct, the passage of a charged particle through the chamber during the pulse will cause current to flow along the path of ions in its wake, leaving a visible spark track. The chamber normally has a clearing field between pulses to eliminate all free ions from the gaps. The bubble-chamber expansion must be initiated before the particles enter, in contrast to spark chambers, which can be and are triggered after the particles have passed through. Thus bubble chambers are normally used for exploratory experiments (in conjunction with an accelerator because one knows definitely a beam pulse is coming) and everything that occurs in the chamber is recorded; in contrast, a spark chamber is normally used for a more concentrated study, of a certain, preselected, type of interaction determined by the "logic" built into the system.

A permanent record of the passage of charged particles can be obtained directly in photographic emulsions, a technique developed about 1947 by an English group led by C. F. Powell. Charged particles produce ions in the emulsion which cause black grains to appear in the tracks when the film is developed. Heavy particles, such as protons, produce much more dense tracks than electrons, because they are able to ionize the emulsion more effectively, and alpha particles or ionized heavy nuclei produce denser tracks than protons. Furthermore, as the particle slows down, it ionizes more of the emulsion in a given distance. Thus the direction and speed of a particle in a nuclear emulsion can be estimated after an initial calibration is made.

Various particle detectors which record the presence, but not the tracks, of charged particles are also used. Almost all are based on the fact that energetic charged particles will excite or ionize atoms. Geiger used a scintillation detector, which consists of a screen that has been painted with a material that gives off light (photons) when struck by an energetic particle. (This is the same sort of interaction that produces light on a television screen.) By examining a small part of the screen with a microscope, it was possible to count the number of alpha particles from a radioactive source that hit that part of the screen in a certain length of time. Today, scintillation detectors are used, but usually with the addition of a photomultiplier. The photons that are given off by the scintillation screen are allowed to strike a second screen, this one yielding electrons when hit by photons. A plate at a lower potential is situated near the second screen, accelerating the electrons produced and releasing more electrons by secondary emission; these are attracted to another plate at an even lower potential. The process is repeated several times (Fig. 42.2) until a reasonable number of electrons is produced. Thus each particle that hits the scintillator can produce a sizable current of electrons, which can then be recorded automatically. A typical photomultiplier yields about 10^3 to 10^4 electrons per incident photon. The amount of light produced in the scintillation, to which the total number of multiplied electrons is proportional, is related to the total energy lost by the high-energy particle in the scintillating material. Therefore the size of the electron pulse is often used to measure this energy loss.

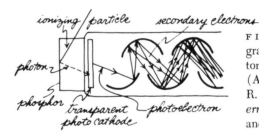

FIG. 42.2. Schematic diagram of a scintillation detector and photomultiplier tube. (After R. T. Weidner and R. L. Sells, *Elementary Modern Physics*, rev. ed., Allyn and Bacon, Boston, 1968)

Various other types of counting devices are sensitive to the arrival of high-energy charged particles. The famous *Geiger counter,* which clicks so faithfully in the movies, works as shown in Fig. 42.3.

We have only suggested the variety and ingenuity of the technology that has been developed to detect the presence of nuclear particles. Because nuclear events are not completely controllable, a major part of the problem is to pick the ones of interest out of the mass of events that occur. A single chamber a few feet in diameter can produce let us say 100,000 possibly interesting events a year. Analyzing all these pictures and picking out only the right ones presents a tremendous problem of data processing. The number of people and the amount of time necessary limit the experiments that can be done. To enlarge the range of possible experiments, computer systems have been designed to map the tracks, cal-

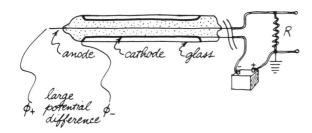

F I G. **42.3**. The Geiger-Mueller counter is usually a glass tube filled with a gas and containing two electrodes. One might be a metal cylinder and the other a thin wire mounted coaxially with the cylinder across which a large potential difference (say 1000 volts) is applied. Ions are swept out of the region between the cylinder and the wire by an electric field. If this electric field is made very intense, the tube can be brought to a critical condition, where the introduction of a single charge will initiate a discharge. Electrons from the original (primary) ion pair produce more ions by collision; the secondary ions produce still more ions. Thus a single electron may result in a cascade of a million or more electrons. When this avalanche strikes the central electrode, the drop in potential is easily detected electronically. (After I. Kaplan, *Nuclear Physics*, 2nd ed., Addison-Wesley, Reading, Mass., 1963)

culate momentum, and so on, of the various particles to determine if the event is one that is being looked for.

43

HOW ARE THEY MADE?

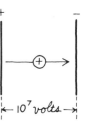

To produce a nuclear reaction, a proton must penetrate the Coulomb potential. For bromine, for example ($Z = 35$), this requires an energy of about 10 million electron volts. Most nuclear reactions of interest require energies of the order of millions of electron volts and higher. To produce a proton with 10 Mev of energy one might let it fall through a potential difference of 10 million volts (household voltages: 110 or 220 volts). Since one cannot easily produce such potential difference without having lightning discharges, it becomes a major preoccupation in the investigation of subnuclear events to produce particles energetic enough to be of interest.

A few natural sources of energetic particles exist: Radioactive nuclei emit alpha, beta, and gamma rays with energies of the order of 1 Mev. It was the alpha particles emitted from polonium that Rutherford used in his early research. High-energy nuclear particles, known as cosmic rays, rain in on us from somewhere in the solar system or the galaxy (energies as high as 10^{13} Mev—1 joule!). This was the source that produced the positron that Anderson observed in 1932. Thus, among the various natural processes that are available, one has the possibility for the production of a range of nuclear and subnuclear particles of various energies, and these, of course, were the first particles with which any experiments were done. But, as with most phenomena of nature, their range and variety is not entirely what one might want—thus man invents.

One of the first machines to produce laboratory accelerated particles was the Van de Graaff generator, an electrostatic accumulator that produces large potential differences essentially by rubbing. In 1929, Van de Graaff built a pilot machine capable of generating 80,000 volts, and in 1931 a 1.5-million-volt machine was built at Princeton which the authors were proud to note had been "constructed at a total cost for materials of only about $100." (It is still used at Princeton for demonstrations. The recently authorized 200-Bev accelerator proposed for Weston, Illinois, will cost an estimated $300 million.)

In England, Cockcroft and Walton constructed what is called a linear accelerator using a high-voltage a-c source (Van de Graaff had used a d-c source) to accelerate protons through 600,000 volts. In 1932 they first used these particles to bombard lithium and split it into two helium nuclei ($_0H^1 + {}_3Li^7 \rightarrow {}_4Be^8 \rightarrow {}_2He^4 + {}_2He^4$). The Cockcroft-Walton accelerator and the Van de Graaff generator have since been refined and replaced as sources of very high energy particles; however, they retain their importance in two respects: (1) they provide injection systems for higher-energy circular accelerators, and (2) they have been improved and extended to provide higher energy linear accelerators.

Another approach to the production of high-energy particles was introduced by Lawrence. He succeeded in substituting a circular race track for the straight-line track of the linear accelerator, and a succession of small pushes for the few larger pushes, thus requiring smaller voltages. "The experimental difficulties go up rapidly with increasing voltage," he and Livingston wrote in 1932 in describing the first *cyclotron*. This first device, barely 1 ft across, produced 1,200,000-volt protons.

The idea was to use a magnetic field to make the charged ion move in a circular path. Each time the particle went around it could be accelerated a little bit thus attaining energies above that used in previous accelerators with comparatively low voltages. The idea came to Lawrence when, by chance, in 1929, he saw an article on the subject written by Wideroe, a German engineer. Lawrence did not understand German, but in the diagrams he found the germ of the idea for the cyclotron (an argument for—or against?—the requirement of a reading knowledge of German for a Ph.D. in physics).

As long as the cyclotron magnetic field is constant, a nonrelativistic particle makes one revolution in the same period of time, no matter what the radius of the orbit. In that case, the accelerating field can oscillate at

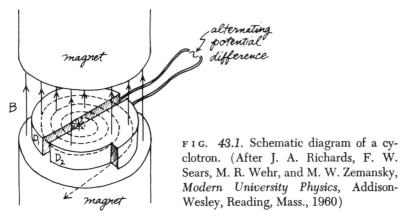

FIG. 43.1. Schematic diagram of a cyclotron. (After J. A. Richards, F. W. Sears, M. R. Wehr, and M. W. Zemansky, *Modern University Physics*, Addison-Wesley, Reading, Mass., 1960)

a constant frequency. (One reason that Lawrence could not use electrons in his accelerator was that even a 50,000-volt electron shows a relativistic 10 percent increase in mass and its period of revolution changes as its mass changes.) A means of solving this problem is to decrease the frequency of the oscillating field as the particle becomes more massive—the principle behind the synchrocyclotron.

In 1941 D. W. Kerst invented the *betatron*, so-called because it was used to accelerate beta particles. Whereas the cyclotron accelerates par-

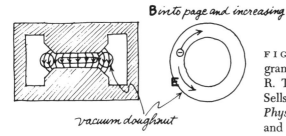

FIG. 43.2. Schematic diagram of a betatron. (After R. T. Weidner and R. L. Sells, *Elementary Modern Physics*, rev. ed., Allyn and Bacon, Boston, 1968)

ticles with an electric field produced by a change in potential, the accelerating electric field in the betatron is produced by the Faraday effect of a changing magnetic field (changing magnetic fields produce electric fields). The field is changed at just the right rate so that the radius of the orbit is always the same, even though the momentum of the particle is continually increasing. Usually ejection of the beam is accomplished by perturbing the equilibrium orbit at the end of the acceleration cycle using either electric or magnetic fields. For some other applications, targets are rammed into the equilibrium orbit at appropriate times during the acceleration cycle. Typically, a beam of 50-kev electrons from a Van de Graaff generator is injected into the machine when the magnetic field is increasing and ejected at high energy when the field reaches its maximum. The betatron, like the cyclotron and the synchrocyclotron (Plate 43.1) before it, uses a magnet that fills the entire orbit. Such huge magnets make the growth of these machines beyond the 1-Bev mark economically prohibitive. (The iron and electricity cost too much.)

In 1945 V. Veksler in Moscow and E. M. McMillan in Berkeley independently introduced the principle of the *synchrotron*, a fixed-radius circular accelerator. The idea was to vary both the electric field frequency and the magnetic field at the same time. Veksler and McMil-

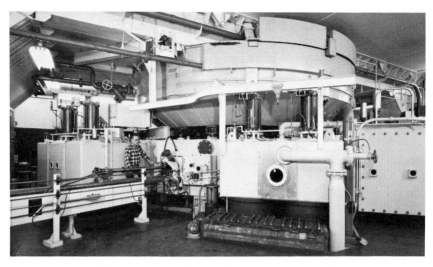

PLATE *43.1.* The 184-in. synchrocyclotron, which accelerates protons to 720 Mev. The lower pole of the cyclotron magnet is below the floor surface and not visible. (Courtesy of the Lawrence Radiation Laboratory, University of California)

lan demonstrated that the orbits could maintain their stability in this process.[*] The magnets of the synchrotron consist of huge *C*-shaped magnets which, since they need be placed only at the fixed radius of the particle orbit, are relatively much smaller than the magnets of varying-radii machines. Generally, the magnet is separated into four sections and the gaps are used for injecting, accelerating, and targets. As the particle's energy (and hence its speed) increases, the frequency of the accelerating field is increased and the magnetic field is made stronger. When relativistic speeds are reached, the frequency can remain constant (although the energy may increase, the speed of a particle moving at 0.99 *c* cannot change by much), and the magnetic field can be increased as the momentum of the particle rises on each successive turn.

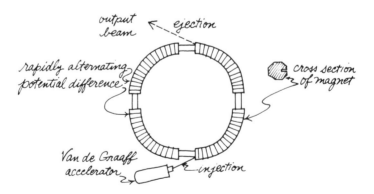

FIG. *43.3.* Schematic diagram of a synchrotron. (After R. T. Weidner and R. L. Sells, *Elementary Modern Physics,* rev. ed., Allyn and Bacon, Boston, 1968)

[*] They showed that a particle, if it got a little out of orbit for one reason or another, would automatically be forced back into orbit rather than farther out. This stability condition is required if any significant number of particles is to be accelerated and the beam not be lost during acceleration.

In circular accelerators, such as the synchrotron, serious limits on the energy that electrons can attain are posed by radiation losses.* The fact that the particle is moving in circles means that it is constantly accelerating even when there is no gain in speed. At very high energies the radiation loss due to the acceleration of the charged particle is sizable. It appears that these losses will limit electron energies, even in the synchrotron, to about 10 Bev. Linear accelerators, in which the particles always move in straight lines, do not suffer such severe radiation losses and can be expected to produce higher electron energies if made long enough (2 miles, 20 Bev, at Stanford).

The largest accelerator in operation as of 1967 is a 33-Bev synchrotron at Brookhaven National Laboratory (Plate 43.2); CERN (European Council for Nuclear Research) has a 28-Bev accelerator in Geneva. A 70-Bev synchrotron is scheduled to be completed in the Soviet Union at Serpukhov between 1968 and 1970. Larger accelerators are planned for the future, most notably, the 200-Bev proton synchrotron proposed for Weston, Illinois, by the early 1970s. Protons in this projected accelerator would receive their initial acceleration in a 750-kev Cockcroft-type generator. From there they would be accelerated to 200 Mev in a linear accelerator and then be injected into an 8-Bev synchrotron. When the protons reached 8 Bev they would be sent on their final journey into the large 200-Bev synchrotron. The main ring would be 0.86 mile in diameter and would be buried in 25 ft of earth for shielding. In addition, a 300-Bev machine is being talked about for CERN.

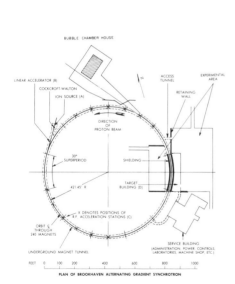

PLATE 43.2. Plan of the 33-Bev alternating gradient synchrotron at Brookhaven National Laboratory and an aerial view of this synchrotron and its ancillary buildings. On the right side of the magnet ring in the aerial view is the 80-in. liquid hydrogen bubble-chamber building; upper left: the cosmotron building; lower left: the graphite research reactor building. (Courtesy of the Brookhaven National Laboratory)

* Radiation losses are also strongly mass-dependent, and losses for protons are negligible. Thus it is practical, at least from the point of view of radiation losses, to build a 200-Bev proton synchroton.

The synchrotron, although cheaper per Bev than the betatron, is a reasonably expensive machine. The accelerator for Weston, Illinois, is estimated to cost several hundred million dollars and to have an annual budget of not quite a hundred million. These costs have wrought certain changes in research. The physics is no different from that of Faraday; however, when it costs a half billion dollars to build and operate a machine, congressmen as well as scientists participate in the decisions. It has been said that scientific projects are pork barrel like everything else—and no doubt it's true. But would Faraday have given us electricity if he had first had to convince the prime minister that an expenditure for twenty pounds of copper wire was justified? ("No, I can't do with ten," we can hear him say.)

44

WHAT HOLDS THE NUCLEUS TOGETHER?·

For much of atomic physics the nucleus could be thought of as a massive charge; any further consideration of its internal structure was almost irrelevant. However, when it became clear that nuclei could disintegrate or that one nucleus could decay into another—that nuclei could be understood as bound combinations of protons and neutrons: hydrogen, one proton; helium, two protons, two neutrons; and so on—the question that arose with increasing persistence and that was increasingly difficult to avoid was, What is it that binds the nucleons together to form a nucleus?

Of the forces already available, those between charged particles repelled rather than attracted, gravitation was much too weak, and the newly proposed force associated with beta decay, whose magnitude could be estimated from the rate at which these beta decays occurred, was of the order of 10^{-11} smaller than electromagnetic forces. Thus it was necessary to introduce some new force to bind these nucleons (protons and neutrons) together. These new forces must be very strong because they hold together nucleons against the force of repulsion of electrical forces with an energy so enormous as to make a nucleus relatively indestructible. At the same time nuclear forces must have a very short range, since they cannot normally be seen in atomic or molecular interactions.

H. Yukawa proposed that this new force is produced by the exchange of a heavy quantum, analogous with the way the electromagnetic force is produced by the exchange of a photon. The mass of the heavy quantum can be estimated from the proposed range of the new force. Assuming a range of 10^{-12} cm, one obtains for the mass of the heavy quantum a value about 250 times larger than the electron mass.

Yukawa wrote:

> As such a quantum with large mass and positive or negative charge has never been found by the experiment, the above theory seems to be on a wrong line. We can show, however, that, in the ordinary nuclear transformation, such a quantum cannot be emitted into outer space.[1]

With a certain modesty he agreed that he might be on the wrong line, but he then proceeded to show that in those nuclear transmutations observed before there was no reason to expect to see the new heavy quantum, and he was at least permitted to hope that if the proper conditions were created, the new quantum would be seen.

Since the new quantum had never been observed previously, one might expect that it wasn't lying around in ordinary matter and that it didn't live a very long time. It does not have to live very long to accomplish its purpose as the carrier of the nuclear force. Roughly speaking, if it interacts strongly with other nuclear particles and if it lives long enough to complete its journey from one nucleon to another, or, to be more secure, if it lives long enough to make several round trips, then it will do its job. But the time required to complete a trip is about

$$\triangle t \simeq \frac{\text{range of nuclear forces}}{\text{speed of particle}} \simeq \frac{10^{-12} \text{ cm}}{10^{10} \text{ cm/sec}} \simeq 10^{-22} \text{ sec} \quad (44.1)$$

(where to make this crude estimate we have used a speed one third that of light).

The Yukawa particle, discovered later and described below, is now known to have a lifetime of several hundred-millionths of a second. Very short by the standards of an ordinary watch, but from the point of view of its purpose for nuclear interactions sufficient to make more than a trillion round trips, in this respect a very long-lived particle indeed.

Nuclear transmutations of the type observed up to Yukawa's time, let us say beta decay or alpha emission, accompanying the natural radioactivity of uranium, involved energy changes within the nucleus of the order of a few million electron volts. However, the rest energy of the new quantum, if its mass is about 250 electron masses, is of the order of 125 Mev. This much energy must be added to the nucleon-nucleon system to produce an observable heavy quantum. Thus Yukawa concluded that in the ordinary nuclear transmutation seen to that time there had never been enough energy involved to produce one of the new quanta.

Almost at the same time and unknown to Yukawa, C. D. Anderson and S. H. Neddermeyer were making an extensive study of the charged particles in cosmic rays. They concluded that certain new types of positive and negative charged particles were present, particles with a mass intermediate between the masses of the electron and the proton; it was difficult to resist the conclusion that they were the particles predicted to exist by Yukawa, the particles responsible for nuclear forces. In a letter to Millikan written in 1938, Niels Bohr said:

The story of the discovery of these particles is certainly a most wonderful one and the cautiousness for which I gave expression during the discussions in the unforgettable days in the spring before last in Pasadena was only dictated from the appreciation of the great consequences of Anderson's work, if the evidence of the new particles was really convincing. At the moment I do not know whether one shall admire most, the ingenuity and foresight of Yukawa or the tenaciousness with which the group in your institute kept on in tracing the indications of the new effects.[2]

However, the situation did not turn out that simply; there followed the kind of surprise that has characterized the study of the interaction of these nuclear particles ever since. In 1947, M. Conversi, E. Pancini, and O. Piccioni published the results of an experiment on the interaction of the intermediate mass cosmic-ray particles with atomic nuclei. The results, as emphasized by Fermi, Teller, and Weisskopf, indicated that the interaction of these particles with the nuclei rather than being very large, as would be expected if they were the particles that transmitted the nuclear force, were extraordinarily small, probably one hundred billion times weaker than electromagnetic forces.

This led to a minor crisis of short duration and produced a flurry of theoretical activity. In particular, the proposal of S. Sakata, T. Inoue, H. A. Bethe, and R. E. Marshak suggested that the intermediate mass particles already observed were intermediate mass particles all right, but were not the particles proposed by Yukawa. Thus presumably intermediate mass particles were not rare, or at least not unique, and the hasty conclusion that any intermediate mass particle would be the Yukawa particle apparently was incorrect. Almost surprisingly these suggestions turned out to be right. The intermediate particle, already observed, to be called soon after the μ meson (now the muon, a lepton, not a meson), was not the strongly interacting intermediate mass particle suggested by Yukawa, but its daughter.

About that time C. F. Powell and his co-workers in Bristol, studying such elementary interactions in nuclear emulsions, observed the two remarkable events shown in Plate 44.1. One could conclude from the figure on the left side of Plate 44.1 that a particle entered from the bottom near the left, slowed down and stopped near the corner at the right bottom, and there disintegrated into another charged particle which went upward, leaving a track and one or more neutral particles that disappear, leaving no tracks. To the entering particle, the primary, was given the name π meson, and to its daughter the name μ meson. Later, when more sensitive photographic emulsions became available, it was found that the μ meson, the daughter, disintegrated also. One of its progeny could be identified as an electron (Plate 44.2).

Measured in units of the electron mass, the masses of the π and μ mesons are now known to be 273 and 207 electron masses. The π meson has since been identified as Yukawa's meson. The quantity of matter a π meson will traverse before suffering a violent nuclear interaction is relatively small, whereas its daughter, the μ meson, can traverse vast quantities of ordinary matter (being found, for example, deep underground in mines or in subway shafts) without suffering any violent nuclear collisions. Its interactions are almost wholly due to its charge.

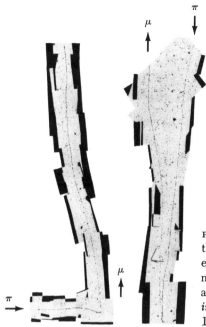

PLATE *44.1.* Powell's photographs of the decay of the π meson. *Right:* π meson enters top right, decays at bottom; μ meson leaves top left. (C. F. Powell and G. P. S. Occhialini, *Nuclear Physics in Photographs,* Clarendon Press, Oxford, 1947)

The entire process has since been identified to go as follows. Beginning, for example, with the positively charged π^+ meson, one finds

$$\pi^+ \to \mu^+ + \nu_\mu \qquad (44.2)$$

(π meson disintegrates into a μ meson + a muon neutrino), and then

$$\mu^+ \to e^+ + \nu_e + \bar{\nu}_\mu \qquad (44.3)$$

(μ meson distintegrates into a positron + an electron neutrino + a muon

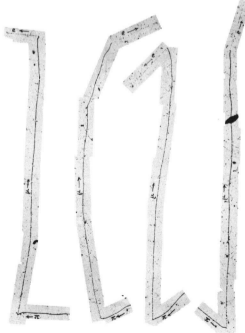

PLATE 44.2. Photographic emulsion records of the decay of the π meson, followed by the decay of its daughter, the muon. (C. F. Powell, *Report on Progress in Physics,* **13**, 1950, p. 384)

antineutrino). The meaning of these (more like a chemical process than an equation) is that a certain collection of properties, known as the π meson, disintegrates into a collection of properties known as the μ meson, plus a collection known as the neutrino. The μ meson, in its own right, disintegrates again into a positive electron, plus an electron neutrino, and a muon antineutrino. Since only the charged particles leave a trail, the existence of uncharged particles such as the neutrino must be deduced by balancing energy momentum and other conserved quantities at each point of disintegration. For example, in those photographs in which a π meson was seen to decay into a μ meson, it seems clear that an uncharged particle must have gone off in the direction opposed to that of the μ meson, because otherwise momentum could not be conserved in the disintegration.

The over-all motion of these particles, as long as they live, seems no different from the motion of other quanta, let us say electrons or protons, already well known. It is their internal properties in which one is particularly interested and in which one finds the characteristic concern of particle physics: What are those collections of properties that exist in the world? How long do they live? With what do they interact? And into what do they disintegrate?

45

STRANGE PARTICLES

Even after the discovery of the π meson a list of known particles was sufficiently small so that one did not object to calling them elementary. However, during the 1950s, evidence for numerous new particles of an almost bewildering variety was uncovered with surprising rapidity. As early as 1944, Leprince-Ringuet in Paris had observed events that could not be easily explained with known particles. In 1947, in Manchester, G. D. Rochester and C. C. Butler found among their cloud-chamber photographs the two shown in Plate 45.1. Because of the characteristic forked tracks, all these newcomers were at first called V particles. It was soon clear, however, that they came in various charges $(+, -, 0)$, widely differing masses, and that some were meson-like, whereas others were nucleon-like.*

They were not isolated events. Soon others were found, so many

---

* The first nucleon-like V particle soon was named the Λ^0 (spin ½, mass 1115 Mev). The first meson-like Vs are now known as K mesons, having for a while taken the names θ and τ.

others that it quickly became clear that what was being observed was common rather than rare when the energy of the bombarding particles was high enough. From the number of such events, one could estimate that they were due to an interaction of roughly the same strength as the strong (π-nuclear) interaction. In a typical such event as shown, the in-

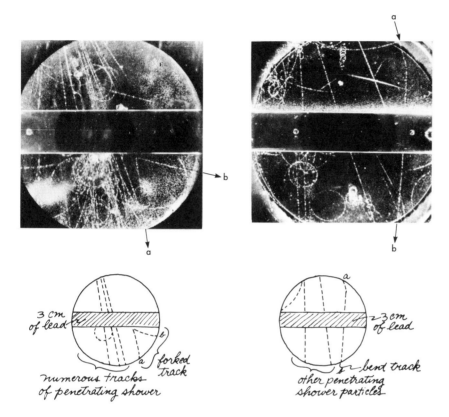

PLATE 45.1. Cloud-chamber photographs obtained by Rochester and Butler, interpreted as showing the decay in flight of V particles. *Left:* A neutral V particle decays into two charged particles (a) and (b). *Right:* A charged V particle (a) decays into a charged particle (b) and a neutral particle. The drawings are schematic diagrams of the two events. (Photographs courtesy of Professor G. D. Rochester, University of Durham. Drawings after A. M. Thorndike, *Mesons: A Summary of Experimental Facts,* copyright 1952; used by permission of McGraw-Hill Book Company, New York)

coming π^- disappears at A. The tracks of the two particles produced at B are identified with π^- and proton, a total mass of 1078 Mev. The empty space between A and B, which asks to be interpreted as a neutral particle, was, however, not to be identified with any particle then known. Further, the lifetime of the invisible stranger could be estimated to be of the order of 10^{-10} sec. (From an analysis of the other tracks its speed could be determined. Thus the length of time it lives can be calculated by measuring the distance it traverses before it decays:

$$\frac{\text{distance between } A \text{ and } B}{\text{speed}} \simeq \text{lifetime})$$

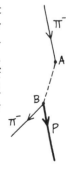

The decay of nuclear particles into other particles should typically take of the order of 10^{-23} sec if they proceed via the strong (nuclear) interactions. [The precise value of the lifetime would depend presumably on a variety of details. However, there is a general agreement that a nuclear particle which lives for 10^{-10} sec does not decay by strong interactions (one does not like to believe that details will account for a factor of 10 trillion). If there were only one instance, there would be no strong resistance to the belief in some rather accidental cancellation; but the phenomenon is quite common.] Because of their copious production and slow decay, these new objects presented an immediate and somewhat strange puzzle. The name remained, and we have with us today the strange particles or the new quantum-number strangeness.

Why, for example, should the Λ^0 particle, which was produced via nuclear interactions, not decay in the same way (and thus live only about 10^{-23} sec—not long enough to leave a 1-cm track)? Perhaps, A. Pais suggested, the Λ^0 particle is not produced alone but with a "little brother."

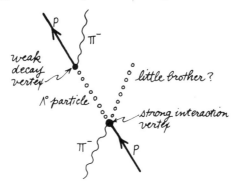

Together, the production proceeds as a strong interaction, but separated; the decay proceeds as a weak interaction. Thus was born the idea of associated production and the implied new quantum number for these particles. This conjecture was confirmed when, very soon later, it was observed that the Λ^0 particle was always accompanied by another new and equally strange little brother (Plate 45.2).

Now one could propose the following. Imagine that the strong interactions conserved some new quantity (call it S), that the Λ^0 particle carried -1 of the S, and that its little brother carried 1, and that further all "ordinary" particles, π mesons or nucleons, carried zero. Then the interaction

$$\pi^- \;+\; p^+ \;\rightarrow\; \Lambda^0 \;+\; \text{little brother}$$

S: 0   0 -1 $+1$ (45.1)

total S $= 0$ total S $= 0$

could be a strong interaction, since S was conserved [$S = 0$ for $(\pi - p)$ and for $(\Lambda^0,\ \text{little brother})$] but the interaction

$$\Lambda^0 \;\rightarrow\; \pi^- \;+\; p^+$$

S: -1 0   0   (45.2)

total S $= -1$ total S $= 0$

could not be strong, since S would not be conserved.

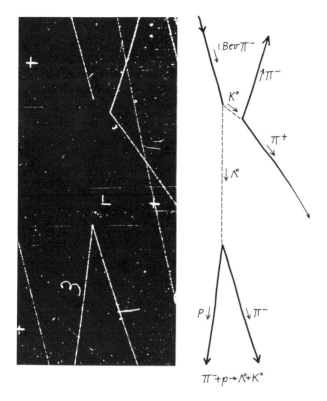

PLATE 45.2. Bubble-chamber photograph of the associated production and decay of a Λ^0 and K^0. The event is analyzed in the diagram. (Courtesy of the Lawrence Radiation Laboratory, University of California)

If one assumed that the strong (nuclear) interactions must conserve S, but that the weak (decay) interactions did not, then one could understand the copious production and the slow decay. This new quantum number (named *strangeness*) was incorporated by Gell-Mann and Nishijima into a scheme relating it to the electric charge of the particles, as discussed next.

Particle		Source of radiation	Specific behavior or measurement	Instrument used for detection
Electron	e^-	Discharge tube	Ratio of e/m	Fluorescent screen
Positron	e^+	Cosmic rays	Ratio of e/m	Wilson cloud chamber
Muons	μ^+ μ^-	Cosmic rays	Absence of radiation loss in passage through Pb (also decay at rest)	Wilson cloud chamber
π mesons	π^+ π^- π^0	Cosmic rays Cosmic rays Accelerator	π-μ decay at rest Nuclear interaction at rest Decay into gamma rays	Nuclear emulsion Nuclear emulsion Counters
K mesons (First meson-like V particles)	K^+ K^- K^0	Cosmic rays Cosmic rays Cosmic rays	$K_{\pi 3}$ decay Nuclear interaction at rest Decay into $\pi^+ + \pi^-$ in flight	Nuclear emulsion Nuclear emulsion Wilson cloud chamber
Neutron	n	Polonium plus beryllium	Mass determination from elastic collisions	Ionization chamber
Antiproton	$\bar{p}$	Accelerator	e/m measurement plus detection of annihilation	Counters
Antineutron	$\bar{n}$	Accelerator	Detection of annihilation	Counters
First nuclear-like V particle	Λ^0	Cosmic rays	Decay in flight into $p^+ + \pi^-$	Wilson cloud chamber
Antilambda	$\bar{\Lambda}^0$	Accelerator	Decay in flight into $\bar{p} + \pi^+$	Nuclear emulsion
Some other hyperons	Σ^+ Σ^- Σ^0 Ξ^- Ξ^0	Cosmic rays Accelerator Accelerator Cosmic rays Accelerator	Decay at rest Decay in flight into $\pi^- + n^0$ Decay in flight into $\Lambda^0 + \gamma$ Decay in flight into $\pi^- + \Lambda^0$ Decay in flight into $\pi^0 + \Lambda^0$	Nuclear emulsion Diffusion chamber Bubble chamber Wilson cloud chamber Bubble chamber

SOURCE: Modified from C. F. Powell, P. H. Fowler, and D. H. Perkins, *The Study of Elementary Particles by the Photographic Method,* Pergamon Press, New York, 1960.

46

CHARGE, INTERNAL SPIN, AND STRANGENESS

Because of their similar masses, spins, and nuclear interactions, the proton and neutron had early been considered perhaps two states of the same particle differing only by their electric charge. Heisenberg had proposed that perhaps these two states could be thought of as the two projections of an internal spin (that is to say, as spinlike states but not connected with space-time). The internal (isotopic) spin quantum number of the nucleon would be

internal (isotopic) spin quantum number $T = \frac{1}{2}$

$$T_z \text{ (the } z \text{ component)} = \begin{cases} \frac{1}{2} & \text{proton} \\ -\frac{1}{2} & \text{neutron} \end{cases} \tag{46.1}$$

[This is in complete analogy with ordinary spin or angular momentum. There if the spin quantum number is $\frac{1}{2}$, the two z projections are also $\pm\frac{1}{2}$. Ordinary spin is interpreted as an intrinsic angular momentum (whose origin however is unknown) and because it is an angular momentum it can be related with the dynamics of ordinary angular momentum. The internal (isotopic) spin formally gives a doubling of levels and a degeneracy associated with the symmetry of the two directions of T_z, but the interpretation of this new quantum number is still unresolved.]

For the proton and neutron the electric charge could be related to T_z by the charge quantum number

$$Q = T_z + \frac{1}{2} = \begin{cases} 1 & \text{proton} \\ 0 & \text{neutron} \end{cases}$$

$$\text{charge} = eQ = \begin{cases} e & \text{proton} \\ o & \text{neutron} \end{cases} \tag{46.2}$$

so that instead of proton and neutron one would have a single object with two different quantum states.[*]

[*] In zero magnetic field the two spin projections have the same energy; the degeneracy is removed and the two states separate when a magnetic field is turned on. In an analogous manner, two states of the same isotopic (internal) spin (for example, proton and neutron) in the absence of the electromagnetic interaction should have the same energy (mass). The observed mass difference is attributed to the electromagnetic interaction.

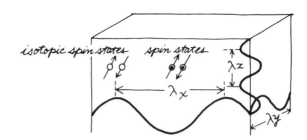

A nucleon in a cubical container has associated with it a wave function characterized by the three de Broglie waves λ_x, λ_y, and λ_z. Its spin quantum number $m_s = \pm\frac{1}{2}$ and its internal spin $T_z = \pm\frac{1}{2}$. When $T_z = +\frac{1}{2}$, the nucleon state has the charge quantum number $Q = \frac{1}{2} + \frac{1}{2} = 1$ and is the proton; when $T_z = -\frac{1}{2}$, $Q = -\frac{1}{2} + \frac{1}{2} = 0$, the neutron.

This simple classification could be extended to the newly discovered π mesons if they were assigned an internal (isotopic) spin quantum number 1:

$$T_z = \begin{cases} 1 & \pi^+ \\ 0 & \pi^0 \\ -1 & \pi^- \end{cases} \tag{46.3}$$

$$Q = T_z$$

The relation between charge quantum number and the z component of the isotopic spin is

$$Q = \begin{cases} T_z + \frac{1}{2} & \text{nucleon} \\ T_z & \text{meson} \end{cases} \tag{46.4}$$

The constant ($\frac{1}{2}$ for the nucleons, 0 for the mesons) seemed a gratuitous addition.

It was generally believed after the discovery of the Λ^0 that like the nucleon it would be a charge doublet (isotopic spin quantum number $\frac{1}{2}$) and that its positive twin would be found in time. Perhaps to be contrary, Gell-Mann and Nishijima explored the possibility that the Λ^0 was a charge singlet. Then its internal (isotopic) spin would be zero, and thus for the Λ^0,

$$Q = T_z + 0 = 0 \tag{46.5}$$

In this they found room for the new quantum-number strangeness.

For the nucleons

$$Q = T_z + \frac{1}{2} \tag{46.6}$$

while for the Λ^0 particle

$$Q = T_z + 0 \tag{46.7}$$

The extra constant in the relation between the charge quantum number and the z component of the isotopic spin was conjectured to be a new quantum number: $Y/2$ (hypercharge divided by 2), which itself could be defined* in terms of the two quantum numbers "strangeness," S, and

* The various quantum numbers, Q, S, Y, T_z, . . . , are dimensionless numbers and thus can be added as above. To obtain, let us say, the charge from the charge quantum number Q, we multiply by the unit charge:
$$\text{charge} = eQ \quad \text{esu}$$

"heavy particle number," B:

$$Q = T_z + Y/2$$
$$Y = S + B \tag{46.8}$$

The quantum-number strangeness (S) is related very simply to the quantum-number hypercharge (Y), which is a little easier to use, by

$$\text{hypercharge} = \text{strangeness} + \text{heavy-particle number}$$
$$Y \quad = \quad S \quad + \quad B \tag{46.9}$$

The heavy-particle number is zero for mesons and leptons (electrons, muons, and neutrinos), 1 for nucleons and hyperons, and -1 for antinucleons and antihyperons. Thus for mesons

$$Y = S \tag{46.10}$$

for nucleons and hyperons (heavy nucleons)

$$Y = S + 1 \tag{46.11}$$

and for antihyperons and antinucleons

$$Y = S - 1 \tag{46.12}$$

In all cases

$$\text{charge quantum number} = z \text{ component of isotopic spin} + \text{½ hypercharge} \tag{46.13}$$
$$Q \quad = \quad T_z \quad + \quad Y/2 \tag{46.14}$$

Now one could propose that the hypercharge (or the strangeness) quantum number was conserved in strong and electromagnetic interactions but not conserved by the weak (decay) interactions, so that processes which conserved strangeness or hypercharge, such as

$$\pi^- \quad + p^+ \longrightarrow \Lambda^0 \quad + \quad \text{little brother}$$
$$Y = 0 \quad Y = 1 \quad Y = 0 \qquad Y = 1 \tag{46.15}$$
$$S = 0 \quad S = 0 \quad S = -1 \qquad S = 1$$

would proceed rapidly via the strong interactions,[*] whereas processes that cannot conserve hypercharge, such as

$$\Lambda^0 \longrightarrow \pi^- + p^+$$
$$Y = 0 \quad Y = 0 \quad Y = 1 \tag{46.16}$$

[*] A process such as $\pi^- + p \to \Lambda^0 + \pi^0$, when energetically possible, is allowed via the weak interaction, but the probability of its occurrence is negligible compared with the above.

proceed 10^{-13} times more slowly via the weak (decay) interactions, yielding a lifetime of 10^{-10}, rather than 10^{-23}, sec.*

Since the first observations of the Λ^0, dozens of new particles have been seen (bosons spin zero and 1 and fermions spin $\frac{1}{2}$ and $\frac{3}{2}$ and higher) with lifetimes ranging from 10^{-8} sec to shorter than 10^{-20} sec. Among the more notable successes has been that of giving them names. Some of them are listed in Table 46.1. It has been possible to assign a strangeness to all the new particles so far observed (their isotopic spin is determined by what kind of a charge multiplet they are), which is consistent with the rapidity of their production and decay and various other properties.

Consider, for example, among the particles classed as baryons† the three Σ particles Σ^-, Σ^0, and Σ^+. Since there are three charge states, this is an internal (isotopic) spin triplet and its isotopic spin is 1 (number of degenerate levels $= 2T + 1$). Its hypercharge can be deduced to be zero. To decay into a nucleon, we might have

$$\Sigma^- \longrightarrow \pi^- + n^0$$

$$\Sigma^0 \longrightarrow \begin{cases} \pi^0 + n^0 \\ \pi^- + p^+ \end{cases}$$

$$\Sigma^+ \longrightarrow \begin{cases} \pi^+ + n^0 \\ \pi^0 + p^+ \end{cases} \tag{46.17}$$

In all these decays the hypercharge on the left side is zero while on the right side it is 1, because the hypercharge of the π meson is zero while the hypercharge of the nucleon is 1. Thus these decays must proceed via the weak interactions, the transition probability is small, and the lifetime of the Σ (if these were the only processes) would be of the order of 10^{-10} sec.

But the decay of a Σ into a Λ^0 plus a photon or a π meson could conserve hypercharge because both Σ and Λ have hypercharge zero.

$$\begin{array}{cccc} & \Sigma & \rightarrow & \Lambda \quad + \text{ meson or photon} \\ \text{hypercharge:} & Y = 0 & Y = 0 & Y = 0 \end{array} \tag{46.18}$$

However, the mass of the Σ is about 1195 Mev, the mass of the π meson is about 140 Mev, and the mass of the Λ is 1115 Mev:

$$\text{mass } \Lambda^0 = 1115 \text{ Mev}$$

$$\text{mass } \pi^+ = 140 \text{ Mev}$$

$$\text{mass } \Lambda^0 + \pi^+ = 1255 \text{ Mev}; \quad \text{mass } \Sigma^+ = 1190 \text{ Mev}$$

---

* The idea is that, analogous to momentum or energy, the initial and final hypercharge or strangeness are equal. Thus, for a two-particle interaction, energy conservation might be phrased
$$E_1 + E_2 = E'_1 + E'_2$$
and hypercharge conservation
$$Y_1 + Y_2 = Y'_1 + Y'_2$$
† The baryons are all heavy particles: protons, neutrons, and heavier with heavy-particle number 1. The nucleons include proton and neutron; the hyperons are all baryons except for the nucleons. Thus baryons = nucleons + hyperons.

TABLE 46.1

Particle		Electric[a] charge	Mass in Mev[b]	Spin[c]	Strangeness	Hypercharge	Mean lifetime in seconds	Common disintegration products	Antiparticle[a]	
Baryons										
Ξ^-	(xi minus)	$-e$	1319	½	-2	-1	2×10^{-10}	$\pi^- + \Lambda^°$	$\overline{\Xi}^+$	(antixi plus)
Ξ^0	(xi zero)	0	$\sim$1311	½	-2	-1	$\sim 2 \times 10^{-10}$	$\pi^0 + \Lambda^°$	$\overline{\Xi}^0$	(antixi zero)
Σ^-	(sigma minus)	$-e$	1196	½	-1	0	1.6×10^{-10}	$\pi^- + n$	$\overline{\Sigma}^+$	(antisigma plus)
Σ^0	(sigma zero)	0	1192	½	-1	0	$\approx 10^{-20}$	$\gamma + \Lambda^°$	$\overline{\Sigma}^0$	(antisigma zero)
Σ^+	(sigma plus)	$+e$	1190	½	-1	0	0.8×10^{-10}	$\pi^+ + n$ or $\pi^0 + p$	$\overline{\Sigma}^-$	(antisigma minus)
$\Lambda^°$	(lambda)	0	1115	½	-1	0	2.5×10^{-10}	$\pi^- + p$ or $\pi^0 + n$	$\overline{\Lambda}^°$	(antilambda)
n	(neutron)	0	940	½	0	$+1$	1.0×10^3	$e^- + \overline{\nu}_e + p$	$\overline{n}$	(antineutron)
p	(proton)	$+e$	938	½	0	$+1$	stable	—	$\overline{p}$	(antiproton)
Bosons										
K^0	(K zero)	0	498	0	$+1$	$+1$	(e)	(e)	$\overline{K}^0$	(anti-K zero)
K^+	(K plus)	$+e$	494	0	$+1$	$+1$	1.2×10^{-8}	$\mu^+ + \nu_\mu$, $\pi^+ + \pi^0$, or others	K^-	(K minus)
π^+	(pi plus)	$+e$	140	0	0	0	2.6×10^{-8}	$\mu^+ + \nu_\mu$	π^-	(pi minus)
π^0	(pi zero)	0	135	0	0	0	$<10^{-15}$	$\gamma + \gamma$	itself	
γ	(photon)	0	0	1	0	0	stable	—	itself	
Leptons										
μ^-	(mu minus)	$-e$	106	½	undefined	undefined	2.26×10^{-6}	$e^- + \nu_\mu + \overline{\nu}_e$	μ^+	(mu plus)
e^-	(electron)	$-e$	0.511	½	undefined	undefined	stable	—	e^+	(positron)
ν_e	(e neutrino)	0	0	½	undefined	undefined	stable	—	$\overline{\nu}_e$	(e antineutrino)
ν_μ	(μ neutrino)	0	0	½	undefined	undefined	stable	—	$\overline{\nu}_\mu$	(μ antineutrino)

[a] $e = 4.8 \times 10^{-10}$ esu.

[b] Mev = million electron volts = 1.6×10^{-6} erg.

[c] Spin = angular momentum in units of $\hbar$, $\hbar = 1.05 \times 10^{-27}$ erg sec.

[d] A particle and its antiparticle have the same masses, spins, and lifetimes. They have equal but opposite electric charges, equal and opposite strangenesses. The disintegration products of an antiparticle are the antiparticles of the disintegration products of the particle. For example, $\overline{\Xi}^- \longrightarrow \pi^+ + \overline{\Lambda}^°$. Compare, however, note below.

[e] K^0 and $\overline{K}^0$ share the same kind of disintegration products: $\pi^+ + \pi^-$, $\pi^0 + \pi^0$, $\pi^+ + \mu^- + \overline{\nu}_\mu$ and others. They both have *two* mean lifetimes $= 1 \times 10^{-10}$ sec and 6×10^{-8} sec. Each of the other particles in the table has only one mean lifetime.

SOURCE: Modified from C. N. Yang, *Elementary Particles: A Short History of Some Discoveries in Atomic Physics*; copyright © 1962 by Princeton University Press; reprinted by permission of Princeton University Press.

Therefore, the decay $\Sigma^+ \longrightarrow \Lambda^0 + \pi^+$ does not occur as a real process because energy cannot be conserved.

The photon decay is energetically possible, but Σ^+ and Σ^- cannot decay into Λ^0 by emitting a photon and conserve charge. (A charged particle must be emitted.) Only the Σ^0 can decay into Λ^0 by emitting a photon:

$$\Sigma^0 \longrightarrow \Lambda^0 + \gamma$$
$$\text{hypercharge:} \quad Y = 0 \quad Y = 0 \quad Y = 0 \tag{46.19}$$

In this process, which is energetically possible and which has been observed, hypercharge is conserved. Thus the decay proceeds via electromagnetic interactions 10^{11} times stronger than the weak interactions.

As a result, the lifetime of the Σ^+ and Σ^- particles is of the order of 10^{-10} sec, the dominant decay made being

$$\Sigma^\pm \longrightarrow \text{nucleon} + \pi \text{ meson} \tag{46.20}$$

The lifetime of the Σ^0 is 10^{-20} sec, the dominant decay mode being

$$\Sigma^0 \longrightarrow \Lambda^0 + \gamma \tag{46.21}$$

Although it is not clear that all the necessary concepts have been introduced, it seems reasonable to believe that the classification according to the strangeness and isotopic spin quantum numbers will be retained in one way or another in whatever theory is developed in the future.

Presented with several dozen heavy bosons and fermions and the seeming necessity of introducing hundreds of particle interactions, the minimum that can be said is that the economy present in electrodynamics (photons, charged particles, and one fundamental interaction) is gone. To bring some order to the mass of data involving particles, decays, interactions, and so on, in the absence of any detailed dynamics, what does not happen has become almost more important than what does. A dominating idea in the analysis of these events has been the assignment of quantities (quantum numbers) that are conserved in some but not all interactions. The four broad classes of interactions conserve the internal quantum numbers, isotopic spin, and strangeness or hypercharge as in Table 46.2.

TABLE 46.2

Interaction	Relative strength	Conserves the internal quantum number
Strong (nuclear)	1	Isotopic spin; hypercharge
Electromagnetic	10^{-2}	Hypercharge
Weak	10^{-13}	
Gravitational	10^{-38}	?

In addition to the classical conservation laws—momentum, energy, and angular momentum—all events have been found to be consistent with the conservation of electric charge, heavy particle number, and lepton number [see Eq. (46.28)].

The conservation of electric charge is essentially the classical observation that the sum of electric charges in any region remains constant if no charges pass in and out of the boundary. If charges are created, they are created in pairs—thus photons produce electron-positron pairs:

$$\gamma + p^+ \longrightarrow p^+ + e^- + e^+$$
$$0 \quad 1 \qquad\quad 1 \quad\; -1 \quad 1$$

$$\underbrace{}_{\text{total charge } e} \quad \underbrace{}_{\text{total charge } e} \tag{46.22}$$

The idea of heavy-particle conservation is a perhaps sophisticated way of repeating the obvious—the universe is stable. One heavy particle can decay into another,

$$\Lambda^0 \longrightarrow p^+ + \pi^- \tag{46.23}$$

the π^- meson decays,

$$\pi^- \longrightarrow \mu^- + \bar{\nu}_\mu \tag{46.24}$$

and the μ^- meson decays,

$$\mu^- \longrightarrow e^- + \nu_\mu + \bar{\nu}_e \tag{46.25}$$

Thus the final products are

$$p^+, e^-, \nu_\mu, \bar{\nu}_\mu, \text{ and } \bar{\nu}_e \tag{46.26}$$

all of which are stable.

If the proton could decay—for example, if

$$p^+ \longrightarrow e^+ + \nu_? + \nu_? \tag{46.27}$$

then eventually* all the matter in the universe would disintegrate into electrons, neutrinos, and light. As far as we know, this does not happen, and a convenient way of expressing this is to postulate a principle of heavy-particle conservation: In any process the number of baryons minus the number of antibaryons remains constant (see Plate 46.1).

A less evident and newer conservation rule is that of leptons (light —nonnuclear fermions: electrons, muons, and neutrinos). The analysis of known events is consistent with the conservation of leptons.

In a process the number of leptons minus the number of antileptons is constant. One writes

$$n \longrightarrow p^+ \;+\; e^- \;+\; \bar{\nu}_e$$
$$\text{lepton number:} \quad 0 \;=\; 0 \;+\; 1 \;+\; -1 \tag{46.28}$$

As they originally came out of Newtonian theory, the conservation laws—momentum, energy, angular momentum—were relatively special theorems concerning the properties of systems subject to forces with particular symmetries. The relation between the symmetry of the force system and the conservation laws is preserved and becomes even more powerful in quantum theory. One observes in hindsight that an analysis of what does not happen (for example, transitions that do not occur) among the hydrogen levels could have been used to guess at some of the properties of the force between the electron and the proton if it had

---

* In less than a second if via the decay interactions.

been unknown.* In the analysis of events among these new particles, where the forces are unknown and the dynamical analysis, if they were known, is almost impossibly difficult, one has tried by observing what does not happen to find selection rules, quantum numbers, and thus the symmetries of the interactions that are relevant. What has gradually come out of this analysis is that in addition to the quantities, which are always thought of as being conserved (momentum, energy, heavy particle number, and so on), there are quantities such as strangeness and isotopic spin, which are conserved by some interactions and not others. No one knows why.

---

* For example, the general pattern of degenerate levels $(2l + 1)$ and the transitions $\Delta l = \pm 1$ (that is, $2P \rightarrow 1S$, not $2S \rightarrow 1S$) could have been used in retrospect to identify the interaction between electron and proton and the electromagnetic field as invariant under rotations (having the symmetry of a sphere).

47
πρώτη ὕλη

The flood of data and unanswered questions about the new "particles" has resulted in so much activity that to a large extent the concern of physics has turned to this area. Here the major preoccupation is to understand why some collections of properties (spin, mass, charge, and so on) hold themselves together for brief periods of time and why others do not; why they interact and decay the way they do; why the various interactions possess different symmetries; and so on. Whether these questions are related, whether they can be answered in any traditional form, whether the framework of quantum physics is large enough to encompass the answers must certainly be included among the questions.

The analogy between the various particle states and their decays, to atomic states and their decays, has not been overlooked. And one possibility that has been explored is that the various "particles" are states of some fundamental system that decay into one another, just as the various levels of the hydrogen atom are excited states of the electron-proton system that make transitions into one another with the emission of photons.

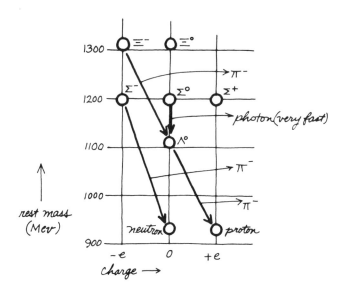

F I G. 47.1. The baryon "octet" with some of the observed decays from one state to another. The decay $\Sigma^0 \to \Lambda^0 + \text{photon}$ is very fast; the others are relatively slow.

On thinking about the hydrogen atom we realize that an infinite number of different levels characterized by

$$n = 1, 2, 3, \ldots$$
$$l = n - 1, n - 2, \ldots \qquad (47.1)$$
$$m_l = l, l - 1, \ldots, -l$$

are interpreted as the excited states of the electron-proton system. The emission of light in an atomic transition, which might, for example, be written

$$2P \to 1S + \gamma \qquad (47.2)$$

is just possibly no different in principle from the transition

$$\Lambda^0 \to p + \pi^- \qquad (47.3)$$

We thus ask ourselves if we can consider the entire array of "particles": baryons and mesons of all spins charges and masses as the various levels of one or several "fundamental" systems. This possibility has suddenly developed in a very promising (and possibly revolutionary) way recently.

A recurring theme in theoretical efforts to organize and classify the new particles has been the attempt to guess the symmetry of the interactions between these particles from observations of their degeneracy structure. The argument often goes as follows: Imagine that one did not know the Coulomb force between electron and proton, but suppose one observed the degeneracy structure for the various angular momentum levels of hydrogen* and other atoms:

groups of $(2l + 1)$ levels: $\quad l, l - 1, \ldots, -l \quad$ for $l = 1, 2, \ldots \quad (47.4)$

From the observation of this degeneracy one might guess that the equi-

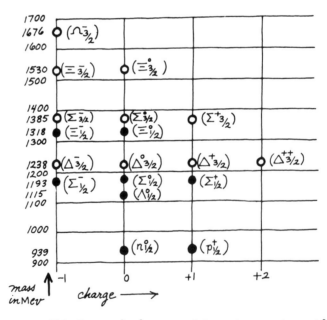

FIG. 47.2. ● are the baryon octet: proton, neutron, Λ^0, Σ (+, 0, −), and Ξ (0, −), all of spin $\frac{1}{2}$. The ○ are the baryon decuplet spin $\frac{3}{2}$. Repetition of Greek letters (e.g., $\Sigma_{3/2}$ and $\Sigma_{1/2}$) indicates that these states have the same hypercharge.

* One is thinking of a hypothetical spinless electron.

potential surfaces were spheres and that the force between electron and proton was invariant under rotations. For the hydrogen atom this is an exercise in hindsight, but among the new particles such an analysis might reveal the symmetry properties of some of the relevant interactions in the absence of a detailed knowledge of what they are or how to calculate their consequences if we knew.

In the last few years attention has been focused on the level spacing of the baryons and mesons. In particular, among the lowest-lying baryons, one can pick out a family of eight particles of spin $\frac{1}{2}$, and above these in average energy another family of 10 particles of spin $\frac{3}{2}$ (Fig. 47.2). To call the eight levels of spin $\frac{1}{2}$ degenerate when they are separated by almost 400 Mev in energy and when other levels supposedly not from the same degenerate-level structure lie among them requires a lively imagination. But it is not impossible. One might say that these levels would be degenerate if it were not for an interaction (supposedly weaker than the dominant interaction) that possesses a different symmetry and which splits the degenerate levels. The situation (always referring to the hydrogen atom, which we understand) for hydrogen would be the same if the external magnetic field were somehow part of the fundamental interaction between electron and proton and if it were very strong.

splitting of l levels in a very strong magnetic field

From this type of consideration directed both at the hyperons and the mesons Gell-Mann and Ne'eman proposed that the internal symmetry was a symmetry under the exchange of three objects such that the strongest interaction did not distinguish between them. This internal symmetry characteristically gives degeneracy levels of 1, 8, 10, . . . ; the eight baryons were identified as one such set of levels, the eight mesons π^+, π^-, π^0, K^0, $\bar{K}^0$, K^+, K^-, and η^0 as another.

Two identical objects (with the labels 1 and 2 visible only to us) among which interactions do not distinguish give four ($2 \times 2 = 4$) states with the same energy:

①①
①②
②①
②②

For convenience they are arranged (using superposition) as

①①
②②   these three (called a triplet) go
①② + ②①   into themselves if $1 \leftrightarrow 2$
                    (1 and 2 are interchanged)

① ② − ② ①   this lone one (singlet) goes into
minus itself if $1 \leftrightarrow 2$

One sometimes writes: $2 \times 2 = 3 + 1$.

Three identical objects

① ① ①   give $3 \times 3 \times 3 = 27$ states with
① ① ②   the same energy
⋮
③ ③ ③

These arrange themselves in a manner similar to that above, into

$$3 \times 3 \times 3 = 1 + \underbrace{8 + 8}_{\text{octets}} + \underbrace{10}_{\text{decuplet}} = 27$$

In particular, the set of levels lying above the eight spin-$\frac{1}{2}$ baryons (short-lived levels of spin $\frac{3}{2}$, of which nine were known at the time) were tentatively identified as $\frac{9}{10}$ of a tenfold degenerate level. The tenth member, named in anticipation the Ω^-, because of its predicted mass and hypercharge, should be relatively long lived ($\simeq 10^{-10}$ sec). Thus nine members of the decuplet had been seen; the tenth, needed to fill out

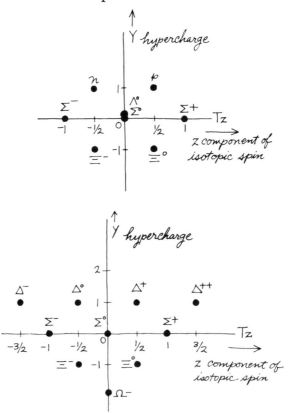

FIG. 47.3. *Above:* the spin-$\frac{1}{2}$ baryon octet, and *below:* the spin-$\frac{3}{2}$ decuplet, classified according to T_z and Y. The repeated Greek letters indicate particles with the same hypercharge.

the little monad, could be hoped to be there if looked for. An intensive search for the predicted particle resulted in its discovery in 1964; Pythagoras could have done no better.

The predicted parameters of the Ω^- were: mass 1676 Mev, hypercharge -2, isotopic spin 0: therefore a singlet particle with charge quantum number

$$Q = T_z + \frac{Y}{2} = 0 + \frac{-2}{2} = -1 \qquad (47.5)$$

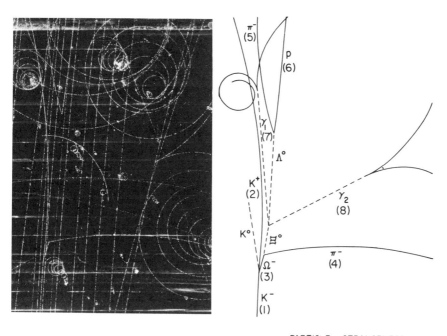

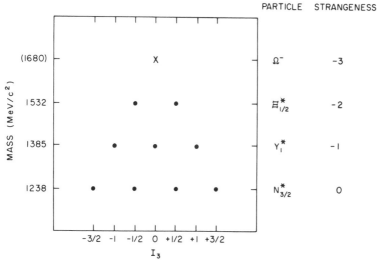

PLATE 47.1. *Above:* photograph and line drawing of first observed decay of the Ω^-. *Below:* Decuplet of $3/2^+$ particles plotted as a function of mass versus third component of isotopic spin. (V. E. Barnes, et al., "Observation of a Hyperon with Strangeness Minus Three," *Physical Review Letters,* vol. **12,** no. 8, February 24, 1964; courtesy of the Brookhaven National Laboratory)

giving a charge $-e$. Because of its hypercharge $Y = -2$ and because all the baryons below it in mass have hypercharge -1, 0, or $+1$, a decay of the Ω^- cannot conserve hypercharge and energy. For example, the most favorable decay that conserves hypercharge (a decay to the lowest mass baryon of hypercharge -1, the Ξ) is

$$
\begin{array}{lccc}
 & \Omega^- & \to & \Xi^- \;+\; \bar{K}^0 \\
Y & -2 & & -1 \quad -1 \\
\text{mass} & 1676 & & 1319 + 498 = 1817
\end{array}
\qquad (47.6)
$$

but since the end products are more massive than the Ω^-, the decay cannot proceed this way. (The problem is that there are no mesons of hypercharge -1 less massive than the $\bar{K}^0$.) Thus if the Ω^- is to decay, it must do so via a process that does not conserve hypercharge—via a weak interaction. Therefore, it is relatively long lived.

The discovery of the Ω^- set off a frenzy of theoretical activity; in particular, the question that was hard to avoid was what were those three objects (subnucleons) whose combinations give rise to the meson and baryon systems. Among the various proposals, perhaps the most startling but the most economical and at this moment the most successful is that of Zweig and again Gell-Mann. The three fundamental entities (masses unknown), called aces by Zweig and quarks[*] by Gell-Mann, have the properties shown in Table 47.1. If anything the proposal was

TABLE 47.1

Quarks	Baryon number	Hyper-charge Y	T_z $\left(\begin{array}{c}z\ component\\of\ istopic\ spin\end{array}\right)$	Charge $Q = T_z + Y/2$
q_p (proton-like)	$\frac{1}{3}$	$\frac{1}{3}$	$\frac{1}{2}$	$\frac{2}{3}$
q_n (neutron-like)	$\frac{1}{3}$	$\frac{1}{3}$	$-\frac{1}{2}$	$-\frac{1}{3}$
q_λ (lambda-like)	$\frac{1}{3}$	$-\frac{2}{3}$	0	$-\frac{1}{3}$

not ordinary. No known objects have anything other than integral charge or integral baryon number. However, to form the proper degeneracies (octet, decuplet), one needs a triplet of objects and each must have the same baryon number. Thus if three of them combine to give baryon number 1, each must have baryon number $\frac{1}{3}$, resulting in charges $\frac{2}{3}$, $-\frac{1}{3}$, and $-\frac{1}{3}$. But this means that the objects are unlike anything ever seen and quite possibly anything that ever will be seen. The lowest mass quark and its antiparticle (antiquark) should be as stable as a proton, because they cannot decay into anything and conserve baryon number or charge in the process.

Various properties of these quark systems that are calculated on the basis of remarkably simple assumptions seem to give reasonable agreement with observations. For example, if one assumes that the mass of the $Y = -\frac{2}{3}$ quark (q_λ) is larger than the masses of the two $Y = \frac{1}{3}$ quarks,

[*] Gell-Mann's name comes from *Finnegan's Wake*, "Three quarks for Muster Mark!" and, because of its firmer literary roots, has been generally accepted.

$$\text{mass}(q_\lambda) = m_0 + \delta$$

$$\text{mass}(q_p) = mass(q_n) = m_0$$

then the masses of the various hyperons in the spin-$\frac{3}{2}$ hyperon decuplet should be split by δ; the separation between the various levels should be proportional to their hypercharge. This is compared with the observed masses in Table 47.2.

Various other properties, such as the lifetimes and magnetic moments of the baryons and mesons, seem to be qualitatively consistent when one makes even the simplest assumptions about the dynamics of

TABLE 47.2

		Observed mass	δ (experimental)
Mass of $\Delta_{Y=1}$	$= m_0$	1238	
Mass of $\Sigma_{Y=0}$	$= m_0 + \delta$	1385	147
Mass of $\Xi_{Y=-1}$	$= m_0 + 2\delta$	1530	145
Mass of $\Omega_{Y=-2}$	$= m_0 + 3\delta$	1674	144

SOURCE: Rosenfeld, et al., *Rev. Mod. Phys.*, **39** (1967), 1. The last decimal place is not certain.

the quarks and their mode of binding to one another. This qualitative success of simple assumptions made about bizarre objects may turn out to be accidental or might possibly herald that kind of revolution physicists hope for but nervously resist.

If quarks "really" exist, then the lowest-mass quark (let us say it is q_p) would be completely stable in the presence of ordinary matter (it requires a $\bar{q}$ to annihilate a q). This means that if quarks are ever found and can be made, it will be possible to store them in a container of ordinary matter (let us say in the form of quark-oxygen):

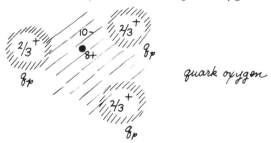

quark oxygen

There is no apparent reason such a molecule could not be kept in an ordinary bottle. (Would it be a solid, liquid, or gas? Would it react with the walls of the container?) Precisely the same argument* would be ap-

---

* Although the quark mass is at present a complete conjecture, it is felt that if they are really here they are likely to be heavier than the nucleon. If the mass of the quark is larger than that of the nucleon, then antiquarks would be stable in the presence of ordinary matter, because the process

$$\bar{q} + \text{nucleon} \longrightarrow q + q$$

for example,

$$\bar{q}_p + (q_p\,q_p\,q_n) \longrightarrow q_p + q_n$$

which conserves baryon number, charge, and hypercharge could not conserve energy, because mass of antiquark + mass of (q_p, q_p, q_n) is smaller than $2 \times$ (mass of quark),

$$m_{\bar{q}} + m_N < m_q + m_q$$

plicable to $\bar{q}_p$ (charge $-\frac{2}{3}$); another container could hold antiquark hydrogen:

antiquark hydrogen

Mixing the contents of the two containers would give

$$H_2(\bar{q}_p)_3 + O(q_p)_3 \rightarrow H_2O + \text{perhaps 10 billion ev of energy}^{\circ}$$

One does not have to think very hard to imagine a variety of practical applications. But it would possibly be more interesting if such quarks did not "really" exist. What then would be the sense in which three objects that do not exist could by their combinations produce everything that does? The answer to that question (the meaning of the question) could be very interesting indeed.

. . .

Even a casual perusal of the current journals reveals what can only be called passionate differences of opinion about what is the useful thing to do next. When one is groping for order, almost nothing seems evident. Cruel April brings its flood of new ideas, autumn the burial of the tired old. And one can only contrast—somewhat wistfully—the finished proportions of Newton's mechanics, the graceful arches of Maxwell's electrodynamics, with the rubble-strewn workshop that is particle physics —a column here, there a half-completed frieze, and everywhere the chaos of broken and discarded stone.

If anyone still had to be convinced that science is more than the collecting of facts, he need only regard the bookshelves full of data that accumulate every year in particle physics alone and the eagerness with which any new idea (even a mnemonic device that enables one to re-member some part of the mass of figures) is grasped and developed to be, perhaps, convinced that the data alone are neither very satisfying nor what we call science. "The scientist," said Poincaré, "must order. One makes science with facts as a house with stones; but an accumulation of facts is no more science than a pile of stones is a house."

But the process by which order is imposed on experience is not as systematic as the order being sought; it contains risks and hazards. To believe that there will be no more surprises is perhaps least interesting and least likely to be true. If the data and conjectures that are physics in development (particle physics now) seem confusing—at once diffuse, speculative, or even sensational—perhaps we can be forgiven as Tyndall forgave Faraday:

$^{\circ}$ And no smog-producing hydrocarbons.

Let those who ponder his works seek to realize the object he set before him, not permitting his occasional vagueness to interfere with their appreciation of his speculations. . . . It must . . . always be remembered that he works at the very boundaries of our knowledge, and that his mind dwells in the "boundless contiguity of shade" by which that knowledge is surrounded.[1]

APPENDICES

THESE APPENDICES are not intended to be self-contained but should be read with the indicated relevant sections of the text—in particular Chapters 2, 3, and 4. There the physical ideas which underlie the mathematical concepts elaborated here are introduced. Where possible, examples have been taken from the body of the text to illustrate the various techniques.

1

NUMBERS, ALGEBRA

The first mathematics to which we are exposed is that of the ordinary numbers. We are drilled early and thoroughly in their use ($3 \times 4 = 12$, $3 + 6 = 9$, . . .), with such emphasis on the mechanics that the structure the numbers form is often lost. The symbol 3, among other things, designates something that is common to all collections of three objects. It could as well be written III (as was done by the Romans) or

$$\text{ΤΤΤ}$$

(as was done by the Babylonians);

$$3 + 6 = 9$$

just as

$$\text{III} + \text{VI} = \text{IX}$$

or for that matter as

$$∴ \; + \; ∵ = ∷$$

The early Egyptians used hieroglyphic symbols to represent powers of 10. A vertical staff | represented 1; a heel ∩ , 10; a scroll 𝟡 , 100; a lotus 𝟙 , 1000; and so on. These were used in an additive system, so that the number we know as two thousand three hundred and thirty-nine was written

$$\text{𝟙𝟙 𝟡𝟡𝟡 ∩∩∩}|||$$

and was read as

$$𝟙 + 𝟙 + 𝟡 + 𝟡 + 𝟡 + ∩ + ∩ + ∩ + | + | + | + | + | + | + | + | + |$$

In Old Babylonian cuneiform (1800–1600 B.C.) the numbers 1 through 10 were written

$$\text{Τ ΤΤ ΤΤΤ 𝍩 𝍪 𝍫 𝍬 𝍭 𝍮 ⟨}$$

The signs for 20, 30, 40, and 50 were

$$\text{⟨⟨ ⟨⟨⟨ 𝍲 𝍳}$$

while 60 was written Γ . It was by its position in the number group that Γ as 60 was distinguished from Γ as 1. Thus, ⟨Γ was 10 + 1, or 11, and K was 60 + 10, or 70. The Babylonian system made it possible to express the powers of 60 with the same set of symbols, whereas the Egyptians had to use a different symbol for different powers of 10.

In the Hindu-Arabic system, a decimal system based on positional notation, and from which was developed the numbers we use today, the use of the zero eliminates ambiguities that occur in earlier systems. (It is used in the middle of numbers, as in 201; or after integers, as in 10, 100, 1000, . . . , to denote the correct power of 10; it also provides the answer to such subtractions as $2 - 2$ and $15 - 15$.) In this system there are 10 integers 0, 1, . . . , 9. These and their placement are used to construct the numbers. For example, 932 means $9 \times 100 + 3 \times 10 + 2 \times 1$. We are thus able, with a limited number of digits, to conveniently write a very large range of numbers.

Among the devices used to simplify the manipulation of large numbers, one of the most useful is the exponential. Multiplying a number like 100 by 1000 gives

$$100 \times 1000 = 100{,}000$$

One followed by *two* zeros multiplied by 1 followed by *three* zeros gives 1 followed by *five* zeros. Such multiplications can be simplified by introducing the idea of an exponent:

$$10^5 \text{ means } 10 \times 10 \times 10 \times 10 \times 10$$

$$10^2 \text{ means } 10 \times 10$$

$$10^{-3} \text{ means } \frac{1}{10 \times 10 \times 10} = \frac{1}{10^3}$$

Thus

$$100 \times 1000 = 100{,}000$$

can be written

$$10^2 \times 10^3 = 10^5$$

In general,

$$10^a \times 10^b = 10^{a+b}$$

or

$$10^c \times 10^{-d} = 10^c \times \frac{1}{10^d} = 10^{c-d} \qquad \text{etc.}$$

It is easy to verify that

$$10^5 \times 10^7 = 10^{12} = 1{,}000{,}000{,}000{,}000$$

or that

$$10^6 \times 10^{-2} = 10^6/10^2 = 10^4 = 10{,}000$$

Since the diameter of the universe is said to be about

$$10{,}000{,}000{,}000{,}000{,}000{,}000{,}000{,}000{,}000{,}000{,}000{,}000{,}000$$

times the diameter of an electron, it is convenient to have available a quick way of writing such a large number. We would usually write 10^{40}. It is easier to write and handle, just as 27 is more convenient than 11111111111111111111111111111.

From the basic rule for manipulation of exponentials any exponential may be defined. For example, $10^{2/3}$ is that number such that

$$10^{2/3} \cdot 10^{2/3} \cdot 10^{2/3} = 10^{2/3 + 2/3 + 2/3} = 10^{6/3} = 10^2 = 100$$

Thus $10^{2/3}$ is the cube root of 100. The number $10^{1/2}$ is that number such that

$$10^{1/2} \cdot 10^{1/2} = 10^{1/2 + 1/2} = 10^1 = 10$$

or

$$10^{1/2} = \sqrt{10}$$

In general, $10^{a/b}$ is that number which when multiplied by itself b times gives 10^a

$$10^{a/b} \cdot 10^{a/b} \cdots 10^{a/b} = 10^{a/b + a/b + \cdots + a/b} = 10^{(a/b)b} = 10^a$$

Thus it could be called the bth root of 10^a.

The various properties of the numbers we know almost by instinct:

$$3 + 4 = 4 + 3 = 7$$
$$3 \cdot 4 = 4 \cdot 3 = 12$$
$$2(5 + 3) = 2 \cdot 5 + 2 \cdot 3 = 10 + 6 = 16 \qquad \text{etc.}$$

Expressed more generally:

$$a + b = b + a$$
$$a \cdot b = b \cdot a$$
$$a(b + c) = a \cdot b + a \cdot c \qquad \text{etc.}$$

We can regard the numbers as forming one special system that possesses these properties, or we can regard these properties as an abstraction of the properties possessed by the positive whole numbers—the road by which they were arrived at historically.*

Having abstracted the rules from the behavior of the whole numbers, we find that objects other than the whole numbers also satisfy them. Further, these objects are extremely convenient to join to the whole numbers, as otherwise many permissible operations among the whole numbers take us out of the domain of the whole numbers. Thus, zero, fractions, and negative numbers are introduced.

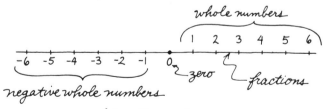

* Because the rules for the manipulation of algebraic symbols and for the manipulation of numbers are identical, we can remind ourselves of the rules by displaying them with numbers:
$$(a - b)(a + b) = a^2 - b^2$$
$$(7 - 6)(7 + 6) = 49 - 36 = 13$$

Fractions make permissible such operations as 1 divided by 3, zero an operation such as $2 - 2$, and negative numbers an operation such as $3 - 4$.

The symbols of algebra, a, b, . . . , x, y, . . . , because they are not explicit numbers, can be used to express the fact that some statements are true of all numbers—not just some particular ones. For example:

$$3 \cdot 4 = 4 \cdot 3$$

is true, as well, for

$$7 \cdot 6 = 6 \cdot 7$$

or, in general,

$$a \cdot b = b \cdot a$$

(the order in which numbers are multiplied does not change the result). However, if we write $3x = 12$, this statement will be true only for certain numbers (that is, $x = 4$).

Perhaps the single most important algebraic relation is that of equality. It satisfies the axioms as stated by Euclid, from which follow most of the manipulations of algebra: equals added to equals give equals, things which are equal to the same thing are also equal to each other, and so on. For Aristotle an axiom was equivalent to a "common saying," that is, an agreement as to what the language was to mean. The axioms satisfied by the relation of equality are neither necessary nor self-evident. They make explicit what the relation means. The answer to the question: Why do equals added to equals give equals? is not that this is self-evident, but that it is part of the meaning of the symbol $=$, that if $2 \times 3 = 6$,

$$2 \times 3 + 7 = 6 + 7$$

It is not true in general. [For example, * added to * gives * is not a true statement for arbitrary substitutions for * (consider "odd number" for each *).]

The various techniques commonly grouped together as algebra often involve the manipulation of algebraic expressions to convert them from one form to another. They follow from the fact that certain relations are maintained under certain operations. (For example, expressions equal to one another remain equal to one another when multiplied by, say, 7.) The reason for the manipulation is to reduce expressions to standard or more recognizable forms or to make explicit relations that are obvious in one form and not in another. To add two fractions, for example

$$a/b + c/d$$

one often rearranges them so that they have a common denominator. This can be done by multiplying the first by

$$d/d = 1$$

and the second by

$$b/b = 1$$

The result is

$$\frac{a}{b} \cdot \frac{d}{d} + \frac{c}{d} \cdot \frac{b}{b} = \frac{ad + cb}{bd}$$

$$\left[\text{i.e., } \frac{2}{7} + \frac{3}{4} = \frac{2}{7} \cdot \frac{4}{4} + \frac{3}{4} \cdot \frac{7}{7} = \frac{8 + 21}{28} = \frac{29}{28} \right]$$

We now analyze one series of operations that occurs in Chapter 23 to illustrate some of the principal techniques.[*] There we obtain

$$t_{\text{forward}} = \frac{\ell}{c - v}$$

$$t_{\text{return}} = \frac{\ell}{c + v}$$

and we defined T as the sum

$$T = t_{\text{forward}} + t_{\text{return}} = \frac{\ell}{c - v} + \frac{\ell}{c + v}$$

There is no objection to leaving T in this form, but the relation of T to v becomes somewhat more transparent and, further, the expression takes a standard form (and so is more readily comparable with others) if we rewrite it as below. First,

$$\frac{\ell}{c - v} \cdot 1 = \frac{\ell}{c - v} \frac{c + v}{c + v} \qquad \text{as } \frac{c + v}{c + v} = 1$$

and equals multiplied by equals give equals.

$$(c - v)(c + v) = c(c + v) - v(c + v)$$

$$= c^2 + cv - vc - v^2 = c^2 - v^2$$

as, in general, $a(b + c) = ab + ac$; for example, $3(7 + 2) = 3 \cdot 7 + 3 \cdot 2 = 27$. Further, $cv - vc = 0$. Thus

$$\frac{\ell}{c - v} = \frac{\ell c + \ell v}{c^2 - v^2}$$

We can also write

$$\frac{\ell}{c + v} = \frac{\ell}{c + v} \frac{c - v}{c - v}$$

and since

$$(c + v)(c - v) = c^2 - v^2$$

we obtain

$$\frac{\ell}{c + v} = \frac{\ell c - \ell v}{c^2 - v^2}$$

[*] We are not concerned, of course, with the meaning of the expression here but only with the manner in which it can be manipulated.

Now

$$T = t_{\text{forward}} + t_{\text{return}} = \frac{\ell c + \ell v}{c^2 - v^2} + \frac{\ell c - \ell v}{c^2 - v^2}$$

The two expressions contain the same denominator (common denominator) and can be added simply. We then get

$$T = \frac{\ell c + \ell v + \ell c - \ell v}{c^2 - v^2} = \frac{2\ell c}{c^2 - v^2}$$

which is one convenient form. If we divide numerator and denominator by c^2 (which is equivalent to dividing by 1), we obtain

$$T = \frac{2\ell c/c^2}{(c^2 - v^2)/c^2} = \frac{2\ell}{c} \frac{1}{(1 - v^2/c^2)}$$

another standard and convenient form.

2

FUNCTIONS

We say that a function $y = f(x)$ [$y = f(x)$ is read y is a function of x] is defined if there is a rule which associates a definite y with each definite x. There is no limit to such functions. Imagine a somewhat highly tuned Ferrari which passes a point on a highway at a time we call zero. If we mark the distance it moves every second or every few seconds, we would obtain two columns of numbers as, for example, below—one for time, the other for distance.

Time, sec	Distance, ft
0	0
1	200
2	400
3	600
.	
.	
.	

We could say that the distance is a function of the time. Notice how arbitrary this is: If the car were going faster, the distance would be a different function of the time; if it were standing still, it would be still another function of the time. In fact, in the absence of any other information, the distance is a completely arbitrary function of the time. As mathematicians, nothing could concern us less. We are not troubled to say that this is one of a myriad of situations in which one can establish myriad of situations in which one can establish a functional relationship.

The fact that this is a physical relation (distance, time), however, implies that there are some functions that will not occur. We know, for example, that even with the latest gasolines the car will not move 20 miles in a second. But this restriction is a statement about the physical world (which, as physicists, is just our interest).

In the table above, we have displayed an explicit illustration of a function. One can sometimes do it this way. Another method, which has the advantage that pictures have over words, is to mark the time out horizontally, the distance vertically, and represent the function by a line; this is known as a *graph*.

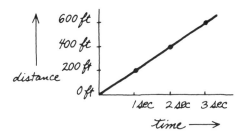

Properly, our observations are only the points that correspond to those pairs of numbers (time, distance) taken from the table above. But we are inclined to connect them with the line, making the assumption that between two and three seconds the distance was between 400 and 600 ft. Of course, this does not have to be so. However, we feel that if we record enough points, we can finally connect them with a continuous curve—a straight line in this case. Again it does not have to be so, but the language we use inclines us to believe it. Later we may be in trouble because of it.

We notice also that, having drawn a line through the observed points, we have related an infinite number of times with an infinite number of distances. This could hardly be tabulated. What is required is some rule so that for any stated time the distance traveled can be determined. The rule for this particular case is

distance in feet is equal to 200 multiplied by the time in seconds

or, in the symbolic notation that says so much in so small a space,

$$d = 200t$$

where the units feet and seconds are implied but not put in explicitly. This is called a *linear equation,* because its picture (or graph) is a straight line.

The most general equation whose graph is a straight line is

$$d(t) = at + b$$

which means: distance as a function of time is given by a multiplied by the time plus b, where a and b are constant numbers. In the example above, $a = 200$ and $b = 0$.

To associate a clear meaning for a and b, we draw a picture of the function.

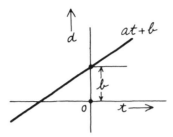

When the time t is equal to zero, we obtain

$$d = a \cdot 0 + b = b$$

Thus b is the point at which the curve crosses the distance line (axis), that is, b is the distance at time $t = 0$; we might therefore call it d_0, so that it more readily gives up its message. The only other undetermined quantity is the angle the line makes with the time line (axis). This is related to the number a, which is equal to the speed (the magnitude of the velocity); a therefore may be called v.

The equation would then conventionally be written

$$d = vt + d_0$$

Its full meaning is: The distance traveled is a linear function of the time, the constant speed is v, and the distance at $t = 0$ is d_0.

Clearly this is a very restricted type of function. There are many nonlinear functions (corresponding to all the curves one might draw). The same Ferrari accelerating from rest easily achieves a uniform acceleration of 10 ft/sec² which results in a relation between distance and time as below:

$$d = \tfrac{1}{2}at^2 = 5t^2$$

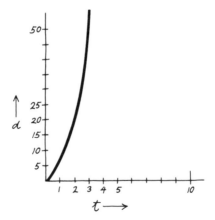

We would now say that the distance was a quadratic function of the time (a function of the time squared) and the resulting curve is a parabola.

Much of what is called high school algebra is the study of such linear and quadratic equations; they are technically simple to manipulate and one can play endless games finding the roots of quadratic equations.[*] However, this almost always obscures the significant fact that they are only two of an infinite class of functions and their only special interest is their simplicity.

3

SOLUTIONS OF EQUATIONS

[To be read after Chapter 2]

It was Descartes who introduced the convention of using letters at the end of the alphabet (x, y, or z) to represent unknowns and letters at the beginning (a, b, or c) to represent constants. We are, for example, asked to solve the equation

$$3x = 6$$

which has the meaning: 3 multiplied by some number x is equal to 6. Thus

$$x = \frac{6}{3} = 2$$

or

$$ax = b$$

which has the solution

$$x = \frac{b}{a}$$

Such algebraic equations may be regarded as conditions on the "unknowns" and there may be one or several possible solutions (that is, values of the unknowns which satisfy the conditions).

We often speak of finding the solutions for one or several equations. The principle is not entirely different from ordinary usage. If we want an enclosure that keeps out the rain, either the Palais de Versailles or a cave will do. If we want heating in addition, the Palais de Versailles is eliminated. If we want to construct a house in which four people can live, which can be heated, is attractive, and so on, there are many pos-

---

[*] The general quadratic function can be written
$$d = at^2 + bt + c$$
where a, b, and c are constants; an unpleasant memory recalls to us that when $d = 0$,
$$t = \frac{-b \pm \sqrt{b^2 - 4ac}}{2a}$$

sibilities. If we add the requirement that the house not cost too much, we may arrive at a situation where no "solution" exists. What this means is that there is no object in this world that meets all the requirements desired.

When we write

$$3x - y = 0$$

we ask for all those pairs of numbers (x, y) such that the relation $3x - y = 0$ is satisfied (that is, $x = 1$, $y = 3$; $x = 2$, $y = 6, \ldots$). If we make a graph of all such number pairs, we find a straight line (light line in the figure), thus a linear relation.

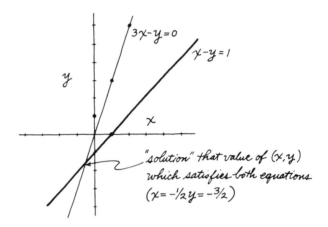

When we write

$$x - y = 1$$

we ask for those values of x and y such that the relation $x - y = 1$ is satisfied (that is, $x = 1$, $y = 0$; $x = 2$, $y = 1, \ldots$). The graph of all such number pairs again gives a straight line (heavy line in the figure). That number pair which satisfies both of these equations $(x = -\frac{1}{2}, y = -\frac{3}{2})$ occurs at the intersection of the two lines.

It might happen that two equations have no "solutions":

$$y = x$$

$$y = x + 1$$

which has the meaning that there is no value of (x, y) which satisfies both; the two lines are parallel. Or the two equations might have many solutions:

$$y = x$$

$$2y = 2x$$

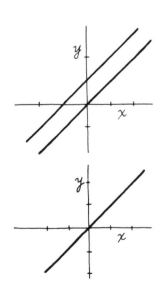

[All values of (x, y) that satisfy the first satisfy the second.] In this case the lines are the same. [That either one or all values of (x, y) are solutions corresponds to the Euclidian postulate that between two points only one straight line can be drawn. When Descartes associated the points of geometry with the number pairs (x, y) and the lines with the linear equations $y = mx + b$ (where m and b are constant numbers), he created what is known as analytic geometry. To every proposition of geometry he could associate an algebraic proposition.]

The "solution" of an equation often requires that we "invert" the relation. Imagine that we are given distance as a function of time

$$d = 16t^2$$

and are asked to find the time required for a body to fall 64 ft. The solution follows: when $d = 64$, $t^2 = 4$, and $t = 2$. In general it is required that we find t as a function of d. In this case this can be done by standard algebraic manipulation:

$$16t^2 = d$$

$$t^2 = \frac{1}{16} d$$

Therefore,

$$t = \frac{1}{4} \sqrt{d}$$

In general, though the inversion might not be simple at all, the principle is always the same. It is perhaps easiest to visualize the process graphically. All the pairs of numbers (t, d) that satisfy $d = 16t^2$ produce the curve shown (a parabola). Inversion is then no more than the process

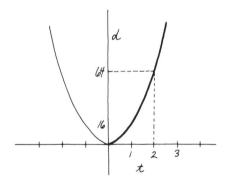

of turning the graph on its side. From one point of view we are given t and ask what is d; from the other we are given d and ask what is t.

There is a subtle point. The parabola has two halves. For each value of d there are two values of t that satisfy the relation $d = 16t^2$, because

$$t = \pm \frac{1}{4} \sqrt{d}$$

Thus the relation $t \rightarrow d$ is single-valued [that is, given t there is only one d such that (t, d) satisfy the equation]. However, given d larger than zero, there are two ts that satisfy the equation.

What this means is usually evident. The problem might be defined only for t larger than zero (a cannonball released from the mast of Galileo's ship 64 ft above the deck at time $t = 0$). Therefore, the "physical region" would be those times for which t was larger than zero. However the extension to negative times might be assigned a meaning. Imagine, for example, that the ball was fired upward from the deck at $t = -2$ sec with just the right speed. Then at time $t = 0$ it would be at the top of the mast and would begin to fall. Thus it would be a distance of say 16 ft from the top of the mast at

$$t = \pm \frac{\sqrt{16}}{4} = \pm 1 \text{ sec}$$

When $t = -1$ it would be on its way up; when $t = 1$ it would be on its way down.

Imagine that Galileo's ship (p. 19) is moving with a speed of 10 ft/sec. From the point of view of the sailor on the dock, how far does the ball move horizontally before it strikes the deck? The ball falls vertically according to

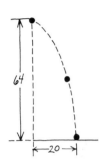

$$d_v = \tfrac{1}{2}gt^2 = 16t^2$$

and moves horizontally according to

$$d_h = 10t$$

It reaches the deck in 2 sec, as

$$64 = 16(2)^2$$

and in this time moves horizontally

$$d_h = 10 \cdot 2 = 20$$

Thus it moves 20 ft horizontally before striking the deck.

As another example, consider the accelerated Ferrari trying to catch a Volkswagen, moving uniformly with the speed of 50 ft/sec, which passed the Ferrari's point of departure at $t = -2.4$ seconds. For the Volkswagen

$$d = 50t + 120$$

while for the Ferrari

$$d = 5t^2$$

The Ferrari overtakes the Volkswagen at that value of (t, d) which is a solution of both equations of motion—that value of (t, d) for which both the equations above are satisfied. At this value of (t, d)

$$5t^2 = 50t + 120$$

or

$$t^2 - 10t - 24 = 0$$

the famous quadratic equation whose solutions are given by

$$t = \frac{10 \pm \sqrt{100 + 96}}{2} = \frac{10 \pm 14}{2} = +12 \text{ or } -2$$

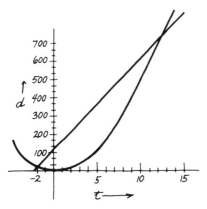

Thus the Ferrari overtakes the Volkswagen at $t = 12$ sec. But what is the meaning of the second solution $t = -2$? This has no "physical" meaning if the Ferrari starts from rest at $t = 0$. However it could be interpreted as follows. Imagine the Ferrari moving in the opposite lane toward the Volkswagen. When the Volkswagen is sighted on the horizon the Ferrari begins to slow down uniformly so that

$$d = 5t^2$$

It then passes the Volkswagen at $t = -2$, stops at $t = 0$, reverses its direction, starts off in pursuit, and overtakes at $t = 12$. If the left branch of the parabola is associated with a physical phenomenon, the intersection at $t = -2$ can thus be interpreted.

4

RATES OF CHANGE, LIMITS

Those who follow the stock market are already familiar with changing functions and their graphs. We often see profits as a function of the month, the Dow-Jones average as a function of the day, or the gross national product as a function of the year. Such functions unfortunately are known only for those points in time that are already past; any extension into the future has a certain risk attached which the investor would be willing to corroborate.

The behavior of the stock market might be called its dynamics. If anyone has unraveled its inner workings he has not revealed it, though there never seems to be a shortage of explanations after the events. The dynamics of the motion of planets and atoms have been unraveled and, because of the generosity and openness of science (possibly because of its small profitability), are available to all. Thus we can, for example, write the position of a planet as a function of time and give the form of that function for all times—past and future. Anyone interested in the position of Venus at 11:31 P.M., April 8, 2066, can with a few telephone calls actually find out where it will be.

A question of great interest both to the mathematician and the investor is how fast and in which direction is a function changing. In the case of a company's capital, the increase or decrease by the year is often as much information as we need. However, there are situations in which we are interested in the change in equity on a day-to-day basis (if we were near bankruptcy for example). The average rate of change of a function in time is obtained by taking its value at one time, subtracting it from that at another, and dividing by the time interval:

$$\text{average rate of change of function in time} = \frac{\text{change in function}}{\text{time interval}}$$

This is accomplished as follows. We consider a function of time $f(t)$. At the time $t_0 + \Delta t$, the function has the value $f(t_0 + \Delta t)$. At the time t_0 the function has the value $f(t_0)$. In the time interval Δt (which is supposed to be small—as small as we wish but *always larger than zero*), the change in the function (which we write Δf, meaning: the change of f) is

$$\Delta f = f(t_0 + \Delta t) - f(t_0)$$

and the average rate of change is defined as

$$\text{average rate of change} = \frac{\Delta f}{\Delta t} = \frac{f(t_0 + \Delta t) - f(t_0)}{\Delta t}$$

Δf and Δt are conventional ways of writing change in f or change in t; the average is just the ratio of the change in f for a given change in t. [When the change in t becomes infinitesimally small, Δt is conventionally written dt and the corresponding change in f, df. What is known as the derivative or the "instantaneous" rate of change of a function is just the above average rate of change in the limit as Δt becomes very small (never zero but as close to zero as we wish).]

What is being done can be pictured as shown.

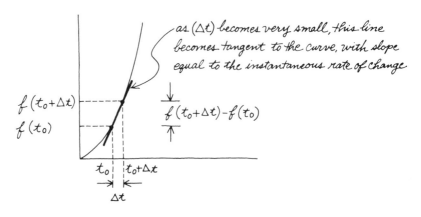

Consider the function $f(t) = 16t^2$. Suppose we take $t_0 = 1$ and $t_0 + \Delta t = \sqrt{2}$. Then $f(t_0) = 16$
and

$$f(t_0 + \Delta t) = f(\sqrt{2}) = 16(\sqrt{2})^2 = 16 \cdot 2 = 32$$

so that

$$\Delta f = f(\sqrt{2}) - f(1) = 32 - 16 = 16$$

and
$$\Delta t = \sqrt{2} - 1 \simeq 1 \cdot 41 - 1 = 0.41$$

Therefore,

$$\frac{\Delta f}{\Delta t} = \frac{16}{0.41} \simeq 39$$

If we interpret $f(t)$ as the distance fallen by a cannonball released from rest from the top of a mast,

$$d = f(t) \doteq 16t^2$$

Then

$$\frac{\Delta f}{\Delta t}$$

is the average speed over the time interval Δt. Thus, for the time interval between $t_0 + \Delta t = \sqrt{2}$ and $t_0 = 1$ the average speed is about 39 ft/sec.

We illustrate the method for obtaining the value of the instantaneous speed with one example. Consider again

$$f(t) = 16t^2$$

Then

$$f(t + \Delta t) = 16(t + \Delta t)^2 = 16t^2 + 32t(\Delta t) + 16(\Delta t)^2$$

The ratio becomes

$$\frac{f(t + \Delta t) - f(t)}{\Delta t} = \frac{\overbrace{16t^2 + 32t(\Delta t) + 16(\Delta t)^2}^{f(t + \Delta t)} - \overbrace{16t^2}^{f(t)}}{\Delta t} = 32t + 16(\Delta t)$$

The essence of the argument is now to say that as Δt becomes as small as we wish (but always larger than zero), the second term approaches zero and can be discarded.[*]

We therefore have the result that the limit of $\Delta f/\Delta t$ as Δt becomes as small as we wish (instantaneous speed) $= 32t$ or $v = at$, the definition of uniformly accelerated motion. At $t = 1$ the instantaneous speed is

$$v = 32 \text{ ft/sec}$$

At $t = \sqrt{2}$ the instantaneous speed is

$$v \simeq 45.3 \text{ ft/sec}$$

Both the change of the function

$$\Delta f = f(t_0 + \Delta t) - f(t_0)$$

and the interval Δt are thought to become very small. But in no case is Δt allowed to become zero. Therefore, Δt can always be treated as a

---

[*] Whether this is exactly true, whether such terms can accumulate to distort the result, were questions that troubled Newton's contemporaries. They were not resolved until the nineteenth century, when it was shown that there is a sense in which the argument is exact.

number unequal to zero, and we can divide and multiply by Δt as we can by any nonzero number. To emphasize this we have always written

$$\frac{\Delta f}{\Delta t}$$

If

$$\frac{\Delta f}{\Delta t} = 32t$$

then this means in the small time interval Δt the change of the function Δf is given by

$$\Delta f = 32t(\Delta t)$$

(which is exact for very small time intervals).

The idea of the limit as Δt goes to zero is subtle and very important. The point is roughly as follows. Consider Zeno's analysis of Achilles' pursuit of the tortoise. Imagine for simplicity the tortoise is at rest; further, let Achilles run at 15 miles/h = 22 ft/sec and start 22 ft behind the tortoise.

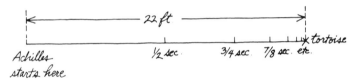

Achilles of course reaches the tortoise in 1 sec. However, we follow Zeno. In the first half-second Achilles goes 11 ft; in the next $\frac14$ second he goes $5\frac12$ feet; in the next $\frac18$ second he goes $2\frac34$ feet; and so on. Dividing the time interval in half at each step, Achilles advances only one half of the remaining distance to the tortoise—so, concludes Zeno (is this really his conclusion?), Achilles never reaches it.

What makes the paradox is that the time interval becomes so small that an infinite number of these intervals add up to a finite time:

$$\frac12 + \frac14 + \frac18 + \frac1{16} + \cdots = 1 \text{ sec}$$

At the same time the distance traveled in these intervals also becomes so small that an infinite number add up to a finite interval:

$$11 + 5\frac12 + 2\frac34 + \cdots = 22 \text{ ft}$$

Thus to follow the time intervals or the distance intervals by themselves is confusing—the heart of the paradox. However, the ratio is very simple. In the first interval Achilles runs 11 ft in $\frac12$ sec, or

$$\frac{11 \text{ ft}}{\frac12 \text{ sec}} = 22 \text{ ft/sec}$$

In the second interval he runs $5\frac12$ ft in $\frac14$ sec, or

$$\frac{5\frac12 \text{ ft}}{\frac14 \text{ sec}} = 22 \text{ ft/sec} \qquad \text{etc.}$$

The ratio $\Delta d/\Delta t$ is just the speed with which he runs: 22 ft/sec.

We encounter many occasions on which it is difficult to follow two quantities that become very large or very small.* However, the product or the ratio of the quantities remains under control.

5

GEOMETRY

After the numbers, geometry is the most familiar mathematical structure. We are taught very young to accept the convention that our space is Euclidian. We do not say what straight lines or points are, but recognize and indicate them. We hope the objects we designate as straight lines will remain straight as they are moved about in space (whether or not massive objects are near by) or that a triangle will remain the same triangle when its position is altered—in short, that the objects we designate as lines, points, and so on, will satisfy Euclid's postulates. If this is so, then all Euclid's theorems will hold among these objects, and the entire structure of geometry follows.

We here present a brief review of some of the more useful definitions and theorems of Euclidian geometry. Two triangles are said to be congruent if each of their sides and each of their angles is equal. A number of theorems are designed to prove that two triangles will be congruent if they are the same in some more limited aspects (for example, if each of their sides is equal).

An *equilateral* triangle is one all of whose sides are equal.

An *isosceles* triangle is one two of whose sides are equal.

A *right* triangle is one that contains a right angle.

For right triangles one has the famous theorem of Pythagoras:

$$a^2 + b^2 = c^2$$

Two triangles are said to be *similar* if the angles of one are equal to the angles of the other.

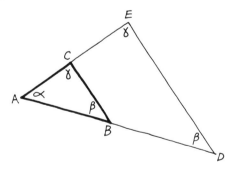

The triangle *ABC* is similar to the triangle *ADE*.

---

* Another example is the force as a function of time in a collision. There the force grows large over a small time interval. However, the product $F\Delta t$ remains under control.

A circle is defined as that curve each of whose points is equally distant from a center.

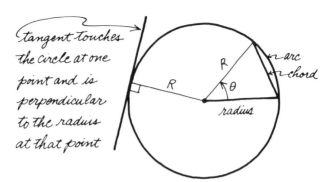

The circumference of a circle is related to its radius (a problem that dates from antiquity) via the famous number π:

$$\text{circumference} = 2\pi R$$

$$\pi = 3.141 \ldots$$

$$\text{area of circle} = \pi R^2$$

The angle between lines that meet at a point (as the angle θ between the two radii above) can be defined as

$$\text{angle (in radians)} = 2\pi \, \frac{\text{arc subtended}}{\text{circumference}}$$

The angle measured this way is given what are called *radians*. The relation between radians and degrees can be easily obtained. A *straight angle* subtends 180°, or one half of the circumference of a circle. Therefore,

$$\text{angle (radians)} = 2\pi \, \frac{\pi R}{2\pi R} \qquad \text{or} \qquad \pi \text{ radians} = 180°$$

and

$$1 \text{ radian} = \frac{180°}{\pi} \simeq 57.3°$$

This particular measure for angles is convenient because then

$$(\text{angle in radians}) \times (\text{radius}) = \text{arc subtended}$$

The special functions sine, cosine, and tangent can be defined by referring to the sides of the right triangle formed as shown from the lines defining the angle θ:

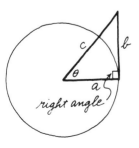

$$\text{sine } \theta \text{ (written } \sin \theta) = \frac{\text{opposite}}{\text{hypotenuse}} = \frac{b}{c}$$

$$\text{cosine } \theta \text{ (written } \cos \theta) = \frac{\text{adjacent}}{\text{hypotenuse}} = \frac{a}{c}$$

$$\text{tangent } \theta \text{ (written } \tan \theta) = \frac{\text{opposite}}{\text{adjacent}} = \frac{b}{a}$$

From the definition we see immediately, for example, that for

$$\theta = 0 \qquad \sin\theta = \frac{b}{c} = 0$$

$$\theta = \frac{\pi}{2} \qquad (90°,\ \tfrac{1}{4}\text{ of the circle}) \qquad \sin\theta = \frac{b}{c} = 1$$

$$\theta = \pi \qquad (180°,\ \tfrac{1}{2}\text{ of the circle}) \qquad \sin\theta = \frac{b}{c} = 0 \qquad \text{etc.}$$

In a similar manner, for

$$\theta = 0 \qquad \cos\theta = \frac{a}{c} = 1$$

$$\theta = \frac{\pi}{2} \qquad \cos\theta = \frac{a}{c} = 0$$

$$\theta = \pi \qquad \cos\theta = \frac{a}{c} = -1 \qquad \text{etc.}$$

When the circle has been completed, $\theta = 2\pi$, we begin again. These yield the periodic* curves:

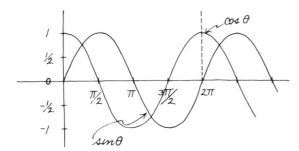

From Pythagoras' theorem various relations between these functions follow immediately. For example,

$$\sin^2\theta + \cos^2\theta = 1$$

Proof:

$$\left.\begin{array}{l} \sin\theta = \dfrac{b}{c} \\[2ex] \cos\theta = \dfrac{a}{c} \end{array}\right\} \ (\text{definition})$$

$$\sin^2\theta + \cos^2\theta = \frac{b^2}{c^2} + \frac{a^2}{c^2} = \frac{b^2 + a^2}{c^2}$$

but

$$b^2 + a^2 = c^2 \qquad (\text{Pythagoras' theorem})$$

Therefore,

$$\sin^2\theta + \cos^2\theta = \frac{c^2}{c^2} = 1 \qquad\qquad \text{Q.E.D.}$$

* Curves which repeat themselves. In this case $\sin(2\pi + \theta) = \sin\theta,\ \cos(2\pi + \theta) = \cos\theta.$

The various standing waves in the theory of light or quantum theory can be written as sine or cosine functions. For example, the standing wave with 2 nodes (Fig. 31.1) on the line ℓ is

$$\sin \frac{2\pi}{\ell} x$$

when

$$x = 0 \qquad \sin \frac{2\pi}{\ell} 0 = \sin 0 = 0$$

$$x = \frac{\ell}{4} \qquad \sin \frac{2\pi}{\ell} \frac{\ell}{4} = \sin \frac{\pi}{2} = 1$$

$$x = \frac{\ell}{2} \qquad \sin \frac{2\pi}{\ell} \frac{\ell}{2} = \sin \pi = 0$$

$$x = \frac{3\ell}{4} \qquad \sin \frac{2\pi}{\ell} \frac{3\ell}{4} = \sin \frac{3\pi}{2} = -1$$

$$x = \ell \qquad \sin \frac{2\pi}{\ell} \ell = \sin 2\pi = 0$$

6

VECTORS

[To be read after Chapter 3]

It is often convenient to "resolve" a vector into two perpendicular components. From the basic rule for vector addition a vector can always be made equal to the sum of two others.

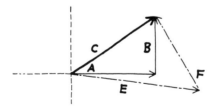

There is, of course, no necessity that the two vectors into which the original is "resolved" be perpendicular. The vector **C** above can be resolved into two perpendicular vectors:

$$\mathbf{C} = \mathbf{A} + \mathbf{B}$$

or, just as well, two that are not perpendicular:

$$\mathbf{C} = \mathbf{E} + \mathbf{F}$$

very much as

$$7 = 5 + 2$$

or

$$7 = 3 + 4$$

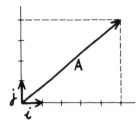

However, there is a particular convenience if the vectors into which the original is resolved are perpendicular (or orthogonal). In a two-dimensional space (a plane, for example) all vectors can be written as a weighted sum (linear combination) of two perpendicular vectors of unit length (denoted conventionally by **i** and **j**). For example, the vector **A** here can be written

$$\mathbf{A} = 5\mathbf{i} + 4\mathbf{j}$$

Any vector in the plane can be written

$$\mathbf{V} = x\mathbf{i} + y\mathbf{j}$$

where x and y are the (x, y) components.

Written this way, the addition or subtraction of any number of vectors becomes particularly convenient. For example, if

$$\mathbf{A} = \quad 5\mathbf{i} + 4\mathbf{j}$$

and

$$\mathbf{B} = -2\mathbf{i} + 3\mathbf{j}$$
$$\overline{\mathbf{A} + \mathbf{B} = \quad 3\mathbf{i} + 7\mathbf{j}}$$

and

If the sum of any number of vectors is required to be zero, then the sum of their **i** and **j** components must both be zero.

SCALAR PRODUCT

[To be read after Chapter 7]

In the definition of such quantities as work, we have used the idea of the product of the component of a vector in a certain direction ($F_{\parallel}$) multiplied by the distance moved in that direction. Both the force and the distance moved can be associated with vectors, whereas the final quantity of interest—the work—is a scalar (a number: magnitude but no direction).

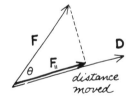

Because this type of operation occurs quite frequently, it is convenient to define a type of product called the *scalar,* or *inner product,* of two vectors. For two vectors **F** and **D** the inner product is defined as

$$\text{scalar or inner product:} \quad \mathbf{F} \cdot \mathbf{D} = F_{\parallel} D \quad \text{(definition)}$$

Since

$$F_{\parallel} = F \,\frac{F_{\parallel}}{F} = F\,\frac{\text{adjacent side}}{\text{hypotenuse}} = F \cos \theta$$

the scalar product can also be written

$$\mathbf{F} \cdot \mathbf{D} = FD \cos \theta$$

From this and the properties of cos θ,

$$\cos \theta = 1 \qquad \theta = 0$$
$$\cos \theta = 0 \qquad \theta = \frac{\pi}{2}$$
$$\cos \theta = -1 \qquad \theta = \pi$$

we obtain in a compact form the results

$$\mathbf{F} \cdot \mathbf{D} = \begin{cases} FD & \text{if parallel} \\ 0 & \text{if perpendicular} \\ -FD & \text{if antiparallel} \end{cases}$$

VECTOR PRODUCT

[To be read after Chapter 18]

Often two different vector quantities interact (for example, in electromagnetic theory where a velocity and a field combine) to yield a third vector (in this case a force):

$$F = q\mathbf{E} + \frac{q}{c} \, \mathbf{v} \times \mathbf{B}$$

The vector (cross or outer) product of **v** and **B** yields a new vector. Its magnitude is defined by

$$\mathbf{v} \times \mathbf{B} = vB \sin \theta \qquad \text{(magnitude)}$$

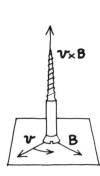

Its direction is perpendicular to the plane defined by **v** and **B**. It remains to define the sense of the new vector. Does it go up or down? The sense of **v** X **B** can be defined by what is known as the *right-hand rule*.

The vector **v** in the first position of the product is rotated through the smallest angle that will bring it into the direction of **B**. The sense of **v** X **B** is in the direction that a right-hand screw (the standard screw in the United States) would move if rotated in the same direction. One could just as well define the sense of the cross-product in the opposite way (left-hand rule, British screws). It is only required that we be consistent.

From the definition it follows immediately that

$$\mathbf{v} \times \mathbf{B} = \begin{cases} vB \ (\text{maximum magnitude}) & \text{if } \mathbf{v} \text{ is perpendicular to } \mathbf{B} \\ 0 & \text{if } \mathbf{v} \text{ is parallel to } \mathbf{B} \end{cases}$$

that

$$\mathbf{v} \times \mathbf{B} \text{ is perpendicular to both } \mathbf{v} \text{ and } \mathbf{B}$$

and that

$$\mathbf{v} \times \mathbf{B} = -\mathbf{B} \times \mathbf{v}$$

The concepts of angular momentum can be constructed using the vector product. We can write for the angular momentum of a particle about a point

$$\mathbf{L} = \mathbf{r} \times \mathbf{p}$$

If **r** is perpendicular to **p**, then

$$\mathbf{L} = \frac{rp \;(\text{magnitude})}{\text{direction and sense as shown}}$$

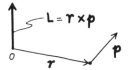

7

SOME COMMENTS ON UNITS

[To be read after Chapters 4 and 18]

Using centimeters ∼ length, grams ∼ mass, and seconds ∼ time as the primitive (undefined) elements—the elements we associate directly with objects or events in the physical world (things we can point to):

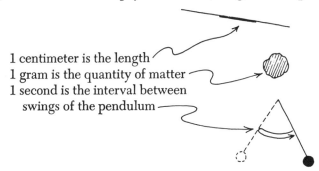

1 centimeter is the length
1 gram is the quantity of matter
1 second is the interval between
   swings of the pendulum

we can from these construct all the other quantities needed for all known physical theories. (Just as, for example, circles, triangles, and so on, can be constructed from the primitive objects, points, lines, . . . of geometry.)

Thus, for example,

$$\text{speed is defined as}\quad \frac{\text{change in distance}}{\text{time interval}} \sim \frac{\text{cm}}{\text{sec}}$$

$$\text{acceleration is defined as}\quad \frac{\text{change in speed}}{\text{time interval}} \sim \frac{\text{cm}}{\text{sec}^2}$$

Newton's second law

$$\mathbf{F} = m\mathbf{a}$$

can be used to define a force. Thus a 1-g mass that undergoes an acceleration of 1 cm/sec² is acted upon by a force—by definition 1 dyne: One dyne is that force which acting on an object whose mass is 1 gram produces an acceleration of 1 cm/sec². Therefore,

$$1 \text{ dyne} = 1 \text{ g} \times 1\frac{\text{cm}}{\text{sec}^2}$$

or
$$\text{dyne} \sim \frac{\text{g–cm}}{\text{sec}^2}$$

This has the meaning that we can always substitute g–cm/sec² for dyne, because if dyne appears, it means force, and by the second law $F = ma$.

All the other quantities constructed in mechanics can be measured in terms of the basic units of length, mass, and time:

$$\text{work} = (\text{force}) \times (\text{distance}) \sim \text{dyne-cm} \sim \frac{\text{g–cm}^2}{\text{sec}^2} \sim \text{ergs}$$

$$\text{kinetic energy} = \tfrac{1}{2}mv^2 \sim \frac{\text{g–cm}^2}{\text{sec}^2} \sim \text{ergs} \qquad \text{etc.}$$

Electrical quantities can also be constructed from the mechanical entities: length, time, and mass. In the CGS system, the charge is defined from Coulomb's law:

$$\mathbf{F} = \frac{q_1 q_2}{r^2} \qquad (\text{magnitude})$$

in such a way that 1 esu of charge on two objects 1 cm apart produces a force on each of 1 dyne.

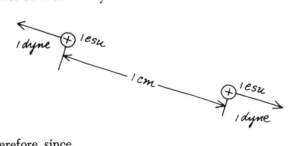

Therefore, since

$$1 \text{ dyne} = \frac{(1 \text{ esu})(1 \text{ esu})}{\text{cm}^2}$$

$$\text{dyne} \sim \frac{(\text{esu})^2}{\text{cm}^2}$$

or
$$\text{esu} \sim (\text{dyne–cm}^2)^{\frac{1}{2}} = (\text{g–cm})^{\frac{1}{2}}\frac{\text{cm}}{\text{sec}}$$

Current defined as

$$\text{current} = \frac{\text{charge}}{\text{time}}$$

in the CGS system is given in

$$\text{current} \sim \text{esu/sec} \qquad (\text{called statamperes})$$

Since, however,

$$\text{esu} \sim (\text{g--cm})^{\frac{1}{2}} \frac{\text{cm}}{\text{sec}}$$

the current can also be written

$$\text{current} \sim (\text{g--cm})^{\frac{1}{2}} \frac{\text{cm}}{\text{sec}^2}$$

In this way every electrical quantity can be expressed in units of length, time, and mass so that these need be the only primitive (undefined) elements of the theory.

For example, the electrical potential energy of a point charge q_1 a distance r from another q_2 is given by

$$V_{\text{electric}} = \frac{q_1 q_2}{r}$$

It thus has the units

$$V_{\text{electric}} \sim \frac{(\text{esu})^2}{\text{cm}}$$

or, since

$$\text{esu} \sim (\text{g--cm})^{\frac{1}{2}} \frac{\text{cm}}{\text{sec}}$$

$$V_{\text{electric}} \sim (\text{g--cm}) \frac{\text{cm}^2}{\text{sec}^2} \frac{1}{\text{cm}} \sim \frac{\text{g--cm}^2}{\text{sec}^2} \sim \text{erg}$$

Thus happily, the electric potential energy, as all other energies, is measured in ergs.

The electric potential due to the point charge q is

$$\phi = \frac{q}{r}$$

which then has the units

$$\phi \sim \frac{\text{esu}}{\text{cm}} \sim (\text{g--cm})^{\frac{1}{2}} \frac{1}{\text{sec}} \sim \frac{\text{erg}}{\text{esu}}$$

This is sometimes called the statvolt. From the relations above we obtain immediately such results as

$$(\text{charge}) \times (\text{electric potential}) \sim \text{energy}$$

$$(\text{length}) \times (\text{electric potential}) \sim \text{charge}$$

Quantity	Common symbol	CGS system	MKS "practical" system	Conversion between the two
primitive terms				
length	ℓ	centimeter	meter	1 meter = 100 centimeters
mass	m	gram	kilogram	1 kilogram = 1000 grams
time	t	second	second	1 second = 1 second
force	$\mathbf{F}$	dyne $\sim \dfrac{\text{gram–centimeter}}{\text{second}^2}$	newton	1 newton = 10^5 dynes
velocity	$\mathbf{v}$	centimeter/second	meter/second	
acceleration	$\mathbf{a}$	centimeter/second^2	meter/second^2	
momentum	$\mathbf{p}$	gram–centimeter/second	kilogram–meter/second	
work, energy	W, E	erg $\sim \dfrac{\text{gram-centimeter}^2}{\text{second}^2}$	joule	1 joule = 10^7 ergs
power $\left(\dfrac{\text{work}}{\text{time}}\right)$	P	erg/second	watt	1 watt = 10^7 erg/second
charge	q	electrostatic units (esu)	coulomb	1 coulomb = 2.998×10^9 esu
current	I	esu/second	ampere	1 ampere = 2.998×10^9 esu/second
electric potential	ϕ	statvolt $\sim \dfrac{\text{erg}}{\text{esu}}$	volt	1 volt = (1/299.8) statvolts
electric field	$\mathbf{E}$	$\dfrac{\text{statvolt}}{\text{centimeter}} \sim \dfrac{\text{dyne}}{\text{esu}}$	volt/meter	
resistance	R	second/centimeter	ohm	1 ohm = 1.139×10^{-12} second/centimeter
electric flux	Φ	esu	coulomb	
magnetic field	$\mathbf{B}$	gauss $\sim \dfrac{\text{dyne}}{\text{esu}}$	tesla	1 tesla = 10^4 gauss
magnetic flux	Φ	esu	coulomb	

SOURCE NOTES

CHAPTER 1

1. Aristotle, *On the Heavens,* W. K. C. Guthrie, trans., Loeb Classical Library, Harvard University Press, Cambridge, 1939.
2. Thomas S. Kuhn, *The Copernican Revolution,* Harvard University Press, Cambridge, 1957.
3. Aristotle, *Physics, The Works of Aristotle, Translated into English,* Vol. 2, W. D. Ross, ed., Oxford University Press, New York, 1930.
4. Lucretius, *On the Nature of Things,* Vol. 2, pp. 230–239, Cyril Bailey, ed. and trans., Oxford Classical Texts, New York.
5. René Descartes, *Discourse on the Method of Rightly Conducting the Reason and Seeking Truth in the Sciences,* John Veitch, trans., Open Court, La Salle, Ill., 1949, p. 34.
6. *Ibid.,* p. 35.
7. *Ibid.,* p. 36.
8. *Ibid.,* p. 46.

CHAPTER 2

1. Galileo Galilei, *Dialogues Concerning Two New Sciences,* Henry Crew and Alfonso de Salvio, trans., Macmillan, New York, 1914, p. 153.
2. *Ibid.,* pp. 153–154.
3. *Ibid.,* pp. 160, 162.
4. *Ibid.,* p. 162.
5. *Ibid.,* p. 178.
6. *Ibid.*
7. *Ibid.,* p. 179.
8. *Ibid.,* p. 250.

CHAPTER 4

1. Alexander Pope, "Epitaph Intended for Sir Isaac Newton."
2. *Sir Isaac Newton's Mathematical Principles of Natural Philosophy and His System of the World,* Andrew Motte, trans., University of California Press, Berkeley, 1962, Vol. 1, p. 1.
3. *Ibid.,* p. 1.
4. *Ibid.,* p. 2.
5. *Ibid.,* p. 13.
6. *Ibid.*
7. *Ibid.,* pp. 13–14.

8. *Ibid.,* p. 14.
9. *Ibid.,* pp. 21–22.
10. Christian Huygens, *Horologium Occillatorium.*
11. Isaac Newton, *Principia,* Book I, Proposition IV, Theorem IV.

CHAPTER 5

1. John Donne, "The First Anniversary," lines 251–258.
2. Copernicus, *De Revolutionibus.*
3. *Ibid.*
4. *Ibid.*
5. *Ibid.*
6. Joshua 10:12–13.
7. William Shakespeare, *Hamlet,* Act II, scene 2, lines 116–119.
8. John Donne, "The First Anniversary," lines 205–208.
9. Johannes Kepler, *De Harmonia Mundi.*
10. Thomas Browne, *Religio Medici,* Part II, Section IX.

CHAPTER 6

1. *Sir Isaac Newton's Mathematical Principles of Natural Philosophy and His System of the World,* Andrew Motte, trans., University of California Press, Berkeley, 1962, Vol. 1, pp. 3–4.
2. Isaac Newton, as quoted by Charles C. Gillispie, *The Edge of Objectivity,* Princeton University Press, Princeton, N.J., 1960, p. 119.
3. *Ibid.,* p. 119.
4. *Ibid.,* p. 137.
5. *Ibid.*

CHAPTER 7

1. *Sir Isaac Newton's Mathematical Principles of Natural Philosophy and His System of the World,* Andrew Motte, trans., University of California Press, Berkeley, 1962, Vol. 1, p. 6.
2. Galileo Galilei, *Dialogues Concerning Two New Sciences,* Henry Crew and Alfonso de Salvio, trans., Macmillan, New York, 1914, pp. 293–294.

CHAPTER 9

1. Francis Bacon, *Novum Organum*, 1620.
2. Benjamin Thompson (Count Rumford), *Collected Works*, Vol. II, Essay II (VII).
3. *Ibid.*
4. *Ibid.*
5. *Ibid.*, Essay IX.
6. *Ibid.*
7. Lucretius, *De Rerum Natura* (*On the Nature of Things*), R. E. Latham, trans., Penguin Books, Baltimore, 1951.
8. Hermann Helmholtz, "Uber dei Erbaltung der Kraft," John Tyndall, trans., *Scientific Memoirs, Natural Philosophy*, 1853.
9. *Ibid.*

CHAPTER 10

1. *The Second Law of Thermodynamics*, W. F. Magie, ed., New York, 1899.
2. *Ibid.*

CHAPTER 12

1. Eric Shipton, quoted in J. Bronowski, *Science and Human Values*, Harper & Row, New York, 1965, p. 30.
2. *Works of Aristotle*, Vol. I, Book IV, "Historia Animalum," D'Arcy Wentworth Thompson, trans., Oxford University Press, New York, 1962, pp. 554–555, Book V, pp. 23–24.
3. Aristotle, "De Generatione e Corruptione," Book I, Chap. 11.
4. O. Neugebauer, *The Exact Sciences in Antiquity*, 2nd ed., Brown University Press, Providence, R.I., 1957.

CHAPTER 13

1. Henry David Thoreau, *Walden*.

CHAPTER 14

1. Isaac Newton, in Charles C. Gillispie, *The Edge of Objectivity*, Princeton University Press, Princeton, N.J., 1960, pp. 122–123.
2. Isaac Newton, reprinted by permission of the publishers from William F. Magie, *A Source Book in Physics*, Harvard University Press, Cambridge. Copyright 1935, 1963 by the President and Fellows of Harvard College, p. 300.
3. Thomas Birch, *The History of the Royal Society of London*, A. Millar, London, 1757, Vol. 3, p. 14.
4. *Ibid.*
5. Isaac Newton, in Magie, *op. cit.*, p. 302.
6. *Ibid.*
7. Diogenes Laertius, R. D. Hicks, trans., quoted from Milton C. Nahm, ed., *Selections from*

Early Greek Philosophy, Appleton-Century-Crofts, New York, 1947, p. 165.
8. *Isaac Newton's Papers and Letters on Natural Philosophy*, I. Bernard Cohen, ed., Harvard University Press, Cambridge, 1958.
9. *Ibid.*
10. *Sir Isaac Newton's Mathematical Principles of Natural Philosophy and His System of the World*, Andrew Motte, trans., University of California Press, Berkeley, 1962, Vol. 2, p. 547.

CHAPTER 16

1. Thomas Young, reprinted by permission of the publishers from William F. Magie, *A Source Book in Physics*, Harvard University Press, Cambridge. Copyright 1935, 1963 by the President and Fellows of Harvard College, p. 309.
2. *Ibid.*, p. 305.
3. *Ibid.*
4. *Ibid.*, p. 306.
5. *Ibid.*, pp. 308–309.
6. *Oeuvres d'Augustin Fresnel*, Paris, 1866–1869, Charles C. Gillispie, trans., in *The Edge of Objectivity*, Princeton University Press, Princeton, N.J., 1960, p. 407.
7. Magie, *op. cit.*, p. 310.
8. *Ibid.*, p. 310.
9. *Ibid.*
10. *Ibid.*, pp. 310–311.
11. *Ibid.*, p. 311.
12. J. C. Maxwell, *Encyclopaedia Britannica*, article.
13. *Ibid.*

CHAPTER 17

1. Charles François de Cisternay Du Fay, *Philosophical Transactions*, Vol. 38 (1734), p. 258.
2. William Gilbert, in Magie, *op. cit.*, p. 417.

CHAPTER 18

1. Hans Christian Oersted, reprinted by permission of the publishers from William F. Magie, *A Source Book in Physics*, Harvard University Press, Cambridge. Copyright 1935, 1963 by the President and Fellows of Harvard College, p. 437.
2. *Ibid.*, p. 438.
3. William Gilbert, in *ibid.*, p. 390.

CHAPTER 19

1. Michael Faraday, reprinted by permission of the publishers from William F. Magie, *A Source Book in Physics*, Harvard University Press, Cambridge. Copyright, 1935, 1963 by

the President and Fellows of Harvard College, p. 473.

2. *Ibid.*, pp. 473, 474.
3. *Ibid.*, p. 474.
4. *Ibid.*

CHAPTER 20

1. *The Origins of Clerk Maxwell's Electric Ideas, As Described in Familiar Letters to William Thompson,* Sir Joseph Larmor, ed., Cambridge University Press, New York, 1937, p. 3.
2. J. C. Maxwell, "On Faraday's Lines of Force," Part I, *Transactions of the Cambridge Philosophical Society,* Vol. 10 (1856), pp. 27–83.
3. *The Scientific Papers of James Clerk Maxwell,* W. D. Niven, ed., Cambridge, 1890.
4. *Ibid.*
5. J. C. Maxwell, letter to H. R. Droop, Dec. 28, 1861.
6. Maxwell, *loc. cit.*
7. *Ibid.*
8. *Ibid.*
9. *Ibid.*
10. J. C. Maxwell, "A Dynamical Theory of the Electromagnetic Field," *Philosophical Transactions,* Vol. 155 (1865), pp. 459–512.

CHAPTER 21

1. J. C. Maxwell, "On Physical Lines of Force," Part 3, *Philosophical Magazine,* Jan.–Feb. 1862, Proposition 16.
2. J. C. Maxwell, letter to William Thomson (Lord Kelvin), Dec. 10, 1861.
3. J. C. Maxwell, reprinted by permission of the publishers from William F. Magie, *A Source Book in Physics,* Harvard University Press, Cambridge. Copyright 1935, 1963 by the President and Fellows of Harvard College, p. 537.
4. C. J. Monro, letter to J. C. Maxwell, Oct. 23, 1861.

CHAPTER 22

1. Lucretius, *De Rerum Natura,* C. J. Monro, trans., Cambridge, 1886, Vol. 3, p. 23.
2. Giordano Bruno, "On the Infinite Universe and Worlds," Dorothea W. Singer, trans., in *Giordano Bruno,* Schuman, New York, 1950, p. 249.
3. *Ibid.*, p. 254.
4. *Sir Isaac Newton's Mathematical Principles of Natural Philosophy and His System of the World,* Andrew Motte, trans., University of California Press, Berkeley, 1962, Vol. 1, p. 6.
5. *Ibid.*, p. 8.
6. Isaac Newton, *The System of the World,* Hypothesis I.

7. J. C. Maxwell, *Encyclopaedia Britannica,* article.
8. *Ibid.*

CHAPTER 23

1. A. A. Michelson and E. W. Morley, *American Journal of Science,* Vol. 34 (1887), p. 333.

CHAPTER 24

1. Albert Einstein, "On the Electrodynamics of Moving Bodies," in *The Principle of Relativity,* Methuen, London, 1923, pp. 37–38.
2. Hermann Minkowski, "Space and Time," address, 1908, in *The Principle of Relativity,* Methuen, London, 1923, p. 81.
3. H. A. Lorentz, *Proceedings of the Academy of Sciences of Amsterdam,* 1904.
4. Minkowski, *op. cit.*, pp. 82–83.
5. Einstein, *op. cit.*, p. 41.
6. *Ibid.*, p. 81.
7. *Ibid.*, p. 37.
8. *Ibid.*, pp. 39–40.
9. *Ibid.*, p. 40.
10. *Ibid.*, pp. 42–43.
11. *Ibid.*, p. 48.
12. *Ibid.*, pp. 37–38.
13. *Ibid.*, p. 41.
14. *Ibid.*, p. 65.

CHAPTER 25

1. Albert Einstein, "Does the Inertia of a Body Depend upon Its Energy-Content?" in *The Principle of Relativity,* Methuen, London, 1923, p. 71.
2. *Ibid.*

CHAPTER 26

1. Albert Einstein, "On the Electrodynamics of Moving Bodies," in *The Principle of Relativity,* Methuen, London, 1923, pp. 37–38.

CHAPTER 27

1. R. V. Pound and G. A. Rebka, *Physical Review Letters,* Vol. 4 (1960), p. 337.
2. Malcolm Lowry, *Under the Volcano,* Vintage Books, New York, 1958.

CHAPTER 28

1. Isaac Newton, reprinted by permission of the publishers from William F. Magie, *A Source Book in Physics,* Harvard University Press, Cambridge. Copyright 1935, 1963 by the President and Fellows of Harvard College, p. 298.
2. *Ibid.*, p. 298.

3. *Ibid.*, p. 354.
4. *Ibid.*, pp. 354–355.
5. *Ibid.*, pp. 355–356.
6. *Ibid.*, p. 354.
7. *Ibid.*, p. 361.
8. *Ibid.*, p. 601.
9. *Ibid.*
10. *Ibid.*, p. 612.
11. *Ibid.*, p. 616.
12. Albert Einstein, "Does the Inertia of a Body Depend upon Its Energy-Content?" in *The Principle of Relativity*, Methuen, London, 1923, p. 71.

CHAPTER 29

1. Lucretius, *De Rerum Natura*.
2. Sir Joseph Thomson, reprinted by permission of the publishers from William F. Magie, *A Source Book in Physics*, Harvard University Press, Cambridge. Copyright 1935, 1963 by the President and Fellows of Harvard College, p. 583.
3. *Ibid.*, p. 589.
4. *Ibid.*, pp. 596–597.
5. Joseph John Thomson, *Recollections and Reflections*, Macmillan, New York, 1937, p. 334.
6. Magie, *op. cit.*, p. 384.
7. *Ibid.*, pp. 384–385.
8. *Ibid.*, p. 385.

CHAPTER 30

1. Ernest Rutherford (Lord Rutherford), "The Scattering of α and β Particles by Matter and the Structure of the Atom," *Philosophical Magazine*, Vol. 21 (1911).
2. *Ibid.*
3. *Collected Papers of Lord Rutherford of Nelson*, J. Chadwick, ed., Wiley, New York, 1963.
4. *Ibid.*
5. Rutherford, *op. cit.*
6. *Ibid.*
7. *Ibid.*

CHAPTER 31

1. Niels Bohr, *Journal of the Chemical Society*, Feb. 1932, p. 349.
2. *Ibid.*
3. Ernest Rutherford, "The Scattering of α and β Particles by Matter and the Structure of the Atom," *Philosophical Magazine*, Vol. 21 (1911).
4. Max Planck, *Scientific Autobiography and Other Papers*, Frank Gaynor, trans., Philosophical Library, New York, 1949, p. 44.
5. *Ibid.*
6. *Ibid.*, pp. 44–45.

7. Niels Bohr, "On the Constitution of Atoms and Molecules," *Philosophical Magazine*, Vol. 26 (1913).

CHAPTER 32

1. C. J. Davisson and H. Germer, *Physical Review*, Vol. 30 (1927), p. 705.

CHAPTER 34

1. Max Born, *Atomic Physics*, 7th ed., Hafner, New York, 1962, Chap. IV.

CHAPTER 37

1. Pieter Zeeman, reprinted by permission of the publishers from William F. Magie, *A Source Book in Physics*, Harvard University Press, Cambridge. Copyright, 1935, 1963 by the President and Fellows of Harvard College, p. 385.

CHAPTER 39

1. Ernest Rutherford, "The Scattering of α and β Particles by Matter and the Structure of the Atom," *Philosophical Magazine*, Vol. 21 (1911).
2. Muriel Haworth, *The Life Story of Frederick Soddy*, New World Publications, London, 1958, p. 184.
3. Rutherford, *op. cit.*
4. A. van der Broek, *Nature*, Vol. 92 (1913), p. 372.
5. Frederick Soddy, *Nature*, Vol. 92 (1913), p. 399.
6. *Ibid.*, p. 400.
7. Ernest Rutherford, "Collision of Particles with Light Atoms. IV. An Anomalous Effect in Nitrogen," *Philosophical Magazine*, Vol. 37 (1919).
8. George Gamow, "Quantum Theory of the Atomic Nucleus," *Zeitschrift für Physik*, Vol. 51 (1928), Henry A. Boorse and Lloyd Motz, trans., in *The World of the Atom*, Basic Books, New York, 1966, Vol. II, pp. 1129–1130.
9. Lise Meitner and O. R. Frisch, "Disintegration of Uranium by Neutrons: A New Type of Nuclear Reaction," *Nature*, Vol. 143 (1939), pp. 239ff.

CHAPTER 40

1. P. A. M. Dirac, "A Theory of Electrons and Protons," *Proceedings of the Royal Society* (London), Vol. A128 (1930).

2. J. Robert Oppenheimer, "On the Theory of Electrons and Protons," *Physical Review,* Vol. 35 (1930), pp. 562ff.
3. Carl D. Anderson, "The Positive Electron," *Physical Review,* Vol. 43 (1933), p. 491.

CHAPTER 41

1. Isaac Newton, *Opticks,* I. B. Cohen, ed., Dover, New York, 1952, p. 400.

CHAPTER 42

1. C. T. R. Wilson, *Le Prix Nobel,* 1927.
2. *Ibid.*

CHAPTER 44

1. Hideki Yukawa, "On the Interaction of Elementary Particles, I," *Proceedings of the Physico-Mathematical Society of Japan,* Vol. 3, No. 17 (1935), p. 53.
2. R. A. Millikan, *Electrons, Protons, Photons, Neutrons, Mesotrons, and Cosmic Rays,* University of Chicago Press, Chicago, 1947.

CHAPTER 47

1. John Tyndall, *Faraday as a Discoverer,* London, 1868.

QUESTIONS AND PROBLEMS

ON THE PROBLEM OF MOTION

2. A Very New Science

Questions

1. If the average speed of a body is zero over a certain time interval, must it have stood still?

2. Does a greater speed imply a greater acceleration? Give a counterexample.

3. We may have heard that an arrow aimed and shot by Paris at Achilles would never reach Achilles, for the arrow must reach the midpoint of the distance between him and the bow; then it reaches the midpoint of the remaining distance, and so on. This process is endless, because, no matter how short the distance remaining to be covered, there always exists a midpoint. How can we explain Achilles' swollen heel? (See the Appendices, p. 486.)

4. J. L. Godard takes pictures with a movie camera of the motion of a stone thrown straight upward. What would the motion of the stone be like if he ran the film through a projector backward? Give another example of reversed motion that can occur in nature.

5. In a mathematical formula, the units of the right side should equal those of the left side. Confirm this with the formulas that appear in this chapter. Suppose you derived a formula that did not have the same units on both sides. What would be your conclusion?

6. The acceleration of gravity on the moon's surface is one sixth of that on earth. Show that if a man's leg muscles give him the same initial velocity when he leaves the surface, he can jump six times as high on the moon as on the earth.

Problems

1. To test the speedometer of his automobile, a driver travels 5 miles on a marked highway

NOTE: Answers to selected odd-numbered questions and problems appear on pp. 526–528.

in exactly 5 min. Throughout the trip his speedometer indicates a speed of 60 miles/hr. Is the speedometer correct?

2. Draw a speed vs. time graph to show that the distance covered by a car traveling at a speed of 60 miles/hr for 1 hr is the same as that covered by a car traveling at a speed of 50 miles/hr for 1 hr 12 min.

3. During the storm K. Lear saw a flash of lightning. After 10 sec had elapsed, he heard thunder. The speed of sound in air at room temperature is about 1100 ft/sec (or about 3.4×10^4 cm/sec). How far away from him did the lightning originate?

4. How many hours does it take for a car moving with a speed of 60 miles/hr to cover a distance of 300 miles? How long does it take for a car moving with a speed of 50 miles/hr to cover the same distance?

5. A car travels on a highway at a speed of 50 miles/hr for 2 hr, and then at 40 miles/hr for 15 min. What distance does the car cover during the 2 hr and 15 min? What is the car's average speed over the 2 hr and 15 min?

6. A car traveling along a straight highway is uniformly accelerated from an initial speed of 30 miles/hr to a final speed of 60 miles/hr in 1 min. What is the value of the uniform acceleration?

7. A stone is released from a height of 490 m. How long will it take to reach the ground and with what speed will it strike the ground?

8. A stone is thrown straight upward. After 20 sec have elapsed, it drops back to the ground. What height does the stone reach, and what is its initial upward speed? What is its speed upon returning to the ground?

9. A stone is thrown horizontally from a height of 64 ft with a speed of 40 ft/sec. How long does it take for the stone to reach the ground, and how far does it travel horizontally? One second after this stone is thrown horizontally, a second stone is thrown vertically downward from the same height. What initial downward speed

must it have in order to hit the ground at the same time as the first stone?

10. A golf-ball leaves the tee with a horizontal velocity of 100 ft/sec. If the tee is 20 ft above the fairway, how far does the ball travel horizontally before it strikes the ground?

11. The sun is 1.50×10^8 km from the earth. How long does it take sunlight to travel this distance at the constant speed of $c = 3 \times 10^8$ m/sec?

12. The nearest star is 4.3 light years from the earth. (A light year is the distance light travels in 1 year.) What is this distance in miles?

13. A rifle slug emerges from the barrel with a "muzzle velocity" of 400 m/sec. The barrel is 1 m long. Assume constant acceleration.

a. How long does it take the slug to travel through the barrel?

b. What is the acceleration?

14. Suppose a batter hits a home run that travels 5.12 ft in the air.

a. What is the initial speed of the ball as it leaves the bat, assuming it "takes off" at an angle of 45° with respect to the horizontal?

b. What is the maximum height of the trajectory?

c. How long is the ball in the air?

Note: Neglect air resistance.

15. A ball is thrown vertically downward from a cliff and strikes the ground 120 ft below in 2 secs.

a. What is the initial velocity of the ball?

b. What is its velocity at impact?

16. Starting from rest, car A begins to accelerate at 5 ft/sec² at the instant that car B flashes by at 90 miles/hr.

a. How long does it take B to catch up with A?

b. How far does B travel in that time? Give your answer in miles.

17. A ball is thrown vertically upward with an initial speed of 80 ft/sec at a place where the acceleration due to gravity is 32 ft/sec².

a. What is the maximum height?

b. How long does it take the ball to reach this peak?

c. What is the velocity of the ball after 2 sec? After 4 sec? Just before it strikes the ground?

Note: Neglect air resistance.

18. A stone is thrown vertically downward from a bridge with an initial speed of 16 ft/sec.

a. How long will it take the stone to reach the water 192 ft below?

b. What is its velocity just before impact?

19. Starting from rest an automobile reaches a speed of 60 miles/hr in 8 secs.

a. What is the (constant) acceleration?

b. How far does the automobile travel in this process?

3. *What Is a Force?*

Questions

1. When there is no force acting, we can say that the force has zero magnitude. What, then, is the direction of this force?

2. Suppose Russians chose east as the direction of the zero force and Americans chose west. How would this affect their calculations?

3. A weight is hung from the middle of a tightly stretched horizontal cable. Must the cable sag?

Problems

1. A car travels north along a straight highway with a speed of 30 miles/hr.

a. What is its velocity v?

b. What are the magnitudes and directions of 2v, −2v, 2v − v, and v − 2v?

2. Using the parallelogram rule for the composition of forces, one can subtract one vector from another: $\mathbf{A} - \mathbf{B} = \mathbf{A} + (-\mathbf{B})$, where the vector $-\mathbf{B}$ is equal to the vector $\mathbf{B}$ in magnitude, but opposite to $\mathbf{B}$ in direction. Draw the parallelograms for the vectors $\mathbf{A} + \mathbf{B}$ and $\mathbf{A} - \mathbf{B}$.

3. To reach a theater a hiker began by walking 5 miles due south, where he met his girl friend. They then walked 12 miles due west to reach the theater. How many miles was the theater from the hiker's starting point?

4. A man fired his pistol vertically upward and found that the bullet reached a height of 100 m. To what height should he climb and fire his pistol horizontally in order that the bullet hit the ground at an angle of 45°?

5. On a stormy day, a person riding in a car traveling southeast notices through a side window of the car that rain is falling straight down. He knows that the car is traveling at a speed of 30 miles/hr. What would be his estimate of the velocity of wind?

6. One can add or subtract many vectors by using the parallelogram rule successively. Show, by repeatedly using postulate V, that the order of summing the vectors is irrelevant.

7. By decomposing the acceleration vector into components, show that for a body to move without changing its speed, the acceleration vector, if it is not the null vector, must always be perpendicular to the velocity. (*Hint:* Remember the rules of composition. Show that, if at any instant of time the acceleration vector has a component along the velocity, then the speed cannot remain constant.)

4. "The Lion Is Known by His Claw"

Questions

1. Sliding a book on a desk, one finds that the book will not maintain a uniform motion in a straight line. If one used the above example as evidence against Newton's first law, what might be a counterargument for the law?

2. Newton's laws governing the motions of a body do not mention the size, shape, or color of the body. Give an everyday example of a motion that depends on the size of the body. Can Newton's laws be saved?

3. A man is a railway car attempts to verify Newton's laws. He finds that they hold sometimes (for example, when the train is traveling in a straight line at a constant speed) but not other times (for example, when the train accelerates or goes around a curve). Suppose the train is sealed so he has no way to know the velocity or change in velocity of the train. Is there a method by which he may know when to expect Newton's laws to hold, or must he conclude that "they work when they work"?

4. If the weight of a body is denoted by W and the acceleration is denoted by a, Newton's second law might be written $\mathbf{F} = (W/g)a$. Would this expression be valid on the moon?

5. Knowing the acceleration of gravity at the earth's surface and the distance and speed of the moon, can one determine the moon's mass?

6. Consider an automobile making a circular turn at a constant speed. The force required to accelerate the automobile is provided at the contact between the tires and the road. If on a rainy day the maximum possible force the road can exert on the tire surface (before skidding occurs) is reduced by one half, by what amount must the automobile speed be reduced?

7. If the above automobile makes a turn at 40 miles/hr rather than 30 miles/hr, what increase in force (exerted by the road on the tires) is required to accelerate it around the turn?

Problems

1. What is the magnitude of the momentum of a 2-g mass moving at a speed of 2 cm/sec?

2. What is the magnitude of the momentum of an electron (mass 0.91×10^{-27} g) moving at a speed of 10^8 cm/sec?

3. A ball of mass 5 g is moving with a velocity of 400 cm/sec toward the east. What is the momentum? If the ball bounces off a solid wall and moves toward the west with the speed of 200 cm/sec, what is the change in the momentum?

4. What is the magnitude of the force required to give an automobile weighing 1600 lb a uniform acceleration of 10 ft/sec²?

5. An electron from the cathode of a radio tube is accelerated uniformly in a straight line toward the anode 0.5 cm away. If the speed of the electron (mass = 9.1×10^{-28} g) is finally 4×10^8 cm/sec, what is the magnitude of the force acting on the electron?

6. A car weighing 3200 lb (100 English mass units) goes around a circular curve with radius of curvature of 200 ft at 60 miles/hr (88 ft/sec). What is its acceleration? What force is required? How is this provided?

7. A boy whirls a 2000-g mass about him on a 50-cm chain once per sec. What force must the weakest link be able to withstand?

8. A binary, or double, star known as J. S. Plaskett's star is the most massive star known. Each one of its component stars has a mass of about 10^{35} g. From spectroscopic studies it is known that each star of the pair goes around in a circle of radius 0.4×10^{13} cm every 14.4 days. What is the force on each star?

9. The earth moves around the sun in a nearly circular orbit of radius 1.5×10^{11} m (about 9.3×10^7 miles).

  a. What is the speed of the earth in its orbit?

  b. What is the force that the sun exerts on the earth to keep it in a circular orbit?

10. In the simplified Bohr theory of the hydrogen atom, an electron moves at a uniform speed of 2.19×10^8 cm/sec in a circular orbit of radius 5.29×10^{-9} cm, with a proton at the center of the orbit.

  a. What is the acceleration of the electron?

  b. What is the magnitude of the force necessary to keep the electron in its orbit? (The electron mass is 0.911×10^{-27} g).

  c. Compare the force found in part b with the weight of the electron assuming $g = 980$ cm/sec?

11. A 180-g baseball approaches the plate moving horizontally at 30 m/sec. After contact with the bat the ball reverses its direction and moves off horizontally at 40 m/sec.

  a. Calculate the magnitude of the change in momentum of the ball.

  b. Assuming that the bat is in contact with the ball for 1×10^{-2} sec, find magnitude of the (constant) force that the bat exerts on the ball.

12. A 150-lb horizontal force pulls an 800-lb sled along a rough horizontal plane surface. The motion is opposed by a frictional force of 50 lb.

  a. What is the acceleration of the sled?

  b. How far does it move in 10 sec, starting from rest?

13. A 1000-kg crate is pushed over a rough horizontal plane surface with the result that it accelerates at a steady rate and moves 4 m in 5 sec. If the frictional force is 20 percent of the weight, what is the magnitude of the propelling force? Assume that $g = 9.80$ m/sec².

14. Starting from rest a 4800-lb limousine accelerates at a constant rate and covers ¼ miles in 30 sec. What is the resultant force?

15. A 500-g projectile is launched at a speed of 50 m/sec at angle of 45° above the horizontal. What is its momentum at the apex of its trajectory?

5. *The Music of the Spheres*

Questions

1. From the point of view of an observer on the moon, how would the following appear: the earth, the stars, the sun?

2. In arguing that the earth is round, Aristotle is quoted as saying:

Further proof is obtained from the evidence of the senses. (i) If the earth were not spherical, eclipses of the moon would not exhibit segments of the shape which they do. As it is, in its monthly phases the moon takes on all varieties of shape—straight-edged, gibbous, and concave—but in eclipses the boundary is always convex. Thus if the eclipses are due to the interposition of the earth, the shape must be caused by its circumference, and the earth must be spherical.

Reconstruct the geometry that lies behind this argument. Is it consistent with an earth-centered universe? He says further:

(ii) Observation of the stars also shows not only that the earth is spherical but that it is of no great size, since a small change of position on our part southward or northward visibly alters the circle of the horizon, so that the stars above our heads change their position considerably, and we do not see the same stars as we move to the North or South. Certain stars are seen in Egypt and the neighbourhood of Cyprus, which are invisible in more northerly lands, and stars which are continuously visible in the northern countries are observed to set in the others. This proves both that the earth is spherical and that its periphery is not large, for otherwise such a small change of position could not have had such an immediate effect. For this reason those who imagine that the region around the Pillars of Hercules° joins on to the re-

° The Straits of Gibraltar.

gions of India, and that in this way the ocean is one, are not, it would seem, suggesting anything utterly incredible. They produce also in support of their contention the fact that elephants are a species found at the extremities of both lands, arguing that this phenomenon at the extremes is due to communication between the two. Mathematicians who try to calculate the circumference put it at 400,000 stades.†

Again reconstruct the geometry that leads to the argument. How might the "mathematicians" have used the phenomena described to attempt to measure the circumference? What is the most likely source of their error? Then

From these arguments we must conclude not only that the earth's mass is spherical but also that it is not large in comparison with the size of the other stars.‡

From these arguments can he conclude anything about (a) the size of the stars, or (b) their distance from the earth compared with the size of the earth?

3. Given Aristotle's universe, his natural motions, and imagine that he accepted the second law ($\mathbf{F} = m\mathbf{a}$). What would he be forced to conclude about the mass of the stars?

4. Imagine that Apollonius had observed stellar parallax during the course of a year. How might he have fitted this into an earth-centered universe?

5. Suppose Ptolemy had accepted a sun-centered solar system. Would he have had to abandon all his devices (epicycles, and so on)? How might he have used the epicycle, for example, most profitably?

6. An experimental airplane is designed to rise to a height of 10,000 ft and hover there for a certain time. The earth, meanwhile, rotates on its axis and the craft returns to land when its destination appears below. Thus one could travel from New York to Los Angeles in 3 hr. Will it work?

7. An earth satellite has a circular orbit with radius $2Re$ (Re is the radius of the earth). What is its period?

8. Why does a comet move faster when it is near the sun than when it is far away?

† That is, 9987 geographical miles. Prantl remarks that this is the oldest recorded calculation of the earth's circumference. . . . (The present-day figure in English miles is 24,902.)

‡ *Exploring the Universe*, p. 81, as quoted in Milton Munitz, ed., *Theories of the Universe*, Macmillan, New York, 1957.

Problems

1. The planetoid Ceres goes around the sun once every 2.6 years. What is the radius of its orbit measured in radii of the earth's orbit around the sun? (Assume it is circular.)

2. A satellite is fired into an orbit halfway between the earth and the moon. What is its period?

3. Halley's comet goes around the sun once every 75 years, and at its point of closest approach it is 8.9×10^{10} m from the sun. Its farthest point from the sun cannot be measured because it cannot be seen. Compute the greatest distance of Halley's comet from the sun. (The definition of the average radius, the quantity referred to in Kepler's third law, is one half the sum of the shortest and longest distances to the sun.)

4. Man will go to Mars by taking an elliptic orbit having the earth's orbit (radius 93 million miles) as its closest approach to the sun and the Martian orbit (radius 141.5 million miles) as its greatest distance. How long will such a trip take?

5. How long will it take to reach Venus (orbital radius 67 million miles) from earth?

6. Calculate the orbital speed and the period of a satellite in a circular orbit just above the surface of the earth. Assume that the radius of the orbit is 6.40×10^8 cm. Neglect air resistance.

7. What is the period of a satellite in a circular orbit just above the surface of the moon? Take $R = 1.75 \times 10^8$ cm. The mass of the moon is 7.35×10^{25} g.

6. Newton's System of the World

Questions

1. Kepler's third law yields the same constant for any satellites orbiting the earth. (The satellites can be either natural or artificial.) What is a quick way to calculate this constant?

2. Does Newton's law of universal gravitation imply that there is only one kind of force acting between two bodies, and that it is gravitation?

3. Do Newton's law of universal gravitation and his laws of motion give any information about the cause of the gravitational force or what is its origin?

4. Discuss qualitatively how the orbit of the moon will be distorted by the gravitational pull of the sun. (Assume that the moon's orbit about the earth is in the plane of the earth's ellipse.)

5. After Hooke's failure to measure a vari- ation of the gravitational attraction from the surface of the earth to the top of a mountain, an unnamed physicist became attached to the idea that the gravitational force does not vary with distance. He also agreed that $\mathbf{F} = m\mathbf{a}$. Why did the moon appear in his nightmares?

6. Newton did not know the exact value of G. (Why not?) However, over a cup of coffee he might have been willing to make some estimates. What might he have proposed?

7. Suppose Newton had agreed that the stars, being celestial material, were massless. (How could he accommodate this?) Could he also agree that the moon was massless? What might he have indicated as evidence for the mass of the moon?

8. How might Descartes have greeted the idea that the gravitational force between two bodies does not act instantaneously but propagates with the speed of light?

Problems

1. At what height above the earth's surface will an object have one half the force of gravity on it that it would have at sea level?

2. If the radius of the moon's circular orbit were reduced by one half, what would be the new period of revolution? What would be the ratio of the speed of the moon in the new circular orbit to that in the old orbit?

3. A man weighs 200 lb and wants to weigh less. Should he go to Mars or to Jupiter? How much would he weigh on each?

4. An accurate value of the mass of the earth can be obtained by measuring the altitude and the period of the revolution of an artificial satellite about the earth. How can this be achieved?

5. What is the force on an object that weighs 100 lb on the earth's surface when it is 2 earth radii from the center of the earth?

6. A satellite circles the earth once every 98 min at a mean altitude of 5×10^7 cm (about 330 miles). Calculate the mass of the earth.

7. A body of mass 10^5 g is located on the surface of the earth.

   a. What is the magnitude of the gravitational force that the earth exerts on the body?

   b. What are the magnitudes of the gravitational forces that the sun and the moon, respectively, exert on the body?

   c. Show that the gravitational forces exerted by the sun and the moon on the body are negligible compared with those exerted by the earth. The mean radius of the earth is 6.4×10^8 cm.

	Mass, g	Mean distance from the earth, cm
Sun	2.0×10^{33}	1.5×10^{13}
Earth	6.0×10^{27}	
Moon	7.3×10^{25}	3.8×10^{10}

8. The mean distance between the earth and the moon is 3.8×10^{10} cm (or about 240,000 miles) and the mass of the moon is 7.3×10^{25} g. Is there a point in space at which the pull of the earth and the moon are equal and opposite? If so, locate the point. (Ignore the gravitational pull of all other planets.)

9. Explain why two objects lying on a well-polished table remain in their respective positions even though the gravitational force is exerted between them. If two objects each weigh 20 lb and are separated by a distance of 1 ft, what is the magnitude of the gravitational force that each object exerts on the other?

10. A body moves with speed v around the earth in a circular orbit of radius R.

a. Prove the relation $v^2R = GM_{earth}$, where M_{earth} denotes the mass of the earth. Since the right side is known, one can find the speed of the body if the radius of the circular orbit is known and vice versa.

b. Show that both sides of the above relation have the same units of $(length)^3/(time)^2$.

11. For a model of the hydrogen atom, one can roughly consider it as an electron (mass $= 9.11 \times 10^{-28}$ g) moving around a proton (mass $= 1.67 \times 10^{-24}$ g) with a speed of 2.2×10^8 cm/sec in a circular orbit of the radius 0.53×10^{-8} cm.

a. Find the gravitational attraction between the electron and the proton.

b. What force would have to be exerted on an electron to hold it in the circular orbit?

c. What is the ratio of the magnitudes of two forces obtained in the above? From this can one conjecture that there must exist a new force that is far stronger than the gravitational force between the electron and the proton?

12. How far from the earth is the point in space at which the resultant force of gravity due to the earth and the sun is exactly zero? Express this distance in terms of the radius of the earth, $R_e = 6.37 \times 10^8$ cm.

7. Particles in Collision

Questions

1. Under what assumptions is momentum conserved?

2. Under what assumptions is kinetic energy conserved?

3. An airplane flying through the atmosphere is able to push itself by (so to speak) pushing the atmosphere, using the atmosphere to provide a force. The propellers drive air backward. They push on the air through which they turn, and the air pushes back. The total momentum of the atmosphere—plane system remains constant. Why does an aircraft of this kind run into trouble at very high altitudes where the atmosphere becomes less dense?

4. A spaceship in the near vacuum beyond the thick part of the earth's atmosphere does not have the same opportunity. Since there is no matter to push on, it must provide its own, throwing material out from behind. The change in momentum of the spaceship is equal and opposite to the change in momentum of the material ejected. In order to carry less, one wants to eject as little material as possible. Why is it that the larger the exhaust velocity, the more efficient the spaceship is?

5. One of the practical advantages of having a conservation law is that it will reduce the number of unknown quantities. For instance, for the system where the conservation of kinetic energy holds, the conservation law will reduce by one the number of unknown speeds of bodies in the system. For a collision of two particles how many speeds should be given in order that all others be determined? (Consider this for a one-dimensional collision and then for a collision in three dimensions.)

6. Apply the results of Question 5 to that situation in which a body explodes into two parts.

Problems

1. A 145-g baseball moving with speed of 3×10^3 cm/sec is bounced off a wall and moves directly backward with a speed of 2×10^3 cm/sec.

a. What is the impulse?

b. If the ball is in contact with the wall for 1×10^{-2} sec, what would be the average value of the force acting on the ball and what would be that acting on the wall?

c. What is the impulse acting on the wall?

2. Show that the impulse due to a force of 1 dyne acting 10 sec is equal to that due to a force of 10 dynes acting 1 sec in the same direction.

3. A skier weighing 160 lb grabs a tow rope to tow him back up the hill. In 2 sec he has accelerated from a standstill to 10 ft/sec. Disregarding any friction his skis might have with the snow, what is the average force his arm must withstand?

4. Discuss Problem 1 in the case that the ball has poorer elasticity. Consider, for example, the limiting case when the ball has as poor elasticity as a lump of clay. Given the same mass and initial speed, which is more likely to penetrate a thin wooden fence, the lump of clay or the baseball?

5. What is the magnitude and direction of the total momentum of a system having a 3-g mass moving 10 cm/sec to the east and a 5-g mass moving 20 cm/sec to the north?

6. What is the momentum of a 6400-lb vehicle whose speed is 50 ft/sec? (Assume $g = 32$ ft/sec^2). At what speed will a 10,000-lb truck have the same momentum?

7. A 3-kg grenade is rolling across the ground at 3 m/sec. Upon exploding it misfires and breaks into two pieces. The first shoots forward in the same direction the grenade was rolling, at 30 m/sec. How fast and in what direction does the other half fly?

8. A three-stage rocket causes the following total upward force to be exerted on a 3200-lb capsule. (The weight of the rocket, air friction, and other assorted downward forces are included.) What is the final speed?

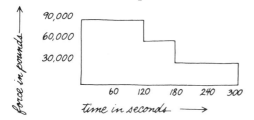

9. A machinegun fires 5-g bullets at a target. The gun fires at a rate of 480 bullets per min, and the speed of the bullets is 400 m/sec. If the target is supported so that it does not move, what average force is required to keep it in position?

8. Mechanical Energy

Problems

1. A satellite circles the earth in an orbit of radius R. What is the increase in the potential energy if the radius of its circular orbit is increased by 1 percent?

2. A spring of natural length ℓ_0 always exerts a restoring force proportional to the change in the length when it is stretched to a length $\ell(\ell \geqq \ell_0)$.

a. Write down the force exerted by the spring.

b. Draw the graph of the restoring force versus the change in the length $(\ell - \ell_0)$ and prove that the magnitude of the work done is one half of the area of the curve: force exerted at the length ℓ as a function of the elongation $(\ell - \ell_0)$. Write down the potential energy.

c. A body of mass m is attached to one end of the spring and the other end is fixed. Discuss qualitatively how the speed of the body will change if the spring is stretched to a length ℓ and released. In this situation, is the mechanical energy of the body conserved?

3. A 10-lb lead ball is thrown so that its upward velocity component is 8 ft/sec. If it is thrown from a shoulder height of 5 ft, what is the maximum height above the ground that it will reach?

4. How much work is done in pulling an 8-lb sled up a 50-ft 30° slope? If the sled is let go at the top of the hill without passengers, what will its speed be at the bottom of the hill, assuming negligible frictional losses? What difference would it make in the final speed if an 80-lb boy were sent down the slope on the sled?

5. A heavy bead slides down a frictionless wire of odd shape. At a certain point it has a speed of 1 ft/sec. How fast will it be going at a point 1 ft lower?

6. A wild goose of mass 3 kg is flying at 3 m/sec at a height of 30 m. What is its kinetic energy? What is its potential energy with respect to the earth's surface?

7. An amusement park roller coaster jumps off the track at the 60-ft-high top of one of the curves. How does the speed with which it hits the ground compare with what its forward speed would have been on the track at the bottom of the run?

8. The mean distance between the earth and the moon is 3.8×10^5 km (240,000 miles) and their masses are 6.0×10^{24} kg and 7.3×10^{22} kg, respectively. What is the potential energy of the moon? If the moon were to stop in its orbit and then plunge into the earth, with what speed would it hit the surface? The mean radius of the earth is 6.4×10^3 km (4000 miles).

9. What is the escape velocity from the moon?

10. How much work is done on the 10-kg object whose motion is described below?

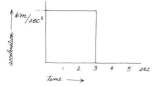

11. A skier who weighs 160 lb is pulled up a 500-ft ski slope. How much work does the ski tow do on the skier?

12. How much work is done in driving a 4-inch nail into a block of wood if each hit of the hammer drives the nail 1 inch with an average force of 50 lb?

13. What is the kinetic energy of a 200-g object moving at a speed of 500 cm/sec?

14. How many feet will it take to stop a car going 60 miles/hr if a car going 45 miles/hr can stop in 150 ft?

15. The wheels on a 1600-lb automobile exert an average forward force of 100 lb over a distance of 100 ft.

a. What is the work done by the force?

b. What is its velocity after it has gone that far?

16. A 100-g stone thrown north at 300 cm/sec hits and becomes embedded in a 1500-g ball of clay thrown east at a speed of 50 cm/sec. What is the velocity (magnitude and direction) of the clay after the collision? How much kinetic energy is dissipated?

17. A 4-g steel bullet moving horizontally at 500 m/sec strikes a 1-kg steel block initially at rest. Following the collision, the bullet rebounds at a speed of 400 m/sec in the opposite direction. What is the final speed of the block? (Neglect friction.)

18. Imagine that one wants to stop a rapidly moving object in a given limited distance, call it d (say an airplane landing on the deck of an aircraft carrier). The maximum force one dares exert on the airplane by a device such as a cable is determined by the strength of the plane, the fragility of the pilot, and so on. Call it F_{max}. Then the maximum work that can be done on the plane is $F_{max}d$. If we assume that the maximum force is limited by what can be exerted on the pilot (say he can withstand an acceleration of $5g$), then for a 4200-lb plane this would mean a force of how much? If the length of the carrier deck is 500 ft, what is the maximum landing speed?

19. Ball A of mass 6 kg collides with ball B of mass 2 kg. Ball B was initially at rest and the encounter takes place on a horizontal frictionless surface. After the collision A moves at 2 m/sec west and B at 6 m/sec south. What was the magnitude and direction of A's velocity before the collision?

20. Give an example to show that a body may have more kinetic energy but less momentum than another body.

21. A 2-kg ball moving to the right at a speed of 6 m/sec collides headon with a 5-kg ball at rest. After the impact the 2 objects rebound and the heavier ball moves to the right at 2 m/sec. Find the velocity of the lighter ball after the impact.

9. Conservation of Energy (*The First Law of Thermodynamics*)

Questions

1. Describe qualitatively how the first law of thermodynamics can be applied to (a) sliding a book on a horizontal table, (b) rubbing hands in winter to warm them up, (c) boiling a saucepan of water on an electric stove.

2. What happens to the kinetic energy of the accelerated particles in an accelerator when they hit a thick target? Why is it necessary to use water cooling on some types of targets?

3. Vast amounts of energy (mental) have been expended in the attempt to design a "perpetual motion" machine. Under what circumstances might such a machine be possible? (Consider, for example, the solar system.)

4. From the point of view of the caloric theory how might one explain the fact that it requires 80 cal of heat to melt 1 g of ice?

Problems

1. Normal body temperature is 98.6° F. Convert this to centigrade degrees.

2. Dalence introduced a scale in 1688 in which the melting point of ice was −10° and the melting point of butter was 10°. Find a formula that converts Dalence degrees to centigrade degrees. What is room temperature on the Dalence scale? The melting point of butter is 31° C.

3. The specific heat of copper is 0.092. How much heat is needed to raise the temperature of 5 g of copper by 10° C?

4. An insulated glass contains 100 g of water and 20 g of ice, both at 0° C. A silver spoon (specific heat 0.056) weighing 30 g and at 100° C is put into the glass. What is the temperature of the water after it has reached equilibrium? Does any ice remain, and, if so, how much?

5. A jug containing 2000 g of water and 2000 g of ice gains 10 cal/sec through its insulation. How long will it take for the ice to melt?

6. Calculate the difference in temperature of water at the top and the bottom of Victoria Falls (height 360 ft). Assume that no heat is lost to or gained from the surrounding air.

7. The heat of vaporization is the amount of heat required to change 1 g of a solid or liquid into a vapor at the same temperature. How much heat will be required to vaporize 15 g of mercury? (Heat of vaporization = 70 cal/g.)

8. A horse can do about 750 joules of work per second. Rumford, in his experiment, used a horse-driven lathe turning inside a metal cylinder that was immersed in water. He knew, from known values of the specific heats of the water and metal, that it would take 1.2×10^6 cal to make the water boil. It took 2½ hr for the water to boil. Joule later used these facts to obtain a rough estimate of the heat-work equivalence. What result did he get?

9. A lead bullet is stopped by a wooden plank and melts upon impact. If its temperature before hitting the plank was 30° C and it melts at 327° C with a heat of fusion of 5.86 cal/g, at least how fast must it have been going? (Specific heat of lead = 0.031 cal/g-C°.)

10. An electrical heater supplies 2400 joules to 80 g of water in an insulated 400-g metal container. As a result, the temperature of the water rises by 5° C. Find the specific heat capacity of the metal.

11. A flow calorimeter is used to measure the specific heat capacity of a liquid. Heat is added at a known rate to the liquid as it flows at a steady rate through a calorimeter. The specific heat of the liquid can be determined from a knowledge of the rate of flow and the temperature difference between the input and output points. **Problem:** A liquid with a density of 1.25 g/cm³ flows through a calorimeter at a rate of 8 cm³/sec. Heat is added at a rate of 300 joules/sec. Under steady state conditions the temperature difference between the input and output points is 20° C. Find the specific heat capacity of the liquid.

12. Calculate the heat required to change 1 kg of ice at −40° F into steam at 212° F. The specific heat capacity of ice is 0.5 cal/g-C°.

13. Steam enters a "radiator" at 120° C and leaves as hot water at 90° C. How much heat does the device supply per g of water? The specific heat capacity of steam is 0.48 cal/g-C°.

EXPERIENCE, ORDER, AND STRUCTURE

Questions

1. Aristotle, who was committed to observation for the verification of any theory, said among other things that

> If all men are mortal and if Socrates
> is a man, then Socrates is mortal.

Might he have hesitated to say this if Socrates had still been alive?

2. Since so many thousands of words are attributed to Aristotle, since he spoke with a lisp, and since he mentions falling bodies so few times and so peripherally, suppose we imagine that when he wrote (or said) those famous words concerning falling bodies he was thinking of something other than Galileo's interpretation. Of what might he have been thinking? How might he have defended the statement?

3. What evidence might one propose for the existence of a chair, for the force of gravitation, for the mass of the sun, for the existence of the sun?

4. The sun also rises. It has always risen. What evidence do we have that it will rise again?

5. Suppose we agree that the structure of chess does not depend upon the material out of which the pieces are made, and suppose also that Bobby Fischer maintains that his game is better when he plays with his favorite ivory set. Are the two statements inconsistent?

ON THE NATURE OF LIGHT

14. The "Phaenomena of Colours"

Questions

1. What might be a way to measure the speed of light? To measure differences in the speed of light?

2. Occasionally there occurs in the heavens an explosion producing what is known as a supernova, in which a star suddenly becomes many times brighter than before. We see the explosion as a bright white light, not as a series of different colors arriving at different times. What does this show about the speed of light of different colors in vacuum?

Problems

1. How long does it take for light from the sun to reach the earth? (The distance is 93 million miles.)

2. According to a recent comic strip, visitors from outer space traveled from Pluto to earth in less than 1 day, although even the sun's rays take 6 hr to reach that far. About how far away is Pluto from the sun? If the "visitors" could reach the sun in 1 day, how fast would they be traveling?

3. A light-year is a unit of length. It is the distance a light beam can travel in 1 year. How many centimeters is it?

4. An observer sees lightning strike a building 12 miles away. What is the time interval between the event and his observation of it?

5. A radar pulse, which travels at the speed of light, is sent out, reflected by an airplane, and received, all in an interval of 5.12×10^{-6} sec. How far from the radar unit is the airplane?

6. How long would it take for a radar pulse to travel from the earth to the moon and back? (The distance between the earth and the moon is 240,000 miles.)

7. The index of refraction at an air–glass boundary is 1.6. If the ray in air is incident at an angle of 30°, what is the angle of refraction?

8. Prove that the assumption that during reflection the horizontal component of the velocity of light particles remains constant while the vertical component is reversed in direction implies that the angle of incidence equals the angle of reflection.

15. Waves

Questions

1. Judging from one's daily experience, a light wave travels approximately in a straight line, so that it cannot go around a wall, whereas a sound wave can. Explain why. The frequency of audible sound is about 16 cycles/sec to 2×10^4 cycles/sec and the speed at room temperature is 3.4×10^4 cm/sec. The frequency of visible light is 4.3×10^{14} to 7.5×10^{14} cycles/sec.

2. How do you suppose the amplitude of a circular wave might decrease as the wave moves outward? (Relate amplitude to radius.)

3. Knowing that sound is a wave, can you think of an explanation for the "rumble" in thunder?

4. Can you give a reason why you can only get water sloshing back and forth in a tub by pushing it at certain frequencies? Can you think of an explanation, on the basis of this fact, for the unusually high tides at the Bay of Fundy (which has 50-ft tides)?

5. What is "echo" and how does it arise?

6. Explain why you can hear a sound from a car moving far away from you better on a quiet summer night.

7. A source moving in a straight line with speed v on the surface of a lake produces spherical waves.

   a. Draw the wave fronts for the case where the speed of the source is zero.

   b. Draw the wave fronts for the case where $v \ll c$. (c is the speed of the wave.)

   c. Draw the wave fronts for the case where $v \gg c$, and show that the wave fronts are enclosed within a cone of angle such that $2 \times$ its sine is equal to c/v. (This cone is called the Mach cone.) In the case where the emitted wave is a sound wave, there will be a finite pressure difference across the Mach cone and this pressure difference gives rise to the sonic boom.

Problems

1. What are the possible wavelengths of standing waves in a wire stretched between two rigid supports that are 40 cm apart?

2. An organ pipe of length ℓ which is closed at one end and open at the other has the ability to contain standing waves whose wavelengths λ satisfy the equation

$$\lambda = \frac{4\ell}{2n - 1} \text{ for } n = 1, 2, 3, \ldots$$

On a day when the speed of sound in air is 1080 ft/sec, what are the three lowest frequencies at which a 16-ft organ pipe can support standing waves?

3. The wavelength of the orange light emitted by krypton-86 is 6.06×10^{-5} cm. What is its period? What is its frequency?

4. The frequency of International A, the note to which all instruments are tuned, is 440 cycles/sec. What length of organ pipe would give International A as its fundamental tone? See Problem 2.

5. Assume operating at a constant frequency of 2 cycles/sec produces a wave that passes from one medium to another. If the velocity in medium 1, 0.2 cm/sec, is twice that in medium 2, what are the wavelengths in the two media?

6. a. In a ripple tank, when one pulse is sent every $\frac{1}{10}$ sec, we find that λ is 3 cm. What is the speed of propagation?

   b. In the same medium we send two pulses, the second one $\frac{1}{2}$ sec after the first. How far apart are they?

16. Light as a Wave

Questions

1. Would you expect to be able to hear a note of frequency $v = 200$ per sec at the side of a 3-ft-wide door?

2. Would you expect to be able to see a light beam of frequency 6×10^{14} per sec at the side of a 0.1-cm hole in your window shutters?

3. Much of the Gulf of Mexico has only one tide a day. Can you think of an explanation for this?

4. From the point of view of the wave theory, how does one understand the colors, the behavior of a prism, and the propagation of light in a straight line in vacuum?

5. In Young's experiment, the distance between the two slits, for best results, should be several times λ, the wavelength of the light. What happens if this distance is much larger or much smaller than λ?

6. Why is it necessary that two portions of light in Young's experiment be derived from the same origin and that they arrive at the same point by different paths?

Problems

1. Coherent green light having a wavelength of 5.5×10^{-5} cm strikes an opaque screen with two slits separated by a distance of 2.5×10^{-2} cm. A diffraction pattern is produced on a parallel screen placed 1 m behind the opaque screen. Find the distances from the central bright line to the next two maxima. One can use the fact that for very small angles θ the difference between $\tan \theta$ and $\sin \theta$ is negligible, for $\cos \theta$ is then very close to 1.

2. Solve Problem 1 with the green light replaced by red light of wavelength 6.2×10^{-5} cm.

3. Solve Problem 1 with the slit separation changed to 0.2 mm.

4. Light of wavelength 5.5×10^{-5} cm passes through a single slit and falls on a screen 90 cm behind it. The first minimum is 0.14 cm from the center of the diffraction pattern. What is the width of the slit?

5. The average television channel has a frequency width of 4×10^6 cycles/sec. How many such channels would be available in the range of visible light, assuming that optical masers or lasers could be used for transmission? Take the range of the visible spectrum as 4 to 7×10^{-5} cm.

ELECTROMAGNETIC FORCES AND FIELDS

17. *Electrostatic Forces: Charges at Rest*

Questions

1. An electroscope is constructed of a metal ball, a conducting wire, and two very fine gold leaves, as shown. The gold leaves are enclosed in

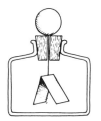

a bottle to keep drafts and fingers away. When the ball is touched with a charged insulating rod the two leaves, which normally hang down together, spring apart and stay in the position shown. Why? If a wire is now connected between ground (or a water pipe) and the ball, the leaves will again collapse. Why?

2. When the charged rod is not touched to the ball, but merely brought close, the leaves will move apart as long as the rod is held there and collapse when it is taken away. Why?

3. This time we bring the charged rod close enough so that the leaves separate, hold it there, touch the ball with the grounded wire, remove the wire, and finally remove the charged rod. After this sequence of operations the gold leaves remain apart. Why?

4. Grass seeds when suspended in a liquid and subjected to an electric field will line up parallel to the field. Why might that happen?

5. Imagine two electrons in orbits about a nucleus made of two protons. Would the Coulomb force between the two electrons produce a smaller or larger perturbation of the orbits than the gravitational force between two typical planets—say earth and Mars? Make an estimate.

Problems

1. What is the charge on a quantity of material containing 6.02×10^{23} molecules (1 mole) if one electron is removed from each molecule? (This quantity of charge has a special name: the faraday.)

2. A charge of 5 statcoulombs is situated 3 cm from a charge of −8 statcoulombs. What is the force between the two charges? Is it attractive or repulsive?

3. In Problem 2, where would a charge of 10 statcoulombs have to be placed so that the charge of −8 statcoulombs would feel no net force?

4. Two pith balls, having equal charge and weighing 200 dynes each, hang on strings that are 20 cm long. The vertex angle *between the two strings* is 10°. What is the charge on each pith ball? (Solve this graphically.)

5. If the pith balls in Problem 4 are not equally charged but, instead, one of them has a charge of 240 statcoulombs, what is the charge on the other? (Solve this graphically.)

6. Three charges of 5 statcoulombs each are placed at the corners of an equilateral triangle that has sides 3 cm long. What is the force on any one of these charges? (Solve graphically.)

7. The electric field strength between two charged plates is 3.4×10^4 dynes/statcoulomb downward. What charge on an oil droplet of mass 5×10^{-8} g would be needed to balance out its weight? This represents how many excess electrons?

8. What is the acceleration of an electron in a field of 10^6 dynes/statcoulomb? Express this in terms of g.

9. If the electron in Problem 8 were initially at rest, how long would it take it to attain one tenth the speed of light? What distance would it travel?

10. Given two parallel plates that are 0.8 cm apart, what potential would the plates have to be at to produce the 3.4×10^4 dynes/statcoulomb field required in Problem 9?

11. A charge of 6 statcoulombs is placed in an upward-directed uniform electric field of 1.5 dynes/statcoulomb. What is the work of the electrical force when the charge is moved (a) 45 cm to the right? (b) 80 cm downward? (c) 260 cm at an angle of 45° upward from the horizontal?

12. A potential difference of 2000 volts is established across parallel plates that are separated by paraffined paper which becomes conducting at field strengths of 5×10^7 volts/m. What is the minimum separation of the plates that will keep a current from flowing?

13. The air becomes conducting when the electric field intensity gets to about 3×10^6 volts/m. How close to the ground would a cloud with a potential 10^9 volts have to be for lightning to arise? (Assume, as a rough approximation, that the ground and the cloud act as two charged plates.)

14. A small pith ball of 0.2 g hangs by a thread of length 4 cm between two parallel plates 5 cm apart. The charge on the pith ball is 6 statcoulombs. What potential difference between the plates will cause the thread to assume an angle of 30° with the vertical? (Solve graphically.)

15. The electric field strength between two plates is 5000 dynes/statcoulomb. What is the force on a 10-statcoulomb charge? How much work is done on the charge if the field moves it 3 cm? What is your answer in electron volts?

16. What is the separation between two protons if the Coulomb force between them is just equal to the weight of one proton? The proton charge is $+e = +4.80 \times 10^{-10}$ esu and its mass is 1.67×10^{-24} g. Take $g = 980$ cm/sec².

17. Find the ratio of the Coulomb force of repulsion between two electrons to the gravitational force of attraction.

18. In the Bohr model of the hydrogen atom, the electron and proton are separated by 5.29×10^{-9} cm. At what separation would the gravitational force equal the Coulomb force at $5.29 \times$

10^{-9} cm? The mass of the electron is 9.11×10^{-28} g and the mass of the proton is 1.67×10^{-24} g.

19. What kinetic energy would the electron need in order to escape from the electrical attraction of the proton in the hydrogen atom? What is the corresponding speed?

20. In the hydrogen atom a proton and an electron are separated by a distance of 5.29×10^{-9} cm. The proton charge is $+e = 1.60 \times 10^{-19}$ coulomb $= +4.80 \times 10^{-10}$ esu and the electron charge is $-e$.

a. What is the Coulomb force on the electron?

b. What is the electric vector **E**?

21. In the hydrogen atom, what is the electric potential at the location of the electron? What is the electric potential energy of the electron?

18. Magnetic Forces: Charges in Motion

Questions

1. In a beam, the electrons are all moving in the same direction. Give a relation between the amount of mass flow and the electric current.

2. For a given potential difference between two ends of a wire, show that the greater the electric current that flows in the wire, the smaller is the value of the resistance of the wire.

3. From Ampere's law concerning the force between two current-carrying wires, explain why two electrons separated by a distance R and moving in the same direction repel each other less strongly than two electrons the same distance apart moving in opposite directions.

4. What is the work done by a magnetic field acting on a charged particle?

5. An electron is moving on a circular orbit under the influence of a uniform magnetic field acting in the direction perpendicular to the plane of the orbit. When one switches on a uniform electric field in the direction perpendicular to the uniform magnetic field, what would be a rough sketch of the electron trajectory? Consider the trajectory for the cases where the uniform electric field is (a) very weak, (b) of intermediate strength, (c) very strong.

6. When a conductor is placed in a magnetic field perpendicular to the direction of current flow, a potential difference is produced across the specimen in the direction perpendicular to both the current and the magnetic field (the Hall effect). Explain why.

7. If magnets are produced by current loops internal to the material, can one ever separate the north pole from the south pole?

Problems

1. Two long parallel wires are separated by 10 cm. If a current of 1 amp flows in each in the same direction, what is the force per unit length attracting one toward the other?

2. Suppose the two currents above are in opposite directions. What is the magnitude and direction of the force per unit length between them?

3. A light bulb having 3.6 ohms electrical resistance is connected directly across the terminals of a 12-volt battery. What is the current flow in the bulb?

4. To construct a microscopic picture of electric current, consider a uniform conductor of cross-sectional area A and length ℓ with a potential difference v between its ends. For a model of a conductor, one may consider it as a system of n_0 free electrons in a unit volume, each of which has the electric charge $-e$ and the mass m. Owing to impurities in the conductor, these electrons suffer collisions with the impurity atoms. The average time between two successive collisions is denoted as τ (the collision time). Assume for simplicity that at each collision the electron stops; then the average drift speed v_d is given by

$$v_d = (\text{acceleration acting on electron}) \times (\text{collision time})$$

a. Show that $v_d = \dfrac{e}{m}\dfrac{v\tau}{\ell}$.

b. Show that the current in the conductor satisfies Ohm's law and is given by

$$I = \frac{e^2 n_0 \tau A}{m\ell} v$$

so that the resistance R of the conductor is given by

$$R = \frac{m\ell}{e^2 n_0 \tau A}$$

c. Discuss the dependence of the resistance R on the length and the area of the conductor.

d. Show that the loss in potential energy in unit time is IV. If this loss in energy is converted into heat, show that the heat developed (joule heat) per unit time is $I^2 R$.

5. In an ordinary metal there are on the order of 10^{22} electrons per cubic centimeter. A current of 1 amp flows through a metallic wire with a diameter of 0.1 cm (about 0.04 inch) and a length of 1 m (about 3.3 ft).

a. What is the order of magnitude of the drift velocity?

b. How long does it take an electron to drift from one end to the other?

c. Try to explain why it does not take more than an instant to switch on household electric equipment even though it takes a very long time for an electron to move from one end of the wire to the other end.

6. A particle of mass m and charge e moves with speed v in the plane perpendicular to the uniform magnetic field B.

a. Show that the particle moves in a circular orbit of the radius mvc/eB.

b. Give the period of revolution and show that the frequency is $eB/2\pi mc$.

c. Show that the faster the particle moves, the larger the radius is, and that the period of revolution is independent of the speed of the particle or the radius of the orbit.

7. Calculate the force exerted on a 10-cm wire 2 cm from an infinite wire, each carrying a current of 1 amp (a) in the same direction and (b) in the opposite direction.

8. An electron (charge 4.80×10^{-10} statcoulomb) moves along a straight line with a speed of 1.00×10^8 cm/sec.

a. What is the current produced by the electron?

b. If a long straight wire carrying a current of 50.0 amp is placed parallel to the trajectory of the electron making a perpendicular distance of 10 cm, what is the force on the electron?

9. Show that 10^4 gauss = 1 tesla.

19. Induction Forces: Charges and Changing Currents

Questions

1. An electron with a speed v moves in a circular orbit under the influence of a uniform magnetic field B. Show that the magnetic flux through the orbit is

$$\Phi = \frac{\pi}{B}\left(\frac{mcv}{e}\right)^2$$

2. A magnet is dropped through a loop of wire. Neglecting air resistance, will the acceleration of the falling magnet be constant?

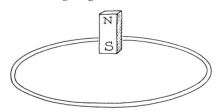

3. Why does the crank on a generator have to be pushed harder if current is being drawn (to heat toasters, for example)?

4. Why was the sudden opening or closing of the switch so important to the experiment of Faraday and Henry?

5. If for some reason their switches turned the current on and off very gradually, what would they have seen? (Use Faraday's law.)

6. Why might a small electric-potential difference develop across the wings of a plane (especially when it dives)?

Problems

1. A uniform magnetic field of 10^3 gauss is applied in the direction perpendicular to the plane of a circular loop of radius 10 cm. What is the flux through the loop?

2. In Problem 1, what is the flux through the loop if the magnetic field is parallel to the face of the loop?

3. What is the potential produced across the ends of the loop if the uniform magnetic field in Problem 1 is increased to 2×10^3 gauss in 1 sec?

4. A circular coil of radius 10 cm with 10 turns is rotated at 60 cycles/sec in a uniform magnetic field of 10^4 gauss. What is the potential set up across the ends of the coil?

5. An electron is constrained to move in a frictionless circular tube of radius 100 cm. A uniform magnetic field perpendicular to the plane of the circular tube increases at the rate of 10^3 gauss/sec. What force acts on the electron?

20 and 21. Electromagnetic Theory and Electromagnetic Radiation

Questions and Problems

1. What is a source for electromagnetic radiation?

2. What evidence is there that sound is not electromagnetic radiation?

3. Is change in current at a position felt by another charge or current at some distance away instantly? How fast is the change felt?

4. The electromagnetic radiation emitted by an electron in a circular orbit about a proton according to Maxwell's theory has the same frequency as the frequency of the electron's orbit. Imagine an electron in a circular orbit of radius $R = 2 \times 10^{-8}$ cm falling gradually into an orbit of radius $R = 10^{-8}$ cm as it emits energy by radiation. What range of wavelengths of radiation would be emitted? To what colors would they correspond? Give a qualitative description of what might be seen.

5. One end of a straight copper rod of length L rotates in a circle about the other end at a rate of N turns/sec. The rod is in a uniform B-field perpendicular to the plane of the circle.

a. What is the magnitude of the voltage difference between the ends of the rod, in terms of B, L, and N?

b. Evaluate this expression for $B = 10,000$ gauss, $L = 100$ cm, and $N = 10$/sec.

SPACE AND TIME REEXAMINED

22. Absolute Motion, Absolute Rest

Questions

1. Consider Newton's first law. How does it change from the point of view of an observer in uniform motion?

2. In what way is it different for an observer moving in a straight line with uniform acceleration?

3. Do Maxwell's equations change for observers in uniform motion with respect to one another? Consider the following two cases:

a. The force between two charges in the rest frame seen by two observers, one at rest, the other in a frame that moves in a straight line with a uniform velocity with respect to the rest frame.

b. The force on a charge in a uniform magnetic field when the charge is moving in a straight line with a uniform velocity. (The observer is riding on the charge.)

4. According to classical conventions what are the speeds of light seen by an observer moving with a speed v (a) along the same direction in which the light propagates, and (b) along the opposite direction in which the light propagates?

5. Why, from the point of view of Newtonian physics, must the center of the universe or the absolute motion of a body be legislated?

6. In what way does the situation seem to be changed with the addition of Maxwell's electrodynamics?

23. The Michelson-Morley Experiment

Questions and Problems

1. Suppose one wanted to use a Michelson type of interferometer with sound waves to measure the speed of an airplane. What would be measured, air speed or ground speed?

2. For an airplane moving at one half the speed of sound, what would be the difference of

transit times if $\ell_{\parallel} = \ell_{\perp} = 150$ cm. (Let the speed of sound be 30,000 cm/sec.)

3. To how many wavelengths does the time difference of Problem 2 correspond?

4. Let us assume that material bodies contract in the direction of motion when they are moving and the foreshortened length in the direction of motion is given by

$$\ell_{\parallel} = \ell \sqrt{1 - \left(\frac{v}{c}\right)^2}$$

where ℓ denotes the length when the body is at rest and v is the speed of the body in motion (Lorentz-Fitzgerald contraction). Show that if this is so, the Michelson-Morley experiment cannot detect any effect of the relative motion of the earth to the ether.

5. How can the results of the Michelson-Morley experiment be explained by assuming that the ether is dragged along with the earth as is the earth's atmosphere?

6. Explain how the above assumption of the ether being dragged with the earth can be tested by performing the Michelson-Morley experiment at two different altitudes above the surface of the earth. (The experimental results give the ether being not dragged with the earth.)

7. A river of width 100 m flows steadily with a uniform speed of 1 m/sec. A boat always travels with speed of 10 m/sec relative to the water.

a. What is the time required to cross the river and return to the starting point in the shortest possible distance?

b. What is the time required for the boat to travel the same distance downstream and return to the starting point?

c. What is the time required for the boat to travel the same distance in still water?

24. The Principle of Relativity

Questions

1. In what sense are Einstein's two postulates "apparently irreconcilable"?

2. What precisely are the suppositions that seem to be contradicted by the observed constancy of the speed of light?

3. If the Michelson-Morley experiment is not to be explained by ether drag, shrinking arms, and so on, and if one does not want to assume that the apparatus is at rest in an absolute sense, how can it be understood by the postulate that the speed of light in empty space is always c (independent of the motion of the observer measuring that speed)?

4. If the speed of light were infinite, in what way would all of Einstein's considerations be altered?

5. Looking at the form of the relativistic equations, what happens to them when c becomes infinite?

6. A particle at rest is found to decay in the time τ_0. When its speed is v relative to the laboratory, what is the lifetime of the particle measured by an observer in the laboratory?

7. Heracles is reputed to have had a pair of scissors 100,000 miles long. Prior to Einstein he could close them on a piece of paper in ½ sec, the scissors cutting the paper at 200,000 miles/sec, and in this way transmit information (the cutting of the paper) faster than the speed of light. In what way is the post-Einstein Heracles limited?

8. How is it that according to Einstein's conventions two observers moving uniformly with respect to each other can both say that the interval between two physical points is measured to be too small by the other and not be in contradiction?

Problems

1. How fast would a rocket ship have to be going relative to an observer for its length to be contracted to 90 percent of its at-rest length?

2. A spaceship is moving with respect to earth at one half the speed of light. Before it left it was provided with a good metal meter stick and a fine Swiss watch. From the point of view of an observer on earth:

a. How long is the meter stick measured to be?

b. How many minutes does the spaceship's watch record in 1 hr of earth time?

3. A student is given an examination to be completed in 50 min (by the professor's clock, always). The student and professor are moving at $0.98c$ with respect to each other. How much time has elapsed on the professor's clock, as measured by the student, when the professor says, "Time is up"?

4. In the book *Mr. Tompkins in Wonderland*, Mr. Tompkins enters a world in which the speed of light is only about 30 miles/hr. If a bicyclist is riding at a speed of 15 miles/hr, how short will his 6-ft bicycle be measured by a man standing still? (Nobody in Wonderland uses automobiles; you cannot go over 30 miles/hr and even to go much faster than 20 miles/hr cuts down on the gas mileage tremendously.)

5. An engineer wants to place a particle detector at the distance from his source where the most particles will die. He knows the average

lifetime of his particles is approximately 10^{-10} sec and that they move with a speed of $0.99c$. How far from the source will he place the detector?

6. In Wonderland, a special high-speed ($29\frac{2}{3}$ miles/hr) subway has been built to "save the commuter's time." Although riding an ordinary subway may cause people to age prematurely, this subway compensates by its time-dilating speed. If Mr. Tompkins rides the subway for 30 min (according to the clocks at rest in the station) each day, how much less will he have aged after 10 years?

7. A ball rolls eastward at a speed of 2 cm/sec, another westward at a speed of 2 cm/sec. Two protons move in opposite directions at speeds of 2×10^{10} cm/sec each. In each case, what is the speed of one particle with respect to the other?

8. One day, when Mr. Tompkins was riding the subway, he looked for the first time at the train on the other track going the other direction. Why did he think it was having trouble?

9. A man in an interstellar spaceship is going at a speed of $0.8c$ with respect to the earth. He is approaching a star group which, with respect to his spaceship, he is closing in on at $0.6c$. With what speed is this star group receding from the earth?

10. A rocket ship leaves earth and travels at $v = \frac{98}{100} c$ to a star which is 3×10^{18} m from earth as calculated from observations made on earth. The observers on the rocket ship can make their own observations.

    a. What is the distance between the earth and the star according to the astronauts?

    b. According to observers on the earth how long does the journey take?

    c. According to the astronauts, how long does the trip take?

11. As seen in a laboratory, a beam of protons traverses a pipe 12 cm long in 5×10^{-10} sec.

    a. What is the speed of the protons in terms of c?

    b. According to an observer moving with the protons, how long is the pipe?

    c. How much time is required for the protons to move through the pipe?

12. A beam of particles moving at 96 percent of the speed of light is found to have an average lifetime of 1×10^{-9} sec, according to an observer in the laboratory. What is the average lifetime of these particles when they are at rest?

13. Suppose the observed average lifetime of a beam of particles is found to be τ when the speed is $v_1 = 3/5 \ c$. Find the speed v_2 at which the observed lifetime is increased to 2τ.

25. The Union of Newton's Laws with the Principle of Relativity

Questions

1. Show that conservation of momentum holds for the relativistic theory if one assumes action and reaction are equal.

2. Discuss what happens to mass, velocity, and momentum if a constant force is exerted on a particle for a long time.

3. What happens to the relativistic modification of Newton's equations if the speed of light becomes infinite?

4. What is the minimum energy of a particle of mass m in the absence of forces from the point of view of the relativistic theory?

5. If a particle has zero rest mass (such particles are presumed to exist), what is its speed? (This can be obtained from the relativistic relation between speed and momentum.)

6. What is the energy-momentum relation for a particle with zero rest mass?

Problems

1. A muon has the same character as an electron except that its rest mass is 207 times larger than the rest mass of the electron and it is unstable. What is the mass of a muon moving with speed $(\sqrt{3}/2)c$?

2. An electron has rest mass 9.11×10^{-28} g. What is the mass and the momentum of an electron moving with $0.5c$?

3. Using the relation $E = m_0 c^2$, one can express the mass of a particle in terms of a unit of energy. In particle physics, a typical unit for specifying the energy is the electron volt ($1 \text{ eV} = 1.60 \times 10^{-19}$ joule). Other units related to the electron volt are:

kiloelectron volts	$= 1 \text{ keV}$	$= 10^3 \text{ eV}$
million electron volts	$= 1 \text{ MeV}$	$= 10^6 \text{ eV}$
billion electron volts	$= 1 \text{ BeV}$	$= 10^9 \text{ eV}$

What are the rest masses of an electron, a muon, and a proton in units of MeV?

4. What is the speed of a particle when the relativistic mass m exceeds the rest mass m_0 by 1 percent?

5. Find the kinetic energy of a proton when its relativistic mass m exceeds its rest mass m_0 by: (a) 1 percent, (b) 10 percent, (c) 100 percent.

6. What is the speed of a 10-MeV electron? Of a 100-MeV electron? By what factor do their speeds differ?

7. What is the speed of a 10-MeV proton? Of a 100-MeV proton? By what factor do their speeds differ?

8. If a constant force is applied for 1 sec, what is the ratio of the forces needed to stop electrons going $0.8c$ and $0.9c$?

9. In burning a quantity of firewood, 5000 joules of heat were obtained. How much mass was lost?

10. Oil can produce 4×10^5 joules/kg. What percentage of mass is turned into energy?

11. One type of nuclear reactor is able to produce 10^{13} joules per kilogram of fuel. What efficiency is this?

12. What is the kinetic energy of an electron if its relativistic mass is (a) twice the rest mass m_0? (b) ten times m_0?

13. What are (a) the kinetic energy and (b) the momentum of an electron moving at $v = 0.980c$? Express the momentum in MeV/c.

14. Find an algebraic expression for $B = v/c$ in terms of the total relativistic energy E and the rest mass energy E_0, where $E = m_0c^2$.

15. It is convenient to express the momentum of a relativistic particle in the units MeV/c.

a. Express $p = 1$ MeV/c in CGS units.

b. What is the kinetic energy of an electron with $p = 1$ MeV/c?

c. What is the kinetic energy of a proton with $p = 1$ MeV/c?

16. What is the kinetic energy of a particle whose momentum p is: (a) $10^{-1} E_0/c$, (b) E_0/c, (c) $10 E_0/c$? Express your answer in terms of $E_0 = m_0c$.

17. What is (a) the kinetic energy and (b) the momentum of a proton moving at a speed of $v = \frac{4}{5}c$? Express the momentum in MeV/c.

18. At the earth's surface the total energy radiated from the sun is 1350 joule/sec per m^2. The mean distance between the earth and the sun is $\frac{1}{50} \times 10^{11}$ m; the mass of the sun is 2×10^{30} kg.

a. Find the total energy radiated by the sun in 1 sec.

b. Find the mass of the sun transformed into radiation in 1 sec.

c. Find the fractional loss in the mass of the sun in 3×10^9 yr (the approximate age of the earth).

26 and 27. The Twin Paradox and The General Theory of Relativity

Questions and Problems

1. For a spaceship moving at two thirds the speed of light, how much does the astronaut twin age in 20 earth-years?

2. Is there any way that the twin in the spaceship could determine that he was not in an inertial frame?

3. From the point of view of forces and changes of motion (that is, from Newton's point of view), can one understand at least qualitatively why light bends as it passes near the sun?

4. If the frequency of a light particle is related to its energy by $E = h_v$, can one explain why the color of a photon is shifted toward the blue as it falls from the top of a building toward the earth?

5. Starting from the earth, an astronaut wants to reach the star Alpha Centauri in 0.43 years of his lifetime. As seen from the earth the distance to the star is 4.3 lights years. (Give the speed in terms of c.)

a. At what speed must the astronaut travel?

b. What is the minimum energy needed to achieve this result? Assume the combined mass of the astronaut and his command module is 1000 kg.

c. Express this energy in kilowatt hours. (1 kilowatt-hr $= 3.6 \times 10^6$ joule.)

6. What is the relative change in frequency of a proton as it moves from the sun to the earth.

7. The strongest known gravitational fields exist at the surface of white dwarf stars. The star 40 Eridani B is a typical white dwarf. Its mass is 43 percent of the mass of the sun and its radius is 1.6 percent the radius of the sun. What is the relative change in frequency of a proton as it moves from the star to the earth? (The actual observation made in 1954 gives a figure some 20 percent to 25 percent above the calculated value.)

STRUCTURE OF THE ATOM

28. Silver Threads

Questions and Problems

1. What would be a quick way to determine whether the contents of a small jar were sodium or strontium chloride?

2. What might the basis be for a claim that there is iron in the cool gas surrounding the sun?

3. How might one decide from spectroscopic observations whether the moon and the planets shine by their own light or by reflected light from the sun?

4. How might one distinguish those Fraunhofer lines in the spectrum of sunlight that are due to absorption in the sun's atmosphere rather than to absorption by gases in the earth's atmosphere?

5. What would be the wavelength of the fifth hydrogen line $(n = 7)$ on the basis of Balmer's formula? Why wasn't it found along with the first four?

6. If a 1-MeV (10^6 electron volts) γ ray is emitted from a radioactive substance, by what amount has the mass of that substance changed? How does this compare with the mass of an electron?

7. At the Cambridge Electron Accelerator, particles are accelerated to energies of 6 billion electron volts (6 BeV). Show that, at those energies, their speed is extremely close to the speed of light. α particles are usually emitted from radioactive materials with kinetic energies of a few MeV. Would relativistic connections be appreciable for them?

8. In what way might one verify that X rays are electromagnetic radiation?

29. Discovery of the Electron

Questions and Problems

1. Thomson attributed the absence of lasting deflection to leakage of gas into the tube. How might such gas interfere with the cathode rays? (Consider the possible electrical constitution of matter.)

2. Current theoretical fantasy has envisaged the existence of quarks—heavy (i.e., much heavier than a hydrogen atom) charged particles one of which has charge $\frac{1}{3}e$. After measuring the charge to mass ratio, e/m, of the quark, what value might Thomson have guessed for the mass of the quark? Suppose that a quark with the same mass but charge $-\frac{2}{3}e$ had also been observed. How might Thomson have tried to fit these particles into a theory of the structure of the atom?

3. Assume Thomson maintained a magnetic field perpendicular to the direction of motion of the cathode particles of 500 gauss. What electric field would be required to produce undeflected passage of those particles through the crossed fields?

4. How can we justify Thomson's neglect of the effects of gravitational forces on the cathode rays? Compute the total deflection by gravity for an electron traveling the length of a 2-m tube at a typical speed of 10^9 cm/sec?

5. Suppose protons, electrons, and alpha particles have all passed through a velocity selector and enter a region of uniform magnetic field. The magnetic field is 250 gauss and the velocity

of the particles is 10^8 cm/sec and perpendicular to the field. What is the radius of the circular paths followed by these particles? Draw a diagram showing the path.

6. Millikan measured the charge on the electron by charging oil drops and watching them move under the influence of gravitational and electrical forces. Consider oil drops of mass 10^{-14} g that become charged by absorbing radiation emitted from radioactive materials. The drops are then suspended between the plates of a parallel plate condenser by adjusting the voltage so that the force of gravity is balanced. Assuming the minimum charge on these drops is that due to Thomson's particle, what is the maximum electric field required to suspend these drops?

7. A particle with charge q and mass m travels in a straight line through a crossed field region R_1 where $B_1 = 200$ gauss and $E_1 = 4.80 \times 10^4$ volts/m. The particle emerges into region R_2 where $B_2 = 1000$ gauss and $E_2 = 0$. In region R_2 the particle follows a trajectory with radius $\rho_2 = 25$ cm.

a. What is the speed of the particle?

b. Find an algebraic expression for q/m in terms of E_1, B_1, B_2, and ρ_2.

8. An electron traveling north in a horizontal plane at a speed of 2×10^6 m/sec enters a region where uniform **E** and **B** fields are combined so that the electron passes through undeflected. The **E** field of 3×10^4 volts/m is directed vertically downward. What is the direction and magnitude of **B**?

30. Rutherford's Nuclear Atom

Questions and Problems

1. Alpha particles of 1 MeV kinetic energy are shot at a sheet of gold foil.

a. What is the distance of closest approach to the nucleus?

b. How close would a proton of the same energy approach?

2. On what basis could Rutherford ignore collisions with electrons in large-angle scattering of alpha particles?

3. What is the minimum kinetic energy (in electron volts) required for an alpha particle to approach a *point* charge $Q = 79e$ to within a distance of (a) 10^{-8} cm? (b) 10^{-12} cm?

4. A proton with kinetic energy of 1 MeV is incident on a gold target and is reflected back along the incident direction. Assume the radius of the gold nucleus is about 10^{-13} cm.

a. What is the value of momentum transfer in this case?

b. At what point along the trajectory is its kinetic energy equal to its potential energy?

c. At what point on its trajectory is the speed zero?

5. The day Rutherford completed his calculations, he went to the theoretical physicist, C. G. Darwin, and asked him to check the results. At the same time, he asked Darwin how close the 1.6×10^9 cm/sec alpha particles could approach a point nucleus if the force of repulsion were $2Ze^2/r^3$. What was Darwin's result?

6. An alpha particle of energy 4 MeV is scattered by a gold nucleus. The initial momentum of the particle is along a line 10^{-8} cm from the center of the gold nucleus.

a. Will this value change?

b. Will the speed of the alpha particle ever be zero along its trajectory?

31. Origins of the Quantum Theory

Questions and Problems

1. What in our ordinary experience indicates that there must be something fundamentally wrong with the analysis that leads to the "ultraviolet catastrophe"? Almost none of the energy in a cavity can leave through the hole in its side; therefore, most of the radiation going into the cavity will not be able to get out. With this in mind, explain why a cavity is similar to a body coated with carbon black.

2. Why from the point of view of 1905 was Einstein's treatment of light in a container as a gas of particles a radical departure from the current doctrine? What is the energy of a radio-wave photon that has a wavelength of 200 cm?

3. What is the minimum frequency photon that can be absorbed by a hydrogen atom in the ground state? Is there a maximum?

4. It takes 2.0 eV to produce photoelectric electrons in potassium. What is the minimum wavelength radiation that will produce these electrons?

5. Silver bromide (AgBr) is a light-sensitive substance used in some types of photographic film. To cause exposure of the film, light must be available with enough energy to dissociate the molecule—an energy of 1.04 eV. What is the longest wavelength that will suffice?

6. The longest wavelength that will produce photoelectrons in sodium is 5400 Å.

a. Calculate the binding energy of these electrons.

b. What kinetic energy will the photoelec-trons have if they are excited by using 2000-Å light?

7. What is the minimum voltage that must be applied across the cathode and anode of an X-ray tube to produce X rays of wavelength 10^{-8} cm?

8. In the upper atmosphere of the earth, molecular oxygen, O_2, is dissociated into two oxygen atoms by photons from the sun. The maximum photon wavelength that permits this process to occur is 1750 Å. What is the binding energy, in electron volts, between two oxygen atoms forming a molecule? (This phenomenon has been proposed as a possible source of energy for high-altitude flying. The solar radiation dissociates the oxygen molecules into atomic oxygen, and then energy is released as the atoms recombine to form molecule.)

9. From the point of view of the Bohr atom, what is the orbital radius of the first excited state of hydrogen?

10. What is the wavelength emitted in making a transition from the second excited state in hydrogen to the first excited state? What part of the spectrum is this line in?

11. Assuming no electromagnetic radiation losses, what is the speed of an electron in a circular orbit 10^{-8} cm from a proton? How many revolutions per second does it make? If it emitted light in this orbit, would it be in the visible spectrum?

12. The lifetime of an excited state of a typical atom is 10^{-8} sec. From the point of view of the Bohr theory, how many revolutions about the nucleus are made by an electron in the first excited state of hydrogen before it makes a transition to the ground state?

13. What would be the minimum size of the hydrogen atom if $h = 1$ erg-sec? How does this compare with the size of the earth?

14. Following Bohr, compute the radius of the ground-state orbit of He^+. At what speed must the electron be circling the nucleus? Compare the spectra of the helium and hydrogen.

15. The mu meson (muon) is a fundamental particle similar to the electron, except that its rest mass is 207 times larger. Suppose a muon is captured by a hydrogen nucleus to form a mu-mesonic atom. Compare the energy levels with those of a normal hydrogen atom. What frequency is emitted when the muon jumps from the first excited state to the ground state? What type of radiation is this?

16. What is the shortest wavelength emitted by normal hydrogen? (It is the *series limit* for the Lyman series.)

17. Find the wavelength (in Ångstrom units) of (a) a 1-MeV photon, (b) a 1-KeV photon, (c) a 1-eV photon.

18. In terms of h, λ, and c, what is the relativistic mass of a photon?

19. What is the wavelength (in Ångstrom units) of a photon whose energy is equal to the rest mass energy of an electron?

20. The human eye is most sensitive to yellow-green light (5500 Å). What is the energy in electron volts of a 5500-Å photon?

21. The human eye can detect a flash of yellow-green light (5500 Å) with an energy of about 1×10^{-18} joules incident on the retina. What is the corresponding number of photons?

THE QUANTUM THEORY

32. *The Electron as a Wave*

Questions and Problems

1. What is the de Broglie wavelength associated with an electron moving at 3×10^9 cm/sec?

2. What is the de Broglie wavelength associated with the earth ($m = 6 \times 10^{24}$ kg) moving at 3×10^6 cm/sec about the sun? For a 75-kg man walking at 0.5 m/sec?

3. Most television sets use a potential difference of about 20,000 volts to accelerate electrons in the picture tube. What is the de Broglie wavelength of such an electron?

4. What would be the necessary separation between the slits of a plane grating that would diffract through 30° a rifle bullet having a mass of 1 g and velocity of 3×10^4 cm/sec? (sin 30° = ½.)

5. Performing an electron-diffraction experiment with a crystal whose atomic spacing is 10^{-8} cm, the first diffraction maximum is found at an angle of 30°. What is the speed of the electrons?

6. If the radius of a hydrogen atom is 10^{-8} cm, what is the speed of an electron in the first Bohr orbit?

7. (a) Write an expression for the wavelength λ of a particle moving at a speed $v = \beta c$. Express the result in terms of β, E_0, and various fundamental constants. (b) Find λ for an electron moving at $v = \frac{4}{5}c$.

8. Find an algebraic expression for the kinetic energy K of a relativistic particle in terms of its wavelength λ, its rest mass energy E_0, and various fundamental constants.

9. Find an algebraic expression for the wavelength λ of a relativistic particle. Express the result in terms of E_0, K, and certain constants.

10. Find the wavelength of (a) a 1-MeV photon, (b) a 1-MeV electron, (c) a 1-MeV proton.

33. *Schrödinger's Equation: The Law of Motion for Quantum Systems*

Questions and Problems

1. What would be some of the differences between our universe and one where Planck's constant equals 1 erg-sec?

2. If an electron were held in the nucleus of an atom ($r = 10^{-12}$ cm), what is the minimum kinetic energy it could have? Compare this with the Coulomb potential of two opposite charges whose centers are held 10^{-12} cm apart.

3. In which Bohr orbit about the sun is the planet Mercury? (Distance from the sun is 6×10^{12} cm; mass is 3×10^{33} kg; velocity is 5×10^6 cm/sec.)

4. If the value of h were 1 erg-sec, what would be the uncertainty in speed of a 1-g ball bearing contained on a line 10 cm long?

5. Imagine a black box that contains free electrons instead of photons. What is the minimum kinetic energy an electron could have if the inner dimensions of the box were 2 cm?

6. The singly ionized anion of a common type of organic dye is shown below:

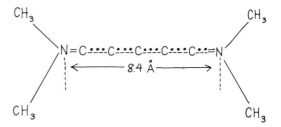

It is a good approximation to assume that the electrons in the conjugated double bonds in the structure above are able to move from one end of the molecule to the other in a constant-potential-energy well bounded by infinite walls at either end. Using this "one-dimensional box" approximation, calculate the energies of the first four energy levels. The color of the dye is due to a transition from level 4 to level 3. What color is the dye? (*Ref.*: W. J. Moore, *Physical Chemistry*, 3rd ed., Prentice-Hall, Englewood Cliffs, N.J., 1962, p. 609.)

7. What is the *minimum* kinetic energy K of a proton confined to a spherical region with a radius of 10^{-12} cm? Compare this result to Problem 2.

8. Find (a) the kinetic energy K_p of a proton with $\lambda = 10^{-12}$ cm, (b) the kinetic energy K_e of an electron with $\lambda = 10^{-12}$ cm.

9. In a nickel crystal the atomic spacing is $d = 2.15$Å. If the preferred direction for electron scattering from a single crystal is $45°$, (a) what is the electron wavelength λ? (b) what is the electron kinetic energy K in electron volts?

10. Find the energy (in electron volts) at which the following objects have a wavelength $\lambda = 1$Å $= 10^{-8}$ cm: (a) photon, (b) electron, (c) proton.

34. What Is the Associated Wave?

Questions and Problems

1. If electrons are shot through a slit one at a time, in what sense do they form a diffraction pattern?

2. What kind of experimental result might revive interest in the interpretation of the square of the wave function as a matter density?

3. Consider an electron confined to move in a one-dimensional box of length ℓ. If the electron were in the first excited state $(n = 2)$, where would the wave function of the electron be at a minimum? Where would the probability of finding the electron be highest? What would the answers to these questions be for the second excited state $(n = 3)$?

4. One of the problems in the everyday operations of a gambling casino is keeping the roulette wheels balanced so that the probability that the ball lands on any particular number is $1/37$. How might a knowledge of quantum physics help?

35. On the Consistency of the Quantum View

Questions and Problems

1. What is the uncertainty in momentum of an electron confined to a nucleus $(r = 10^{-12}$ cm$)$? What uncertainty in velocity does this correspond to?

2. What is the uncertainty in velocity of a pinball that has a mass of 300 g and can be localized to a thousandth of a centimeter?

3. What is the uncertainty of the pinball in Problem 2 if the value of h were 1 erg-sec?

4. In an experiment we localize a proton to 10^{-12} cm. What is the minimum energy it can have?

5. An electron beam impinges on a slit 10^{-6} cm wide with a velocity of 10^9 cm/sec. What is the uncertainty in the direction perpendicular to the slit and direction of motion of the beam at a distance of 1 m from the slit?

6. If one localized a particle so that $\triangle x = 0$, then the uncertainty in momentum must be infinite. Discuss the consistency of this statement with the fact that one knows, by relativity theory, that the speed must be smaller than c.

7. A radioactive isotope of silver (^{108}Ag) decays in about 2.4 min. What is the uncertainty in its energy?

8. Suppose we localize an electron with an accuracy of 10^{-8} cm, how much kinetic energy must it have? Compare this with the binding energy of hydrogen. How is this related to the possibility of following an electron as it traces out the first Bohr orbit of hydrogen?

9. Some elementary particles (or resonances) have uncertainties in their rest mass as large as 200 MeV. What is the lifetime of such a particle?

10. Express Planck's constant in units of electron volt sec $\equiv$ eV-sec.

11. Assume that a photon with $\lambda = 6000$ Å is produced by a transition with a mean decay time of 10^{-8} sec. (a) What is $\triangle \nu$, the uncertainty in the frequency of the photon? (b) What is $\triangle \nu / \nu$, the *relative* uncertainty in the frequency?

12. Consider a photon with wavelength λ and frequency ν. Show that $\triangle \lambda / \lambda \simeq - \triangle \nu / \nu$.

THE QUANTUM WORLD

37. The Hydrogen Atom

Questions and Problems

1. What is the angular momentum quantum number of a 1-g ball being twirled at 2 revolutions per second on the end of a 20-cm-long string? What would this quantum number be if the value of h were 1 erg-sec?

2. The angular momentum of a system is found to be equal to $6\sqrt{2}\,\hbar$. What is the angular momentum quantum number?

3. What is the maximum difference in wavelength due to the splitting in the hydrogen $3S \rightarrow 2P$ emission line due to a 10,000-gauss magnetic field? (The normal wavelength of this line is about 6500 Å.) Would it be possible to see this shift as a change in color without the use of instruments?

4. What is the wavelength of the $3D \rightarrow 2P$ transition in hydrogen? Is this visible to the unaided eye? If so, what color is it?

5. A hypothetical one-electron atom is observed to emit wavelengths of 1200 and 1000 Å when excited. Assume these are the lines corresponding to the $2P \rightarrow 1S$ and $3P \rightarrow 1S$ transitions.

a. If a beam of electrons with an energy of 11 eV were incident upon a gas of these atoms, what energy electrons would you expect to see coming out?

b. If the beam energy were increased to 13 eV, how would your answers change?

6. If the energy emitted in the $2P \rightarrow 1S$ transition of a one-electron atom is known to be 7.8 eV, what is the minimum speed a striking electron needs to excite the atom from the 1S to the 2P state?

7. Laser light is produced when a great number of atoms in an excited state return to the ground state almost simultaneously. Why would one expect the light emitted to be much more monochromatic than that due to ordinary atomic transitions (in, say, a heated gas)?

8. The two D lines in sodium, at 5890 Å and 5896 Å, are produced by an internal magnetic field B_{int} which removes the electron-spin degeneracy. What is the magnitude of B_{int}?

38. Many-Particle Quantum Systems

Questions and Problems

1. Construct a periodic table for the first eight elements using *spinless* electrons. What would the qualitative chemical properties be?

2. If either radium or radioactive strontium were swallowed, where would these elements concentrate in the body? To answer this, consider the chemical properties of the elements in the groups or columns in the periodic table.

3. Why was helium for so long undiscovered? (It was not discovered until the late 1800s.)

4. Can you suggest why francium is so much more reactive than sodium?

5. How is it that a mu meson can fill the lowest orbital state (1S) in an atom, where all the lowest electron orbitals are already filled?

6. (a) What is the total energy E of a system of 10 identical fermions confined within a one-dimensional container of length ℓ? (b) If the particles are electrons and $\ell = 10^{-8}$ cm, what is E in electron volts?

39. The Atomic Nucleus

Questions and Problems

1. What are the number of neutrons and protons in the nuclei $_{10}\text{Ne}^{20}$, $_{82}\text{Pb}^{208}$, and $_{92}\text{U}^{238}$?

2. To what properties of the nucleus might a mu-mesonic atom be particularly sensitive?

3. Is there a mass defect in chemical reactions? What is its approximate magnitude?

4. The mass of an isotope of beryllium $_4\text{Be}^8$ is 13.2880×10^{-24} g. Calculate the binding energy.

5. The unstable isotope $_{84}\text{Po}^{213}$ decays through an intermediate atom to the stable isotope $_{83}\text{Bi}^{209}$. Write down the decay scheme.

6. In the process beryllium$^8 \rightarrow$ two alpha particles, what is the kinetic energy of the two final alpha particles?

7. How much energy is released in the fusion reaction $_1\text{H}^3 + _1\text{H}^1 \rightarrow _2\text{He}^4$?

8. What could one say about the respective energies of the alpha particles emitted in the decays $_{90}\text{Th}^{230} \rightarrow _{88}\text{Ra}^{226} + _2\text{He}^4$ and $_{84}\text{Po}^{214} \rightarrow _{82}\text{Pb}^{210} + _2\text{He}^4$? (Refer to Table 39.1, p. 410.)

9. The energy received from the sun at the earth is 8×10^7 erg/cm²-min on a surface perpendicular to the rays of the sun. At what rate, in tons per minute, must hydrogen be consumed in the reaction $4(_1\text{H}^1) \rightarrow _2\text{He}^4$ to provide this radiated energy?

10. Calculate the height of the Coulomb barrier of the $_{92}\text{U}^{235}$ nucleus for (a) protons and (b) alpha particles.

11. Why are there no stable isobars (elements with the same atomic weight) that have their atomic numbers differing by only 1?

12. As a rough estimate of the energy released in the fission of $_{92}\text{U}^{235}$, calculate the electrostatic potential energy of the $_{56}\text{Ba}^{146}$ and $_{36}\text{Kr}^{90}$ fragments at the instant the $_{92}\text{U}^{235}$ nucleus begins to split. Assume that the initial separation is of the order of 10^{-12} cm.

40. Dirac's Relativistic Electron

Questions and Problems

1. Suppose electrons had a spin of $\frac{3}{2}$. How would the allowed levels of the hydrogen atom be affected? Note that all particles with odd half-integer spin obey the exclusion principle.

2. In a relativistic theory, how many internal states would an electron of spin $\frac{1}{2}$ have?

3. Give examples of negative energy states that exist in nature. How do these differ from the negative energy states of Dirac particles discussed in Chapter 40?

4. An energy of 3 MeV is put into the Dirac vacuum and an electron with kinetic energy of 1 MeV is created. What else is created? What is its energy? Describe the process in terms of Dirac's ideas. Draw energy-level diagrams for the system before and after the excitation. How is the "hole" observed experimentally?

5. What is the minimum energy photon that can cause pair formation of an electron-positron pair? (*Hint:* Assume that all the energy is converted into an electron and a positron at rest.)

6. What is the minimum energy required to create a proton-antiproton pair?

7. Show that conservation of energy and momentum require that at least two photons of equal energy be emitted when an electron and a positron moving in opposite directions annihilate each other.

8. Consider the annihilation of an electron-positron pair into one photon. For an observer stationary with respect to the center of mass of the two particles, $p_1 = p_2$. For this observer:

a. What is the initial energy of the pair?

b. What is the total momentum of the system?

c. If energy and momentum are conserved, compute the "mass" of the photon produced in the annihilation using the Einstein relation $E^2 = m^2c^4 + p^2c^2$. Does this diagram represent a real process?

FIRST MATTER

Chapters 41–47.

Questions and Problems

1. What is the kinetic energy carried away by the neutrino in the decay of a charged pion at rest?

2. If one wished to study the interactions of a high-speed particle that has a relatively small probability of interacting with the matter it is passing through, would one be more likely to use a cloud chamber or a bubble chamber?

3. A 600-MeV K^+ meson (rest mass 494 MeV) is observed to decay after leaving a 100-cm-long track in a hydrogen bubble chamber.

What is the approximate lifetime of the K^+ meson? What assumptions have you made that may affect the accuracy of your answer? Would you say that the K^+ meson decays via a strong interaction?

4. Which of the following production processes are not possible and why? Assume the incoming particle has sufficient kinetic energy to allow the process to proceed.

    a. $\gamma + p \rightarrow \pi^+ + \pi^- + \pi^0$
    b. $\gamma + p \rightarrow n + \pi^+ + \pi^0$
    c. $\gamma + p \rightarrow K^+ + \Lambda^0$
    d. $p + p \rightarrow \Xi^0 + K^0 + \pi^+$
    e. $\pi^- + p \rightarrow \pi^0 + \Lambda^0$
    f. $\pi^- + p \rightarrow n + \pi^0$
    g. $\pi^+ + p \rightarrow \pi^+ + \mu^+ + \bar{\nu}$

5. Which of the following decays are not possible and why?

    a. $\Xi^0 \rightarrow \Sigma^0 + \pi^0$
    b. $\Sigma^+ \rightarrow \pi^+ + \pi^0 + \pi^0$
    c. $K^0 \rightarrow \pi^+ + \pi^- + \pi^+$
    d. $\pi^0 \rightarrow \Xi^\gamma + \pi^\gamma$
    e. $\Lambda^0 \rightarrow n + \pi^0$
    f. $n \rightarrow \pi^+ + e^- + \nu + \bar{\nu}$

6. Which of the following decays would have a lifetime typical of a strong or electromagnetic interaction and which that typical of the weak?

    a. $\Xi^+ \rightarrow \pi^+ + \Lambda^0$
    b. $K^- \rightarrow \mu^- + \nu$
    c. $\Sigma^0 \rightarrow \Lambda^0 + \gamma$
    d. $\Omega^- \rightarrow \Xi^0 + \pi^-$
    e. $\Xi^- \rightarrow \pi^- + \Lambda^0$

7. Place the following decays in order of their lifetimes.

    a. $\Delta^{++} \rightarrow p + \pi^+$
    b. $\Sigma^0 \rightarrow \Lambda^0 + \gamma$
    c. $\Xi^0 \rightarrow \Lambda^0 + \pi^0$

8. A Van de Graaff generator can be used to accelerate electrons, protons, deuterons, and alpha particles. If the generator potential is 6.0×10^6 volts, what energies can each of the above particles attain? What speeds will they have?

ANSWERS TO SELECTED
ODD-NUMBERED
QUESTIONS AND PROBLEMS

Chapter 2

  1. Yes
  3. 1.1×10^4 ft, 2.1 miles, 3.4×10^5 cm, 3.4 km
  5. 110 miles, 48.8 miles/hr
  7. 10 sec, 98 m-sec
  9. 2 sec, 80 ft, 48 ft/sec
  11. 2.58 sec
  13. a. 5×10^{-3} sec; b. 8×10^4 m/sec^2
  15. a. 28 ft/sec; b. 92 ft/sec
  17. a. 100 ft; b. 2.50 sec; c. 16 ft/sec; -48 ft/sec; -80 ft/sec
  19. a. 11 ft/sec^2; b. 352 ft

Chapter 3

  1. a. 30 miles/hr; b. north
  3. 13 miles
  5. 30 miles/hr from the northwest

Chapter 4

  1. 4 g-cm/sec
  3. 2×10^3 g-cm/sec east; 3×10^3 g-cm/sec west
  5. 1.44×10^{-10} dyne
  7. 3.94×10^6 dyne
  9. a. 2.99×10^6 cm/sec; b. 3.57×10^{27} dyne toward the sun
  11. a. 12.6 kg-m/sec; b. 1.26×10^3 N
  13. 2280 N
  15. 21.2 kg-m/sec

Chapter 5

  1. 1.89 Re
  3. 5.4×10^{12} m
  5. ·0.4 years
  7. 6.57×10^3 sec

Chapter 6

  1. 0.41 Re
  3. Mars; 80 lb on Mars, 490 lb on Jupiter

  5. ¼ of its weight on earth, or 25 lb
  7. a. 9.8×10^7 dynes; b. 5.9×10^4 dynes by the sun; c. 3.4×10^2 dynes by the moon
  9. 1.3×10^{-8} lb
  11. a. 3.6×10^{-42} dyne; b. 8.34×10^{-3} dyne toward the proton; c. about 2.3×10^{39}

Chapter 7

  1. a. 7.25×10^5 dyne-seconds acting along the direction in which the ball is bounced off the wall; b. 7.25×10^7 dynes in both cases; c. 7.25×10^5 dyne-sec opposite in direction to the impulse exerted on the ball
  3. 25 lb
  5. 104 g-cm/sec directed 17° from the north toward east
  7. 24 m/sec in the opposite direction
  9. 16×10^5 dynes

Chapter 8

  1. about 1 percent
  3. 6 ft
  5. 8.1 ft/sec
  7. it is the same
  9. 2.4×10^3 m/sec (or 5300 miles/hr)
  11. 80,000 lb-ft
  13. 2.5×10^7 ergs
  15. a. 10,000 lb-ft; b. 20 ft/sec
  17. 3.60 m/sec
  19. 2.83 m/sec, southwest
  21. 1 m/sec

Chapter 9

  1. 37.0° C
  3. 4.6 cal
  5. 4 hr 27 min
  7. 1050 cal
  9. 3.55×10^4 cm/sec
  11. 0.359 cal/g-C°
  13. 559.60 cal

Chapter 14

1. 8.3 min
3. 9.4600×10^{17} cm
5. .48 miles
7. 18°

Chapter 15

1. 80, 40, 80/3, 20, . . . , $80/n$, . . . (cm)
3. 2.0×10^{-15} sec, 4.95×10^{14} sec^{-1}
5. 0.1 cm; 0.05 cm

Chapter 16

1. 0.22 cm, 0.44 cm
3. 0.28 cm, 0.55 cm
5. 80 million

Chapter 17

1. 2.89×10^{14} esu
3. 4.24 cm from the charge of -8 statcoulombs, directly opposite the 5-statcoulomb charge
5. 0.89 statcoulombs
7. -14.4×10^{-10} esu, 3 electrons
9. 5.7×10^{-15} sec, 8.55×10^{-6} cm
11. a. 0; b. -720 ergs; c. 1656 ergs
13. 3.33×10^2 m
17. 4.16×10^{42}
19. 27.2 ev; 3.09×10^8 cm/sec
21. 9.07×10^{-2} statvolt/esu; -43.5×10^{-12} erg

Chapter 18

1. .2 dyne
3. 3.3 amp
5. a. 7.95×10^{-2} cm/sec; b. 126×10^3 sec $=$ 21 min
7. a. 0.1 dyne, attractive; b. 0.1 dyne, repulsive

Chapter 19

1. 3.14×10^5 gauss-cm^2 $= 3.14 \times 10^{-3}$ weber
3. 1.05×10^{-5} statvolts $= 3.15 \times 10^{-3}$ volt
5. 8×10^{-16} dyne

Chapter 24

1. $.44c$
3. 4 hr 12 min
5. 21 cm
7. 4 cm/sec, 2.8×10^{10} cm/sec
9. $0.39c$
11. a. ⅘c; b. 7.20 cm; c. 3×10^{-10} sec
13. $\sim 0.92c$

Chapter 25

1. twice its rest mass
3. 0.511, 106, and 938
5. a. 9.38 Mev; b. 93.8 Mev; c. 938 Mev
7. 4.24×10^9 cm/sec, 1.28×10^{10} cm/sec, ratio $= 3.03$

9. 5.56×10^{-11} gm
11. .011 percent
13. a. 2.06 Mev; b. 2.52 Mev/c
15. a. 5.33×10^{-17} g-cm/sec; b. 0.612 Mev; c. 5.32×10^2 ev
17. a. 625 Mev; b. 1.25×10^3 Mev/c

Chapters 26 and 27

5. a. $0.995c$; b. 8.10×10^{20} joules; c. 2.75×10^{14} kWh
7. 5.7×10^{-5}

Chapter 28

1. Place a sample in a flame
5. 3969.7 Å; it is not in the visible part of the spectrum

Chapter 29

3. 500 v/cm
5. 43.1 cm; .0235 cm; 86.2 cm
7. a. 2.40×10^6 m/sec; b. $E_1/B_1B_2\rho_2$

Chapter 30

1. a. 4.32×10^{-11} cm; b. 2.16×10^{-11} cm
3. a. 2.3 kev; b. 23 Mev
5. 1.46×10^{-6} cm

Chapter 31

3. min $= 2.46 \times 10^{15}$; max $= 3.29 \times 10^{15}$
5. 11,900 Å
7. 12,400 v
9. 2.12×10^{-8} cm
11. a. 1.6×10^8 cm/sec; b. 2.54×10^{15}; c. no
13. 2×10^{44} cm
15. 5.86 Å; X rays
17. a. 0.0124 Å; b. 12.4 Å; c. 12,400 Å
19. 0.0243 Å
21. $\sim$ 3 photons

Chapter 32

1. 2.43×10^{-9} cm
3. 6.2×10^{-9} cm
5. 5.15×10^8 cm/sec
7. a. $\sqrt{(1 - \beta^2)/\beta^2}\ hc/E_0$; b. 1.82×10^{-10} cm
9. $(hc/K)1/\sqrt{1 + 2E_0/K}$

Chapter 33

3. 8.5×10^{82}
5. 1.75×10^{-27} erg
7. $\sim$ 21 Mev $<< E_0$
9. a. 1.52 Å; b. 65.25 ev

Chapter 34

3. $x = 0, \ell/2, \ell; x = \ell/4, 3\ell/4; x = 0, \ell/3, 2\ell/3, \ell; x = \ell/6, \ell/2, 5\ell/6$

Chapter 35

1. 6.6×10^{-15} g-cm/sec; about 3×10^{10} cm/sec
3. 3.3 cm/sec
5. 0.7 cm
7. 4.6×10^{-30} erg
9. 2.1×10^{-23} sec
11. a. 10^8 H_z; b. 2×10^{-7}

Chapter 37

3. 0.6 Å; no
5. a. 11 ev, 0.7 ev; b. 13 ev, 10.9 ev, 2.7 ev, 0.6 ev

Chapter 39

1. 10, 10; 82, 126; 92, 146

3. Yes; 13.6 ev for hydrogen ($\simeq 3 \times 10^{-32}$ g)
5. $_{84}Po^{213}$ $_{82}Pb^{209}$ $_{83}Bi^{209}$
7. 16.2 Mev
9. 4×10^{10} tons/min

Chapter 40

5. 1.022 Mev

Chapters 41–47

1. 30.4 Mev
3. 10^{-8} sec
5. Violates: a. energy conservation; b. baryon conservation; c. charge conservation; f. baryon conservation
7. a, b, c

INDEX

Italicized numbers refer to pages on which definitions appear.

Light (*Continued*)
wave theory of, 144, 145, 170 ff.
Line spectra, 302–306
Line waves, *162*, 163
Linear accelerators, *438*
Liquids, superheated, 434
Loadstone, 217, 218
Local time, 257, 263, 273
Localization of wave functions, 364–373
Loops of conductors, magnetic fields produced by, 212
Lorentz, H. A., 255–257
Lorentz force, *209*, 210, 213, 214, 232
Luminiferous ether, *see* Ether
Lyman series for hydrogen atoms, 335

MKS system of units, *39*
electrostatic, *191, 192*
Magnetic effects of electric currents, 205–208
Magnetic field lines, 219
Magnetic fields, 210–214
effects of, on atomic energy states, 389–390
effects of, on spectra of hydrogen, 390–391
produced by electric currents, *230–233*
produced by loop conductors, 212
produced by solenoids, 212–213
produced by straight conductors, 211
right-hand rule for, 210, 211
uniform, effect of, on moving charged particles, 214–217
moving conductors, 225–227
Magnetic flux, 223–225
Magnetic forces, 204 ff.
Magnetic poles, 219
Magnetic quantum numbers, 386, 387
Magnetism, permanent, 218
Magnets, 217–219
Mass, *32*
of the earth, 61
of electrons, *313*
and wave functions, 353 ff.
equivalence to energy, 284–286
relativistic, *282*
rest, *282*
units of, *38*, 39
Mathematics, relation to physics, 123–128
Matter, electrical properties of, 313–314
equivalence to energy, 284–286
nature of, waves or particles, 353 ff.
Maxwell, James Clerk, 228 ff.
Maxwell's equations, 231–232, 233–235
and Bohr's atomic model, 335–336

and Rutherford's atomic model, 323–324
Measurements in physics, 123–128
Mechanical energy, 77 ff.
conservation of, 86, 87, 88
Mechanical equivalent of heat, 100–104
Mechanics, relativistic, 280 ff.
Media for waves, dispersive, *156*
nondispersive, *155*
Melting, 96–97
Mercury (planet), perihelion advance and relativity, 295–296
Mesons, 444
decay of, 444–446
interactions of, with baryons, symmetry of, 460–464
K, *446*
mu, *444*
pi, *444*
Metal model, container with enclosed quantum particles, wave functions of, 395–397
Metals, alkali, quantum physics aspects of, 400–401
Meter (unit), equivalents of, 39
Metric system, *39*
Michelson-Morley experiment, 247 ff., 258–260, 270, 272
Momentum, *32*, 36
angular, *see* Angular momentum
conservation of, 71 ff.
of quantum particles, 347–353, 366–373
relativistic aspects of, 282–286
total, *74*
Moon, motion of, 57
Motion, of bodies, Aristotle's ideas about, 3–8
Galileo's ideas about, 7 ff.
history of ideas about, 3–10
circular, uniform, *see* Uniform circular motion
of freely falling bodies, 7, 13–18
Newton's laws of, *33, 34*, 36, 37
of projectiles, 18–20
uniform, *11*
uniformly accelerated, *13–18*
Multidimensional waves, 161 ff.
Multiplets, *389*
Muons, 444

Negative charges (electric), 185–186
Negative energy states in Dirac's electron theory, 418–423
Neutrinos, 408
formation of, 429
in nuclear reactions, 445
Neutrons, *403*
decay of, 428–429
mass of, 405
Newton, Sir Isaac, 31 ff.
laws of motion of, *33, 34*, 36, 37
and optics, 133 ff., 144, 145
place in society of, 111–113

and planetary motion, 56 ff.
and space concepts, 242
Newton (unit), equivalents of, *39*
Newton's laws, 68
and relativity, 273–274, 280 ff.
Newton's laws (third), 69 ff.
Nickel crystals, diffraction of electrons by, 344–346
Noble gases, quantum physics aspects of, *400*
Nondispersive media, *155*
Nuclear barriers, 408–409
Nuclear bombs, 412
Nuclear fission, 411–413
Nuclear forces, 404–407, 442 ff.
Nuclear fusion, 412–414
Nuclear particles, decay of, 410, 447–449
properties of (table), 450
Nuclear processes, 407–410
Nuclear reactions, 401 ff., 437, 446–449
conservation laws for, 456–458
and cosmic rays, 443–444
interaction strengths of, 456
and quanta, 442–444
Nuclear reactors, 412
Nuclear tunneling, 409
Nuclei, binding forces in, 442–446
bombardment of, with alpha particles, 403
Coulomb forces in, 406, 408, 413
Rutherford's model of, 321, 322–323
size of, 323
stability of, 409–410
structure of, 401–407
Nucleon processes, 443–446
Nucleons, *442*
internal spin of, 451
Numbers, review of (Appendices), 471 ff.
Numerical functions, review of (Appendices), 476 ff.

Observations, scientific, 117–123
Octets and degeneracy, 464
Oersted, Hans Christian, 204
Ohm, Georg Simon, 203–204
Ohm's law, 203–204
One-dimensional waves, at boundaries, 157–159
periodic, 154–157
properties of, 148 ff.
Optics, Newton's theories of, 144, 145
Orbits, of atomic electrons, Bohr's model of, 332–336
standing-wave relationship for, 342–343
of planets, types of, 63, 64

Parallelograms in force diagrams, 34
Particle accelerators, 437–442
construction and cost of, 441–442
Particle physics, state of, 466

Quantity		Value
speed of light in vacuum	*(c)*	3.00×10^{10} cm/sec
charge on the proton	*(e)*	4.80×10^{-10} esu
Planck's constant	*(h)*	6.63×10^{-27} erg-sec
gravitational constant	*(G)*	6.67×10^{-8} dyne-cm²/g²
rest mass of the electron		9.11×10^{-28} g
rest mass of the proton		1.6725×10^{-24} g
rest mass of the neutron		1.6747×10^{-24} g
Avogadro's number	*(N₀)*	6.02×10^{23}
Boltzmann's constant	*(k_B)*	1.38×10^{-16} erg/°K

Astronomical

Earth

acceleration of gravity at surface *(g)*	$\simeq 980$ cm/sec²
radius of orbit	1.49×10^{13} cm
average radius of earth	6.37×10^{8} cm
mass	5.98×10^{27} g
period of revolution (year)	3.16×10^{7} sec
period of rotation (day)	8.64×10^{4} sec

Moon

radius of orbit	3.84×10^{10} cm
radius	1.74×10^{8} cm
mass	7.34×10^{25} g
period of revolution ($\simeq$ month)	2.36×10^{6} sec

Sun

radius	6.96×10^{10} cm
mass	1.99×10^{33} g

SOME CONVENIENT NUMBERS

1 light year = 9.46×10^{17} cm
1 mile = 5280 feet = 1.61×10^{5} cm
1 angstrom (Å) = 10^{-8} cm
Bohr radius of the ground state of hydrogen = 5.29×10^{-9} cm
1 radian = 57.3 degrees (57°20′)
60 miles/hour = 88 ft/sec
1 calorie = 4.18 joules
rest energy of the electron (m_0c^2) = 5.11×10^{5} electron volts
1 electron volt (eV) = 1.6×10^{-12} erg
ratio of mass of proton to mass of electron (m_p/m_e) = 1836

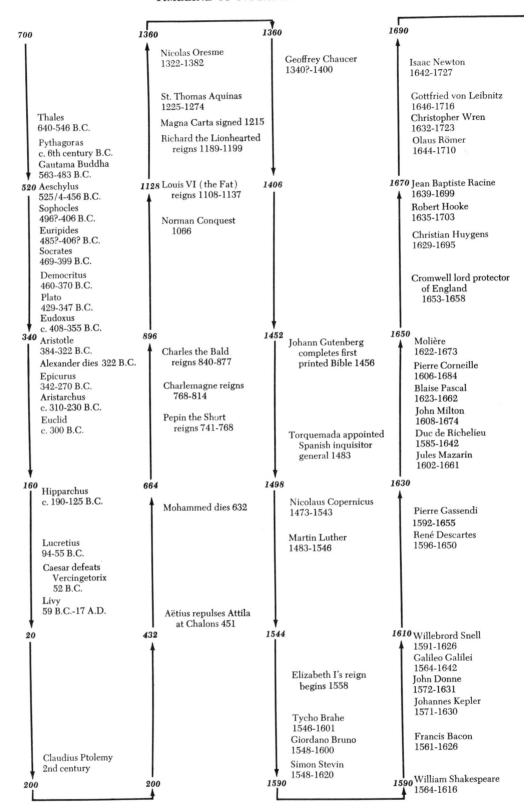

700

1360
Nicolas Oresme
1322-1382

1360
Geoffrey Chaucer
1340?-1400

1690
Isaac Newton
1642-1727

Thales
640-546 B.C.

Gottfried von Leibnitz
1646-1716
Christopher Wren
1632-1723
Olaus Römer
1644-1710

St. Thomas Aquinas
1225-1274
Magna Carta signed 1215
Richard the Lionhearted
reigns 1189-1199

Pythagoras
c. 6th century B.C.
Gautama Buddha
563-483 B.C.

520 Aeschylus
525/4-456 B.C.
Sophocles
496?-406 B.C.
Euripides
485?-406? B.C.
Socrates
469-399 B.C.

1128 Louis VI (the Fat)
reigns 1108-1137

Norman Conquest
1066

1406

1670 Jean Baptiste Racine
1639-1699
Robert Hooke
1635-1703
Christian Huygens
1629-1695

Democritus
460-370 B.C.
Plato
429-347 B.C.
Eudoxus
c. 408-355 B.C.

Cromwell lord protector
of England
1653-1658

340 Aristotle
384-322 B.C.
Alexander dies 322 B.C.
Epicurus
342-270 B.C.
Aristarchus
c. 310-230 B.C.
Euclid
c. 300 B.C.

896
Charles the Bald
reigns 840-877

Charlemagne reigns
768-814

Pepin the Short
reigns 741-768

1452
Johann Gutenberg
completes first
printed Bible 1456

Torquemada appointed
Spanish inquisitor
general 1483

1650
Molière
1622-1673
Pierre Corneille
1606-1684
Blaise Pascal
1623-1662
John Milton
1608-1674
Duc de Richelieu
1585-1642
Jules Mazarin
1602-1661

160 Hipparchus
c. 190-125 B.C.

664
Mohammed dies 632

1498
Nicolaus Copernicus
1473-1543

Martin Luther
1483-1546

1630
Pierre Gassendi
1592-1655
René Descartes
1596-1650

Lucretius
94-55 B.C.
Caesar defeats
Vercingetorix
52 B.C.
Livy
59 B.C.-17 A.D.

Aëtius repulses Attila
at Chalons 451

20

432

1544
Elizabeth I's reign
begins 1558

Tycho Brahe
1546-1601
Giordano Bruno
1548-1600
Simon Stevin
1548-1620

1610 Willebrord Snell
1591-1626
Galileo Galilei
1564-1642
John Donne
1572-1631
Johannes Kepler
1571-1630

Francis Bacon
1561-1626

Claudius Ptolemy
2nd century

200

200

1590

1590 William Shakespeare
1564-1616

Each individual's position on this calendar is determined by the midpoint of his life.

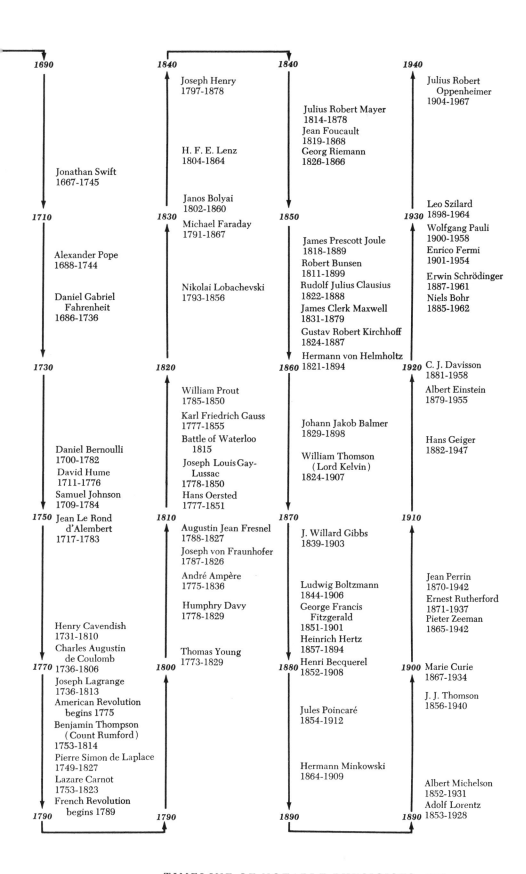

1690

Jonathan Swift
1667-1745

1710

Alexander Pope
1688-1744

Daniel Gabriel
Fahrenheit
1686-1736

1730

Daniel Bernoulli
1700-1782
David Hume
1711-1776
Samuel Johnson
1709-1784

1750 Jean Le Rond
d'Alembert
1717-1783

Henry Cavendish
1731-1810
Charles Augustin
de Coulomb
1770 1736-1806
Joseph Lagrange
1736-1813
American Revolution
begins 1775
Benjamin Thompson
(Count Rumford)
1753-1814
Pierre Simon de Laplace
1749-1827
Lazare Carnot
1753-1823
French Revolution
1790 begins 1789

1840

Joseph Henry
1797-1878

H. F. E. Lenz
1804-1864

Janos Bolyai
1802-1860
1830 Michael Faraday
1791-1867

Nikolai Lobachevski
1793-1856

1820

William Prout
1785-1850
Karl Friedrich Gauss
1777-1855
Battle of Waterloo
1815
Joseph Louis Gay-
Lussac
1778-1850
Hans Oersted
1777-1851

1810
Augustin Jean Fresnel
1788-1827
Joseph von Fraunhofer
1787-1826
André Ampère
1775-1836
Humphry Davy
1778-1829

Thomas Young
1800 1773-1829

1790

1840

Julius Robert Mayer
1814-1878
Jean Foucault
1819-1868
Georg Riemann
1826-1866

1850

James Prescott Joule
1818-1889
Robert Bunsen
1811-1899
Rudolf Julius Clausius
1822-1888
James Clerk Maxwell
1831-1879
Gustav Robert Kirchhoff
1824-1887
Hermann von Helmholtz
1860 1821-1894

Johann Jakob Balmer
1829-1898

William Thomson
(Lord Kelvin)
1824-1907

1870
J. Willard Gibbs
1839-1903

Ludwig Boltzmann
1844-1906
George Francis
Fitzgerald
1851-1901
Heinrich Hertz
1857-1894
Henri Becquerel
1880 1852-1908

Jules Poincaré
1854-1912

Hermann Minkowski
1864-1909

1890

1940

Julius Robert
Oppenheimer
1904-1967

Leo Szilard
1930 1898-1964
Wolfgang Pauli
1900-1958
Enrico Fermi
1901-1954
Erwin Schrödinger
1887-1961
Niels Bohr
1885-1962

1920 C. J. Davisson
1881-1958
Albert Einstein
1879-1955

Hans Geiger
1882-1947

1910

Jean Perrin
1870-1942
Ernest Rutherford
1871-1937
Pieter Zeeman
1865-1942

1900 Marie Curie
1867-1934

J. J. Thomson
1856-1940

Albert Michelson
1852-1931
Adolf Lorentz
1890 1853-1928

TIMELINE OF NOTABLE PHYSICISTS 537

University Press of New England publishes books under its own imprint and is the publisher for Brandeis University Press, Brown University Press, University of Connecticut, Dartmouth College, Middlebury College Press, University of New Hampshire, University of Rhode Island, Tufts University, University of Vermont, and Wesleyan University Press.

Library of Congress Cataloging-in-Publication Data

Cooper, Leon N.
    Physics : structure and meaning / Leon N. Cooper. — New ed.
        p.   cm.
    Rev. ed. of: An introduction to the meaning and structure of physics. Short ed. 1970.
    Includes index.
    ISBN 0–87451–592–0
    1. Physics. I. Cooper, Leon N. Introduction to the meaning and structure of physics. II. Title.
QC21.2.C67   1992
530—dc20   92–19283